BEYOND UNCERTAINTY

HEISENBERG, QUANTUM PHYSICS, AND THE BOMB

BEYOND UNCERTAINTY

HEISENBERG, QUANTUM PHYSICS, AND THE BOMB

David C. Cassidy

Bellevue Literary Press
New York

First published in the United States in 2009 by
Bellevue Literary Press
New York

FOR INFORMATION ADDRESS:
Bellevue Literary Press
NYU School of Medicine
550 First Avenue
OBV 640
New York, NY 10016

This book was published with the generous support of
Bellevue Literary Press's founding donor the Arnold Simon Family Trust,
the Bernard & Irene Schwartz Foundation and
the Lucius N. Littauer Foundation.

Library of Congress Cataloging-in-Publication Data

Cassidy, David C., 1945-
Beyond uncertainty : Heisenberg, quantum physics, and the bomb /
by David C. Cassidy.
p. cm.
Includes bibliographical references and index.
1. Heisenberg, Werner, 1901-1976. 2. Physicists—Germany—Biography.
3. Atomic bomb—Germany—History—20th century. I. Title.
QC16.W518C37 2008   530.092—dc22 [B]   2008039885

Book design and type formatting by Bernard Schleifer
Manufactured in the United States of America
ISBN 978-1-934137-13-0
1 3 5 7 9 10 8 6 4 2

CONTENTS

FOREWORD

BEYOND UNCERTAINTY: HEISENBERG, QUANTUM PHYSICS, AND THE BOMB DRAWS UPON, YET in many ways transcends, the detailed account provided in its now-out-of-print predecessor, *Uncertainty: The Life and Science of Werner Heisenberg.*

The predecessor first appeared in a limited edition in 1991. It was the product of a dissertation and six years of research in Germany and other nations while I was a fellow of the Alexander von Humboldt Foundation in Stuttgart and an assistant professor in Regensburg, Germany. Several more years of research and writing in the United States followed.

My goal in *Uncertainty* was to attain the most comprehensive biography of Heisenberg possible at the time, and to write it primarily for a highly educated, even scholarly, audience in both science and history. My models were not works of science history but eminent literary biographies, such as those of Henry James (Leon Edel), James Joyce (Richard Ellmann), and Fyodor Dostoyevsky (Joseph Frank). Not only were these exhaustive of the person and his work and times, but the life and work were closely integrated, while both were understood as expressions of the culture and the times, as well as the accidents of upbringing and personality. In this way the subject was seen at once as both a highly creative individual and a member of a community in a specific place and time.

My hope in *Uncertainty* was to enable readers of the late twentieth century to comprehend in a very fundamental way two of the most significant events of that century. The first was the truly remarkable achievement of one of the premier scientific breakthroughs of the century, the invention of quantum mechanics, followed by its further development in the contemporary sciences of atoms, nuclei, particles, and solids. Quantum mechanics and the sciences it has spawned have brought us profoundly new and remarkable understandings of the workings of nature and of our universe, and have transformed our daily lives through such technologies as lasers, medical imaging, and the transistors at the base of the computer and digital revolutions of today. Heisenberg was one member of the small band of young people and their mentors who helped bring about the quantum revolution during the 1920s and who helped push it forward during the decades to follow. More than many other scientific advances, it was a community effort, extending beyond any one individual. It was also centered at first primarily in Germany, but gradually extended throughout and beyond the European continent. What did Heisenberg and his colleagues actually do in creating this revolution? How did they do it? What larger forces made it possible? What has been its impact? And how does the quantum journey continue today?

The second significant event of the twentieth century entailed the world's first encounter with advanced industrial totalitarian and genocidal dictatorships, the Nazi dictatorship in particular. How did this happen, and in Germany of all places, the leading cultural and industrial nation at that time? As a member of the non-Nazi upper academic stratum of German society, the product of the best culture and education that Germany could offer, Heisenberg provides a valuable insight into these questions as he, and many others like him, encountered and eventually found accommodation with the new regime. This raises a host of further questions. What events of their past informed their response to the new regime, and why did their efforts at opposition fail? How could Heisenberg remain in Germany and lend his prestige to that society as one of its most prominent remaining scientists? How could he become a representative to occupied nations? How could he work on nuclear fission, and potentially on an atomic bomb, for such a regime at war? As we know from the experiences of others in similar situations, the answers are not as straightforward as they might appear.

Since the fall of the Third Reich, other problems have arisen, many associated with cold-war fear, new weapons of warfare, the threat of terrorism, and the dislocations brought about by globalization. My hope was and is that the lessons of this encounter with a totalitarian, genocidal regime, as it came to power within a democracy and consolidated its hold on the minds of its subjects, will heighten our sensibilities and our resolve whenever similar tendencies and even regimes emerge today.

A lot has happened since *Uncertainty* first appeared. The cold war had just ended, new perspectives on the Nazi era were developing, and many documents that had been classified or sequestered until then were suddenly brought to light. Among these were captured German war documents in formerly Soviet archives. Some have argued that these documents suggest the detonation of some sort of rudimentary nuclear device in Germany at the end of the war. In addition, many new documents relating to Nazi science policy and antiscientific propaganda were made available in the former East Germany, and in other nations. Also, Heisenberg's family decided to publish many of his private letters to family members through 1945 in a volume that appeared in 2003 and on the Web. Such letters make a comprehensive biography possible. During my research in Germany for *Uncertainty*, I had seen only some of these new materials and only briefly.

In addition, in response to the popular and widely debated play by Michael Frayn, *Copenhagen*, in 2002 the Niels Bohr Archive in Copenhagen released a series of previously withheld drafts of unsent letters from Bohr to Heisenberg, starting in 1957. These draft letters contained Bohr's unflattering recollection of Heisenberg's visit with Bohr in German-occupied Copenhagen in 1941. During that meeting they had discussed in some way the prospect of a German atomic bomb.

Finally, a thirty-year-long effort to gain the release of the Farm Hall transcripts culminated in their declassification and release from British and American archives in February 1992. These were transcripts of secretly recorded conversations among ten of the captured German nuclear scientists, including Heisenberg, while held in Allied captivity at the British estate of Farm Hall. These transcripts offer new and important insights into the scientists' fission work, their reasons for doing such work, the formation of a postwar rationale for their work, and their plans for reestablishing postwar German science.

As nearly every batch of new documents became available after the publication of *Uncertainty*, it set off a new round of debate within scholarly circles over such questions as: Was Heisenberg really intent on building a bomb for Hitler? If so, why did the German project make such little progress? If not, why not? Was Heisenberg actually intent on building the bomb but inept as a nuclear scientist and scientific head of the project, or were the war circumstances against rapid progress, or did he secretly sabotage the effort out of moral scruples? What does an overall view of his life and times reveal about his wartime behavior? These are questions that have been hotly, even emotionally, debated, and books have appeared that argue practically every side of this debate.

Finally, during the years since 1991, like many others I have become increasingly concerned about the state of science education in the United States and elsewhere. Science is not just a body of abstract, mathematical concepts invented and manipulated by a small scientific elite, but a living part of human culture and experience, a product of the unending human quest to understand our world and ourselves in relation to it, an adventure that real people with real faults but also enormous determination and creativity have pursued in bringing us to where we are now, and will continue to do into the future. Having firsthand experience with the limited appreciation of this quest and its results at the college level, the elementary level (through my spouse), across academe, and among the general public, I have become increasingly intent on bringing this wonderful story to students, nonscience academics, and the general public. This has been one motivation of my teaching of physics for nonscience students. It has found expression in a textbook for such students that I co-authored recently with Gerald Holton and James Rutherford. In both venues we have attempted to view the science as the product of the historical human quest to understand our world, and as a carefully developed body of knowledge about the workings of our physical world.

All of these developments have now motivated the effort to reach "beyond uncertainty": to draw upon much of what is still valid in *Uncertainty* but to transcend it where appropriate by incorporating new material, new perspectives, the lessons of recent debates, and the insights that a new century, with new problems, now affords us.

Most importantly, my purpose is not to write "primarily for a highly educated, even scholarly, audience in both science and history." Such an audience has recourse to the original work and to many of the recently available technical secondary sources, references to some of which appear in the notes. Instead, my purpose here is to reach beyond technically trained readers to a more general audience, especially one that has little or no experience with quantum physics. Although I will discuss some of the technical details of the physics, the purpose is to provide readers with a general appreciation of the scientific problems that Heisenberg and his colleagues were trying to solve, how they were trying to solve them, and the intensive, often frustrating work this required even of these brilliant scientists. The same struggle continues today.

By the same token, I am attempting here to reach beyond scholars of German history and the bomb project to a more general audience whose members may be less familiar with the details of the history or of that nature of the Nazi dictatorship and totalitarian thinking. My purpose is likewise to provide today's readers with an appreciation of how difficult it was for a "mere scientist" to respond to a regime for which he and they were completely unprepared.

Now that the Nazi regime is over 60 years and the Soviet Union over 20 years behind us, we are in a position to gain not only a new perspective on that era but a new understanding of how it happened so that, hopefully, we will be better prepared whenever and wherever some of the same arguments and same thinking reappear. In science, in political history, and in the moral and ethical behavior of the individual, as both individuals and members of a community and a culture, we can now begin to move beyond *uncertainty* in the story of Heisenberg from the vantage point of a new century with new challenges.

CHAPTER 1

THE EARLY YEARS

ON NOVEMBER 11, 1901, AUGUST HEISENBERG, A SECONDARY-SCHOOL TEACHER OF classical languages, presented a formal lecture to the faculty of the University of Würzburg—the last step in his candidacy for qualification as a university lecturer. Three weeks later, his wife, Annie, gave birth to her second child, a boy. Like his older brother, Erwin, born in Munich, the infant arrived in the Heisenberg home, at Heidingsfelderstrasse 10 in the elegant Würzburg suburb of Sanderau. His birth certificate lists not only his name and date of birth but also the exact time: Werner Karl Heisenberg, born on Thursday, December 5, 1901, at 4:45 PM. Before the month was out, the Bavarian Interior Ministry added to the joys of the proud father its approval of his appointment as a lecturer at the university, a position he held in addition to his duties at the secondary school.

The coincidence between Werner's birth and his father's appointment hinted at three essential elements in the child's future development: his timing with respect to important events, the high academic and cultural level of the family into which he was born, and the rapid upward social and academic momenta the family had attained by the time of Werner's birth. August Heisenberg came from a family of middle-class craftsmen. Within a decade of Werner's birth he would reach the top of the social and academic ladders as Germany's only full professor of middle and modern Greek studies.

The Heisenberg family's social mobility is evident in a carefully constructed family tree preserved in Werner Heisenberg's private papers. The tree, rather a typed pedigree replete with certificates of birth and baptism, owes its origin to the search by Nazi authorities for a Jewish ancestor in the scientist's past. It traces the Heisenbergs back five generations to one Heissenberg in Heidenoldendorf, a village in the northern region of Westphalia. The eighteenth-century ancestor is succeeded by a brandy burner, a master cooper, and a locksmith. The locksmith, Wilhelm August Heisenberg (1831–1913), dropped the second *s* in his name and moved north to Osnabrück, then in the realm of Hannover, where he raised three daughters and two sons, one of whom was Werner's father.

After Werner's grandfather learned the locksmith trade, he set out on a "wander year," a common rite of passage in those days. He obviously did well: on his return he purchased his master's business, barn, and house. With business, property, and title (master locksmith), he easily rose to the rank of official Bürger of Osnabrück, a voting member of the town's middle class. In 1858 he ensured his status by marrying the daughter of a prosperous local farmer. The two complemented each other well. Wilhelm Heisenberg is remembered as a quiet, cerebral man, an impression confirmed in a surviving photograph. His wife, Anne Marie, is remembered for her strong will and keen intelligence.

Werner Heisenberg's middle name was that of his father's younger brother, Karl, the black sheep of the family. The fifth of the five children, he became a *Tunichtgut*, a ne'er-do-well. Always in trouble, he once stole a sum of money from one of his sisters, whereupon his father handed him 200 marks more and packed him off on the next ship for America—in those days, the end of the world. The clever young man quickly fulfilled the American dream: he opened a factory for uniform buttons in Flushing, New York, and soon became the richest Heisenberg. The exile's dollars proved invaluable to his German relatives during the inflation after World War I, as did his American connections to Werner after World War II.

Werner's father, Kaspar Ernst August Heisenberg, was born in Osnabrück in 1869, less than two years before the unification of the German Reich under Kaiser Wilhelm I. Hannover, then a Prussian territory, was already subject to Wilhelmian rule when Wilhelm's chancellor, Otto von Bismarck, induced the recalcitrant southern states to join a united Germany. A period of enormous industrial, commercial, and technological expansion ensued throughout the Reich, matched by the rising nationalism of the middle and upper classes and the increasing solidification of a social and political hierarchy centered on the Kaiser and his chancellor. Like many others of his generation, August (the name he used) came to maturity under the Bismarckian monarchy, and, like many other German academics, he came to idolize Bismarck and the empire. Evidence suggests that August joined other academics in his allegiance to the National Liberal Party, a party on which Bismarck had greatly relied in unifying the Reich. Liberals believed that the best route to civil reform, and the advance of their own social status, lay in national unity under Prussian leadership, the predominance of secularized Protestantism, and rapid commercial expansion—ideals that August later tirelessly impressed on his children.

August recalled a happy childhood in Osnabrück among "numerous siblings."[1] At the age of 10 he entered a nine-year course of study at the local gymnasium, the first step in the German education system toward university education and an academic or professional career. Only graduates of a gymnasium, which brought its students to the equivalent of the early junior year of a modern American college, could pursue higher education leading to the professions. August's study at a gymnasium, instead of apprenticeship to a craftsman, required a fundamental family decision, since it constituted the first break in the family tradition of producing middle-class craftsmen. With the complete support of his family, August would attempt to reach the next social stratum via the uncertain route of an academic career. When his father died in 1913, August, a lately appointed professor, wrote of the "sincerest and most trusting relationship" that he had enjoyed with his father "continuously from childhood on, until the very last. . . . In everything that I achieved in life, he stood by me with his counsel, and whatever I succeeded in doing pleased me, because it pleased him."[2]

Compared with a modern American professor, a German professor within the stratified world of Wilhelmian Germany enjoyed far more prestige and power. During the decades following the Napoleonic conquests, which ended for Germany in 1815, German administrators attempted to rebuild the nation's prestige through the promotion of German culture as one of the pillars of national strength. Administrators and scholars regarded scholarship as an essential component of the cultural pillar, and neohumanist studies of Greek works of the heroic age—exemplars for a heroic new Germany—as its crowning achievement. Because of this, a German university professor, especially one in classical Greek philology (the study of culture and language), ranked in status along with other nonpropertied "bearers" of the Bismarckian state—judges, army officers, industrialists, higher bureaucrats—among the upper-middle-class elite. Just above them stood the nobility and the propertied upper class; below stood the middle class of craftsmen, farmers, lower civil servants, and gymnasium teachers. Unskilled industrial laborers occupied the bottom of the social scale.

August's talents and Germany's economics encouraged the family strategy of social advance through academic achievement. By 1879 the German industrial revolution was in full swing. Master craftsmen were finding it difficult to compete with mechanized industries and their growing pools of cheap labor. Although August Heisenberg's attempt to compete in the academic rather than the economic world entailed a grave risk—only the best students could attain a professorship—the Heisenberg family was not at all unusual in taking it. The expansion of industry and empire required more administrators, jurists, and professors than could be supplied by simple replication. Sons of middle-class and lower-class families were increasingly recruited. (Women were not considered for such positions.) According to one study, during the period of August's education fully two-thirds of Prussian Greek philology students originated from the middle class.[3] More than a quarter of such students came from the families of skilled craftsmen, small businessmen, and innkeepers—the families who could best afford to finance the long years of study. As more of the middle class attained the coveted title *Herr Professor Doktor*, they viewed themselves more consciously—for protection and privilege—as a group apart, an academic class defined and established not by title or inheritance but by education and culture.[4]

Two years out of gymnasium, August headed south to Bavaria, attracted to the southern province by the Wagnerian music of the Bavarian capital, Munich, and by its enthusiasm for the glories of ancient Greece. Even more attractive were the efforts of Bavarian state officials to raise the cultural level of the rural province through generous funding of education and the importation of famous Prussian scholars, the so-called northern lights. August was drawn to one of these beacons at the University of Munich, Karl Krumbacher, a lecturer who soon founded Germany's only chair, or research group, for Byzantine studies (middle and modern Greek philology). Heisenberg immediately converted to the promising yet nearly untouched field, assured of bright career prospects in the rural southern province.

August Heisenberg completed his doctorate under Krumbacher in 1893, passed the difficult teacher-qualifying examination, and soon became a teacher trainee at the prestigious Maximilians-Gymnasium in Munich under its learned and powerful rector, Nikolaus Wecklein.[5] Gymnasium teachers were expected to hold a doctorate and to engage in publishable research. Two years later August abruptly left for his required year of military training, which he performed not in Bavaria but back home in Osnabrück with an infantry regiment under Prussian command. An unfortunate romance with the elder of Wecklein's two daughters, Annie, apparently precipitated the sudden move. Her esteemed family flatly disapproved of her unestablished suitor.[6]

August returned to Munich as a reserve army officer—still dedicated to national unity and Prussian predominance—and soon came under Wecklein's scrutiny in a required pedagogical seminar. Wecklein's doubts apparently dissipated, for August's romance with his daughter revived. Heisenberg remained in Munich only six months before taking off to the Bavarian hinterland, this time to a Latin school in Lindau on Lake Constance. But three days before the start of his appointment he telegraphed his father the good news: he and Annie were engaged.[7]

Little is known of Werner's mother, Annie. Neither she nor her sister received a university education: German universities were closed to women, as a rule, until 1895, and Munich did not admit female students until 1903. Nor are there informative state personnel files on which to rely: German civil careers were open only to men. Both Wecklein girls no doubt attended one of the segregated girls' middle schools, which typically offered training in the fundamentals—math, history, and literature—and prepared their pupils for their future roles as genteel wives and cultivated mothers of educated sons.

Like Wecklein, August's father, Wilhelm, conferred his blessing on the union. But the marriage was postponed for over two years while the groom attained the acceptable status of gymnasium teacher, with corresponding salary, and a transfer back to Munich. During those two years August obtained a state research grant to prepare for academic advance after marriage. While August rummaged for artifacts in Greece and Italy, Rector Wecklein arranged his promotion and transfer to the Luitpold-Gymnasium in Munich (which its most famous pupil, Albert Einstein, had left only recently), and Annie changed her religion from Roman Catholic to August's Lutheran faith in order not to risk opposition by the Catholic Church, the most influential church in Bavaria. In January 1899, August, then in Rome, submitted his required official request to the Interior Ministry for permission to marry. The Interior Minister personally concurred, after assuring himself of the bride's moral reputation (a state employee could not bring dishonor upon his employer).[8] With all in readiness, August returned in the middle of May 1899, and within a week the happy couple were wed at the Erlöserkirche in the upscale Munich suburb of Schwabing.

Nikolaus Wecklein had also married within the higher social stratum to which he had risen through academic achievement. The son of a long line of farmers in the northern Bavarian province of Mittelfranken, Wecklein was married in 1870 to

Three generations of the family in chronogical order. Nikolaus Wecklein at the top, Werner at the bottom right, below his grandmother and mother.

Magdalene (Magda) Zeising, whose ancestors all had served with royal titles at the court of the Duke of Bernburg in the Harz Mountains between Halle and Magdeburg. After the failed liberal revolution of 1848, Magda's father, Dr. Adolph Zeising, an educator and poet, had gone to Munich to study the aesthetics of the Bavarian king's Greek statues. He died there in 1872 of a painful illness he contracted shortly after Magda's marriage to Wecklein.[9]

Wecklein owed his rise from humble beginnings to the top of the Bavarian school system to his scholarly command of classical Greek and to the helpful encouragement of powerful superiors whom he had impressed as a student. With his long white beard and stern demeanor, he had the look in later years of a patriarch, and when he wore his visored cap, he looked a little like an old sea captain. After the marriage of his elder daughter (his younger never married) and the birth of her two sons, Wecklein often did function as the patriarch of the family and overseer of the professional advance of his son-in-law and grandsons. The grandsons entered the Max-Gymnasium while it was still under Wecklein's direction. Photographs of the extended family during their frequent Sunday outings display the familiar pyramidal arrangement of the era. One photograph shows the patriarch perched on the craggy summit of a hill with the succeeding generations dispersed beneath him in chronological order.

Wecklein had aimed early for a teaching chair (as professorships were called) in Greek philology. He wrote a dissertation on the Greek sophists, qualified as a university lecturer with a treatise on Greek grammar, and became a leading authority on

Greek tragedy.[10] But the year of his marriage marked the end of his advance and the beginning of a profound disappointment: his failure to obtain a university professorship. The Munich faculty refused to appoint him to an opening after it judged his lectures on Aeschylus too dull. The dull lecturer found himself instead on a fast track into the education hierarchy when the Bavarian Interior Ministry for Church and School Affairs, encouraged by Wecklein's professors and his Liberal Party credentials, appointed him to gymnasium administration. Wecklein eventually rose to influential positions on the Bavarian school board and the academy of sciences, while rectoring the famed Maximilians-Gymnasium.

Still, without the prestige and standing of a university chair, Wecklein never felt his life complete.[11] When he retired from the top of the school system in 1913 at the age of 70 with all sorts of titles and awards, his family barely managed to keep him from becoming a lowly "private lecturer" (Privatdozent) once again. A lecturer, having officially qualified for a chair, lectured for student fees until appointed by the state to a professorship. In his last years, Wecklein must have been comforted in the knowledge that his son-in-law had been granted that exalted status, that his two grandsons had by then obtained doctorates, and that Werner had habilitated (qualified for a professorship) and had even recently substituted for a physics professor (Max Born) in Göttingen. After paying a good-bye visit to Werner in 1926, Wecklein died before seeing him occupy a chair in Leipzig less than a year later.

Even without a university chair, Wecklein was well positioned by the turn of the century to play a crucial role in the family strategy for social advance. When Krumbacher proved unable to secure the habilitation of Wecklein's new son-in-law in Munich, Wecklein's connections saved the day. After the birth of his first son, Erwin, in March 1900, and a return to Italy to complete his habilitation treatise (a dissertation beyond the doctoral thesis), August Heisenberg learned through Wecklein that Wecklein's alma mater, the University of Würzburg, would consider him. Shortly after submitting the treatise to Würzburg, August also learned through Wecklein of a vacancy at Würzburg's Altes Gymnasium. Should the habilitation succeed, which it did, he could lecture privately at the university while teaching (with sufficient salary) at the gymnasium. Nearly all gymnasium teachers held doctorates, and many engaged in original research. A request for transfer received the immediate approval of Wecklein's school board, and, after Heisenberg completed his annual six weeks of military exercises in the summer of 1901, he and his family moved to Würzburg, about 250 miles northwest of Munich. August began teaching in September, while his wife prepared to give birth to her second child.[12]

In contrast to Munich and other German cities, Würzburg during the early 1900s remained a remarkably quiet, rural, traditional, provincial town. The family's two-story suburban house near the Main River, with the hilly Franken vineyards along its banks and the nearby fields and woods, was ideal for the two growing Heisenberg boys. Despite the immigration of rural families into Würzburg and the sudden jump in gymnasium pupils in the year Dr. Heisenberg began teaching, the social and economic structure of Würzburg had remained nearly untouched.[13] Civil servants, merchants, landed nobles, and especially university professors retained control of the political hierarchy.[14] The son of a skilled craftsman began teaching at an annual base salary of 2,820 marks, double that of a skilled worker in Würzburg. By 1906 it was nearly triple. The family social strategy had paid off handsomely.

August Heisenberg is remembered by his family, superiors, and pupils as a rather stiff, tightly controlled, authoritarian figure. A former student recalled that the schoolmaster demanded "unbending fulfillment of duty, absolute self-control, and meticulous precision."[15] "He treats his pupils with propriety but tolerates no lazy boys in his class," his Lindau rector noted.[16] August must have applied the same standards in raising his own two boys, who grew up in a family structure typical of Bürger families at the turn of the century: father-centered, authoritarian, hierarchical. In the male-dominated, monarchical society of Wilhelmian Germany, it is not surprising that families were organized in the same fashion, or that men considered it an obligation to preserve such an organization.

By the same token, a German woman of that era, no matter what her interests or talents, regarded as her obligation being an obedient wife and a self-sacrificing mother.[17] As the daughter of a gymnasium rector, Annie knew when she married August that her self-realization and recognition would be achieved only in ensuring the success of her husband and the well-being of their children. She excelled brilliantly. Indeed, she made it possible for her husband to discharge at an outstanding level the almost incredible workload he carried during the Würzburg years. Despite the exclusion of women from higher education, she had obviously sought and received advanced instruction, probably from her father, for it was she who graded the daily homework of her husband's pupils. She even learned Russian in order to translate research papers for her husband's use—all this, of course, in addition to caring, no doubt without much help, for the two growing boys, who surely required coaxing to keep them from disturbing their busy father.

As a gymnasium teacher, Dr. Heisenberg at first taught 14 hours per week of Latin, German, and geography to large classes (35 to 40 pupils) of 9- and 10-year-old boys in the first and second grades. After promotion to gymnasium professor, he took charge of the more demanding sixth grade of 14-year-olds.[18] Throughout that period, he also offered three two-hour courses per week at the university on Byzantine philological topics, was deeply involved in the political affairs of the local Gymnasium Teachers Association, and generated scholarly writings at a prolific rate. His bibliography contains an astonishing 56 titles published while in Würzburg.[19] Among them was a two-volume account in 1908 of his own archaeological research in Constantinople—the foundation for his appointment to university professor two years later.[20]

Dr. Heisenberg's capacity for teaching and research also astonished his Würzburg gymnasium rector, who consistently gave him the highest marks as both teacher and scholar. In his evaluation for 1902, he noted in the man "a restless drive to expand and deepen his narrower professional knowledge." Yet Heisenberg's teaching did not suffer: "The lessons are carefully and thoroughly completed; the needs of the class are closely followed; the interest and encouragement of the pupils are close to his heart. Because of this, he exerts a stimulating and permanent influence on the education of the mind and morale of his pupils."[21]

Yet August's academic life took a heavy toll on him and his family. With the tremendous pressures exerted by his work, his restless drive, and his rigid role as family provider, it is little wonder that the professor is also remembered for his stormy temperament and rapid oscillations from pleasantness to depression. His wife, the daughter of an equally authoritarian and probably equally bad-tempered schoolmaster, had learned to handle such behavior with a carefree, childlike disposition and a quiet, pleasant, even-tempered manner. One visitor to the Heisenberg home recalled her as "a small, dear woman, concerned, kind, but not very conspicuous." Her husband, on the contrary, whether present or not, served as the center of attention and authority in all family matters.[22]

Children were at the bottom of the Heisenberg family hierarchy. As Werner grew into adolescence, what he saw and felt from that position must have increasingly distressed him. Like any other turn-of-the-century Bürger family, the Heisenbergs cherished the appearance of genteel respectability, social grace, and allegiance to nationalist trappings. This was especially so for German academics and their families, whose public deportment and professional demeanor were expected to reflect their superior social station and the virtues of their Christianity-dominated, monarchical state. Respectability—frugality, devotion to duty, and restraint of the passions— writes one historian, defined and maintained the precarious position of the upper bourgeoisie during the late prewar period.[23] Bavarian gymnasium rectors, the educators of the future upper classes, were accordingly required to report on the ethical deportment of their subordinates. August Heisenberg's rector always wrote "impeccable" and usually added: "His family circumstances are the best imaginable."[24]

Werner eventually saw the respectability as a facade. Children and adolescents easily perceive and condemn hypocrisy in their elders, and it is usually nowhere more evident than in religious matters. With the recently increased role accorded churches in Bismarckian state affairs, no sharp separation between church and state had developed as it had in other countries. Both Heisenberg boys were duly baptized and confirmed in the German Lutheran Church, and the family adhered closely to prescribed religious practices and to the requirements of Christian ethics. This combination of social, religious, and cultural virtues in the context of the monarchical bureaucracy produced that paragon of Prussian virtue, the upright Wilhelmine professor, the decent and dutiful civil servant, exemplified by the Munich-educated Berlin physicist Max Planck. According to Planck's biographer, "respect for law, trust in established institutions, observance of duty, and absolute honesty—indeed sometimes an excess of scruples—were the hallmarks of Planck's character."[25] Certainly August Heisenberg's character came out of the same mold.

But how solid was the religious foundation of such scrupulous rectitude? Both Heisenberg parents were probably unusual among intellectuals in admitting to their sons the lack of any private religious beliefs. European culture smacked in those years of bourgeois hypocrisy and duplicity.[26] Dr. and Mrs. Heisenberg judiciously left the matter of belief to the boys' personal preferences—as long as those preferences did not conflict with public norms. Years later Heisenberg wrote to his parents that, as he saw it, for them Christianity was "just an empty form" used merely for appearances.[27] He told one interviewer: "My parents were far away from the Christian religion as far as the dogmas were concerned, but they would always stick to the Christian ethics. They would accept the rules of how to behave and to live, and say that we can take them from the Christian religion, but we cannot accept literally all these old stories."[28]

The bourgeois ambivalence of Heisenberg's childhood may have played a role in his own adult ambivalence toward the sweeping claims of every system of thought and belief, including science. At middle age and again near the end of his life, Heisenberg declared science and religion to be "complementary" aspects of reality, each with its own language and symbolism and each with its own limited realm of validity. Different religiously or intuitively apprehended truths should be viewed as different sides of the same truth, while rational science—his own profession—should be viewed as just one among a variety of ways of perceiving reality.[29]

While young Heisenberg reveled in ambivalence about ultimate reality, his brother Erwin became a convinced follower of a religious-philosophical system, anthroposophy, that enjoyed considerable popularity in Germany during the early decades of the twentieth century. Heisenberg once recalled a heated argument with his brother in which Erwin declared that he knew his soul existed but was not so sure about the existence of his body or matter, an essential anthroposophic position. Teenaged Heisenberg was sure that his body existed but was not so sure about his soul.[30] "If someone were to say that I had not been a Christian, he would be wrong. But if someone were to say that I had been a Christian, he would be saying too much," Heisenberg remarked to his longtime colleague and confidant Carl Friedrich von Weizsäcker shortly before his death.[31]

While the problems of ambivalence and hypocrisy became especially acute during Heisenberg's adolescence and early adult life, he emerged from his childhood already endowed with his family's recently acquired cultural station and infused with his father's tremendous drive for recognition and success in academic affairs. The indoctrination took place in the close context of the four-member family.[32] Erwin is remembered as his father's favorite son, the shy and retiring "Wernerle" his mother's favorite. Werner's allergies may have encouraged the favoritism. At 5, he nearly died of a lung infection, which must have increased his mother's protectiveness immensely. Disabling allergies and illnesses recurred throughout Heisenberg's life, while the quiet and even temperament encouraged by his mother's loving care became a permanent fixture of his personality.[33]

A family outing.

In snapshots of Heisenberg as a young man, he always appears radiant, confident, alert, and pleasant. But in photographs as a child, the slightly built youngster, with his close-cropped blond hair, freckled face, and typical Bavarian lederhosen, invariably appears uncomfortable in posture and expression, even hurt and withdrawn. His brother, in the same photographs, always has a mischievous look in his eye. The two boys were in continual competition.

The competition early focused on gaining the attention of their busy father, and to do so, the sons had to excel in academic and cultural skills. Among these skills was the playing of a musical instrument, another essential activity of cultured Germans. Like impeccable deportment, the enjoyment and playing of classical music served the cultured classes as recognizable expressions of a common cultural outlook. All cultured children were thus taught a musical instrument and learned to play the great classical works on it. Music served Heisenberg throughout his life as a significant and sometimes crucial social vehicle.

August Heisenberg and his sons reportedly practiced together daily without the boys' mother, who, avoiding the male competition, claimed complete lack of musical talent.[34] The father, endowed with a fine operatic voice, was accompanied by Erwin on the violin and Werner on the cello. Later Werner played the piano. One can easily imagine the spirited father filling the house with arias accompanied by the halting notes of the budding musicians, while wife and mother retreated to her husband's homework.

Already inspiring ambition by example, August strongly encouraged the competition between his sons, especially in the academic realm and within that in mathematics. Years later Heisenberg recalled: "Our father used to play all kinds of games with [us]. . . . And since he was a good teacher, he found that the games could be used for educating the children. So when my brother had some mathematical problems in his schoolwork . . . he tried to use these problems as a kind of game and find out who could do them quickly, and so on. Somehow I discovered that I could do that kind of mathematics rather quickly, so from that time on I had a special interest in mathematics."[35]

Early and intense rivalry, deliberately stoked by August, coupled with August's "restless drive" and the family's upward momentum, must have engendered Werner's own enormous lifelong drive to excel in everything he did—mathematics, music, even table tennis. His teacher at the Max-Gymnasium in Munich often observed in his grade reports: "The pupil is also extraordinarily self-confident and always wants to excel."[36] But the rivalry also must have engendered the increasing dissonance and eventual disharmony between Werner and his brother. As boys, the two often fought fierce battles with each other. As they grew older, they fought even more frequently and intensely. Finally, after one particularly bloody fight—in which they beat each other with wooden chairs—they called a truce and went their separate ways. After that, they had little to do with each other, except for occasional family visits as adults.

After 1916, Erwin was hardly at home, anyway; he was away first on military duty, then later in Berlin where he studied, married, and settled. Werner, meanwhile, moved increasingly outside his family as he became involved in youth-movement activities. Werner never mentioned his brother to his youth-movement comrades, nor do they recall ever having met him. Erwin, who became a chemist, is conspicuously absent from most of Heisenberg's public recollections. Even after World War II their relationship remained cool, especially after Erwin, now an official of the anthroposophic movement, unsuccessfully tried a second time to convert his brother. Erwin died in 1965 and was buried on the grounds of the movement's headquarters in Basel, Switzerland.

For Heisenberg, competitive achievement outside the family seems to have served more as a personal challenge than as a means of impressing or subordinating others. Although he was a poor loser, he often settled for acceptance and recognition by his peers. The personal challenges that he continually set himself and the hard work required to achieve them were clearly noted by Heisenberg's later youth-movement comrades. He was a *Willensmensch*, a man of will, one of them recalled. This was especially evident in athletics, a favorite pastime of the youth movement.[37] Not a natural skier, Heisenberg nonetheless trained himself to ski excellently and over difficult terrain. Nor was he an exceptional runner, yet one former comrade recalls him running laps alone at the school with a stopwatch in hand to improve his time. Whenever there were long-distance running races—supreme challenges to determination—Heisenberg, endowed with incredible endurance, was always among the early finishers. Later he won a *Sportabzeichen*, a state badge for outstanding physical accomplishment, that he proudly saved. Eventually, the driven youth found other challenges to his enormous talents in the beauties of classical piano and, after 1920, in the seemingly insoluble puzzles of quantum physics.

For Heisenberg the child, the family world in which he lived was simple, ordered, predictable, and full of motivations to excel. Family roles and expectations of propriety were clearly defined, even if the bases for bourgeois values were not. Although control lay in the hands of his male elders, Heisenberg could always gain attention through excellence. Despite his father's outbursts, Heisenberg apparently never lost a sense of fairness. In competing with his brother for his father's attention, he was sure of the rules and certain that the competition would be judged fairly. And when he lost, he could always find solace in the arms of his mother—a situation that must have contributed early to his unusual insecurity and the importance he attached to trust.

When Heisenberg entered school in Würzburg at the age of 5 or 6, one of his teachers falsely accused him of some misdeed—at least the boy regarded the accusation as false—and rapped him sharply on the knuckles with a switch. More emotionally than physically hurt, the lad immediately withdrew into himself, broke off all further interaction with the teacher, and refused to cooperate for the rest of the year.[38] The pattern repeated itself to the end of his life. Whenever Heisenberg felt that his freely (even naively) offered trust had been betrayed, instead of confronting the offender he would sever relations irrevocably. He never irrevocably rejected his parents—although he later dismissed their social outlook—but his brother was apparently cut off to the end of his life.

Whenever Werner felt maligned or betrayed, he retreated into the orderly, secure world of his inner thoughts and dreams. Never very imaginative or fanciful, this inner world grew increasingly abstract, logical, and mathematical. Long hours of solitary piano practice, immersed in a world of order and harmony, must have contributed to the shaping of this inner world. His gymnasium teachers noted the distinction: "More developed toward the side of rationality than of fantasy and imagination"; "the pupil . . . appears . . . to be developed mainly toward the side of understanding; feeling for form and imaginative powers seem to be less developed, which is probably related to the fact that he is less outgoing than would be desirable."[39] Among Heisenberg's favorite childhood stories was a collection about Klaus Störtebeker, a fifteenth-century Hanseatic sea pirate and folk hero whose adventures were less fantastic than realistic.[40]

Because of the geographic dispersal of their relatives, Werner and Erwin became familiar with long train rides and even with traveling alone at an early age. They frequently visited their father's family in Osnabrück. August's sisters were especially fond of their two nephews, and Werner was particularly enthralled by the younger one, Aunt Grete. "She led us around on invisible reins," he recalled.[41] Heisenberg always felt at home among his close, warm, and predominantly female Osnabrück relatives, and he always enjoyed the company of his grandfather, the kind and encouraging Wilhelm Heisenberg. Years later Heisenberg still remembered him for his fine, white hands and his lessons in such practical matters as nailing a lid on a box—lessons he surely was not taught at home.[42] Perhaps because of Grandfather Wilhelm the Heisenberg boys became fond of building technical gadgets in their teens. Their masterpiece was a 1.5-meter electric battleship, equipped with remote-controlled steering and electrically fired canons; it was proudly displayed in the Heisenberg home for decades. With electric lighting only two decades old in Bavaria, the ship was no small achievement.

Family interaction of quite a different sort occurred with the academic Weckleins. Munich and Würzburg were separated by a two-and-a-half-hour train ride, but that did not deter the two sides of the family from frequent Sunday strolls together. Grandfather Wecklein was an enthusiastic hiker—a common activity in those days. He devoted every Sunday to the great outdoors, and during holidays he disappeared on long tours through Bavaria, Austria, and Switzerland. He often wandered with a small group of school colleagues, who called themselves the *Alte Herren-Riege*, the old men's club. Among the club's members was the assistant rector of a Landshut gymnasium, Gebhard Himmler, father of the infamous Heinrich Himmler—a connection that later proved of benefit to Heisenberg.[43]

During the Sunday walks with his Würzburg relatives, Wecklein must have taught his grandsons the joys of experiencing the romantic beauty of their German homeland. Strolling with the family schoolmasters, Werner also learned—along with probable drilling in Latin verbs—the academic and social situation of his family. Matters of school policy were often discussed on those outings, for Werner's father and grandfather were thoroughly involved at that time in efforts to preserve the social gains of gymnasium teachers and in continuing their own climb up the social ladder.[44] Although Heisenberg was only a boy, he later claimed that not all these discussions passed over his head: "The problems of the gymnasium of that period are thus well known to me from my youth."[45]

The pressures and idylls of Heisenberg's Würzburg life, and with it his early childhood, came to an abrupt end when his father was suddenly called to Munich to succeed his teacher, Karl Krumbacher. Krumbacher was only 54 years old when, weakened by diabetes and overwork, he collapsed and died on his way to a lecture in December 1909.[46] The search for a successor began immediately. As was common for teaching chairs, the entire program in middle and modern Greek studies rested on the shoulders of one man. The search did not last long. Within nine days of Krumbacher's death, the dean of the philosophical faculty notified the academic senate that, instead of proposing the usual list of three candidates, the faculty had already decided on a single one, August Heisenberg.[47]

The dean had noted the elder Heisenberg's unusual pedagogical and scholarly talents. Even greater accomplishments could be expected with a full-time university appointment. More in his favor, Heisenberg was actually the only university-qualified teacher of middle and modern Greek in Germany.[48] The rector accepted the faculty proposal and passed it to the interior minister. On December 30, 1909, the interior minister, no doubt eager to silence an articulate spokesman for gymnasium personnel, ordered the immediate appointment of the candidate with an annual salary of 6,000 marks, to which were added seminar fees. The candidate accepted readily, and within a month Herr Professor Dr. August Heisenberg began lecturing in Munich as the occupant of the only chair in Germany for Byzantine philology.[49] The rest of the family remained in Würzburg to finish out the school year.

CHAPTER 2

THE WORLD AT WAR

THE HEISENBERGS MOVED IN 1910 INTO A LARGE APARTMENT ON THE TOP FLOOR OF A three-story building at Hohenzollernstrasse 110 in the fashionable Munich district of Schwabing.[1] The house, built at the turn of the century, stood near the intersection of Hohenzollernstrasse with streets named for Isabella and Fallmerayer, almost opposite one named for Joseph Klar—all in existence today. Werner and Erwin shared a bedroom at the rear of the traditionally furnished, dimly lit apartment, heated by a coal-burning stove. Their window looked across a small yard toward the buildings on Fallmerayerstrasse to the north, away from the noise and bustle of the busier Hohenzollernstrasse. When not afloat, their electric battleship lay docked on a dresser in their room. Probably a live-in maid also occupied the third-floor quarters of the Heisenberg home. Servants were as much a social necessity as a practical help for a professorial family of two growing boys.

The boys keenly regretted trading their spacious Würzburg residence for a Munich flat, but they could not have had a more sheltered or cultured setting in which to grow up. Munich, a city of more than a half million people at that time, 75 percent of whom were Roman Catholic, served as the secular and religious administrative center of Bavaria, and Schwabing served as its cultural apex. Dominating the city were the Bavarian king's residence, flying the blue and white checked flag of proud Bavaria; the massive royal ministry buildings; and the medieval town hall, all stretching from the northern side of the center of town, Marienplatz, with its famous cathedral, the twin-towered Frauenkirche (Church of Our Lady). These structures, along with the ornate stone Siegestor (Victory Arch) at the Schwabing border, all gave visual testimony to the administrative power of the "court and residence city" nestled along the Isar River. Yet Munich was also the capital of the rural province of Upper Bavaria, which stretched south to the foothills of the Alps, near the Austrian border. Farm produce arriving daily at the "victuals market," the frequently seen traditional dress of the residents, the flowing beer at the Oktoberfest, and the Föhn wind that swept down from the Alps along the Isar River basin were constant reminders of the city's rural roots.[2]

The Schwabing district presented quite a different aspect of the city. Schwabing lay on the northern outskirts of Munich, just north of the university. At that time it approached the height of its reputation as a center for art, music, and literature. It had also become the preferred neighborhood of the social elite. Nobles, officers, state bureaucrats, and academics shared the neighborhood with artists, writers, and the owners of the numerous shops, pubs, and cafés that served them.

With Professor Heisenberg seated in a university chair and Dr. Wecklein heading the school system and Maximilians-Gymnasium, the Heisenbergs easily mixed with the upper strata of their new neighborhood and city. Those strata had long since eclipsed the monarchy in power and status. With Bavarian monarchs prone to madness, Munich's ministerial officials, merchants, and civil-servant professors formed a "ministerial oligarchy," dominated—despite the particularist Roman Catholic majority—by pro-Prussian, Protestant liberals. This oligarchy, and the Heisenberg family's solid position within it, remained intact until well after the outbreak of war in 1914.

Changes and threats of change were already apparent. Munich, as the leading commercial and cultural center of the south, served as a magnet for both rural and elite outsiders. While the city experienced an overall expansion in those years, Schwabing underwent the greatest development of any Munich suburb. Many of the trees in the area had already fallen victim to lot clearing for future housing, and much construction was under way. A 1911 map of the city shows the Heisenberg home at the edge of development, one of the last buildings on the north side of Hohenzollernstrasse. There were no buildings at all on the opposite side of the street.[3] Barren lots, fields, and mounds of dirt must have made tempting playgrounds for neighborhood children.

The most conspicuous feature of Werner's new neighborhood was a new electric tram that thundered west down Hohenzollernstrasse past the Heisenberg home, before turning south at Kurfürstenplatz.[4] Almost no other motorized traffic ran along the streets. An engineer named Mailer owned one of the few automobiles in the area—a fire-engine-red contraption that yielded only grudgingly to the crank—much to the amusement of the local boys.[5] Save for the tram and an occasional car, horse-drawn wagons and carriages delivered practically everything: mail, ice, beer, milk—and street cleaners. The Heisenbergs could also catch the sounds of the many wandering musicians, watch the gas-lamp lighter, and wave to the policeman stationed on the corner near their home.

But the high point of neighborhood life arrived with the Bavarian soldiers who regularly marched and rode down Hohenzollernstrasse to the rhythms of marching bands and drums, on their way to and from training at the army barracks on the Oberwiesenfeld at the west end of the street. Everyone threw open doors and windows to watch them parade, and neighborhood boys ran cheering after them. In the summer, the boys often followed the soldiers out to their exercises.

The extent to which Heisenberg participated in such neighborhood fun is apparent from his studious nature. He was not a gregarious youngster, nor, when school was in session, did he devote much time to frivolous play. Even though he finished his regular school lessons with ease, he undertook extra studies. In addition, his parents enrolled him in piano lessons with the well-known Munich pianist Peter Dorfinger, who demanded hours of intensive practice. By the time he was a teenager, Heisenberg could play master piano compositions, and he participated in chamber presentations and frequent musical interludes at school ceremonies. No evidence survives of a childhood playmate.

In the fall of 1910, Heisenberg enrolled—for his fourth and last year of primary education—in the Elisabethenschule, several blocks from his home.[6] Erwin, however, having completed primary school, took the entrance exams for his grandfather's gymnasium and spent the next nine years in the B section of each class. Werner followed Erwin to the gymnasium the next year and consistently attended the A sections, which were usually reserved for the brighter pupils. This circumstance and the one-year separation in grades, despite their nearly two-year separation in ages, must have inflamed the boys' already intense rivalry.

While awaiting the completion of renovations and new construction, the Max-Gymnasium occupied temporary quarters in the middle wing of the Damenstift on Ludwigstrasse near the university to the south of the Heisenberg home. The school was still there when Werner entered. Going to school each day, he could join his father and brother on their early morning walk to their respective classes. Not until Werner's second year did the gymnasium reopen in its present location on the corner of Morawitzkystrasse and Karl-Theodor-Strasse, several blocks northeast of the Heisenberg home.

Nine-year-old Werner entered the Max-Gymnasium in September 1911 for a nine-year course of study that prepared its students for entering into advanced study at a university. By then, the school had attained a reputation under Wecklein as an academically and socially elite institution—a "plutocratic gymnasium," one administrator called it.[7] Rector Wecklein's efforts to ensure the scholarly excellence of his teachers made the institution increasingly attractive to the wealthy elite. Among its illustrious pupils was Heisenberg's most noted predecessor, Max Planck, who also briefly taught physics at the school. Since school was not free and pupils were not assigned to neighborhood schools, the elite naturally chose the best for their children. The arrival of the cultured classes in Schwabing reinforced the social composition of the gymnasium. Although a modest fund existed for needy pupils, few took advantage of it. Of the 37 pupils who survived the 1911–1912 school year in Werner's first class, 19 of their fathers held titles that began with Königlicher (royal). Eleven fathers were jurists or state officials. Eight are listed as professors and five as military officers. The remaining thirteen held respectable positions: businessman, craftsman, factory owner, artist. With 576 pupils enrolled (all boys) and 44 men as faculty members, the Max-Gymnasium was second in size only to its brother school, the Luitpold-Gymnasium attended earlier by Einstein.[8]

The Interior Minister for Church and School Affairs decreed the curriculum in detail throughout the Bavarian school system. During the decades preceding Heisenberg's gymnasium study, German gymnasium education had undergone considerable controversy and reform. Neohumanism, the dominant educational ideology, had fostered the notion that only the classics should serve as the foundation for any professional career requiring higher education. By the turn of the century, the more practical demands of commerce, industry, and technology rendered humanistic studies no longer the only route to higher education. With Werner's classicist grandfather on the school board, Bavaria remained one of the last bastions of classical humanism. Even after the promulgation of reforms in 1914, the overall aims of gymnasium education remained the same: moral education on the basis of Christian ethics, "education in the spirit of the fatherland," and preparation for independent academic research, in that order.[9] Gymnasium curricula continued to stress the classics. Of the 263 credit hours decreed over nine years of study, 63 were devoted to Latin, 36 to Greek, and 31 each to mathematics and German. The remaining 133 hours were divided among history, religion, athletics, French, geography, and nature studies, in that order. Physics received the least attention, next to drawing, with 6 hours spread over three years.

During his first three years at the gymnasium, Heisenberg's main subjects were Latin and mathematics, of which he received eight and four class hours per week, respectively. Beginning in his fourth year, he received six hours per week of written and spoken classical Greek. Three years later, he began two weekly hours of physics. In his last year, Heisenberg read Horace and Tacitus in Latin; Homer, Sophocles, and Plato in Greek; and pondered elementary classical mechanics from a single physics text. One can imagine little Werner in his school suit standing at attention next to his desk as he recited a Latin or Greek text from memory, while the teacher reigned ominously over the boys of his class with a switch.[10]

The education ministry's official neglect of science and technology was countered by unofficial interest in these subjects among the pupils. Such exciting developments as the discovery of X-rays and radioactivity, the rapid advance of the electrical and chemical industries, and the Kaiser's personal interest in promoting and financing scientific and technological research captured the pupils' imaginations and opened many new career opportunities. Of the 41 pupils in the graduating class at the end of Heisenberg's first year, 20 intended to enter careers in science, technology, or medicine. The next choices reflected other advancing fields of the day: seven hoped to go into banking and five into the military. Only one brave pupil opted for a classical subject, archaeology.[11]

As with Heisenberg's father, nearly all of the school's teachers possessed doctorates and an interest in scholarly research. The newly renovated Max Gymnasium was now the most modern school in Munich. With it came the most extensive facilities for physics instruction in the city, fortuitously enabling advanced science studies in an otherwise literary setting. Since gymnasium teachers qualified in at least two subjects, mathematics teachers taught physics; and they usually taught in all nine grades. Consequently, a precocious younger pupil could be kept stimulated with material from the science teacher's upper classes, while the teacher, inspired by the school's facilities, would refer often to science in his lower classes. Both were true for Heisenberg and his favorite and most influential teacher, Herr Christoph Wolff.

Little is known of Herr Wolff, whose personnel records were lost in World War II. Apparently Wolff did well on his qualifying examinations in the early years of the century, for board member Wecklein, who kept the best teachers for himself, immediately assigned the young man to his school. Although Herr Wolff never obtained a doctorate—then not necessary for science teachers—he had quickly advanced by 1910 to the top of his profession, gymnasium professor with the title of royal study councilor (Königlicher Studienrat). Wolff taught Heisenberg mathematics during his first three years at the school and both math and physics in his last three years there.[12]

Heisenberg's need for study beyond that in the prescribed curriculum is apparent from the curriculum. For a mathematically inclined pupil accustomed to solving his older brother's math problems, the course work was pretty weak. Until the fourth grade, Heisenberg was taught only simple arithmetic. From grades three to seven he studied plane geometry from a single elementary textbook. Not until the seventh and eighth grades did he finally encounter plane and solid trigonometry.[13] By fourth grade he was restless. His ability to dash off his assignments, fueled by his eagerness for recognition, left him few remaining challenges. His first-grade teacher noted in his grade report: "The thought operations, namely in grammatical and arithmetic questions, are completed rapidly and in most cases without error. Spontaneous diligence, great interest that the subject is done thoroughly, and ambition." At the end of the second grade, his teacher wrote, "He has attained his excellent achievements with playful ease; they have cost him no expenditure of effort." By the fourth grade, the teacher complained: "However, with his ease of comprehension [he is] often careless in his homework assignments."[14] Heisenberg easily mastered his homework in brief bursts of work, then turned to other, more challenging (and enjoyable) activities, such as music—a pattern that persisted throughout his life.

Another influence was soon at play. Heisenberg's fourth school year coincided with Germany's first war year, and his father, still a reserve army officer, absented himself for long periods of active duty. Challenged neither at home nor at school, Heisenberg challenged himself in his own studies of advanced mathematics. By the end of the school year his diligence had brought results: he had moved to the very top of his class, where he remained.

Despite his shyness, Heisenberg was acknowledged and well liked at school, not only for his pleasant manner, mathematical talent, and musical ability, but also as one of the rector's two grandsons—a combination that can make life miserable for a quiet boy. But his classmates' respect for authority and accomplishment apparently prevented serious jealousy.

Heisenberg's family connection brought him a special honor in March 1913 when Bavarian Prince Regent Ludwig came to dedicate the new gymnasium building. Werner's mother, a descendant of literati, wrote a poem for the occasion. Her favorite son, 11-year-old Werner, recited it to the prince during the convocation, which concluded with an enthusiastic singing of the *Regents Hymn*.[15] The delighted prince duly thanked the proud lad with an official certificate and a pair of cufflinks engraved with the letter *L*—a memento that remained among Heisenberg's most cherished possessions.

Heisenberg's encounter with the prince was the culmination for him of an era that would soon fade. On September 1, 1913, Grandfather Wecklein, approaching 71 and having seen his grandsons safely started on their school careers, finally retired. He had by then attained the prestigious title of royal privy court councilor (Königlicher Geheimer Hofrat). Two months later, Bavaria's Wittelsbach dynasty itself underwent a change. Upon the death under suspicious circumstances of insane King Ludwig II in 1886, his uncle Prince Luitpold had served as regent for the rightful heir to the throne, Ludwig II's deranged brother, Otto. Luitpold presided over a decline in monarchical influence until his own death in 1912, whereupon his son Ludwig assumed the regency. A year later, Prince Regent Ludwig deposed Otto and declared himself King Ludwig III, promptly alienating his tradition-minded subjects, who five years later made Ludwig the last of the Wittelsbachs to rule Bavaria. Both transitions, Wecklein's retirement and Ludwig's coronation, marked a transition in gymnasium education, in German history, and in Heisenberg's life.

A year after Heisenberg's recital before the prince, the world erupted in war. Although they would later face bitter disillusion, all sides greeted the outbreak of war with jubilation—a feeling that emerged less from the snapping of international tensions than from an explosion of nationalist fervor, social unity, and war romanticism. Munich broke out in a patriotic fever. Sandwiched in central Europe between potential enemies on the east and west, Germany had settled on instant mobilization and lightning offense as its secret weapons against potential encirclement and a seemingly hopeless two-front war. The strategy instantly galvanized Munich and the nation at the news of war in the early days of August 1914. "There was plenty of excitement at the Pasing train station," began a story in a local paper on the first day of war. "Almost every minute brought in a new train with reservists waving from the windows, whom people greeted with fluttering handkerchiefs and shouts of 'Hurrah.' . . . On the streets themselves, groups stood around everywhere in front of the notice boards; people surround officers and give them ovations. . . . In front of the Feldherrnhalle, during the changing of the watch, there were stormy outbursts of enthusiasm and constant cheering for the troops."[16] On Kurfürstenplatz, near the Heisenberg home, people gathered daily during the first weeks to celebrate the many early German victories with patriotic songs and to cheer on their boys, who, they were certain, would be home by Christmas.[17]

The Heisenberg family faced a dilemma. They were as much in favor of the war as their compatriots, but they had to face the fact that Professor Heisenberg might have to fight in it. The grim response of Werner's parents in those early days contrasted sharply with the town's jubilation, a contradiction he found bewildering.[18] The professor indeed received a call to duty within days. Wife and sons accompanied him to his regiment in Osnabrück at the end of August. A good-bye photograph of Werner and Erwin on each side of their mustached father, dressed in an officer's uniform with a sword and the traditional pointed helmet, reveals both pride and concern in all three faces. The 12-year-old Werner, dressed in an ill-fitting suit for the occasion, is standing in his frequent defensive posture with his arms crossed and his face set.

The family stayed with their relatives until October, when August's regiment marched into German-occupied Belgium. In Osnabrück, Werner's father received promotion to captain with command of a local infantry company. His primary duty was to station one of his men, armed with a machine gun, on the roof of the tallest building, in order to guard the city against enemy planes. Werner visited the soldier in his perch from time to time, probably a welcome relief for the fellow, who certainly had little to do. At that time, airplanes were used only for reconnaissance. Bombers and fighters were later innovations.

*Erwin and Werner (right) with their newly mobilized father in front of
the University of Munich at the outbreak of World War I.*

During those months in Osnabrück, Werner surely discussed and reflected upon his father's often expressed political outlook and his many activities that resulted from it during the prewar years. Professor Heisenberg's views were typical of his generation and station, and, despite Werner's later rebellion, they were influential on his son's outlook. Studies have shown that, following the unification of the Reich in 1871, socially prominent German professors turned ostensibly apolitical in public affairs. This was intended both to insulate themselves from the mass politics of the lower, working classes—officials elected by the lower classes could not be entrusted to administer their exalted profession—and to establish their status as keepers of German culture. Although they still engaged in politics, they regarded their activities as residing outside the party system and in service to German culture (especially to themselves as "bearers of culture"), so they could believe themselves free of political taint.[19] Many academics joined self-interest pressure groups, such as the Gymnasium Teachers' Association, which promoted German cultural interests in the international arena. They were encouraged by the example of other professionals—bankers, industrialists, military men—who were exercising their own "apolitical" influence in state service.

By 1910, the unified empire had achieved a powerful economy and military, but its diplomacy was weak. A late entrant in the international grab for power and markets, the country had failed to achieve a meaningful alliance with any of its competitors or to establish a large colonial market and source of raw materials for its expanding industries. The German people and their political leaders felt themselves increasingly isolated internationally and surrounded by hostile forces. The Balkan

Peninsula to the southeast, the "powder keg of Europe," provided one outlet and a possible geographic wedge to Germany's ally, Turkey. German classical scholars, including the occupant of Munich's chair for Byzantine studies, provided a connection—a cultural one—to Greece, the southern end of the Balkan Peninsula.

When Werner's father assumed his Munich chair in 1910, he immediately took up the apolitical cultural barrage his predecessor, Krumbacher, had instigated, and he aimed it directly at the Balkans, Greece in particular. Culture could obviously serve many purposes. Although he followed Krumbacher's example on nearly every academic issue, Heisenberg's very first publication as professor contradicted his colleague's stand on the intense controversy then raging over vulgar Greek.[20] Heisenberg and other German scholars lobbied against a proposal to replace classical Greek as the official language of Greece, seeking to bind Greece more closely to Germany through the German preservation of classical Greek language and culture.

The outbreak of war intensified the political efforts of academics on all sides. No nation saw itself as an aggressor; each believed that it was simply fighting to defend home and culture from those who would destroy them. For Germany, lightning offense as a defense justified the invasion of neutral Belgium. This action, together with the unfortunate destruction of Belgian art works, set the seal on other countries' perception of Germany as an aggressor and a destroyer of culture. Germans, on the contrary, convinced themselves of the profound interdependence of the military, nation, and culture. For many the three were almost identical, making it easy to believe that military measures were essential for the preservation of German culture.

The rationalist veneer overlying such fierce nationalistic emotions pervaded professional German society, including humanistic gymnasiums, and even infected supposedly objective scholars. This is nowhere more evident than in the infamous German academic manifesto "To the civilized world," which overtly supported the German cause. More than 4,000 "apolitical" professors—almost the entire German professorate—eventually signed the manifesto (to their later regret).[21] Only Einstein and two others are known to have refused. Of course, German academics were not alone in their folly—numerous equally unscholarly, nationalistic manifestos and countermanifestos were written and issued throughout the war in nearly every land.[22]

Professor Heisenberg's war duties prevented him from joining his like-minded colleagues in the manifesto war, but they did not hinder his enthusiastic participation in the propaganda war. As early as 1912, as Balkan tensions erupted into battles, Heisenberg and his Munich philhellenic colleagues formed the ostensibly studious and objective German-Greek Society.[23] It unleashed an overt propaganda campaign to persuade Greece to remain pro-German, or at least neutral, in the coming war.[24] Unfortunately for them, Bavaria's mad king, Otto, had once ruled Greece, leaving little infatuation with Germany among the Greeks.

The German-Greek effort ended when, in 1917, England and its allies invaded Greece, deposed King Constantine, and induced the new parliamentary government to join the war against Germany. At least Professor Heisenberg could console himself with the thought that "We Germans were well represented with the intellectual leaders of Greece through the activities of our academic circles."[25] When Werner encountered Greek for the first time and his father put his words into action in 1914, the boy's admiration for his father's efforts must have increased greatly.

Unlike most other older academics, Heisenberg's father was not content with waging a mere cultural battle. Although he was of military retirement age, the 45-year-old professor demonstrated the strength of his commitment to his country by taking up arms for her. When his battalion was ordered into occupied southern Belgium in October 1914, Captain Heisenberg led the charge over the border. His regimental commander ordered his company to guard a 26-kilometer stretch of railroad track near Manage près Mons. Captain Heisenberg personally patrolled the track on horseback every day. His only other duty was to issue travel permits to the local populace from the home of a Belgian family that (willingly or not) provided him quarters.[26] But the enthusiastic soldier soon chafed under such tame assignments, and on Christmas Eve, 1914, he transferred to Landwehr Infantry Regiment 56, headed for the front.

To prepare for battle, the middle-aged captain underwent a month of infantry combat training in Belgium. He marched out in command of a company for the trenches of the Argonne Forest near Servon, France. Before leaving, August penciled his last words to family, friends, and colleagues on postcards. Just two weeks later, at the end of January 1915, his exhausted company pulled back for a nine-day respite. The much-sobered captain wrote to a Munich colleague, "If anyone speaks to you about the poetry of trench warfare, then please object, for holding out a grenade attack is nothing but pain, misery and suffering."[27]

By April, the captain had had enough. The courage of his convictions had now evaporated under fire. On April 24 he wrote again to his colleague—this time from Osnabrück: "Despite all of my bellicose inclinations, the longing for peaceful activities had become very alive once more. Therefore, when, after a very difficult and strenuous fight, I was rather at the end of my nerves, I asked for my discharge from the front and my transfer to Area Command I Munich, in order to do garrison duty once again in Munich."[28] Because of his age, his request was readily approved, and within a month the erstwhile warrior was back in Munich delivering Greek lessons to a seminar of four students. His cherished company, many of whose members were about the same age as his students, remained at the front to face the horrors of a war that bellicose professors had helped bring about. Werner, who later remembered that his father had returned from the front because of wounds,[29] must have suddenly seen his father in an entirely new light: either as a tragic hero to a cause that ultimately failed, or as a man now compromised. Either perception would have served as a setup for what would later become total disillusion.

Heisenberg's school immediately felt the effects of war. Within days of the war's outbreak, the Bavarian Army Command seized the school's year-old building for the quartering of newly mobilized troops. The garrison was meant to last only until the men went into battle a few weeks later, but the building remained occupied until shortly before Heisenberg graduated in 1920. With Bavarian troops settled in for the duration, the Ludwig-Gymnasium near Marienplatz, the center of town, hosted the required courses of its brother institution with half-day sessions for each school. Electives in music, art, and languages were curtailed, and, unfortunately for Heisenberg, the physics laboratory was closed. Lack of coal in early 1917 forced the school to cease operations almost entirely. For a month, the pupils simply picked up and turned in their daily homework.[30]

The Bavarian Army vacated the Max-Gymnasium at the end of the war, only to be replaced by a hospital for the wounded. Once the hospital left, the Bavarian regime then in power quartered a unit of local troops in the building. Two Free Corps companies, among the Reich troops dispatched to Munich, replaced the Bavarian troops in the fall of 1919. After a much-needed disinfecting, the building finally returned to classroom use, but a lack of coal again forced pupils onto half-day sessions until March 1920. Heisenberg graduated four months later. Despite these interruptions, or more likely because of them, he progressed far beyond his assigned schoolwork through independent study.

The war also affected the gymnasium's faculty and pupils. Six reserve army officers among the faculty, including Lieutenant Christoph Wolff, were called to duty during the first few months. All returned unharmed, including now-Captain Wolff, who, like Heisenberg's father, quickly headed home after tasting the realities of trench warfare. But the youngsters they sent into the trenches fared far less well. Eager pupils answered patriotic calls to the colors by school, university, and academy officials, who also urged a draft for boys over 17.[31] Of the 452 pupils enrolled in the Max-Gymnasium in the fall of 1914, 74, including the entire graduating class, had joined the army by the end of the year. Eleven never returned. A plaque with their names and those of 22 other pupils and one student teacher killed in the war now hangs in the gymnasium.

In order to foster the identification of culture with the military and to instill "love of the fatherland" in its pupils, the Max-Gymnasium subjected them to heavy doses of patriotic and military indoctrination throughout the war. The celebration of battlefield victories, the birthday of the popular Field Marshal Paul von Hindenburg, and the anniversary of King Ludwig's coronation offered opportunities for patriotic school convocations. Carefully edited letters from former pupils at the front appeared in gymnasium annual reports and were used for instruction. Heisenberg's 1916 composition class wrote an essay on the topic "My participation in the war for Germany's world stature."

For at least two years, from 1916 to 1918, Heisenberg belonged to the Max-Gymnasium's paramilitary *Wehrkraftverein*, the Military Preparedness Association, which institutionalized the military indoctrination. His membership in the association led to many of his postwar extracurricular activities. The association was founded in 1910 by a group of Munich army officers intent on providing early training to gymnasium students, the officers of the future. A year later they convinced Prince Luitpold to grant the association official sanction, and soon a Military Preparedness Department was attached to every Bavarian gymnasium. Initially, the association held only occasional after-school drills for the boys, but with the outbreak of war, it offered more extensive training and indoctrination. At the Max-Gymnasium, Dr. Ernst Kemmer, head of the school's military department, set up a "young storm regiment," one of whose later members was Heisenberg. He joined the regiment probably for fun at first and in anticipation of his own likely entry into service upon turning 17 in December 1918. Fortunately for him, the war ended before his birthday.

Throughout the war, Dr. Kemmer, who had also headed Heisenberg's third-year class in 1913–1914, wrote yearly accounts of his youth training in the gymnasium annual reports. After the outbreak of war, Kemmer's training expanded to twice week-ly, on Saturday and Sunday afternoons, and once monthly for an entire day. In keep-ing with Bavarian war ministry guidelines, Kemmer's training focused not on weaponry but on physical and mental preparation for combat. This included physical fitness and familiarization with military discipline; instruction in marching, map read-ing, and reconnaissance; and such later youth-movement activities as tent making, outdoor cooking, and "the joy of hiking, the love of nature and of homeland."[32]

Ninety pupils descended on Kemmer's unit during the first year of the war. So many youngsters clamored to join that Kemmer had to lower the minimum age to 13. Yet interest in paramilitary training ebbed as the war dragged on. Parents complained that their tired children had little energy left for schoolwork and that the school itself was fast turning into a "prep school for recruits" to an increasingly nasty war. Kemmer argued that his program proved its utility as each graduating class marched into the trenches. The school could not abandon its moral duty, he wrote. The war "has made [the school] responsible for the future of German culture!"[33]

Disillusion with such belligerence quickly spread as German forces bogged down in the west, casualties mounted, and food and coal began to run low. In Munich alone, 13,000 inhabitants died in the war. In the early days, reports by supply experts had encouraged undue optimism. A Munich paper boldly proclaimed, "The danger of a cut-off of food supplies to large cities does not exist, say the experts."[34] The experts obviously had not reckoned with or did not care to inform the public of the possibil-ity of a near-total British naval blockade. It proved unbreakable even after Germany unleashed an unlimited U-boat war, bringing the United States to arms.

In 1915, rapidly inflating food prices led to demonstrations and bread rationing in Munich. Workers at the Krupp armament works, which was practically Bavaria's only heavy industry, were, as elsewhere, growing dangerously restless. In August 1916, the victorious commanders of the eastern front, Hindenburg and Ludendorff, took command of the economy as well as the military—alienating the already anti-Prussian Bavarians. To make the best of shortages, the commanders introduced food rationing and a centralized control system for all food production and distribution. In Bavaria, the rationing of milk, meat, and sugar was administered by weekly ration cards distributed by gymnasium pupils. But even with ration cards, one needed to know a friendly farmer or a black-market source to obtain all the necessities. In late 1916, an early frost wiped out most of the potato crop, a staple of the Bavarian diet.

A coal shortage in the ensuing months made the terrible "turnip winter" of 1916 to 1917 (when turnips were the main staple) the worst of the war. Food and coal ran out in Munich, soup kitchens were set up, and teenagers no longer sported on Hohen-zollernstrasse. "We boys mostly went hungry," one of them wrote. When one of Wecklein's colleagues asked how he was, the old fighter responded with a stiff upper lip, "We're starving and freezing, but we can take it."[35]

Faced with a failed harvest, a hard winter, and a lack of workers, in early 1917 the two military commanders ordered all men between the ages of 17 and 60 who were not in military service to register for agricultural and factory war-assistance work (the *Hilfsdienst*). With little industry in Bavaria, nearly all Bavarian helpers went to the farms. The authorities pressured schoolboys under 17 into volunteering during the spring and summer months and gave them an early final examination before they left for the fields in April. Thirty two pupils from the Max-Gymnasium entered the agriculture service that year, among them Werner's brother and eight of his own classmates.

Pupils who did not enter any service were put to work in vegetable gardens during the holidays. Werner's class worked in gardens at a small factory and at King Ludwig's residence at Leustetten under the direction of their military training master, Dr. Kemmer. "The boys worked diligently and for the most part happily," he wrote.[36] The agriculture service was less successful. Newspapers reported that the poorly organized pupils regarded their early country adventure as "a pleasant summer holiday," and farmers had little patience with a pack of wild city boys rampaging through their fields.

When Heisenberg joined the agriculture service the next time—while his brother entered the army—the program and pupils were better prepared and more tightly controlled. The continuing hardships had rendered the boys much more cooperative. Although the Bavarian agricultural officer reported only moderate success in his personal campaign to recruit volunteers among the military training units, Heisenberg, then a 16-year-old member of Dr. Kemmer's unit, required little prompting.[37] The food shortage had taken its toll on the growing boy, whose parents had yet to locate an independent source to supply the family. Heisenberg grew so weak from hunger that he once fell off his bicycle into a ditch. He and his family decided he would go to the farms that summer.

Learning from past experience, the military overseer of the project divided the nearly 4000 pupils, called "young men" (*Jungmannen*), into squads of 10 to 30 boys, each under the close supervision of a military training leader or an officer, who was in turn under the supervision of the district army corps commander. To encourage better work, the army supplied each boy with a new pair of combat boots. Judging from photographs of Heisenberg's youth group, the footwear found excellent use on outings. The boys were also warned that an evaluation would be placed in their annual grade report. Werner, always the well-behaved pupil, received what sounds like the standard commendation for good behavior: "As a Jungmann he earned the recognition of his leader . . . through his good behavior and his work accomplishments." Soon afterward, Kemmer promoted him to group leader, in charge of a squad in the gymnasium's military unit.[38]

Heisenberg (4th from right) and his comrades on the Bavarian diary farm in 1918.

The home-front commanders posted Heisenberg's farm squad at a large dairy school near Miesbach in the foothills of the Alps in Upper Bavaria, south of Munich. There he worked, save for short leaves, from early May until September 5, 1918, when he returned to school for ten days before leaving again to help with the potato harvest. That summer marked Heisenberg's first extended time away from home. It also marked the start of his frequent correspondence with his parents, a correspondence that lasted until the death of his mother in 1945. (His father died in 1930.) In nearly every letter home that summer, Werner wrote about his family's main concern—food. The diet on the dairy farm, although strange, apparently sufficed for the hardworking teenager. Midday dinner consisted of pancakes and sauerkraut; supper of potatoes, butter, cheese, and milk. Breakfast was at 5:30 AM, and the boys had two breaks during the day for bread and milk. Bread, however, required personal ration cards. Meat was simply not to be had.

From photographs of Heisenberg and his farm companions, one might suppose that they, like their predecessors, led an idyllic country existence that summer. In fact, the farm life was far from easy, and its hardships contrasted sharply with the playful ease of his school experience. "There I learned to work," he would tell his children. "Taking it all together, I think that was one of my most important times, considering my education, because on a farm you really learn to work. You know it's not like in school where you think it's not so important."[39]

The boys worked from 6:00 in the morning until past sundown (as late as 10:00 PM). After he recovered his strength, Heisenberg spent entire days during the first month chopping and sawing wood. This was hard labor, but his only complaint was: "The work is just too boring for me."[40] In June and July he and his companions were sent into the hay fields; during the rainy August they helped around the cow barns.

Heisenberg spent his free time in the evenings playing games of chess, which he always won. He took some mathematics textbooks and a copy of one of Immanuel Kant's *Critiques* with him to study, but, he wrote home, "For school work and things such as reading, etc. we are mostly too tired, i.e., we simply aren't interested."[41] He also brought some sheet music, planning to practice a piece by Liszt to perform on his return home. There was a piano in the farm school's central building, but piano practice lapsed until after the hay harvest in late July.

The ambitious pupil's confinement of his mental activities to chess testifies to the intensity of his physical labor. He also came into more direct contact with the people and life of the laboring world than he had ever done in Osnabrück. Both performing and observing concentrated physical labor may have fostered a similar type of concentration and perseverance in his later mental efforts. Years later, he recalled rather romantically the effect his farm summer had on his postwar years: "Others, including myself, had been working two years earlier as farm hands on farms in the Bavarian Highlands. So the raw wind was no longer alien to us; and we were not afraid to form our own opinions on the most difficult problems."[42]

Two months after the laboring student returned to Munich and his studies, the war came to a long-awaited but, for most Germans, surprising end. Having won a favorable armistice in the east with Bolshevik Russia, in the spring of 1918 Hindenburg and Ludendorff launched an all-out offensive in the west. At first it seemed to succeed, but a counteroffensive in July by massively superior forces, bolstered by newly arrived American doughboys, forced the German army into retreat. The ever-confident commanders couldn't believe they were losing. Only in September did they admit to the kaiser the hopelessness of their situation. By November, the Allies had forced a capitulation and the acceptance of President Wilson's terms for surrender. The German public, thoroughly unprepared for defeat, was appalled. Even the *Münchner Neueste Nachrichten* (Munich latest news), the highbrow newspaper of Munich's cultured classes, had failed to comprehend or to report the situation accurately.[43]

The armistice on November 11, 1918, imposed with the entire German army still in position far beyond the Rhine and occurring at a time of growing labor unrest at home, gave rise to the infamous legend of a "stab in the back," propagated in years to come by fanatic nationalists. On November 8, 1918, as the kaiser held tight to his throne in Berlin, the leader of the Bavarian Independent Socialist Party, Kurt Eisner, declared in Munich an end to the Bavarian monarchy and the establishment of a socialist republic. The next day, a defeated Kaiser Wilhelm abdicated the German throne, and a prominent Social Democrat declared the republic in Berlin—a further "stab in the back" to fanatics, and an end to the imperial era for all.

CHAPTER 3

THE GYMNASIUM YEARS

HEISENBERG GRADUATED FROM MUNICH'S MAXIMILIANS-GYMNASIUM IN THE SUMMER of 1920 and entered the University of Munich that fall. During his first two years of university studies, he published four physics research papers, submitting the first just 18 months after graduating from the middle school.[1] Three of the papers dealt with atomic spectroscopy, one with hydrodynamics. Their publication thrust Heisenberg, at the age of 20, into the forefront of research in quantum atomic physics. This extraordinary achievement was certainly due in large part to the marvelous training he received from his university mentor, physics professor Arnold Sommerfeld. But Heisenberg could not have absorbed so much so quickly if he were not already advanced far beyond the gymnasium curriculum. The disruptions of the war years had encouraged his independent study and accelerated his education.

Heisenberg impressed his gymnasium teachers from the start. On his grade reports they consistently noted not only his spontaneous drive, which always brought "very commendable achievements," but also his sheer intelligence: "The pupil is *very highly gifted*," noted his fifth-grade instructor; "A *highly talented, capable individual*," remarked his eighth-grade professor; "[He] is among the best in the class"—a unanimous opinion.[2] His achievements in grade 8, 1918–1919, were all the more remarkable, his teacher noted, because military activities in Munich had caused him to miss the last crucial weeks of school before the final examinations.

An average of Heisenberg's final grades, weighted according to the number of prescribed hours for the listings available (starting in grade 4), confirms his teachers' assessment. It yields a grade average of 1.22 on a scale of 1 (very good) to 4 (unsatisfactory). His best subjects, for which he always received a 1, were mathematics, physics, and religion. (There was no separation of church and state.) In the main subjects, Greek and Latin, he received only one 2, the rest 1s. His worst subjects were German and athletics. Athletics accounted for his only 3s (received twice), and both subjects prevented his receiving all 1s during his last three years of school. Heisenberg's literate family notwithstanding, his fourth-grade teacher noted, "In essay, clear but dry."

On his graduation examination (Abitur), Heisenberg's lowest grade was in German, his only 2. The theme of his essay, "What Makes Tragedy a Significant Form of Poetic Art?" would have intrigued his grandfather, an authority on Greek tragedy. The polite examiner, a close friend and colleague of Grandfather Wecklein, judged

Werner's essay "a comprehensive, fluently written work that, however, does not always succeed in its argumentation."[3] In mathematics and physics, on the other hand, the pupil was simply "outstanding." The astonished state examiner reported, "With his independent work in the mathematical-physical field he has come far beyond the demands of the school."[4] Three months later, in October 1920, the young man arrived at Sommerfeld's institute.

Although Heisenberg and his classmates were heavily saturated with classical studies and German literature, the young man, like other youngsters of his era, grew interested in science and technology. By early adolescence he was fascinated with technical objects. During his fourth term, 1914–1915, the 13-year-old's preferences had become apparent to his teacher, who wrote, "His interest has turned in a decisive and impressive way to physical-technical things." It was probably in those years that the Heisenberg brothers built their electric battleship.

Reconstructing Heisenberg's intellectual growth beyond technical things during those gymnasium years requires considerable ingenuity. The surviving school reports and curricula do not fully coincide with his scattered recollections, which were committed to paper only late in life. Together, however, they enable considerable insight.

In recalling his budding interest in matters scientific, Heisenberg often referred to his early fascination with technical toys, under the encouragement of his mechanically adept grandfather, as the starting point.[5] This fascination led him, so he claimed, along the path of the geometry of objects into the realm of theoretical physics, especially the mathematical analysis of physical objects and data. But the shy teenager, ill at ease in the socially superficial "real" world, grew ever more fond of mathematics itself—particularly the harmonious, orderly beauties of abstract number theory. He learned differential and integral calculus, essential for physics, in his spare moments. In addition to these independent studies, he later claimed encounters at that time with some of the classic philosophical problems concerning the relationship between mathematics, experimental data, and atomic science within the context of ancient Greek philosophy. With these interests and stimuli, he rapidly advanced far beyond the meager demands of the school's science and mathematics curricula.

Heisenberg was smitten with mathematics even earlier than the recollected stimulus of technical toys. His father had already set him competing with his brother in arithmetic at an early age, and the effects of these skirmishes were clearly evident when he entered the gymnasium. His teacher, Herr Wolff, noted the boy's skill in "calculational problems." Heisenberg's fortunate encounters with Herr Wolff during his first three years at the gymnasium further encouraged the budding prodigy.

"He tried to interest me and give special problems to me. He told me, 'Try to solve that and that.'" But when the class turned from arithmetic to geometry, Heisenberg claims he lost interest. "I felt this to be very dry stuff; triangles and rectangles do not kindle one's imagination as much as do flowers and poems."[6]

The otherwise unimaginative Heisenberg recalled a sudden spark of interest in geometry only when Herr Wolff explained that universally valid propositions can be drawn from geometry, and that these propositions correspond to the transitory world of physical phenomena. The correspondence between mathematics and the physical world "struck me as remarkably strange and exciting," he wrote. He remembered then applying mathematics to his homemade gadgets, fascinated by the notion that mathematics could be used to such ends. "Because of all this, I remained far more interested in mathematics than in science or apparatus during most of my life at school."[7]

The discovery that, as Galileo said, "the book of nature is written in mathematical symbols" comes as a revelation to many idealistically inclined youngsters intrigued by transcendent harmonies. But exactly when (and if) this occurred to Heisenberg in this way is uncertain. Although Heisenberg recalled studying geometry in his third year under Herr Wolff, school records do not list Wolff as teaching the subject at that time. Geometry was a fourth-year subject, and in that year Lieutenant Wolff was at the front. One possibility is that Wolff introduced his intelligent class to geometry before the end of the third year. The following year's school curriculum (1914) recommended this. Alternatively, Heisenberg's remembered independent study of mathematics and its applications to his gadgets may actually have taken place when his father and Herr Wolff were both at the front. It was in that year, his fourth year (1914-1915), that Werner's grade reports note an unusual interest in "physical-technical things."

That geometry can correspond to physical reality, yet transcend it, was likely driven home to Heisenberg most strongly not by his gadgets or even by Herr Wolff but by Einstein. Like many other science-minded youngsters of his day, Heisenberg had heard of Einstein's theory of relativity and of its celebrated difficulty: "That naturally especially fascinated me." He accordingly obtained a copy of Einstein's explication of relativity theory—both the special and the general theories—written expressly for gymnasium pupils.[8] The opening section spoke right to the point: "Physical Content of Geometric Propositions."

From his geometry textbooks, Heisenberg had already learned that "as the name indicates, geometry arose from practical needs" (as Einstein put it), particularly the needs of ancient surveyors. Einstein postulated that pure geometry deals only with logical relationships between concepts, from which logically valid propositions emerge. Determining the "truth" of these concepts and propositions, however, requires a comparison with "real" objects and apparatus, at which point "geometry is then to be treated as a branch of physics." The implication is that, once the "truth" of geometric axioms and propositions is ascertained through physical references, then any propositions derived from them are also likely to be "true." "Freely constructed" mathematical and physical laws are restricted by empirical and logical constraints. If Herr Wolff later followed the ministerial decree for teaching physics, he would have grounded physics even more in "real" data. Physics instruction, the decree ordered, "is to take its starting point from observations and facts of experience and not from mathematical considerations."[9]

Heisenberg later recalled that mastering the mathematics in Einstein's book gave him no difficulty—not surprising since the only mathematics in it was the Lorentz transformation—but that, ironically, he did have trouble with Einstein's physics: "It was extraordinarily difficult for me to think my way into this problem."[10] Einstein's theory thus did not convert him to physics as it did others of his generation, notably Wolfgang Pauli. But Heisenberg soon did manage to think his way into relativity through Hermann Weyl's even more advanced text, *Raum-Zeit-Materie (Space-Time-Matter)*.

In the third edition of his essay, published in 1918, Einstein recommended Weyl's newly published treatise to those seeking an advanced treatment of relativity. Heisenberg, however, did not look at Weyl's text until after he had left gymnasium in 1920. "And that again interested me a great deal," he later said, "so I tried to understand the Einstein relation and the Lorentz transformation and so on. Still I didn't think about [studying] physics." That fall he inquired instead about studying pure mathematics.[11]

In his earlier cited recollection, Heisenberg remembered that he had taken up calculus during the first year of the Great War in order to comprehend the advanced physics of his homemade toys. If true, it was a gigantic leap for a 13-year-old with no apparent interest in physics. But another of his recollections indicates that he did not actually become adept in calculus until as late as 1918, when his parents asked him to help a family friend, Paula Fries, a doctoral candidate in chemistry, prepare for the mathematics portion of her oral examinations. (Women began to gain unhindered admission to Munich universities in 1903.) For three months the 16-year-old gymnasiast tutored the 24-year-old university student in calculus. "And in that time I don't know whether she had learned it, but I certainly had."[12]

Fräulein Fries learned enough to pass her examinations, while her tutor applied what he had learned to the principles of elementary physics. By then he was in his second year of gymnasium physics and was gaining a mild interest in the subject. Although his physics textbook required knowledge only of algebra, by the end of his studies Heisenberg could apply calculus to derive and solve the equations it presented.

On his final orals, Heisenberg volunteered a demonstration of his newly acquired skills. He amazed the examiner with his use of calculus to solve the Newtonian equations of motion: "Heisenberg solved the problem posed with playful ease. Above and beyond that he now treats out of his own volition the problem of free fall and vertical throw with air resistance taken into account, whereby he makes use of infinitesimal calculus and proves that he has already gone far beyond the goal of middle school mathematics."[13]

Heisenberg's father returned from the front in 1915, but he absented himself again a little over a year later. In August 1916 the professor, fluent in Greek, left to serve as liaison officer between the Prussian War Ministry and a corps of captured Greek troops encamped near Görlitz, southeast of Berlin, where he remained until May 1918.[14] The war ended six months later. August's Munich interlude markedly accelerated the pace of his son's independent studies. The momentum occasioned by Werner's anxiety to impress his father helped to carry him through the following years of hardship and disrupted schooling. His teacher noted the obvious effect of his father's presence on the boy's performance: "Attentive and stimulating education at home, with which a good relationship existed, made itself noticeable."[15]

Grandfather Wecklein, rector of the Max-Gymnasium during Heisenberg's first years as a student there.

Professor Heisenberg was naturally most concerned with his son's progress in Greek and Latin. Noting the boy's enthusiasm for mathematics and his requests for mathematics texts from the state library, where the father performed his research, the father sought out mathematics works written in Latin. Since Professor Heisenberg had personally catalogued the classical manuscripts of the state library, he had little trouble locating mathematical texts in Latin, which had been the formal language of mathematics in Germany until well into the nineteenth century. Knowing nothing of mathematics, he brought home everything he could find in that language.

Among the treasures Professor Heisenberg presented to his son that year was a copy of Leopold Kronecker's 1845 doctoral dissertation, *De unitatibus complexis*. In it Kronecker (known to physicists today for his delta function) had attempted, unsuccessfully, to prove Pierre de Fermat's famous "last theorem" in number theory by employing complex number units in cyclotomy, the science of algebraic rings.[16] Heisenberg devoted considerable time to studying Kronecker's thesis, fascinated by Fermat's last theorem and Kronecker's failed proof. Like most budding mathematicians, Heisenberg tried his own proof of the deceptively simple last theorem and, like Kronecker, he failed. Only recently has a proof been found.

Shortly after encountering Kronecker, Heisenberg happened on what is known as Pell's equation in number theory, which certainly does have solutions. His gymnasium mathematics teachers occasionally handed out offprints of current research papers. One such paper addressed the solutions to Pell's equation, which arises from the quadratic representation of integers. For one set of parameters, the equation has an infinity of solutions, one group being the so-called elliptic functions. Kronecker is known for having contributed long treatises on Pell's equation and the properties of elliptic functions.[17] Heisenberg, familiar with Kronecker's work, certainly knew how to manipulate such functions. This was shown by his solution of one of Herr Wolff's special homework problems on the diffraction of light in a vessel of water. Heisenberg executed a long derivation of the diffraction equations, which led him into the realm of elliptic functions. Unfortunately for Heisenberg, Herr Wolff couldn't say whether it was right or wrong!

His interest in technical objects having waned, in 1916 Heisenberg began to devote all of his energies to music and to number theory, fields that are not entirely unrelated. "I was fascinated by the theory of numbers," he later recalled. "It gave me pleasure to learn their properties—to know if they are prime numbers or not and to try to see if they can be represented as sums of quadratic numbers, or finally to prove that there must be infinitely many prime numbers." He liked number theory much better than calculus "because it's clear, everything is so that you can understand it to the bottom."[18] Heisenberg apparently understood the offprint on Pell's equation to the bottom, for he applied the solution to other cases, wrote a short paper on the subject, and tried to get the paper published in a mathematics journal. The journal rejected it but without discouraging its author. In his only surviving school notebook, from about 1917 when disruptions forced the pupils to study independently at home, one finds such typical school exercises (with checkmarks penned by his teacher) as a plot of a trigonometry function in radians and the graphical determination of the roots of a quadratic equation. But on the first pages of the book, one also finds a graph of one example of Pell's equation worked out by hand.[19]

Neither Heisenberg's recollected research paper nor the offprint on Pell have been found, but remnants do exist of the works from which he probably first learned number theory. They were the standard, multivolume series of advanced texts by Paul Bachmann, entitled *Zahlentheorie (Number Theory)*.[20] Crumbling copies of the first two volumes survive in Heisenberg's book collection. The first volume, *Die Elemente der Zahlentheorie (Elements of Number Theory)*, begins with the number system and arithmetic, which were of special interest to Heisenberg. It then leads into a long section on quadratic forms that pays particular attention to Pell's equation. In the second volume, *Die analytische Zahlentheorie (Analytic Number Theory)*, Bachmann notes that solutions of Pell's equation lead to applications of the theory of elliptic functions, but he does not elaborate, referring the reader instead to Kronecker for details. According to a surviving acquisitions list in the Maximilians-Gymnasium library, its collection at the time included a copy of Bachmann's first volume.[21]

In his seventh-grade class, 1917–1918, Heisenberg received a dose of elementary trigonometry and his official introduction to physics. The sole textbook for the three years of physics (1917–1920) was surprisingly good, though elementary.[22] It covered, without calculus, such subjects as elementary mechanics, electricity, magnetism, heat, kinetic theory of gases, optics, and energy conservation. Save for mathematics, it was comparable to a sophomore physics text at a modern American college. Contemporary physics—the relativity and quantum theories—did not exist for the author of this text. But, heeding the education ministry, the author did provide material on other physical sciences, such as meteorology, astronomy, and geography, and offered explanations of such technical devices as the steam engine, water pump, telescope, and telegraph. The 500-page book was crammed with nearly 700 carefully detailed realistic drawings. Yet Heisenberg, although supposedly enthralled with technical apparatus, insisted that he had little interest in physics until his last two years at the gymnasium, beginning in 1918. And even then, he maintained, his curiosity was piqued by his philosophical pondering of the problem of atoms, rather than by any specific desire to study physics.

Heisenberg's pondering, so he frequently claimed late in life, derived from two encounters with atoms at about that time. One involved a drawing of multi-atomic gas molecules in his physics textbook. In it atoms were joined into molecules with little "hooks and eyes." Accustomed to the realistic drawings of technical devices elsewhere in the book, the demanding adolescent was disturbed to find molecules portrayed in what was to him such a superficial, utilitarian manner. "To my mind, hooks and eyes were quite arbitrary structures whose shape could be altered at will to adapt them to different technical tasks, whereas atoms and their combination into molecules were supposed to be governed by strict natural laws. This, I felt, left no room for such human inventions as hooks and eyes."[23]

Heisenberg's prescribed physics textbook did treat atoms in a rather superficial manner. With the influential anti-atomist Ernst Mach having deceased only in 1916 and the education authorities urging empiricism, the author accepted the notion of combinations of atoms in compounds as useful for describing the properties of gases, but he did not pursue the notion beyond that. Nevertheless, a search through Heisenberg's prescribed physics text yields no picture of hook-and-eye atoms. The drawing most similar to his description is a new figure inserted in the seventh edition (1916), showing water molecules constructed of what might be Ping-Pong balls joined together by toothpicks. Either Heisenberg was using another physics book, perhaps in independent study, or he remembered the picture differently. Heisenberg remembered the picture as being toward the end of his book, but the end of his prescribed book is devoted to astronomy. In any case, Ping-Pong ball atoms would have had the same distressing effect on him.

Heisenberg's second remembered encounter with atoms occurred when he read Plato's dialogue *Timaeus* while freed from school by military duty in May and June 1919. The relevant passage involved a fictional attempt by Timaeus to explain to Socrates that the observed properties of the four elements—earth, air, fire, and water—can be attributed to the transcendent properties of ideal geometric "atoms." To each of the four elements, Plato assigned one of the so-called Platonic, or regular, solids. Plato, or one of his followers, had early proved that there exist in nature only five solid bodies composed of equal-sided, two-dimensional geometric shapes.

Timaeus used the properties of four of these solids—cube, tetrahedron (pyramid), octahedron, and icosahedron—in assigning each to one of the elements:

> Let us assign the cube to earth; for it is the most immobile of the four bodies and the most retentive of shape, and these are characteristics that must belong to the figure with the most stable faces. . . . And again we assign the smallest figure to fire, the largest to water, the intermediate to air. . . . We must, of course, think of the individual units of all four bodies as being far too small to be visible, and only becoming visible when massed together in large numbers.[24]

Purist Heisenberg reacted to this passage with astonishment and dismay, as he had to the drawing in his textbook. How could the sagacious Plato believe that atoms are cubes and pyramids? More important: "The whole thing seemed to be wild speculation, pardonable perhaps on the ground that the Greeks lacked the necessary empirical knowledge."[25] Atoms were not to be so rudely treated as objects either of pure speculation or of superficial utility. Certainly, Plato's atoms bore no relevance to modern science. Or did they?

In his 1969 memoir, *Der Teil und das Ganze* (transl.: *Physics and Beyond*), Heisenberg recalled turning to two close friends from his gymnasium's military training unit and the postwar youth movement, Kurt Pflügel and Robert Honsell. As recounted in his memoir, the three young men entered into a Galilean (or Platonic) dialogue on Plato's *Timaeus* soon after Heisenberg had encountered Plato's puzzling passage. In Heisenberg's dialogue, Kurt, a budding engineer, is cast as a crude pragmatist, while Heisenberg assumed the role of the perplexed seeker of enlightenment. Robert, the deep thinker, is given the role of the Platonist—atoms are not things but mental constructs, mathematical ideals or forms as transcendent yet as bound to reality as mathematics itself. Heisenberg's two friends argue their positions as though in a chess match, until Robert finally wins, convincing Heisenberg of the validity of Platonism and helping him to comprehend Plato's geometric atoms.

In another account of his struggle with Plato's atoms, delivered to the Max-Gymnasium in 1949 in defense of classical studies, Heisenberg went so far as to claim that his reading enlightened him to basic notions of atomic physics and that from then on "I was gaining the growing conviction that one could hardly make progress in modern atomic physics without a knowledge of Greek natural philosophy." The illustrator of his physics text "would have done well," he felt, "to have made a careful study of Plato."[26]

Much later in life, when he wrote many of the foregoing recollections, Heisenberg believed that he had found striking similarities between modem elementary particle physics and Platonic idealism—so striking, in fact, that he believed that Platonism provided genuine clues for contemporary physics. However, little corroboration can be found for his retrospective portrait of himself in memoirs as a lifelong Platonist, either in the surviving evidence or in the testimony of one of his closest students and colleagues, Carl Friedrich von Weizsäcker. Moreover, a study of Heisenberg's work and of his own statements from that period suggests that Heisenberg was in fact singularly devoid of any systematic personal philosophy relating to physics. Only the stimulus of his colleagues and the needs of his science encouraged a deeper concern with philosophical matters. Responding in 1925 to a "philosophical letter" from his close colleague Wolfgang Pauli, Heisenberg, by then notorious for his pragmatic, utilitarian physics, remarked, "Unfortunately my own private philosophy is far and away not so clear, but rather a mishmash of all possible moral and aesthetic calculation rules through which I myself often can not find my way."[27]

Heisenberg did entertain a modest, usual interest in philosophical issues born of scientific research. The copy of Kant's work that accompanied him during his farm labor service in 1918 and his later youth-movement debates attest to his concern. But colleague Weizsäcker, whom Heisenberg first met in 1926, reports that what interested Heisenberg most in philosophy—other than what he needed professionally or in defense of his profession—was not the substance of philosophical systems but rather their beauty and their literary poetry; the same beauty and poetry he found in music—and in mathematics and physics.[28] Heisenberg inclined to idealism in science and music, but not in a systematic sense. This orientation became much more pronounced in the context of postwar events and during his early years with Kurt and Robert in the youth movement.

Weizsäcker reports that Heisenberg found neither Kant nor Mach beautiful in the above sense, and that he regarded only a few passages in Plato as worthy of the label. These, it turns out, are the standard passages he learned in school, which were taught precisely because of their poetic beauty as a pedagogical device. They included the first half of the *Phaedo*, the *Apology*, and the banquet scene in the *Symposium*. Among Heisenberg's colleagues, Weizsäcker, both philosopher and physicist, probably best understood his feelings regarding the lack of poetic beauty in philosophy. And Weizsäcker suspected that, for this reason, Heisenberg never actually read Kant, Mach, or even the *Timaeus* in any detail, nor did he study the intricacies of Platonic thought in any depth. However, he did immerse himself in the beauteous harmonies of physics and music, relying instead on his friends and colleagues for philosophical stimulus.

Heisenberg completed his gymnasium education in 1920 with a two-part final examination (Abitur). A successful performance certified one for entrance into any German university. The written exam was administered at the end of June to 31 pupils in the gymnasium's ninth grade. They were joined by two girls from a nearby girls' school.[29] All passed. Heisenberg did so well on the written exam that he was exempted from the subsequent oral section. But he underwent the oral ordeal anyway as a candidate for support from the prestigious Maximilianeum Foundation.

Each year, the Max-Gymnasium entered its best graduating pupils in a Bavaria-wide competition sponsored by the foundation that King Maximilian II had established in 1852. Twenty-six of Bavaria's best students were provided with room, board, and cultural stimulus in the foundation's elegant Munich quarters. Originally, the honor was reserved for those intending to enter state service via the legal profession. Later, students headed for the sciences and other disciplines were allowed to compete—if their number did not exceed one sixth of the total. Naturally, according to foundation rules, the trainees were to be "of outstanding intellectual talent and impeccable moral behavior." As future state servants, they were also expected to be unblemished Christian men; women, Jews, and the physically challenged needed not to apply.[30]

In 1920, the year Heisenberg graduated, the foundation had eleven openings. The Max-Gymnasium entered its top two graduates in the competition—Werner Heisenberg and his classmate Anton Scherer, who planned to study linguistics. Acceptance brought as much prestige to a candidate's school as it did to the candidate himself.

The Bavarian Interior Ministry instructed Johannes Melber—the ministerial examiner assigned to the Max-Gymnasium's Abitur committee, and a close colleague of Dr. Wecklein—to deliver a comprehensive report on the qualifications and suitability of the two nominees. Unfortunately, state records on the 1920 foundation candidates went up in smoke during World War II. Also destroyed was a surely revealing special report by Herr Wolff on Heisenberg's progress in mathematics and physics. A copy of Melber's report on the two Max-Gymnasium candidates does survive in the foundation files.

The examining committee, of which Melber was a member, administered a two-and-a-quarter-hour oral examination to Heisenberg in the gymnasium's seminar room beginning on July 7 at 8:00 AM. In his report, Melber praised the candidate's display of scientific prowess even more enthusiastically than might have been expected from an admirer of the candidate's grandfather: "The examination provided in mathematics and physics really shining examples of the extraordinary and rare ability of the pupil in this field."[31] But Melber was far less impressed with Heisenberg's German essay on his grandfather's specialty, tragedy as poetic art especially when he compared Heisenberg's paper with that written by Scherer, the future linguist. Nevertheless, Melber was convinced that in science, in any case, "he . . . will one day accomplish something first-rate." Melber recommended the nominee "earnestly" for acceptance by the Maximilianeum Foundation, but he still ranked him a clear second "behind his equally extraordinarily gifted fellow pupil Scherer." Both were selected. On the foundation's 1920 list of successful trainees, Scherer is in sixth place; Heisenberg secured the eleventh, and last, position.[32]

Heisenberg declined the foundation's offer of free room and board. He recalled that he preferred to live with his parents. In its records, the Max-Gymnasium notes: "Entitled to support in 1920, but because of his economic situation not supported."[33] With the country in the midst of a raging inflation, others were needier than Heisenberg. In any case, although Heisenberg did occasionally drop by the foundation for polite evenings of tea and music, by then he much preferred the companionship of his postwar youth-movement comrades to the genteel environment of either state or family.[34]

CHAPTER 4

THE BATTLE OF MUNICH

HEISENBERG COMPLETED HIS GYMNASIUM STUDIES DURING ONE OF THE MOST TURBULENT periods in modern German history. Defeat in the world war, collapse of the monarchy, and revolution across the empire ripped away the fragile facade of bourgeois propriety and patriotism, seemingly throwing the entire nation into turmoil. Munich and Bavaria experienced some of the worst of it. The disillusioned 17-year-old went through his own turmoil at the turn of events. The years ahead proved decisive for his adult political orientations. His reactions and orientations reflected political transformations occurring throughout the empire and within his own family.

The seeds of turmoil had already sprouted before the war. With legalization after Bismarck's demise in 1890, the political representative of the working class, the Social Democratic Party (SPD), experienced a steady increase in influence until, in the last prewar election, in 1912, it gained over a third of the seats in the Reichstag (parliament), the largest representation in that body. The Catholic-sponsored Center Party showed equally dramatic gains. Elections to the Bavarian Landtag (state parliament) had parallel results. And in both cases, but especially in Bavaria, where 75 percent of the population was Catholic, the gains occurred at the expense of the mostly Protestant National Liberal Party, to which the Heisenbergs owed closest allegiance. As the party of upper-middle-class professionals, the liberals strongly supported national unity under Prussian leadership as conducive to commercial expansion. Bismarck, chancellor of the Reich, easily gained liberal favor, as demonstrated when August Heisenberg adulated "the creator of Germany" when he visited Munich.[1]

The National Liberal Party achieved perhaps its greatest influence in Bavaria, the only state to force a king from his throne during the liberal revolt of 1848. Liberal Party pressure, including that of onetime party deputy Nikolaus Wecklein, had helped induce the fiercely independent Bavarians to join Bismarck's empire. As Bavarian monarchs succumbed to insanity, liberals grasped control of the Bavarian ministerial oligarchy.[2]

By 1912, the rising Center Party had eclipsed liberal predominance in Bavaria. While socialists and liberals longed to gain or regain control of the government and to press for constitutional reforms, liberal professionals, such as the Heisenbergs, feared the dilution of their recently achieved social status should socialist "proletarianization" occur.

Such tensions and differences were put aside as Germany went to war in 1914. But as the war dragged on and the body count of worker-soldiers mounted, German social-ists came to believe that they had been duped by imperialistic capitalists. Annual Reichstag votes for war financing precipitated a split on the left. A radical minority, the Independent Socialists (USPD), opposed war credits and broke with the Social Democrats (SPD) in 1917. Leaders of both parties remained loyal to the constitution, but the USPD harbored a revolutionary wing, the Spartacus League led by Rosa Luxemburg and Karl Liebknecht. Encouraged by the surprising success of their Russian counterparts, the Spartacists agitated for a German Bolshevik revolution: the establishment of a workers' and soldiers' council—*soviet* in Russian—to implement revolutionary working-class demands. They experienced their greatest success in Munich.

In Bavaria, the worsening food and fuel crisis, combined with the authoritarian rule of Hindenburg and Ludendorff over nearly every aspect of Bavarian life, awakened anti-war and anti-Prussian sentiments. By the terrible winter of 1917 to 1918, Bavarian socialists could count on liberal support—including the support of the Heisenberg fam-ily, whose two sons were in or being prepared for a now-unpopular war.

Bavarian socialists also experienced a split in 1917, but it was one of personali-ty more than politics. The founder of the Bavarian USPD was the very un-Bavarian Prussian, Kurt Eisner, a Jewish writer and intellectual who had gone to Bavaria in self-imposed exile from Berlin. Eisner's Bavarian USPD quickly gained support from the war-weary man in the street. During a nationwide strike in January 1918 for food and peace, several thousand workers demonstrated in Munich under Eisner, the first such defiance in Bavaria since 1848. The demonstration provoked brutal suppression by the Bavarian army and landed Eisner in jail until October 1918.

October was chaotic. The German army was in hopeless retreat, Ludendorff had lost his command, Austria had collapsed, and the Triple Entente—France, Russia, and the United Kingdom—threatened to push north into Bavaria. August Heisenberg's border-guard company hurried south to meet the threat.

The Bavarian Landtag finally promulgated democratic reforms on November 2, 1918, but events had already overtaken it. That night, when a ship was ordered to sail into a hopeless battle against the Entente fleet, the sailors mutinied at the north German harbor at Kiel, took over the city, and established a sailors' and soldiers' council, igniting revolution throughout Germany.

In Munich, socialists of all stripes called for a peace demonstration on November 7 on the Theresienwiese, the site of the Oktoberfest. During the rally, attended by 50,000 citizens of all classes and stations, Eisner seized the podium, proclaimed a socialist republic, and called for the abdication of the king and the establishment of a workers' and soldiers' council. The next day armed soldiers and civilians seized the army barracks, train station, newspaper offices, and Landtag building. King Ludwig III, informed of his overthrow as he strolled in his garden, quietly gathered his family, placed them in his new Mercedes, and drove into the countryside, vacating the Wittelsbach throne forever.

One day later, during a similar demonstration for peace at the Berlin Reichstag building on November 9, Social Democrat Philipp Scheidemann declared a republic just ahead of what he thought would be a similar Spartacus proclamation planned for later that day. He assuaged middle-class and industrial anxieties by professing that much of the old bureaucracy would be maintained and that every attempt by radical socialists to gain political control would be suppressed with any means necessary. The German social democracy would be neither revolutionary nor socialist, but parliamentary, bourgeois, and liberal. It would look for support to the army and to the upper middle class, not to the revolutionary workers' councils or their representatives.

Eisner offered the same promise for the Bavarian council republic, which at first enjoyed wide support among the populace, including the support of Social Democrats, former Liberal Party members, and the newly influential Bavarian Peasants Party. Eisner did not tamper with the social and political structure of Bavaria, nor did he institute proletarian rule, or try to socialize industry. According to one account, Eisner sought little more than the introduction of the constitutional reforms already contemplated before and during the war.[3] But the new regime proved pitifully unequal to its task, completely unable to control the forces it had unleashed. The party recruited its lower officials from among Schwabing coffeehouse radicals, men long on theories and short on practical sense. As the economy declined, Eisner and his regime lost favor with their supporters. Radicals on the left and the right prepared to seize control.

New parties sprang up to replace the old imperial formations. Bavarian liberals, such as the Heisenbergs, gravitated toward either the SPD or the Bavarian faction of the new German Democratic Party (DDP), which entered into a national and local alliance with the SPD. Bavarian Center Party members joined the more conservative, Church-supported, and now, in ostensible reaction to Eisner, more overtly anti-Semitic Bavarian People's Party (BVP). In the first postwar elections, held in January 1919, universal suffrage became a reality in Bavaria, with women voting for the first time. The results were staggering for Eisner's party. The USPD received only 2.5 percent of the votes. The BVP achieved a plurality, with the SPD and the DDP close behind.[4]

As prime minister of Bavaria, Eisner had presided over a socialist government that relied on the Bavarian army for support. The war-weary army, independent of the empire during peacetime, had rallied to Eisner's revolution in the early days. But the support quickly evaporated when—under restrictions imposed by the victorious Entente and in line with his antimilitary sentiments—Eisner forbade the formation of a peacetime army. A failed Spartacist putsch in Berlin and a failed left-wing Munich coup in December 1918 convinced right-wing extremists of the need for a counter-revolutionary militia. Secret protofascist societies, such as the Thule Society, organized private armies to protect against Bolshevism. One aristocratic Thule Society member, Anton Graf von Arco-Valley, apparently eager to demonstrate his anti-Semitic fervor to his comrades (because of a Jewish ancestor), gunned down Eisner in the street on February 21, 1919. Ironically, his victim had been on his way to the Landtag to submit his resignation after losing the election and the support of the middle class.

Chaos reigned in Munich. A gunfight erupted in the Landtag, killing two deputies. Street fighting broke out all over the city, while Werner and other liberal students, now contemptuous of Eisner, burst into jubilant celebration at the news of his demise. The SPD, backed by the DDP, peasants, and the USPD, soon gained control of the government and assumed command on March 18. The government was headed by Johannes Hoffmann, minister for education and culture under Eisner. As one of his first acts, Hoffmann abolished the nobility and its privileges, which alienated gymnasium pupils and the middle- and upper-class majority. Gustav Wyneken, a well-known reform educator, scolded the elite pupils and their teachers in schoolmasterly fashion for their cultural snobbery and resistance to social change. They used their new freedom of opinion, he wrote, "in order impudently, spitefully, and scornfully to turn against the revolution and its leading men . . . and to form something like a silent conspiracy of resistance against the new order of things."[5] Wyneken's chiding only instilled deeper resentment in the students.

The "silent conspiracy" grew louder as events grew even more chaotic. When Hoffmann, no Bismarckian diplomat, attempted to integrate the now rabidly federalist state into the SPD-controlled Weimar Republic that had replaced the German Reich, right-wing extremists gained new support from the Catholic hierarchy and its party, the BVP. At the same time, communist victories in Hungary and Austria further radicalized Bavarian leftists and instilled even greater fear and resentment on the right.

Like Eisner, Hoffmann proved unequal to his task. On April 7, a self-styled Revolutionary Central Council composed of radical USPD members seized control in Munich and proclaimed a new soviet republic to rule Bavaria. The new regime of "coffeehouse anarchists," led by the expressionist poet Ernst Toller, vainly attempted to socialize Bavaria's press and educational system.

A regime newspaper proclamation to gymnasium pupils on April 12 drove them directly into the hands of the increasingly violent opposition: "Pupils! You have experienced the political and economic collapse of Germany; now you will experience the last and greatest collapse, that of her culture." Genteel "Kultur" had been used too long by the upper classes to separate themselves from the uneducated masses. "The collapse of our culture has now become a historical necessity," the central council proclaimed.[6] The next day the city was in revolt.

On orders from Berlin, Hoffmann and his officials fled to Bamberg in friendly northern Bavaria. On Palm Sunday, April 13, his Munich followers unleashed a coup d'état, bringing down the Toller regime. But after a bloody street battle at the main train station and a one-day rule of the Munich garrison, Bolshevik forces gained the upper hand, declared Toller's "pseudo-soviet republic" at an end, and proclaimed a genuine Russian soviet republic. Lenin telegraphed congratulations from Moscow. August Heisenberg and his family now turned to the SPD as the only hope for protection against Bolshevism and for preservation of national unity.

Like their immediate predecessors, the Bavarian Bolsheviks faced a failing economy, a hostile populace, and a severe coal shortage, exacerbated by a brutal cold wave. On April 1 half a meter of snow had lain on the ground. The soviet regime's survival owed solely to the presence of its "red army," inherited from Toller, and to false rumors that red armies from soviet republics in Hungary and Austria were marching up the Danube. The Munich forces, without uniforms but well equipped with weaponry from the demobilized Bavarian army, were also well paid, receiving the highest army wages in Germany—in advance. Munich had little trouble raising a ragtag army of 10,000 to 20,000 men, mostly unemployed workers, front veterans, and former Russian prisoners of war.

Supporting this expensive army proved impossible. Munich's tottering economy collapsed and remained in chaos after a near-total general strike by most of the Munich population during Easter Week, April 14 to 22. The plight of the people, including the Heisenbergs on Hohenzollernstrasse, grew ever more desperate: Hoffmann's forces had set up a total blockade of the city. In this they had the cooperation of the Bavarian peasants, who, though their party had once allied with the Independent Socialists, violently opposed the soviet rebellion. The peasants prevented food and fuel from entering the city. Rejecting the regime's paper money, they refused to sell their produce.

The Heisenberg family found itself again in trouble. The wartime blockade of Germany had earlier deprived the family of food; the new blockade threatened starvation again. But this time Annie Heisenberg managed to locate a sympathetic farmer in Garching, about 15 kilometers north of Munich on the Isar River. The farmer agreed to supply the Heisenbergs with food staples, but only if the family could run the blockade and pick up the food at his farm. Many years later, 60-year-old Heisenberg recounted to his former youth-movement fellows how he, his brother, and his engineering friend, Kurt Pflügel, set out one night to collect their black-market provisions.[7] "Kurtei," one of the debaters of Platonic atoms, was two years behind Werner at the Max-Gymnasium and a fellow member of the school's Military Preparedness unit. He and Heisenberg had become acquainted through the tutoring Heisenberg provided Kurt at the request of Kurt's father, a demobilized major in the Bavarian army and no doubt an acquaintance of Captain Heisenberg.

Warmly dressed against the bitter cold but without their uniforms, which would have raised questions, the three young men tempted death on that freezing all-night journey as they put their military training to good use. Munich at that time was encircled by a massive "white army" ordered into Bavaria by Berlin in preparation for an invasion of Munich. Following the most direct route to Garching, the teenagers slipped through the red army line at one of its strongest points, the Krupp munitions works at Freimann, near the English Garden and the icy Isar River. Since numerous students left the city at that spot to join the white forces, the red guards paid particularly close attention to movement from that direction. Somehow, the three boys made it safely across both lines and reached the farm. When they tried to return, however, the white forces detained them, fearing that the boys might reveal their positions if they were captured.

The Heisenbergs at the end of World War I.

Heisenberg and his companions managed to slip away, thanks to their intimate familiarity with the local terrain. Passing the Krupp works, they went past Aumeister, a beer garden in summer, and over the broad, wind- and snow-swept field near the North Cemetery and into Schwabing. They arrived home safely with knapsacks full of flour, butter, venison, and, so Heisenberg claimed, unbroken eggs. Years later, partly in remembrance of the exploit, Heisenberg, then director of the Max Planck Institute for Physics and Astrophysics, erected a new building for his institute on the very spot near Aumeister (still a beer garden) where the boys had crossed the army lines that night.

The communist regime in Munich had meanwhile made itself thoroughly hated and feared by most of the population, especially the educated upper classes. The closing of the university and newspapers, the confiscation of food and weapons, and the imposition of a "military dictatorship of the proletariat" further traumatized the people, who began to speak of a "red terror" and to equate communism with thievery and disorder.[8] As Heisenberg later put it, "Pillage and robbery, of which I myself once had direct experience, made the expression 'Räterepublik' [soviet republic] appear to be a synonym for lawless conditions."[9]

The red terror reached its zenith as the white army closed in. Red guards rounded up politically suspect persons and seized hostages from among the leading bourgeois and noble families. Adolf Hitler, who was living in Munich at the time, claimed in *Mein Kampf* that he was to have been interned but defended himself with a carbine. Heisenberg's father, like many potential hostages, went into hiding.

Meanwhile, Social Democrat Hoffmann had lost a skirmish at Dachau, just north of Munich, to red-army commander Toller. Hoffmann retreated with his wounded pride to Bamberg, called on Berlin for help, and exhorted the right-wing secret societies, his staunchest supporters in Munich, to mount guerrilla attacks and to prepare for a general uprising.

The societies in their turn recruited the sympathetic gymnasium students to the cause, many of whom were already organized into paramilitary units through the Military Preparedness Association, to the causes. The schools had closed for Easter recess and remained closed until early May 1919, but the pupils relayed the news from friend to friend. Kurtei's father, Major Pflügel, enlisted the preparedness company of the Max-Gymnasium and organized it into an assistance unit under his command. The assistance units were to act as guides for the invading white army troops and as an auxiliary force during and after the invasion. Among those recruited into Major Pflügel's schoolboy unit were his son, Kurt, Werner Heisenberg, their schoolmate Werner Marwede, and probably Heisenberg's brother. All except Erwin later joined Werner's youth-movement group.

Berlin's socialist war minister, Gustav Noske, charged with stamping out radicalism, dispatched a massive force of regular army troops and free corps units to Bavaria. Noske had organized these units, with financial backing from German industry, to accomplish his mission without having to rely on the defeated and unprepared regular army. Because of the postwar turmoil, Noske's free corps units came to consist mainly of adventurous, often ruthless, mercenaries—former officers (usually with royal titles), restless front veterans, and students lusting for action but too young to have fought in the war. They became a fertile breeding ground for right-wing extremism. Many infamous Nazi careers had their start in one of these units.[10]

On April 23, 1919, Noske ordered the onslaught on Bavaria to begin two days later under the command of Lieutenant General Wilfred von Oven.[11] In anticipation of the attack, Heisenberg's unit was assigned to assist Cavalry Rifle Command 11, with 1,500 men, a component of Group Deetjen, commanded by a Colonel Deetjen. In his invasion plan, Noske ordered Group Deetjen to penetrate Bavaria from the north and to secure the northeast sector of a circle around Munich. At Jena, Deetjen assembled the cavalry command, part of a regular army unit that had fought at the front—a move that caused riots and a general strike among the pro-soviet populace in Jena.

Once assembled, the cavalry unit traveled through Bavaria by train on April 28 via Regensburg and Freising, put down minor resistance it encountered along the way, and positioned itself in the vicinity of the Aumeister beer garden, for its final assault. Heisenberg and his companions had first encountered the men they would now support during their midnight expedition to the Garching farm.

By April 30, the north and east sectors of the circle around Munich had been secured, precipitating the final act of red terror. When the invaders captured the main red army base at Dachau on that day (the site of the future concentration camp), the more bloodthirsty of the Munich Reds reacted by murdering ten of their hostages in the basement of soviet headquarters, the Luitpold-Gymnasium. Among the victims were eight aristocratic members of the Thule Society and two prisoners from Hussar Regiment 8, attached to Group Deetjen. The final assault on the city had been set for May 3, but after the killings there was no stopping the momentum of events.

University and gymnasium students slipped through the Red lines that night to inform Captain Hermann Ehrhardt, commander of the Second Naval Brigade in Group Deetjen (and one of the leaders of the right-wing Kapp putsch a year later), of the senseless murders. Early on the morning of May 1, as a light snow fell, the Munich underground spontaneously rose up, stormed the former king's residence and the Feldherrnhalle, seized weapons from the army barracks, and captured the university and the Luitpold-Gymnasium.

Hearing the clamor of battle, troops in the north broke ranks and began pushing toward their targets. Group Deetjen, spearheaded by inflamed Hussars, attacked the Krupp ironworks, where they met and overcame heavy resistance, then crossed in the snow through Aumeister to the North Cemetery. Fighting their way from house to house down Leopoldstrasse and Schleissheimerstrasse, they battled units of Red guards at the Max-Gymnasium and near the Heisenberg home on Hohenzollernstrasse, crossed the Siegestor near the university, and by nightfall had pressed all the way to the Feldherrnhalle near Odeonsplatz, just north of Marienplatz in the center of town. Student units and secret societies guided the troops through the unfamiliar streets, but Cavalry Rifle Command 11 remained as a rear guard in the vicinity of Schleissheim, just outside the city limits. As daylight faded, the forward troops of Group Deetjen withdrew for the night into northern Schwabing and camped at the Max-Gymnasium.[12]

General von Oven decided not to wait for the entire southern flank of the city to be surrounded and ordered a full-scale assault the next day. At dawn on May 2, Group Deetjen smashed its way out of Schwabing toward the former royal residence and the inner city. Local citizens spontaneously joined the invaders in heavy street battles raging at the war ministry, the Luitpold-Gymnasium, and the train station. Kurt Pflügel carried ammunition to his father, who spent the day blasting with a machine gun near the Wittelsbach Fountain. On the evening of May 2, General von Oven established his headquarters in the city and declared the red terror at an end. But the White terror was just beginning.

It flared on May 3. Following Oven's order and incensed by the hostage murders, the ruthless white troops stormed for days through the city once held by the fiercely hated Red guards, summarily shooting anyone who carried a weapon.[13] During the rampage, every home was subjected to a systematic search for weapons and Red Army members and sympathizers. Minor battles flared as residents defended their homes. Anyone captured as a red was usually shot after the sketchiest of courts-martial. Gymnasium units, Heisenberg's among them, guarded the many prisoners awaiting trial and the firing squad.

The frenzy continued unabated until the traumatized populace was finally shocked into sense by the murders on May 7 of 21 Catholic journeymen, shot, bayoneted, and beaten to death by drunken free corps soldiers who mistook their meeting for a red conspiracy. As one writer described it: "At the municipal cemetery where dead White soldiers lay on mortuary slabs adorned with wreaths and white-blue [the Bavarian colors] garlands the remains of the massacred Catholic journeymen were dumped among the workers' corpses on dirty ground in lean-to sheds. Crying women slithered on blood-soaked sawdust trying to identify their husbands by numbered cardboard tags tied to the corpses' limbs."[14]

By May 8, more than a thousand had died in the White terror, including all the Red leaders except Toller. Toller, arrested in a lady's boudoir disguised in women's clothing, escaped summary execution only because of the sport he afforded his captors. In comparison, the White forces, by their own count, lost 58 men in the battle for Munich.[15]

Heisenberg and his unit were stationed at the headquarters of the cavalry command in the Gregoranium, a Catholic seminary on Ludwigstrasse directly across from the university. The boys, dressed in the boots and green uniforms of their school military training unit, carried loaded rifles while on duty. Under pressure from authorities, Heisenberg stayed with his unit for several weeks, even after the Max-Gymnasium reopened on May 9, 1919. He recalled his reading of Plato's *Timaeus* early one morning while relaxing on the seminary roof where "it was nice and warm."[16] Weather records do not report a break in the cold and snowy conditions until the end of the month.[17]

The Berlin forces remained in Munich until July 1 to reestablish "quiet and order"—that is, to suppress every social democratic tendency in Hoffmann's reinstated Social Democratic government. The legitimacy of the once-moderate regime now rested on the antidemocratic right, which made the raising of a loyal Bavarian army a top priority. Socialist war minister Noske designated as the nucleus of the new army one of the more ruthless of his Free Corps units, that led by Bavarian Colonel Franz Ritter von Epp. Eager to send the "Prussians" back north, and believing that Bavaria's troubles were caused by the lack of an army to protect against Bolshevism, Bavarian officials carried on an intensive recruitment campaign for the harmless-sounding "rifle corps" (Schützenkorps). Free Corps students, many from Prussia, eagerly joined the Bavarian cause, while local gymnasium units remained on duty into the summer.

School and university officials aggressively recruited Munich students, who were the most sympathetic to the new counter-left militia. On the second day of the new term, the university prorector informed his students that their studies must take second place to the defense of Bavaria against Bolshevism.[18] To allay any doubts about the new unit, the protector assured the students that "in Free Corps Epp the democratic spirit reigns in a practical sense." Among the champions of "democracy" on Epp's staff were Rudolf Hess, later to serve as Hitler's deputy führer, and Ernst Röhm, soon to command Hitler's storm troopers. Epp himself became Hitler's Reich governor

(Reichsstatthalter) for Bavaria in 1933. Anti-Semitism now came to play an even more open role in German politics.[19]

Erwin Heisenberg, already an army veteran, signed up for military service before school reopened, took an early finishing exam in April, and graduated into Epp's Bavarian rifle corps. Werner, lately of military age and also under pressure by his superiors, would have joined along with his brother; but, the protector notwithstanding, his studies took priority—as they always had. When Berlin needed recruits a year later to suppress another soviet uprising, this time in the Ruhr, Erwin again enlisted—while Werner again remained with his books. Of the 33 pupils in Heisenberg's ninth grade, 14 signed up for the temporary duty and were dispatched to the Ruhr in April 1920. But for Werner, military adventures had paled beside the excitement of his approaching final school exam.

Werner Heisenberg always regarded this period of violent political upheaval as a puzzle: "Why all this happened is no longer quite clear to me," he wrote. And whenever he described his activities during the Red and White terrors of 1919, they seemed little more than youthful fun when enjoyable, otherwise a crushing bore: "Well, I was, you know, a boy of 17, and I considered that a kind of adventure. It was like playing [cops and robbers], and so on. . . . I just had to write things for an officer, and sometimes I had to take the guns somewhere; this was nothing serious at all." Like many teenagers, he and his chums made the most of the opportunity: "We were freed from school, as so many times before, and we wanted to use our freedom to get to know the world from different sides."[20]

Yet the reality of that world was much less lighthearted than Heisenberg made it seem. His future wife later supplemented her husband's reminiscences with two more serious episodes he had recounted to her.[21] During Heisenberg's guard duty, one of his training unit comrades accidentally shot himself while cleaning his rifle and died screaming in agony. On another occasion, his commander ordered Heisenberg to guard a prisoner overnight—a "Red" who was to be tried and executed the next day. Face to face with the enemy, probably for the first time, the teenager asked the man for his life story, which was as ordinary as the history of most other Red soldiers. The majority were much more in need of the high army wages than committed to any political ideology. By morning, Heisenberg was convinced of the man's innocence and managed to have him released. Death, apparently, was something more than a frivolous adventure to the 17-year-old.

Nor should one readily accept that Heisenberg did not fully comprehend the stakes. The political upheavals of the period entailed more than mere shifts in political power. Matters of class, social standing, and cultural recognition were at stake for everyone, and these were of primary concern to the Heisenbergs. The plethora of political parties, each representing a specific economic or religious interest group, suggests the extremely close identification of class interests with political aspirations. The shifting political alliances—even among such diverse groups as peasants and professors—in concert with the shifting economic and political situation, also attest to the relationship.

The Heisenberg family was part of the pattern. Before the war, the liberal, Protestant professionals in Catholic Bavaria naturally supported a unified Reich and readily hoped for the expansion of middle-class rights as well as the protection of their status against the rising influence of masses of industrial workers. The collapse of the old order and the threat and eventual reality of local proletarian regimes bent on destroying the carefully crafted cultural bases of bourgeois academic status pushed the bourgeois liberal family into the seeming irony of supporting the party of lower-class workers, the Social Democrats. But the Social Democrats actually ascribed largely to liberal objectives and to liberal defensive measures against serious working-class challenges.

The more-democratic-than-socialist Weimar Republic that succeeded the Wilhelmine empire now subjected itself to mass party politics and to shifts in cultural and economic policies at the seeming whim of lower-class voting majorities. Yet the Weimar academic continued to maintain the fiction of the older generation's apolitical stance—even to appear fiercely oblivious to political change. Objective scholars and scientists could not be tainted by subjective, self-serving political intrigue—even as they engaged in such intrigue. Most of the established physicists of the period—Arnold Sommerfeld, Max Born, Max Planck, Max von Laue—reacted in this way. The most obvious exception, of course, was Albert Einstein, whose outspoken defense of the Weimar democracy earned him the disapproval of his colleagues.

By the middle of the war, young Heisenberg had internalized his family's social and class allegiances. To the end of his life he always vigorously opposed the federalist tendencies of Bavaria and other German states in favor of a centrally governed nation. By 1919 he had also committed himself to the family goal of gaining and preserving social standing through academic achievement, and he immediately identified with the upper-middle-class academic elite during the Weimar period. He, too, assumed an apolitical stance, but for different reasons. The ideological and emotional allegiances of his elders, born of their station in Wilhelmine society, lost most of their attraction for him on the collapse of Wilhelmine Germany in bitter military defeat, domestic violence, and bickering political factions. The frivolity of his military adventures seems less an indication of adolescent obliviousness than a relief from the intense adult pressures of home and school.

Faced with defeat and revolution, Werner and many other young people reacted with a bitter sense of betrayal and exploitation. The heavy-handed indoctrination they had received into German war aims and the hollow facade of bourgeois gentility now juxtaposed with armed revolution and counterrevolution on the streets of Munich rendered them angry and mistrustful. "We therefore took the right to see for ourselves," he later wrote, "what in this world is valuable and what is worthless, and not to ask our parents and teachers about it."[22]

Heisenberg the physicist would enter the apolitical, bourgeois world of the upper-middle-class academic, but Heisenberg the man would perceive his place within it in terms derived from the emotional and ideological commitments espoused by what he and his friends were now calling a youth movement.

CHAPTER 5

FINDING HIS PATH

"IT MUST HAVE BEEN IN THE SPRING OF 1920. THE END OF THE FIRST WORLD WAR HAD thrown Germany's youth into great turmoil. The reins of power had fallen from the hands of a deeply disillusioned older generation, and the younger one drew together in larger and smaller groups in an attempt to blaze new paths, or at least to discover a new star to steer by."[1]

With these opening words Heisenberg set the stage for his 1969 memoir, *Physics and Beyond* (German title: *Der Teil und das Ganze*). He began not with childhood or adolescence but with the period that most profoundly influenced him as both scientist and citizen—the chaotic years immediately following World War I. And he focused neither on family nor on formal education but on his participation in the postwar German youth movement, the experience that most directly affected his adult values and outlook.

The first chapter of *Physics and Beyond* refers to Heisenberg's diverse, often difficult and confusing experiences during those early postwar years. Between neo-Socratic dialogues on the nature of atoms, Heisenberg discusses his assistance in suppressing the Bavarian soviet republic, his remembered reading of Plato, and his study of textbook atoms. He also recalls debates with his comrades about the lost war, the meaning of social order, the search for order within their own lives, and their developing notions of nature and homeland. One theme emerges clearly from this often confusing account: a yearning for the return of order in all aspects of thought and life. Heisenberg and his friends longed to regain a sense of orderly purpose and belonging—and they found it with each other in the German youth movement.

For Heisenberg, there were added benefits. The youth movement became a vehicle for his adolescent rebellion, adventurous impulses, and budding leadership qualities. It spurred his intellectual independence, taught him how his primary interests—science and music—could transcend the chaos of daily life, and gave him close and secure friendships with his comrades, with whom he formed valuable lifelong relationships.

As Heisenberg wrote in the opening lines of his remembrances, the postwar youth movement grew out of a profound sense of crisis that engendered a spirit of rebellion among bourgeois German youth after the collapse of the old order at the close of the World War. But the roots of rebellion reached back into the prewar decades. Young people increasingly detested the charades of bourgeois propriety and nationalistic saber rattling and felt no desire to pattern their lives on them. By the same token, middle-class society throughout Europe provided little room for adolescence, the crucial transition from childhood to adult roles.[2] Young people, like the children seen in Renaissance paintings, were expected to behave like miniature adults, to prepare for their adult careers and future station in life, and to accept with-- out question the values and ideals handed to them.

The rapid urbanization of Germany at the end of the nineteenth century brought with it the problem of what to do with young people in large cities. Where could they come together outside school? Where could they find the adventure and challenges of being a teenager? Before the war, some urban youngsters literally headed for the hills, seeking to rediscover fundamental values in the romance of nature, music, dance, and Germanic ritual. Groups like the Wandervögel (Migratory Birds) and the Freideutsche Jugend (Free German Youth) embodied the spirit of prewar youthful rebellion in northern Germany, but neither survived the war intact. Of the 11,000 members of the Wandervögel, 7,000 perished in the war; the Freideutsche Jugend fragmented into factions.

For those too young to fight, the state provided youth organizations, paramilitary training, and agricultural assistance work. Youngsters were aggressively indoctrinated with nationalistic values to prepare them for the task their elders set them: to fight and likely die in a brutal war.[3] Those not battling at the front struggled at home with bitter cold, desperate privations, and near starvation. How carefree can a teenager be when he grows so weak from hunger that he falls off his bicycle into a ditch?

The sudden, humiliating defeat of Germany, the loss of friends and relatives, the collapse of the old regime, the political chaos that ensued, and the forced democratization of their schools traumatized middle-class youngsters, leaving them angry and mistrustful. "A gaping hole opened up for us young people," recalls Wolfgang Rüdel, one of Heisenberg's comrades. Their response: "We're going to make something for ourselves instead, without an organization from above."[4]

The situation was particularly acute for bourgeois Bavarian youngsters, many of whom belonged to the only existing youth organization, the gymnasium's Military Preparedness Association. Few had any use for the North German, "Prussian" youth groups, including the tradition-minded Boy Scouts. The Boy Scouts had originated in England and spread to Germany in 1909, where they were called *Pfadfinder* (Pathfinders). Like their English counterparts, German Pathfinders were paramilitary and puritanical, but unlike the English Scouts they focused less on international ideals and more on preparing their young members to fit into the existing German adult social structure.[5] Two years after the First Munich Pathfinder Troop was founded in 1909, it joined the state-sponsored Military Preparedness Association.[6]

At war's end, the adult-led Military Association lost any raison d'être, and Pathfinder units began dropping out. In January 1919, a Pathfinder troop in Regensburg rebelled against the "decadent" adult values that, in their view, had failed to preserve the monarchy. At the same time, they rejected socialist attempts to dilute the cultural elite by democratizing their schools. The Regensburg troop quit the state's

Preparedness Association and pushed for a renewal of all German Pathfinders, and ultimately society itself, through the ideals of the Wandervögel—a genuine *Jugendbewegung* (youth movement) that would replace adult *Jugendpflege* (youth care).

On Easter Sunday 1919, at the height of the soviet republic's power, the equally traumatized Munich troop followed Regensburg's example. A month later, during Hoffmann's socialist restoration, the Preparedness Association changed its name to the more youthful-sounding Young Bavaria League, and in the last months of the school year, during the bloody mopping-up operations in Munich, a group of boys in the Max-Gymnasium's military unit debated their future.

Wolfgang (Wolfi) Rüdel, then 13 years old, had belonged only briefly to the military unit before it changed its name and some of its activities. Under intense pressure from their elders to support the democratic socialist restoration, he and his friends now resisted "youth care" under any name. Wolfi, his older brother Eberhard, and several other boys from the military unit gathered one day during recess at the old fountain in the courtyard of the Max-Gymnasium. They agreed to reject adult youth care but still wanted the guidance of an elder. They decided to seek an older boy of suitable character to replace teachers and adults as their leader. At Wolfi's suggestion they turned to a well-respected older group leader in the Young Bavaria League— Werner Heisenberg.

Heisenberg satisfied every prerequisite: he was an older student, disillusioned with youth care, well liked and well regarded at the school for his mathematical and musical talents, and endowed with intellectual self-confidence, good looks, and leadership qualities. He was also known as "a very great friend of nature," familiar with the mountains and countryside—a perfect choice.[7] Heisenberg was then 17 and in the eighth gymnasium grade. He was just finishing his military guard duties following the suppression of the soviet republic and readily accepted the boys' invitation. By the summer of 1919 he was guiding Wolfi and eight of Wolfi's friends into the postwar world.

Group Heisenberg, as it was known, belonged at first to the Regensburg reform movement within the Young Bavaria League. It became independent in 1921, but remained closely associated with the Regensburg faction of Pathfinders, which it officially rejoined in 1922. Gottfried Simmerding, one of Wolfi Rüdel's classmates, had joined Group Heisenberg in the fall of 1919. He recalls that the group was then part of Troop B18 of the Young Bavaria League, headed by Dr. Kemmer, the gymnasium's former military commander and one of Heisenberg's former teachers. The troop consisted at the time of six or seven groups led by Hans Schlenk, a veteran of brief war service now in grade 9B who was a friend of Heisenberg's and later became a well-known actor. Most of the boys in Troop B18 had previously served in the agricultural assistance service and in Major Pflügel's schoolboy unit during the suppression of the soviet republic.[8]

Besides Heisenberg, the group leaders in Troop B18 included Heisenberg's comrades Kurt Pflügel and Werner Marwede. Marwede's younger brother, Heini (Heinrich), helped found Group Heisenberg. The group met regularly with the other boys in Troop B18 in several basement rooms provided by the Max-Gymnasium; after breaking with the Young Bavaria League, they met in the Heisenberg home.[9]

Just days after the formation of Group Heisenberg, the Regensburg reformers, led by Franz Ludwig Habbel, a wounded war veteran, and Ludwig Voggenreiter, a publisher's son, called a meeting of all Pathfinder leaders interested in founding a genuine youth movement. Held on the weekend of August 1–3, 1919, the meeting took place, appropriately, in a medieval castle, Schloss Prunn, in the Altmühl Valley near Regensburg. Group leader Heisenberg was still in the throes of his own postwar and post-soviet confusion when he encountered a young man his age on Leopoldstrasse near the university who, as he recalled it, told him of the Schloss Prunn meeting in the passionate words of an inspired youth: "'All of us intend to be there, and we want you to come. Everyone should come. We want to find out for ourselves what sort of future we should build.' His voice had the kind of edge I had not heard before. So I decided to go to Schloss Prunn, and Kurt wanted to join me."[10]

On Friday, August 1, young Heisenberg, carrying his knapsack and a guitar, took the train with Kurt to Kelheim at the end of the Altmühl Valley. There they joined a stream of boys hiking the remaining several kilometers to the castle. The valley and castle made an ideal setting for the adolescent adventure. The narrow valley, a prehistoric Danube River bed, is lined by steep cliffs and jutting rocks. The castle, still in existence, perches precariously at the top of one of the cliffs. Above it lies a large wood where the boys pitched their tents.

About 250 Pathfinders found their way from all over Germany and from Vienna, Austria, to the meeting. Gathered in their castle in the sky, the boys were alone at last to debate the questions of the day that concerned them the most: Had the German soldiers fallen in vain, now that the war was lost? How should young people respond to the new political situation following the fall of the monarchy? How should they interpret the Boy Scout ideals of internationalism, self-sacrifice, and tradition? But the crucial questions were those of any reform movement: How was the movement to define itself, and how was it to address the decadent mass society in which it existed? The answers were vital to Heisenberg, who had hoped to discover his own order at the castle—a philosophical, social, even personal harmony. "I myself was much too unsure," he recalled, "to join in the debates, but I listened to them and thought about the concept of order myself."[11]

Incredibly, their discussions were recorded and a transcript later published in *Der Weisse Ritter* (*The White Knight*), the periodical for reform movement leaders.[12] This was to be a meeting that would set the agenda of German renewal for the ages. The transcript and related writings vividly display the German youth rebellion—indeed, the rebellion of German society at large—against the modernity of urban, industrial "civilization" and the bitter sense of loss of common purpose, of meaningful traditions, of well-grounded values with the passing of the seemingly simpler and more orderly era of the monarchy.[13]

According to the transcript, the young men agreed that their society had declined into lifeless mechanism, capitalistic greed, urban anonymity, and personal hypocrisy. Young people had to cut the chains of material and moral decadence. A year earlier, Regensburg reformer Habbel had declared: "The first demand of our conviction is for *truth* and *uprightness*. We must find our way out of the lie and swindle of our time."[14] Two years after the Prunn meeting, Dr. Kemmer, converted in his old age from youth care to youth movement and largely responsible for the Munich youth rebellion, declared without irony in *Der Weisse Ritter*: "The youth movement is a freedom movement. It has freed itself from the soulless mechanism and materialism of modern civilization and has victoriously defended its value and the right of young life against the limitations of tradition and authority."[15]

As their first order of business, the Schloss Prunn conferees aimed to reestablish truth and virtue. For them this meant an embracing of values derived from a revival of German romanticism: escape from the dead city to the living, genuine, fundamental virtues of pristine nature. There a complete renewal of the total man, a mystical revival of each human soul in unison with other souls, would occur through sustained contact with nature, with the cycle of seasons, and especially with each other—"not just an idea or a thought, but rather something internal, something fundamental, a harmony of souls."[16] Transformed and revitalized by immersion in the genuineness of nature, the young men believed they could eventually reclaim decadent German society and establish themselves as the new and incorruptible elite. "We want a new humankind that lives in our lifestyle. We believe that there is something good in everyone, but that it is suppressed in the shabby struggle for daily bread."[17]

Conflict arose with the practical question of when and how to bring about the social renewal. The "radical" reformers at Schloss Prunn embraced political elitism, seeing society and its masses as hopelessly lost and unredeemable by a handful of enlightened youth. They called for a total retreat from "civilization" to await its collapse, out of which a new order would emerge.

The "conservative" reformers rejected elitism and called for the immediate reformation of society. A reformed society would encourage reformed individuals. Their most articulate spokesman declared: "We can't build a world of ideas next to us; we must reckon with the material side. For we have to deal with age, school, party, food, work. We have to remain standing on the ground with both feet, that is, we have to work for the masses." Did it really help the masses to tramp around the countryside or to enjoy good music and literature? The conservatives believed that the group must deal with the world in order to change it, and that meant dealing with parents, dealing with women, and dealing with what some regarded as the lowest form of materialism—political intrigue. "Political struggles are self-evident for anyone who has learned to take pleasure in development, pleasure in the material side, pleasure in numbers, however reactionary that may sound."[18]

It did sound reactionary to the Schloss Prunn radicals, for involvement with politics, or with social activities in general, seemed an embrace of the very faults of youth care that these youth had rejected. They had grown up and been indoctrinated with political ideals that were suddenly shown to be "a lie and swindle." The older generation had exploited those ideals to wage a war that it had lost, plunging society into a chaos of "materialistic" (pecuniary, urban, working-class) regimes, all vying to control upper-middle-class young people and to command their allegiance.

The first issue of *Der Weisse Ritter* championed the sentiments of the antipolitical elitists: "*Der Weisse Ritter* stays away from all efforts to win young people for the party politics of one or the other direction. It sees in these attempts an irresponsible crime against the highest right of young people to decide for themselves their relationship to a state whose crumbling society they themselves will have to rebuild anew."[19]

Yet, for all of their professed hatred of hypocritical bourgeois values, the young rebels rejected not their own middle-class social and economic status and mores but what seemed to them the hollowness of the ideals that sustained them.[20] The antipolitical reformers actually rejected not politics but the party politics of their elders and of the new Weimar democracy. Their approach was merely the pursuit of the political aims and interests of the upper middle class by other means.[21] Ultimately, the young elitists did a great disservice to the society they so zealously sought to reform. In their frustration, they reacted to the immediate situation by removing themselves from the postwar social and political arena, where the future of their country was being decided. In doing so, they nullified the salutary effects their rebellion should have brought.

These highly educated young men could have exerted a democratizing influence by challenging and encouraging their elders, instead of literally heading for the hills. Their elitism left the arena open to those aggressive and ruthless enough to fill the void. Large numbers of conspiratorial secret societies, paramilitary organizations, assassins, and future Nazis emerged in Bavaria in those years just after the war, while the bewildered older generation, too, largely retreated from the political arena. As young people entered universities and embarked on careers, they were unprepared for and easily overwhelmed by social unrest and political extremism. Ironically, in the late 1920s, the apolitical stance of many university members made those institutions fertile breeding grounds for dictatorial demagoguery—the very behavior that had so repulsed the students and made them apolitical in the first place.[22]

Throughout the 1920s, Heisenberg, too, held himself aloof from politics. But his apolitical stance coincided more with that of moderate older academics of the Weimar era than with that of the radical youth movement. Still, his reasoning derived from the latter. For Heisenberg, science and politics simply did not mix. This was not because of overt rejection of democratic politics—he actually considered himself a social democrat—nor because he regarded political intrigue as criminal exploitation, but rather because he came to view his physics—along with nature and music—as belonging to a higher plane of existence and truth that somehow transcended the ephemeral, dirty world of politics. As Bavaria slid deeper into political extremism in 1923, Heisenberg wrote Kurt from Göttingen: "I never thought that I could interest myself in politics, because it seemed to me to be a pure money-business."[23]

Nor was Heisenberg quite the extreme social elitist his comrades were. Simmerding recalls that his views were even regarded as a little taboo, an indication of just how extreme his friends' opinions were. He was "absolutely for the simple, poorer levels of society, certainly very social."[24] This is supported by his brief participation during his early university days in the so-called Volkshochschule (literally: people's high school) movement, an activity that the Schloss Prunn radicals would have rejected.

The Munich Volks school movement had arisen from an idealistic student goodwill program begun by Father Carl Sonnenschein. As had the Social Democrats before the war, the school sought to integrate uneducated workers into the respectable German middle class through cultural education. Social Democrat Heisenberg, perhaps a little guilty over his role in the suppression of the workers' revolt, wanted to make amends—now that the threat was past.[25] In 1920 and 1921, Heisenberg, together with a young woman, conducted evening classes on German opera for the benefit of the workers: "She sang arias and I accompanied her on the piano." Heisenberg also offered a workers' astronomy course. He often went into the fields at night to observe the stars with several hundred workers and their wives. He also accompanied them to the state observatory, where his friend Hans Kienle served as an assistant.[26] Heisenberg's enthusiasm for the uplifting of the masses was earnest, if naive.

At Schloss Prunn, the question of apolitical elitism became so divisive that on Saturday the radicals stalked out of their own conference. The two sides met separately that afternoon and made up their quarrel by evening, but not before the radicals—mainly the Regenburg circle, the Austrians, and their Munich followers—proclaimed an oath of allegiance and gave themselves a new name, the New German Pathfinders (Neudeutsche Pfadfinderschaft). Heisenberg and his group immediately affiliated with this organization and its successor, the League of New German Pathfinders (Bund der Neudeutschen Pfadfinder, or simply Neupfadfinder, New Pathfinders).

Because of the emphasis on virtue, trusty leadership, and group cohesion, Heisenberg, as official leader of his group, exercised absolute authority over it. Because he was chosen by the group, he received their absolute trust and personal respect. For the New Pathfinders, the group leader, or *Führer*, contrasted sharply with the *Erzieher*, the schoolteacher. The latter led only by force, with meaningless rules and laws, and with no personal authority. A Führer, by contrast, led by virtue of his charisma, his natural authority, and his total concern for the group. He had earned the trust of his followers, and he never betrayed it.

In a diary of Group Heisenberg kept during their outings, Simmerding recorded the following episode in 1921. In that year, Simmerding had recruited some younger boys to join their group, among them the "rather thick-headed" son of Colonel Hans Ritter von Seisser, chief of the Bavarian state police.[27] During an outing into the hills of Upper Bavaria, led by Heisenberg, the anti-proletarians managed to eat up all their personal food prematurely, "as if Communism had been declared." With nothing to eat, they fell back on their emergency recipe for a dish they called cement: flour, water, and blueberries. Young Seisser had withheld some food of his own, but when he tried to steal off to eat it alone, Simmerding caught him, confiscated his hoard, and insisted that he eat what the others had concocted. As the group prepared to leave, Seisser remained behind sulking, impervious to all attempts to force him to come along. Then Heisenberg stepped in. Telling the others to go on ahead, he sat down on a log near Seisser, pulled out a letter, and read it silently. He then stood up and, without a word, Seisser accompanied him as he rejoined the group. Trust, not force, carried the day. That, Simmerding exclaimed, was a leader!

Group leader Heisenberg had at last rediscovered the trusting relationships he had known before the war. But now he was at the top of the power structure, where his ambition and talents found new scope and freedom, but where he also found new, lifelong responsibilities. In Heisenberg his young followers found the admirable father figure they needed, the sincere elder who would set rules and challenges for them and in whom they could place their absolute trust. "He demanded a lot of us but not more than we could take," recalls Simmerding.[28] These were serious responsibilities for Heisenberg, but he apparently never failed to meet the needs of his boys, along with his own needs.

Before every outing Heisenberg conducted a "rag parade." Each boy had to take everything out of his knapsack to be inspected and weighed, and, if the contents weighed more than 10 kilograms, he had to discard the excess. At mealtimes, either Heisenberg or one of the other leaders in the troop—now called a tribe (*Stamm*)—designated a space on the ground as the table and marked it off with pine branches. No one was permitted to put his feet on the table, except the cook, and no one could come to the table without being properly dressed. Before every meal, a quotation from German Poet Joseph Freiherr von Eichendorff was read, and proper table manners were always observed.

A strict moral code, essential for the new man, was firmly enforced in Heisenberg's tribe. As in a cult, everyone was subject to judgment by the group. On extended outings, the boys regularly held intensive criticism sessions, called by the Germanic name *Thing*.

Tobacco and alcohol were strictly forbidden; the strongest drink permitted was cocoa, even after the members attained adulthood. But Heisenberg again made a slightly taboo exception. When Niels Bohr visited the group's ski hut in the early 1930s, Heisenberg's followers were shocked to see the local tavern owner arrive with a case of beer ordered by Heisenberg. But when, during a swimming party on the North Sea, some of the more exuberant members of the other all-male groups threw off their swimming suits and frolicked in the waves, Heisenberg strictly forbade such unfettered fun.

Shortly after the Schloss Prunn reformers had solemnly pledged their devotion to their cause, further debates ensued, but by Saturday evening the two factions were friendly enough to join together for an evening of song, music, and playacting in the romantic castle courtyard. Heisenberg, however, watching the debates, had as yet reached no such accord within himself. Partially sympathetic to both sides, he recalled that each seemed to offer only a piece of the true order he sought. He could be swayed by either side. A link to some sort of fundamental "central order" or cohesion was still missing: "The lack of an effective middle ground became more painfully obvious to me the longer I listened," he wrote. "I suffered almost physically under this."[29]

As the young men gathered in the courtyard to sing in the twilight, the impressionable Heisenberg suddenly experienced what he regarded as that mystical central order. "As the song came to an end," the recorded transcript of the meeting reports, "out of the stillness the sound of a violin flows downward longingly from a narrow, mysteriously lit tower window and upward to the eternal stars. . . . The music ends, no wayward sound breaks the solemn silence."[30]

In that moment of silent epiphany, under the summer stars and the fading note in the courtyard of the medieval castle, Heisenberg writes, "All at once, and with utter certainty, I found my link with the center." Suddenly, it seemed, everything fell together for him—music, science, philosophy, religion: "The clear phrases of the Bach Chaconne touched me like a cool wind, breaking through the mist and revealing the towering structures beyond. There had always been a path to the central order in the language of music, in philosophy and in religion, today no less than in Plato's day and in Bach's. That I now knew from my own experience."[31] Science, music, philosophy, and religion all now contained for him a transcendent inner harmony, an eternal truth or validity that did not inhere in the transitory, discordant, everyday world of a postwar Munich student. Somehow, he had at last perceived the stabilizing insight that he so desperately needed during those chaotic adolescent days, and he experienced it only within the context of this faction of the reformed romantic youth movement.

Following the Schloss Prunn experience, Heisenberg traveled to Osnabrück, where he stayed for two weeks while his brother was on leave from antisoviet military service. On August 16, the transformed youth wrote to tell his father that he would not to return to Munich until September; he planned to wander the mountains with his boys for three days, then to tour for a while with Hans Schlenk. He hoped that his father would not be too angered by his refusal to stay at home. "It is indeed probably not right of me," he wrote, "but I simply prefer to be among my young friends than to sit alone and forgotten, and what I have [among my friends] older people could never give me, no matter how well or how kindly they care for me."[32] In his friends and in his experience of the central order, he felt he had found at last his place in the world and the place of thought and music within his life.

During the previous year, Habbel had joined forces with a Prussian Protestant minister named Martin Völkel to form the New Pathfinders. Their Munich representative, Karl Sonntag, a student at the technical college, began recruiting a new group of boys and promptly merged it with those led by Heisenberg's comrades Werner Marwede, Kurt Pflügel, and Wolfgang Ott, to form the Third Munich Pathfinder Troop, which Völkel headed. When the Boy Scouts finally expelled Habbel and Völkel in November 1920, the newly independent New Pathfinders quickly reorganized themselves along more explicitly feudal lines. Völkel named himself Herzog (duke); Sonntag was elevated to Bavarian *Gaugraf* (district baron) and editor of *The White Knight*, while remaining head of the Third Munich Tribe; Karl Seidelmann from Augsburg became *Gaukanzler* (district chancellor) and editor of *Die Spur in ein deutsches Jugendland* (*The Trail into a German Land of Youth*), the magazine for Bund members; Heini Marwede became treasurer after moving to Berlin; and Heisenberg's friend Wolfgang Hurt became führer of the entire Munich contingent, consisting of the three Munich tribes.[33]

Although the details are sketchy, Group Heisenberg remained at first independent of these various pathfinder leagues, but it did maintain close ties with Sonntag's Third Munich Tribe, to which many of their friends from the old Preparedness Association still belonged. As Heisenberg prepared to leave Munich to continue his studies in Göttingen in the fall of 1922, his group decided to join their companions in the Third Munich Tribe in order to gain the security of a larger organization. Heisenberg opposed the move, as he had all along, but he acquiesced in the decision.[34] A year later, when he moved permanently to Göttingen after receiving his doctorate, Heisenberg transferred the leadership of the group to his second in command, Gottfried Simmerding.

Besides their contempt for official organizations, Heisenberg and his boys had several reasons to resist joining the Völkel-Habbel league. Foremost among them, Simmerding later confirmed, was a strong strain of anti-Semitism in the league, even though it was apparently one of the few youth groups on the romantic right that admitted assimilated Jews. Anti-Semitic volkism did not prevent such groups from admitting Jews.[35] Voggenreiter, Habbel's comrade who headed White Knight Publishers (Weisser Ritter-Verlag) in Regensburg and Berlin, published several anti-Semitic pieces during the early twenties. Among them was Hans Blüher's infamous diatribe on Zionism, *Secessio Judaica*, in 1922, which prompted a well-publicized Jewish boycott of the press.[36]

In Voggenreiter's "Affirmation of Adolf Hitler," published in *Die Spur in ein deutsches Jugendland* in May 1933, the still rabidly antipolitical publisher claimed that he had met the Nazis in Munich as early as 1919 and had presented them with copies of *The White Knight*. In 1924, when Hitler went to prison for his Beer Hall Putsch of the previous year, he asked Voggenreiter for a copy of *Secessio Judaica*. Voggenreiter gladly obliged and included a special issue of *The White Knight* containing Völkel's romantic "Hie Ritter und Reich" ("Yon Knights and Empire"). To Voggenreiter's dismay, Hitler made no reply. But on his release, Hitler forged a new Nazi movement by recruiting students, many of whom were already indoctrinated into the ultraromantic notions of Reich and Führer promoted by the New Pathfinders and similar groups.[37]

Anti-Semitism infecting the Munich contingent is evident from a letter to Völkel in 1921. In it, Munich leader Hurt reported that, to his distress, the Munich members had split into pro-Jewish and anti-Jewish factions, the former headed by Walter

Tuchmann. Since some Pathfinders refused to work with Jews, wrote Hurt, "a sort of Jewish counter-pathfinder group has come into existence, whose relationship to me through its leader Walter Tuchmann, a very fine person, is at the same time both sincere and painful."[38]

Tuchmann is the "Walter" who appears in chapter 2 of Heisenberg's memoir, *Physics and Beyond*. Here Heisenberg recalls that he and Rolf von Leyden, the violinist of Schloss Prunn, often met at Walter's home to practice classical chamber music, Heisenberg on Walter's Bechstein grand, Walter on the cello, and Rolf on the violin. Two other Jews, Alfred Neumeyer and Kurt Bloch, were also closely associated with the Third Munich Tribe. Bloch, a group leader in the Second Munich Tribe who remained in Germany after 1933 and survived internment at the Dachau concentration camp, was considered an honorary member of the Third Tribe.[39]

Another reason for the preference of Group Heisenberg to remain independent of the New Pathfinders league was the league's reputation in Catholic Bavaria for romantic, anti-Church, and immoral *Schwärmerei* (dreamy nonsense). Its overt rejection of church traditions—if not of religion itself—resulted in its proscription by some schools and parishes. Even worse, in late 1921 the Bavarian culture minister attempted to ban the entire league as a pack of Bolsheviks, after Reverend Völkel publicly supported former education minister Gustav Wyneken, who, attracted to the league by Blüher's books, had been caught in a homosexual act with two of his pupils and sentenced to three years in prison.[40]

The Wyneken affair occurred just before Heisenberg's group finally joined the League. Bavarian baron Karl Sonntag was caught in the middle, inflaming tensions between himself and Völkel. Sonntag finally had enough and "retired" in 1925, turning over leadership of the Third Tribe to Werner Marwede and the district leadership to Heisenberg's philosophical crony Robert Honsell. After his retirement, Sonntag became the head of a small group of "Altmannen," young men who had grown too old for the Boy Scouts—one of whom was now Dr. Werner Heisenberg.[41]

The New Pathfinders were notorious throughout Germany for their romantic *Schwärmerei*: Teutonic rituals, visions of a white knight, and dreams of a mystical third Reich, most of which derived solely from Völkel's overactive imagination. Although he participated in some of their ceremonies, all of this must have made the more rational Heisenberg a little uncomfortable. Most of the New Pathfinder ceremonies took place in conjunction with their outdoor camps, which, together with field games, distinguished their activities from those of the prewar youth groups. Where the Wandervögel preferred wandering and singing, the New Pathfinders favored the stationary camp, and they spent long periods, sometimes weeks, living in tents far from city and civilization. There they communed with each other and with nature in song, poetry, and ceremony, and they held day-long war games and feudal sporting matches.[42] The most important of the annual encampments—in which Group Heisenberg usually joined—were the August anniversary of the League, the Christmas-New Year camp, the spring camp at Pentecost, and the solstice celebration in June.[43]

After moving out of the gymnasium, Heisenberg's boys usually met as an independent group one evening each week in the Heisenberg home. They spent weekends taking day trips into the countryside—either alone or with Sonntag's Third Tribe. Weekday meetings were devoted principally to culture: music, song, and poetry. Although they were regularly in the Heisenberg home, neither Rüdel nor Simmerding recalls much of Heisenberg's parents; Annie Heisenberg usually disappeared after serving milk and cookies. During these meetings Heisenberg often fulfilled his paternal role, dispensing advice (informed or otherwise) on school problems and adolescent dilemmas. Traditional Boy Scout crafts, practical skills, and merit badges were excluded as being formal holdovers from Victorian "scoutism" and wartime "youth care." Neither were uniforms worn until the group finally joined Sonntag's New Pathfinder tribe in 1922 and adopted its plain grey shirt, devoid of badges or insignia, and blue neckerchief. A photograph of a smiling Dr. Heisenberg in his uniform playing a spirited game of kickball with his boys in 1926 appeared in a post–World War II newsletter of Sonntag's revived New Pathfinder tribe.

During the evening meetings, Heisenberg often read aloud to his boys from the works of romantic German poets and writers—Johann Wolfgang von Goethe, Friedrich Hölderlin, Ernst von Wildenbruch. Heisenberg's favorite work was Goethe's oriental *Westöstlicher Divan* (*West-Eastern Divan*). These authors were so familiar to the boys that they often recited sections from memory on outings, gathered around the campfire at night. After their literature lesson, they turned to music. Each of the boys played at least one instrument, and during meetings they played in a small ensemble. They also experimented—sometimes hilariously—with new instruments and combinations of instruments. Plucking the right chords, Heisenberg once tried to play piano on his guitar! But the high point of musical enjoyment came when Heisenberg turned in earnest to his piano and his beloved Bach and Mozart. Rolf and Walter occasionally accompanied him. Boys from other groups and the Pathfinder tribe were often invited to these musical evenings, for which Group Heisenberg was widely celebrated. Music, poetry, and nature occupied practically their entire thoughts and activities. "That was our world," says Rüdel.[44]

On weekends, the group headed into the hilly countryside south of Munich or to nearby Lake Starnberg, where as a treat Heisenberg rented a sailboat. Sometimes they went out to the English Garden, the large park running through Munich along the Isar River, where they held field games such as "German ball" and "spear throw." The latter was Heisenberg's favorite game, but it was also the most dangerous. The group divided into two teams that tossed a spear back and forth; the spear had to be caught in flight by an opposing team member. A missed catch meant that the intended receiver was "out," technically if not physically.

During school vacations, Group Heisenberg embarked on long and arduous hiking tours. Sometimes they ventured abroad, to South Tyrol or to Finland, but most of the time they remained in Germany itself, whose beautiful countryside they delighted in discovering. Heisenberg's appreciation of the beauties of his homeland, instilled by family and teachers, became a profound attachment during those early youth-movement years.

The competitive Heisenberg often enjoyed a flirt with danger, even in his later years. According to Friedrich Hund, Heisenberg's later colleague and a prewar Wandervogel, Heisenberg and his comrades gained a reputation for engaging in dangerous activities. The boys often hiked the most rugged mountain terrain where other groups did not dare venture. During one tour, one of the boys fell into a glacier crevice. Only after considerable effort was Heisenberg able to pull him out.[45] On another occasion a youngster broke his leg skiing. Heisenberg managed to get him down the mountain and restore him to his nearly frantic parents—who thereafter paid regular visits to the group's ski hut. Heisenberg's mother once wrote to Kurt while the boys were in South Tyrol: "How nice that you have such beautiful weather—but now be more reasonable than Werner and hold him back a little from the high peaks."[46]

During the summer of 1921, after Heisenberg's first year at the university, the group left for an entire month's tour. Under Heisenberg's leadership, they hiked first from Munich to the Harz Mountains in northern Germany to attend the annual celebration of their league, then they made their way back to Munich through the countryside via Jena, Bamberg, and the Thüringen Forest.[47] They used no other means of transportation on their trips and carried no money for food or supplies. With inflation in full swing, they had to fend for themselves, which was just as they wanted it. They led an enviably unfettered life—hiking and singing during the day, then sleeping under the stars or on the sweet-smelling hay in a farmer's barn at night. In the morning, they did chores for the farmer as payment for his hospitality. In the diaries kept during those carefree tours, one finds an occasional drawing of a beautiful nature scene. Art, explained Simmerding, was simply one of the accomplishments expected of them. Heisenberg tried his hand at painting, too, with typical success: a book of quite good watercolors from that period is among his private papers.

Having just offered a brilliant performance on his final gymnasium examinations in the summer of 1920 and about to launch his scientific career, Heisenberg went off with his group on a two-week hiking tour through the Altmühl Valley near Schloss Prunn. One day towards evening they came upon another castle and asked to spend the night. The castle keep consented and showed the boys to mattresses already laid out in a large hall. Little did the boys know that the castle had recently served as a military typhoid ward. As it happened, Heisenberg was the only victim, returning home terribly sick and suffering from a high fever. His father, alone at home, was at a complete loss as to what to do. He promptly put Heisenberg on a train to his mother in Osnabrück. Since all the beds were taken at the Osnabrück house, the feverish Heisenberg was settled, deathly ill and highly contagious, on a couch in the living room. An uncle, Dr. Mutert, was hastily summoned. He prescribed fresh milk and eggs as the only remedy, but both were practically impossible to come by in those early postwar years. Dr. Mutert apparently saved the future physicist's life by daily fetching eggs and milk himself from one of his patients in the country. Fortunately, no one else contracted the disease. By the time university classes began in the fall, Heisenberg had made a complete recovery.[48] Ironically, Heisenberg's father died of the same disease a decade later, contracted while in Greece.

During their long excursions, when not singing or reciting, Heisenberg and his young friends often engaged in philosophical debates or played chess. Heisenberg's skill at the game was legendary. Besides playing during leisure hours on the farm in 1918, Heisenberg had often held chess matches under his desk at school. Even without a queen, he could still win. "That is indeed an accomplishment!" he bragged to his father.[49] Werner and Kurt often played chess in their heads while hiking, and one night when an actual board was upset an hour into play, Heisenberg managed to reconstruct the entire game from memory. After he entered the university, his obsession with chess apparently became so obvious that Professor Sommerfeld finally had to forbid him to play, claiming it was a waste of his time and talents. But in one way it might have had a beneficial effect. As a physicist, Heisenberg possessed an unusual ability to perceive the physical result arising from a mathematical formulation after many complicated moves. Perhaps this skill, vital to any physicist, was enhanced for him by his many hours of complicated chess matches.

Although Heisenberg and his friends came together to rejoice in their youth, to be young and carefree, they had their serious moments as well, as their solemn bonding, their strict moralism, and the intensity of their philosophical debates attest. During his youth-movement days, Heisenberg's favorite partners for serious discussion were the three young men closest to his own age and development: Kurt Pflügel, Karl Sonntag, and Robert Honsell. The first two planned to study engineering and often debated the merits of science with Heisenberg, but it was Honsell who apparently proved most influential. When Heisenberg later identified the three people who had most affected his intellectual development, he named Honsell second, between Niels Bohr and Carl Friedrich von Weizsäcker.[50] Honsell and Weizsäcker were slightly younger than he, and all three first came into his life during the postwar years. Weizsäcker suggests that all three gave Heisenberg, the pragmatist, a much-needed push toward philosophical reflection about his science and an equally necessary critical perspective on his scientific views. Honsell, apparently a well-read young man, may have provided Heisenberg with an education in the Western philosophical tradition, especially the idealist strain, that Heisenberg would not have accepted at that time from a member of the older generation—namely, his father.

Honsell, a group leader in the Third Munich Tribe, is remembered as "a very deep man." Some say he once wrote a book on philosophy but lacked the confidence to publish it. Nor did he display much personal ambition. He was so scholastically advanced for his age that his teachers at the Luitpold-Gymnasium urged him to skip a grade. He is said to have refused with a remark worthy of a movement youngster: "No, I want to enjoy my youth."[51] He later became a district judge.

In keeping with New Pathfinder ideology, the discussions among these youthful thinkers focused almost exclusively on such otherworldly topics as theology and idealist philosophy. Economics and politics received hardly any attention. Their inquiries ranged from the reality of atoms to the bases of religious belief and the existence of the mystical central order. Heisenberg frequently discussed such questions with the theology students in his group. In 1926, Altmann Heisenberg gave a campfire lecture to the youngsters of the tribe, entitled "God and the World." Karl Sonntag recalled a long debate—naive but spirited—on the difference between the concepts *unendlich* (unending) and *endlos* (endless) and another on the abstruse subject of the numbers 3 and 4 as dimensional "forms of the space-time world."[52]

Heisenberg clearly treasured his relationships with his young friends and the many journeys he made either with them or on his own. Once he started his university studies, these activities provided a necessary respite from his intensive research, enabling him to return refreshed to another round of work. Indeed, the almost incredible intensity of his work during the early 1920s leading up to the founding of quantum mechanics was possible only because he could relax completely during these outings. He had few friends or even acquaintances outside his youth-movement circle.

The deeper impact of the youth movement—as emotional, intellectual, and political phenomenon—on Heisenberg as both physicist and adult can be seen in a variety of ways and on a variety of levels. The intensive questioning and rediscovery of traditional values naturally included the value of science, physics in particular. Some blamed science for the supposed mechanistic materialism of the age. "In addition to many other values, we also discovered science anew," he wrote.[53] In a little-known article entitled "Old Values in New Forms," Heisenberg wrote that the youth movement led him to search for something new, to try "new ways of music cultivation," for example. "Even in science our interests concentrated on those fields in which it was not simply a question of the further development of what is already known."[54] In the end, the search reinforced his own intellectual interests and professional ambitions.

Heisenberg soon realized, thanks in part to his fellow physics student Wolfgang Pauli, that he could achieve something new and creative in atomic physics. Classical music, with all its beauties, and even other fields of physics seemed less promising areas in which to invest his talents, despite the urging of his youth-movement friends. In *Physics and Beyond*, Heisenberg recalls a long discussion with Rolf, Walter, and Walter's mother about his decision, soon after entering the university, to make a career of atomic physics rather than of classical music. As Heisenberg tells it, all three strongly disapproved. Walter's mother gave as her reason what was then typical youth-movement hyperbole: "The future of the world will be decided by you young people. If youth chooses beauty, then there will be more beauty; if it chooses utility, then there will be more useful things." Heisenberg remained unmoved. The newness of atomic physics was too tempting: "Here, I believe, we are on *terra incognita*, and it will probably take several generations of physicists to find the definitive answers. It seems to me very tempting to play some part in all this."[55] Heisenberg had, of course, already discovered that he had an exceptional talent for this work. Still, his early efforts reveal a taste for risk taking and a pragmatic toughness that were probably as much products of his youth-movement training as they were fruits of his innate intellectual self-confidence and ambition.

Although his early science seemed relatively free of direct philosophical influence, especially the influence of Platonic idealism, he lived immersed in the romantic, otherworldly notions of the New Pathfinders. Despite his choice of what seemed utility over beauty, Heisenberg was not unaffected by the antiscientific romanticism of the New Pathfinders. For instance, in his middle years he tried to link modern physics to Goethe's poetic worldview and, later in life, directly to Platonism. During his last years, he seems to have supported efforts by some thinkers to connect contemporary quantum physics, including his own contributions, to certain irrational elements in Taoist philosophy.[56] The English translation of his memoirs appeared in 1971, with Heisenberg's consent, as volume 42 of a series of mystical and religious monographs entitled "World Perspectives." Heisenberg's name appears on the board of editors of this series. The general editor's epilogue on the series could have come straight from the antiscientific New Pathfinders: "It is the thesis of *World Perspectives* that man is in the process of developing a new consciousness which, in spite of his apparent spiritual and moral captivity, can eventually lift the human race above and beyond the fear, ignorance, and isolation which beset it today. . . . *World Perspectives* endeavors to show that the conception of wholeness, unity, organism is a higher and more concrete conception than that of matter and energy."[57] It might be argued as well that Heisenberg's support of the introduction of indeterminism and acausality into quantum mechanics owed in part to his immersion in the romantic, antimechanicist ideals of the youth movement throughout that period.

What are we to make of the obvious political dimensions of the New Pathfinders? Did all the talk of a führer and a coming third Reich have any impact on Heisenberg and his group when Führer Hitler arrived proclaiming a Third Reich encompassing a racially cohesive society? Such questions are extremely difficult to answer. Nevertheless, looking to the future, the seeds of Heisenberg's reactions to the first years of Hitler's Reich already seem to have been planted within the formative environment of the youth movement. Heisenberg had already come to regard his science as above and beyond politics, but he had also come to see himself as party to a less transcendent special relationship with his young charges, for whom he had accepted full, all-encompassing, lifelong responsibility. In 1933, the now established, successful, and famous physicist would feel the same way toward his students and his younger colleagues—even though the situation and his responsibilities as a German professor were quite different from those of a youth movement leader.

For Heisenberg, physics and physicists were meant to exist above the mere "money business" of political intrigue, but, as a good Pathfinder leader should, he entered into that world of intrigue for the sake of those for whom he felt responsible. Perhaps even by gathering a small contingent of his closest students and colleagues around him, he could rescue them from the outside world while awaiting a brighter world to come. Thus arises in part a divergence of perceptions that is usually encountered when examining that period. To Heisenberg himself his actions and motives largely coincided with the lofty ideals of duty and responsibility that he, as an exemplary Pathfinder leader, had internalized during his fundamental formative experiences in the youth movement; to others, viewing his responses to the Third Reich, he was so infused with the volkish politics of the New Pathfinders that he capitulated all too easily to a brutal, antiscientific, antihuman dictatorial regime with which he sympathized.

Heisenberg, like many other German intellectuals, did indeed sympathize with what he perceived to be the nationalist ideals of Hitler's Reich, but the realities of Hitler and his Reich fell far short of what he and the New Pathfinders had envisioned. This became clear even during Hitler's rise to power, as the Nazi movement rejected elitism and entered into mass party politics. As Karl Seidelmann told the younger boys in 1931, nationalism itself was admirable, but Nazi nationalism was not. "On the contrary, the Hitler party is a shocking example of what bad leadership can make of a good thing."[58]

For the New Pathfinders, the coming third Reich was to be the culmination of centuries of German history, the final realization of the ideals of the first Reich, the Holy Roman Empire. Numerous petty princes and political parties would happily coexist within one apolitical empire, ruled by a single trustworthy, God-appointed führer. He would ensure the peace and well-being of the German people—especially, of course, of the cultured upper middle class—in the same way a group führer did for his small, tightly knit group of followers.

Such romanticism cast a spell on the New Pathfinders, many of whose members had little trouble joining Nazi youth organizations when all non-Nazi groups were banned in 1933. But among the members of Group Heisenberg only one—or so it was remembered—joined the Nazi Party.[59] Instead, it was the idealistic theology of their imagined third Reich that seems to have exerted the most immediate impact on Group Heisenberg. As they conceived it, the coming third Reich bore a striking resemblance to the Christian concept of the coming kingdom of God, where all Christian believers will live together in peace and harmony under one God-given savior. Such ideals and the otherworldly orientation of their entire activities may account for the strong inclination toward Christian theology in Group Heisenberg. Most of its younger members studied theology and later joined the clergy— both Rüdels became Lutheran ministers, Karl-Heinz Becker a Lutheran theologian, Gottfried Simmerding a Catholic priest, and Otto Heimeran a monk. The rest pursued other unworldly academic or scientific careers, save for Heini Marwede, the tribe treasurer. Much to Heisenberg's displeasure, Marwede went into the money business, banking.

If the transcendence of science and the Reich went hand in hand, then perhaps, when the bad leadership that corrupted the ideals of the coming Reich could be replaced, those ideals would flourish, bringing in the new moral age. Perhaps one needed only patience, one needed only to endure for a short while until a new leader of deeper vision arose. These were hopes that Heisenberg cherished in 1933, and they were hopes that he had apparently forged over a decade earlier, during what he always remembered as "the most beautiful days of my life."[60]

CHAPTER 6

SOMMERFELD'S INSTITUTE

HEISENBERG ENTERED THE UNIVERSITY OF MUNICH IN THE FALL OF 1920. SEVEN YEARS later he was appointed professor of theoretical physics at the University of Leipzig. At the age of 26, he was Germany's youngest full professor. His rapid rise through the ranks owed not only to his brilliance, but also, once again, to his timing. His talents were perfectly suited to work in the still-new field of theoretical physics. During those seven years, quantum theory underwent a profound transformation, from what Max Born called a state of disorder in 1920 to the orderly system of quantum mechanics that Born and Heisenberg together pronounced complete in 1927.[1]

While still a student and junior lecturer, Heisenberg was a prominent member of the small cast of talented young physicists that brought about the transformation to quantum mechanics. His role arose from the felicities of location and colleagues, in addition to timing and ability. During those years, Heisenberg studied and lectured at three of the world's major quantum research centers—Munich, Göttingen, and Copenhagen—under three of the leading quantum theorists of the day: Arnold Sommerfeld, Max Born, and Niels Bohr. He also studied and worked with some of the brightest young theorists of his generation: Wolfgang Pauli, Pascual Jordan, Paul Dirac, H. A. Kramers.

The profound inadequacy of quantum theory in resolving the riddles of the atom was just coming to light in Munich when Heisenberg began his studies. Through his teachers and colleagues, Heisenberg became acquainted with the problems and with the various attempts to resolve them. This groundwork and the training and outlook he received in each of these locations provided the basis for his own contributions over the coming years.

His contributions during that period were fundamental. They included Heisenberg's pivotal breakthrough to quantum mechanics in 1925, his participation in the development of the complete formalism of quantum mechanics from 1925 to 1927, and perhaps his most well-known achievement—the uncertainty principle—in 1927. The last constituted Heisenberg's part in the fundamental "Copenhagen interpretation" of quantum mechanics. Although they remained controversial, quantum mechanics and the Copenhagen interpretation of its formalism completed the quantum revolution first ignited by Albert Einstein, Max Planck, Bohr, and others during the first decade and a half of the century. It has formed the basis of research on the atomic scale to this day, enabling new and profound advances in understanding all aspects of the physical world, from nuclei and quarks to stellar energy and the big bang. These developments exerted profound effects on the world in which we live, in areas from culture and philosophy to the technologies of nuclear reactors, atomic bombs, semiconductors, lasers, medical imaging, and superconductivity. Heisenberg played a leading role in many of the theoretical developments from the very moment in the fall of 1920 that, as an 18-year-old student, he entered the University of Munich.

Heisenberg's work and studies during the 1920s continued to parallel the upheavals of Germany's social unrest. The parallel, in fact, helped to foster his work through financial support from abroad and challenges at home.

As the youth movement so clearly demonstrated, young people and their political views in and around the University of Munich tilted precariously to the far right during the early twenties. Many of the university students were front veterans or members of Free Corps units sent to suppress left-wing uprisings. The University of Munich quickly became a stronghold of right-wing extremism.[2]

The university administration openly encouraged its fanatical students. During the height of their power, the soviet leaders had attempted to subject the elite institution to "proletarian" rule. Armed student revolutionaries had seized control of the university, taken the rector hostage, and set up a revolutionary council to dictate university policy. The university senate promptly closed the institution and sent the faculty into hiding.[3] When the restored Hoffmann regime proclaimed Bavaria a social democratic republic in 1919, the move was denounced almost universally by the monarchical faculty. When in 1920 the courts sentenced Anton Graf von Arco-Valley to death for assassinating Kurt Eisner, the university rector himself joined his students in a demand of clemency for Arco that did not stop short of threatening violence. The threat succeeded. Max Weber, then a professor of sociology, attempted to defend a socialist student who opposed the demand, only to find his lectures constantly disrupted.[4]

The dire political and economic situation, combined with horrendous overcrowding at the university, contributed immensely to student fanaticism during the early twenties. In the fall of 1920, when Heisenberg entered the University of Munich, it had a student population of 6,879, 62 percent of whom were Bavarian. A year later, 9,659 students were enrolled, only 52 percent of whom were Bavarian.[5] Many students came from middle-class families who were among those hardest hit by economic inflation. During Heisenberg's first semester, in the winter of 1920 to 1921, a counselor discovered that a disturbing 61 percent of the students existed on a monthly income that was below the minimum cost of living for a Munich student.[6] The lowest-paid unskilled worker at the university earned more than the minimal budget of a student—a circumstance that further embittered the already resentful students.

Although Heisenberg suffered some economic difficulty, he was far better off than most. He was one of a small number (2.3 percent) of male Bavarian students who lived at home and avoided boarding expenses. His father's income, moreover, was not seriously threatened by inflation until 1923. Professor Heisenberg's salary was already near the top of the civil-servant pay scale (step 12 of 13), and it nearly tripled between July 1920 and October 1921, while the mark decreased in value relative to the U.S. dollar by only slightly more than that amount.[7] Even without the added income of seminar fees, the professor's salary in 1921 was just under the total expenses of an average five-member professorial family in the more expensive city of Berlin.[8]

However, with Erwin pursuing a doctorate in chemistry in Berlin between military stints, the professor found it necessary to put Werner on a budget. In October 1921, the professor paid Werner's way to Jena so that he could attend his first physics conference, at which he met some of the great names in his profession for the first time, among them were Planck and Max von Laue, but, to Heisenberg's dismay, not Einstein, who could not attend. During the conference, Heisenberg decided to impose further on his father's finances and travel by train to visit the Berlin contingent of his youth movement, but only briefly his brother.[9] His parents maintained that they could not afford to give him any more. On Heisenberg's vehement insistence, his parents finally withdrew a sum from a gold-dollar account set up by Heisenberg's rich American Uncle Karl to help his German relatives in emergencies. Independent minded and little concerned with parental dilemmas, Heisenberg longed even more keenly for the day when he would not have to rely on family generosity.[10]

Weimar scientists, especially atomic scientists, reacted to the worsening economic conditions by developing increasingly innovative strategies to obtain research support for themselves and their students. However, in the political sphere, these same scientists reacted to the upheavals of the era by withdrawing even further into their offices and labs. In keeping with the blame cast on Germany for unleashing the world war, an international boycott of German science plagued the profession following the war. German scientists stubbornly refused to allow any outside events to contaminate them or their science. Richard Willstätter, a Munich chemistry professor, deftly dodged machine-gun bullets on his way home for lunch during the battle of Munich—he said he refused to let his soup get cold.[11] During the right-wing Kapp putsch against the Weimar democracy, Born encountered a heavy street battle in Frankfurt; yet, he recalled, "After things had settled a little we went about as if everything were normal."[12]

Heisenberg's future Munich mentor, Sommerfeld, managed to complete two of his most significant scientific papers in the midst of this turmoil. He submitted his well-known but unsuccessful ring model of the atom at the height of the soviet republic in April 1919; his fundamental paper on the splitting of spectroscopic lines in a magnetic field, the inspiration for Heisenberg's early work on spectroscopy, arrived at the publisher within four days of the Kapp putsch.

But even the ivory tower of Heisenberg's new academic home, the Institute for Theoretical Physics. headed by Arnold Sommerfeld, could not protect its members from a collision with the university's fanatics, many of whom were also anti-Semitic. The issue, which arose during Heisenberg's second year of studies, was Einstein. Sommerfeld's students learned of the episode from his correspondence with Einstein, which Sommerfeld often read aloud to his seminar.[13] Einstein, well known in

Germany as a pacifist and democratic socialist, openly supported the Weimar democracy and worked tirelessly in the cause of international understanding. Since his newly confirmed theory of relativity seemed so profound yet so incomprehensible to most lay people, many saw in him a symbol of their own incomprehension of recent events, and in Jews a scapegoat for Germany's troubled situation. Outspoken, captivating, and Jewish, Einstein made an ideal target for both anti-Semitic and antiscientific hate. In 1920, Einstein's opponents unleashed an anti-Semitic campaign in Berlin against the man and his theory.[14] They were supported by several prominent experimentalists, including the Nobel laureate Philipp Lenard. Although the sources of their hatred were myriad, Lenard and his colleagues were especially furious at Einstein's sudden popularity and prestige as a theoretical physicist, which threatened to overshadow them and their field of experimental physics.[15]

During the nineteenth century, German physics and German physicists had established their power and prestige mainly in the field of experimental physics, the gathering and analysis of data. Experimental work also required mathematical methods and the framing of general hypotheses, but by the early years of the twentieth century a new professional discipline had emerged, especially in Germany, that focused in a new way on hypothesis, mathematical analysis, and empirically informed theories of natural phenomena—theoretical physics. By the end of World War I, the startling results and successes of professional theorists such as Planck, Laue, and Einstein provided the new field with enormous popular appeal and prestige in Germany, but it still held a secondary professional status behind the established and supposedly more well-grounded discipline of experimental research. Many still held empirical data to be more fundamental than mathematically construed theories. Because of the lower status accorded theoretical work, Jewish physicists found more opportunities in theoretical physics, but—as everywhere else in German society—they also encountered anti-Semitism, both before and after World War I.[16]

Moderate, non-Jewish Weimar physicists usually considered opposition to anti-Semitism to be a political issue, rather than a moral or an ethical one. Such a view apparently arose from the politicization of anti-Semitism in Germany. Anti-Semitism had already become a plank in the platforms of several major political parties, and those who engaged in anti-Semitism often did so for obvious political ends. Because of this, scientists and academics regarded blatant anti-Semitism and overt opposition to it as engaging in politics and thus to be avoided. As determined opposition failed to develop, implicit anti-Semitism flourished in German academe, at times infecting even Sommerfeld's institute.[17]

Einstein was a special case. While German repute abroad suffered because of the war, Einstein's international fame reflected favorably on the foreign image of German physics. The growing anti-Einstein campaign in Berlin threatened to tarnish that image. Sommerfeld and others, while carefully avoiding political involvement, realized that something had to be done. In 1920, Sommerfeld and several Munich professors formed an Einstein support committee. As their first order of business, they invited him to lecture in Munich. Sommerfeld obtained financing from a sympathetic philanthropist, and Einstein readily accepted, intending to arrive in Munich in November 1921. Heisenberg and his fellow student Pauli, disappointed at having missed Einstein in Jena, eagerly anticipated an encounter with the great man in Munich.[18]

The plan unraveled shortly before Einstein's appearance. An article in a leftist Berlin literary magazine reported an ominous meeting held nearly a year earlier between Sommerfeld and representatives of the Munich student government in the rector's office. Noting the disruptions of Einstein's lectures in Berlin, the rector—now suddenly worried about the reputation of his school—demanded assurances that no such disruptions would occur in Munich. The students responded with objections to Einstein's "person," prompting Sommerfeld to lecture them on the physicist's signif- icance—but without success. During the next student government meeting, extremist representatives, many of whom were members of the "Swastika Majority," refused to give any assurance against disruption. The article reporting these events appeared just before Einstein's arrival in Munich. Einstein immediately canceled his appearance. Sommerfeld pleaded with his colleague to reconsider. Einstein absolutely refused. "There is just no other way," he told the non-Jewish Sommerfeld. "That you must feel yourself."[19] Heisenberg would have to wait three more years before meeting Germany's foremost theorist.

Sommerfeld and his students stubbornly maintained their insulation, both before and after the Einstein affair. These were talented and intense young scientists con- sumed by the demanding intricacies of their discipline. It was very easy for them to relegate social issues to second place, and they welcomed the opportunity to do so. The arduous work and tantalizing promise of scientific research served as a conven- ient antidote to social upheaval—as it has for other scientists in similar situations throughout history.[20] Pauli, for instance, entered the university in the winter of 1918 to 1919, just before the end of the war. The years immediately following witnessed the soviet republic, civil wars in Munich and Vienna, the taking over and closing of the university, rampant inflation, and hateful violence. Yet he notes none of these events in his available correspondence or in any of his published recollections. Years later he wrote, "The war was over, with Sommerfeld I was in my right element. What then were the political and economic situations in Germany and Austria to me as a young man?"[21] Insular concentration fostered successful work, which in turn encour- aged further insulation.

Heisenberg found his own escapes. "My first two years at Munich University were spent in two quite different worlds: among my friends of the youth movement and in the abstract realm of theoretical physics," he wrote. "Both worlds were so filled with intense activity that I was often in a state of great agitation, the more so as I found it rather difficult to shuttle between the two."[22] Heisenberg literally did shuttle between the two. During the warmer months of the summer semester (April through July), he usually camped out with his boys in the mountains at night, then hiked to the nearest train station early the next morning, arriving in Munich in time for

Sommerfeld's 9:00 AM lecture. Constantly moving between the worlds of physics and youth conveniently left him little time for anything else.

Heisenberg had still planned to study pure mathematics when he completed his gymnasium studies in 1920. Fresh from his brilliant Abitur, the ambitious young man intended to launch immediately into an advanced research seminar leading to a doctoral degree. Having passed the Abitur, which brought them to the junior year of a modern American college, students were automatically admitted to the lectures and exercise sessions of the German university of their choice. Most students attended local universities, trying for a doctorate or an intermediate diploma (not available then in Munich). To obtain a doctorate, one had to be accepted by a professor into his research group, usually centered on his advanced seminar. There the student learned the fundamentals of research while working on an independent project. Rather than requiring the general education courses, course examinations, and semester grades of American schools, the German university focused on advanced courses in the student's field and related fields and on independent research through early study with working specialists. Once accepted into a seminar, the student completed a thesis project under the professor's direction. Final approval of the thesis and a grade for the entire study were conferred at the final oral examination administered by professors in the candidate's major and minor fields. The new doctor was now qualified to teach at a gymnasium. A university teaching career, however, required an even higher degree: the habilitation, or qualification, which entailed additional research, oral examination, and approval of the entire faculty. This latter degree was comparable to the tenure process at an American university. Once habilitated, the candidate could be appointed permanently to a full professorship, or teaching chair.

Shortly after recovering from his bout with typhoid in the summer of 1920, Heisenberg had his father arrange an appointment for him with the Munich mathematics professor Ferdinand von Lindemann, a colleague of the elder Heisenberg. Lindemann seemed an ideal candidate for the role of Heisenberg's "doctor father," as advisors were often called. He was well known in Heisenberg's intended field of number theory for his proof of the transcendence of pi (which has an infinity of decimal numbers), and he was co director of the university's "mathematical-physical seminar," composed of four professors and one assistant. It was not a seminar in the sense of a study group; rather, it was somewhat like an American department, but with more independence among its members. It was designed to train future gymnasium teachers of mathematics and physics in the fundamentals of their field and in basic research.[23] Presumably a good researcher would make a good teacher. In 1920, Wilhelm (Willy) Wien, the newly arrived professor of experimental physics, co-directed the seminar with Lindemann. Their two colleagues were Aurel Voss, professor of mathematics, and—in fourth place in the pecking order—the professor of theoretical physics, Sommerfeld.[24]

As a favor to August Heisenberg, Lindemann agreed to meet with his son—but only as a favor. The old gentleman, a longtime chairman of the university's administrative committee, was two years from retirement. He had little patience with first-year students who intruded on him in his office, and none at all with audacious novices who demanded immediate admission to advanced research. The interview ended in disaster.

As Heisenberg recalled it later, Lindemann received him in a dimly lit office, seated behind a desk on which perched his pet poodle. When Heisenberg began to speak, the poodle barked so loudly that the partially deaf professor could barely understand him. Finally, Lindemann asked his young visitor which textbooks he had studied. After mentioning Paul Bachmann's *Number Theory*, Heisenberg volunteered that he had just finished Hermann Weyl's *Space-Time-Matter*. Lindemann, looking for an excuse and perhaps unsympathetic to Weyl's contamination of pure mathematics with physics, abruptly closed the interview with the remark: "In that case you are completely lost to mathematics."[25]

Stunned by the rejection, the 18-year-old returned to his father to seek alternatives. They considered the three remaining seminar professors. Wien, the experimentalist, would not do, and between Sommerfeld and Voss, the former was the more likely choice, since he and August were already well acquainted. The slightly built, balding, broadly mustached Sommerfeld, who always stood so erect that he looked, in Pauli's words, like a Hussar officer, had served as dean of the science faculty during the previous summer semester. During the coming year, he would serve as senator from his faculty. These duties had already brought him into frequent contact with Professor Heisenberg, the university representative to the German College Teachers League (Hochschullehrerbund). Father and son decided to try Senator Sommerfeld.

The physicist proved much more sympathetic than his elderly colleague. His office was well lit and devoid of poodles, and its less imperious occupant gladly received eager students of all levels. Unlike Lindemann, he was elated—and amazed—to learn that Heisenberg had read Weyl. "You are much too demanding," he told his visitor, with good reason. Obviously impressed, the perceptive Sommerfeld admitted Heisenberg provisionally to his research seminar, even before he had completed any advanced courses. "It may be that you know something; it may be that you know nothing. We shall see."[26] Heisenberg was on his way into theoretical physics.

Sommerfeld's approach to his science and his relations with his colleagues typified the state of German theoretical physics at the onset of the quantum revolution. Like many theorists of his generation (he was then 52), Sommerfeld began his career in mathematics. Coincidentally, he was born in Königsberg, East Prussia, the site of the first mathematical-physical seminar in Germany, which originated the Central European branch of theoretical physics. Sommerfeld had attended the local gymnasium with Willy Wien and his cousin Max Wien, both of whom became physicists. In 1886, Sommerfeld began studying mathematics at the local university, attending the mathematical-physical seminar, which was directed by the professor of mathematics —Ferdinand von Lindemann. But, like many other mathematicians, Sommerfeld became intrigued with the mathematical physics of William Thomson (Lord Kelvin), as outlined in his attempt to envision a mechanical model of the electromagnetic field in concert with James Clerk Maxwell's mathematical equations of this field. Sommerfeld promptly switched from Lindemann's number theory to Kelvin's mathematical physics—the study of mathematical applications to physics—and wrote his doctoral dissertation on the subject under Paul Volkmann, professor of mathematical physics in Königsberg.[27] Doubtless Sommerfeld saw something of himself in Heisenberg.

Still a mathematician, in 1893 Sommerfeld headed for Göttingen, then the capital of German mathematics. There he fell under the influence of the famous mathematician Felix Klein, a superb teacher and administrator who at the time pursued a program for mathematizing science and establishing institutes for applied mathematics.[28] Thirteen years later Sommerfeld began teaching theoretical physics in Munich.

Physics research in Munich derived from the university's instrument collection, consisting of experimental apparatus, and the professor who used it. In 1892 the cabinet moved into the university's new Physics Institute, which was headed until 1920 by Wilhelm Röntgen, the discoverer of X-rays, who had previously taught at the University of Würzburg. Röntgen's successor at Würzburg was Willy Wien, Sommerfeld's old school chum. Wien again succeeded Röntgen at Munich, just as Heisenberg entered the university.

Professor Wien was proof of the respect experimental physics commanded in Munich. The Nobel Prize-winning Wien, then 55 years old, would not leave Würzburg unless granted special concessions. He got everything he demanded, despite the grim economics of the day. He received a fat salary, four assistants, three technicians, and six-figure grants to expand and retool the institute.[29] By contrast, Sommerfeld's institute consisted of a lecture hall, three rooms, a modest laboratory, one assistant, and one technician. It was located on the ground floor and basement level of the university building, two floors directly beneath August Heisenberg's office. Aside from seminar fees, the institute received the modest sum of 2,000 marks per year to purchase apparatus and to maintain a small library.[30]

Although Sommerfeld's chair and quarters were located in the university, they were administered as the state's scientific instrument collection, its mathematical-physical cabinet. Hence there were two professors of physics, two independent experimental laboratories, and two very different schools of thought as to how physics should be defined and taught. Ironically, the university's physics professor handled experimental physics; the conservator of the state's experimental instruments pursued theoretical physics with a parallel university appointment. Ludwig Boltzmann had occupied the position until 1894. In 1905, Röntgen, interested in electron theory, appointed Sommerfeld to the post, over the strenuous objections of Sommerfeld's former mentor, Ferdinand von Lindemann.[31]

On arriving at the university in 1906, Sommerfeld divested the cabinet of most of its outmoded instruments and gave it a new name reflecting its new primary focus: Institute for Theoretical Physics. It quickly became a leading center of research in the new relativity and quantum theories. Sommerfeld was reportedly the first professor in the world to lecture regularly on both subjects, and he enjoyed world renown as one of the best and most stimulating teachers of the era. His institute produced a steady stream of first-rate theorists—the largest number of doctorates in the field until the 1930s. Einstein was amazed at its fruitfulness and, prompted by a report on Heisenberg, wrote to Sommerfeld in 1922, "What especially impresses me about you is that you have produced so much young talent, like stamping them out of the ground. That is something entirely unique. You must be able to activate and to cultivate the minds of your pupils."[32]

What particularly distinguished Sommerfeld as a teacher and researcher was not so much the brilliance of his physical insight but rather, as Born put it, his "logical and mathematical penetration of established or problematic theories and the derivation of consequences that might lead to their confirmation or rejection."[33] Sommerfeld combined this talent with an inspiring teaching style and a gradual selection process that served to weed out the weaker pupils. The institute's Munich location helped to ensure a steady supply of talent for the program.

As the state's conservator of apparatus, Sommerfeld was still obliged to allow at least some experimental research—work that he relegated to assistants whom he banished to the basement. Despite their lowly status, theorists Max von Laue and the outcasts made at least one major discovery: proof that X-rays, which were thought to be particles because of their penetration, exhibit electromagnetic wave behavior. Laue received the 1914 Nobel Prize for his discovery, and Sommerfeld's institute received generous grants to continue the research—thereafter in broad daylight.

By the time Heisenberg joined the institute, theory once again eclipsed experiment. Sommerfeld focused his theoretical interests on two topics: hydrodynamics and quantum atomic physics. His interest in hydrodynamics arose with the financial support of the Isar Company in Munich, which had been contracted to channel the Isar River. Work on quantum spectroscopy—the study of the emission and absorption of light by gases as clues to the internal structure of the constituent atoms—grew out of Sommerfeld's concern with the modification of his own models of atoms in the light of new and puzzling data. To Heisenberg's extraordinarily good fortune, Sommerfeld's institute was unique. It was one of only a handful of institutes for theoretical physics in Germany and one of only two or three that performed research on quantum atomic theory. Moreover, it was the only one at that time concerned with theoretical quantum spectroscopy.

Heisenberg was also fortunate in the timing of his entry into Sommerfeld's teaching program. Like mathematics and other sciences, physics was then part of the philosophical faculty, which required of doctoral candidates a minimum of only six semesters (three years) of study. Sommerfeld accordingly arranged the topics of his main lecture in a six-semester cycle, starting with "classical mechanics," the study of matter, motion, and forces founded on Isaac Newton's work in the 1600s. If a student entered in mid-cycle, he could either learn the material out of sequence or spend his first semesters on required minors such as mathematics while waiting for the cycle to begin again. Heisenberg entered the program just at the start of a cycle, in the winter semester of 1920–1921. Sommerfeld had just spent the previous year teaching a tiresome series of makeup semesters for newly entering war veterans and Free Corps volunteers.[34]

Sommerfeld designed his teaching program to satisfy a variety of needs. While Heisenberg studied under him, he offered five main lectures (one each semester for four hours a week) covering nearly all of "classical" (pre-relativity and pre-quantum) theoretical physics. For advanced students, he taught contemporary subjects in a special lecture on current research. He also conducted the research seminar for doctoral candidates and gave an occasional public lecture on modern theories to raise money for the institute. The main lectures were attended by as many as 80 to 100 students from a variety of scientific fields. Students of chemistry and medicine who attended Sommerfeld's lectures on atomic models in 1916 and 1917 encouraged him to write his famous textbook *Atomic Structure and Spectral Lines* (*Atombau und Spektrallinien*), which became for a generation the "bible of the modern physicist."[35]

At each lecture, Sommerfeld assigned homework problems to be turned in during weekly one-hour exercise sessions. An assistant corrected the problems and discussed them with the students during the exercise, which Sommerfeld himself often attended. No grades were given; a student's work spoke for itself. Heisenberg recalled turning in such long and complicated solutions that Sommerfeld's assistant complained.[36]

The assistant was probably Peter Paul Ewald, who held the post when Heisenberg arrived. Ewald remembered Sommerfeld as a "true doctor father."[37] He took a personal interest in his charges, treated them with dignity, and gave them sympathetic fatherly counsel. He set an example for them as a hardworking, intensely active researcher, yet he was always accessible. Heisenberg was often in Sommerfeld's office for an hour or two each morning during his last semesters. On Sundays, Sommerfeld would invite his charges to daylong outings in the countryside. Winter weekends were often spent with other physicists skiing at Willy Wien's country cottage in Mittenwald near the Austrian border. When students felt the pinch of economic inflation, Sommerfeld dipped into his own pocket to help them out. Heisenberg, too, benefitted from his generosity, which further increased the student's admiration for his mentor.[38]

Stimulation and selection began early in Sommerfeld's institute. It was his strategy to involve students at once with research and institute affairs, both to encourage and to test them. The professor, as Heisenberg called him, gave his beginning pupils minor tasks, such as checking his calculations, analyzing newly received data, or correcting galley proofs of articles. Advanced students assisted with revisions of his textbook or with articles for Klein and Sommerfeld's multivolume *Encyclopedia of Mathematical Sciences*. It was in this work that Pauli's famous article on relativity theory, still considered one of the best summaries of the subject, first appeared.[39] Heisenberg recalled that Sommerfeld would often motivate a bright pupil by handing him a small problem with the remark, "Well, I can't solve this problem; now you try it." Based on performance of these tasks and in the exercise sessions, Sommerfeld assessed his pupils' suitability for admission to advanced training.

Sommerfeld offered his two-hour special lecture each semester on a topic that he was currently researching but had not yet fully grasped. When once asked how he could lecture on a subject he did not understand, Sommerfeld replied, "If I knew something about it, I wouldn't lecture on it!"[40] The object was to enable pupils and teacher to grapple with a current problem together and, in the process of searching for a solution (successfully or not), to arrive at a systematic comprehension of the subject. The communal effort made these sessions particularly stimulating. Sommerfeld prepared the special lecture in advance, but he usually tried to re-derive the results at the chalkboard without referring to his notes. One can imagine the animated discussions that must have occurred when a derivation didn't work out. Throughout Heisenberg's studies in Munich, Sommerfeld devoted the special lecture each semester to the major atomic physics problem of the day: quantum spectroscopy. Young Heisenberg was captivated.

Advanced study at the institute revolved around the research seminar, which was attended by all advanced students, assistants, lecturers, and the occasional precocious beginner. Heisenberg was one such beginner; before him, Pauli had also attended during his first semester. Both managed to survive the weeding out that the course entailed. Sommerfeld devoted each semester's seminar to a current topic of research. Each attendee was given a small problem to solve or a large published article to study, and the results were presented to the seminar for critical review. A successful performance was required for permission to write a dissertation on the subject. Heisenberg obviously did well, for his dissertation and several of his first papers grew out of his early seminar projects.

Heisenberg's first-semester registration form indicates that, despite his audacity and ambition in entering Sommerfeld's seminar, his father must have advised restraint: Heisenberg had prudently protected himself in the event of failure by signing up for five hours of mathematics lectures and exercises conducted by Artur Rosenthal, but for only one hour of theoretical physics—the exercise session following the main lecture. This meant that Heisenberg was in fact a guest auditor in Sommerfeld's seminar and main lecture and could withdraw promptly into mathematics should he prove unsuitable for physics. By the second semester, such caution was no longer necessary. Sommerfeld had admitted him without reservation to the program, and Heisenberg filled in his next registration form with all of Sommerfeld's offerings.[41]

Since students in the university's mathematical-physical seminar were required to take Wien's course in experimental physics, Heisenberg, like Pauli before him, also registered for the five-hour lectures in experimental physics (mechanics and optics) his first semester.[42] As a second-semester physics student, he registered, as required, for Wien's tortuous eight-hour beginner's laboratory. Heisenberg continued to study mathematics with the aged Rosenthal and his colleagues Alfred Pringsheim and

Voss—but he avoided Lindemann. Mathematics and astronomy were his two minor subjects, and in each he was expected to register for lectures, exercises, and one seminar. Already rejected by Lindemann, Heisenberg soon discovered that he had lost interest in Lindemann's abstract number theory but had gained interest in Rosenthal's "visualizable" geometry. The budding number theorist was ripening into a theoretical physicist.

Heisenberg's decision to study theoretical physics rather than mathematics caused his father much concern.[43] Public interest in the relativity and quantum theories was certainly strong, as was demonstrated by the large audiences at popular lectures such as Sommerfeld's. Nevertheless, employment opportunities were meager. Mathematics and experimental physics were well-established disciplines that could lead to any number of jobs in industry and gymnasiums, but professional careers in theoretical physics were still restricted to a few university chairs, all of which were already occupied. Although academic positions would increase during the next decade, Professor Heisenberg knew that his son would have to do extremely well, particularly on the doctoral and habilitation exams, in order to obtain a full professorship and thus continue the family's success in producing university professors.

While Heisenberg's abilities were keen, potential problems were already looming. Personal and professional differences had arisen between Sommerfeld and his new and more powerful colleague, Willy Wien. Both were required to sit on the doctoral committee for physics students, and both had to agree on a single physics grade for each candidate. Wien made no secret of his opinion of theoretical work. Although he had, in fact, once done theory himself, he simply regarded experimental work as more fundamental. Any doctoral candidate in physics had to convince Wien of his mastery of experimental techniques. Moreover, Wien insisted on a traditional, rigid program of study, leading gradually to advanced work. This method was the very opposite of Sommerfeld's habit of confronting his pupils early with research, while simultaneously feeding them the fundamentals. By omitting courses outside their major and minor fields, bright students, such as Pauli and Heisenberg, could obtain doctorates under Sommerfeld in as few as three years. The rapid schedule might leave gaps in a student's knowledge, but, to Wien's horror, Sommerfeld assumed students could fill them in themselves. Wien soon discovered that Heisenberg's training did in fact leave him with serious gaps but not apparently with any regrets. Years later, Heisenberg told a group of young people that, regardless of the many years now required to obtain a doctorate in physics, they should be doing original research by the age of 24.[44]

With such fundamental differences between the two Munich physicists, the final doctoral examinations could easily deteriorate into a pedagogical wrangle. To forestall problems, Sommerfeld ordered his pupils to enroll again in one of Wien's laboratory courses before the final orals. Pauli, who took the course in 1921, apparently did not encounter much difficulty with Wien, but Heisenberg—who suffered the course with ill-concealed scorn—did. His father's anxiety over impending trouble is apparent in a remark he made when Heisenberg visited Göttingen for a semester in 1922, a year before his final orals: "How have Herr Professor Born and the other gentlemen received you? Please don't neglect the experimental physics!"[45]

In addition to his formal training, Heisenberg could credit his rapid advancement under Sommerfeld to the stimulus of an extraordinary group of colleagues and companions. Their names read like a *Who's Who* of their generation. When Heisenberg arrived, the principals included the assistant Ewald, lecturers Karl Herzfeld and Wilhelm Lenz, and students Gregor Wentzel, Wolfgang Pauli, and Karl Bechert. Otto Laporte arrived in 1921 from Born's Frankfurt Institute for Theoretical Physics to continue his studies with Sommerfeld, and Adolf Kratzer, a pioneer in the quantum theory of molecules, habilitated in 1921 and served as a lecturer thereafter. Outside the institute, Heisenberg met Hans Kienle, an assistant at the astronomical observatory who became a close associate, and the mathematician Robert Sauer, a fellow student in Rosenthal's lectures. As had Heisenberg's brother Erwin, Sauer entered into a fierce competition with Heisenberg, and in vying with each other to solve the problems presented, they left the other students far behind.

Of the three non-laboratory institute rooms, with their creaky wooden floors, high ceilings, and drab interiors, Sommerfeld used one for himself and designated another the seminar room. It became a forum where the select five to ten advanced students met daily to discuss and debate various problems and papers. Each student had his own desk. When the newly graduated Wentzel replaced Ewald as assistant in 1921, Sommerfeld appointed Pauli, a younger recent graduate, to the unofficial post of deputy assistant. Among his duties was the correction of Heisenberg's homework. Indicative of the position he would take in physics, Deputy Assistant Pauli's desk was perched on a small platform, from which he could oversee everyone's work.

On or off his perch, Pauli proved to be the most influential and vocal of the seminar members and especially so for Heisenberg. Young Pauli had come to Sommerfeld, from Vienna, even more advanced in the study of physics than was Heisenberg. He arrived in Munich with a paper on general relativity ready for publication. Although barely two years older than Heisenberg, Pauli was already in his fifth semester when Heisenberg first met him, in 1920. Born in Vienna as the son of a Jewish university professor, Pauli was baptized a Catholic, as was frequent at the time. He and Heisenberg experienced a similar, well-bred upbringing, and in their personalities had much in common. Both were sensitive, naive, adolescent, personally insecure but academically confident, enormously ambitious, and thoroughly dedicated to theoretical physics. Outwardly, Heisenberg was quiet and friendly, at once retiring and yet almost recklessly daring—in life and in science—while Pauli was outspoken, aggressive, carefully systematic, and often devastatingly critical. The virtuous Heisenberg loved the purity of the outdoors, youthful games, and the sunshine of long summer days. Pauli preferred the city nightlife, risqué cabarets, and the pubs and coffeehouses of Weimar Schwabing. Heisenberg rose early in the morning, worked intensively throughout the day, and sank into depression during long winter nights. Pauli haunted the cabarets by night, worked feverishly until dawn, then slept until noon, missing his morning lectures. Sommerfeld tolerated Pauli's behavior since he was a mere deputy. But Pauli obviously annoyed Born, whom Pauli assisted in Göttingen beginning in 1921. Pauli left Göttingen for a new institute in Hamburg after only six months. "He can't stand life in a small town," wrote Born.[46]

Although Heisenberg and Pauli were together in Munich for only two semesters, the two physicists—so opposite and yet so similar—formed a close professional friendship that lasted to the end of Pauli's life. That association, recorded in their voluminous correspondence, is one of the most important in modern physics. Each was significant, perhaps crucial, to the other's work. Although they never became close personal friends—they used the formal "*Sie*" (you) form of address (rather than the familiar form, "*du*") until as late as 1927—Pauli functioned for Heisenberg in ways remarkably similar to those of Heisenberg's older brother. Pauli was more advanced in physics and offered Heisenberg brotherly advice on research. But as a colleague and grader of homework, he could also issue ruthless criticism that pushed the insecure yet ambitious Heisenberg to try even harder. Heisenberg once told an interviewer, "Pauli had a very strong influence on me. I mean Pauli was simply a very strong personality. . . . He was extremely critical. I don't know how frequently he told me, 'You are a complete fool,' and so on. That helped a lot."[47]

As noted earlier, Pauli was apparently also partly responsible for converting Heisenberg to the study of atoms. Having read Einstein and Weyl, Heisenberg considered work on relativity after abandoning number theory. During his first semester, he solicited Pauli's opinion of his prospects. Pauli was not optimistic. As author of the then-definitive summary of relativity theory, he warned Heisenberg that research opportunities in the field would be meager. But Pauli was also the author of a dissertation on the quantum theory of the ionized hydrogen molecule that proved a failure in the agreement between theory and experiment. He could therefore assure his colleague that research in quantum atomic physics was wide open.[48]

If Pauli was Heisenberg's "brother" at the institute, Sommerfeld was his "father." Heisenberg, rebellious toward his real father yet still searching for new authorities to replace the old, put his education and early career completely in Sommerfeld's hands. When Sommerfeld left to lecture in the United States for a semester in 1922 and 1923, he sent Heisenberg to Born in Göttingen. They had all agreed that Heisenberg would return to Munich to complete his doctorate. During Heisenberg's visit, Born discovered that he needed a new assistant and hoped that Heisenberg might return to Göttingen to habilitate, after receiving his doctorate. When Born asked him about his future plans, Heisenberg responded: "I don't have to decide that! Sommerfeld decides that!" Born had to apply to Heisenberg's guardian for permission to allow him to habilitate in Göttingen.[49] Heisenberg had become by then a valuable commodity.

CHAPTER 7

CONFRONTING THE QUANTUM

THE QUANTUM ENTERED PHYSICS WITH A JOLT. IT DIDN'T FIT ANYWHERE; IT MADE NO sense; it contradicted everything we thought we knew about nature. Yet the data seemed to demand it. For three decades following the turn of the twentieth century, some of the most creative physicists of the century struggled to comprehend and assimilate the quantum into a new understanding of nature at the atomic level. The story of Werner Heisenberg and his science is the story of the desperate failures and ultimate triumphs of the small band of brilliant physicists who, during an incredibly intensive period of struggle with the data, the theories, and each other during the 1920s, brought about a revolutionary new understanding of the atomic world known as quantum mechanics. Together with relativity theory, quantum mechanics set off a profound transformation throughout physical science and was a forceful impetus to many of the technological innovations that have changed our way of life, from lasers and new medical imaging to the transistors that have powered the digital revolution.

Heisenberg and his colleagues worked on the forefront of highly abstract theories, unfamiliar and newly invented fundamental concepts, and advanced mathematical techniques. The physics is, by its nature, highly technical. There are many good accounts of it available.[1] The purpose in this context is not to explore the technical details of their work but to gain an appreciation, through descriptive accounts of what Heisenberg and his colleagues were trying to do and how they were trying to do it. In the process we can begin to appreciate also the intensity of their struggles to understand the quantum; their frequent failures, setbacks, and feelings of despair; and their truly remarkable creativity in overcoming these difficulties in bringing about what became the new quantum mechanics.

By the turn of the twentieth century, understanding of the workings of the everyday physical world had reached a culmination in what is now called "classical mechanics," the study of moving matter and forces going back to Isaac Newton. It was joined by "classical electrodynamics," the study of electricity, magnetism, and light, based on the contributions of James Clerk Maxwell, Heinrich Hertz, and H. A. Lorentz during the last decades of the nineteenth century.

In 1905 Albert Einstein published three papers that shook classical physics. They had to do mainly with extraordinary situations not encountered in everyday life. The first of Einstein's papers was the special theory of relativity, a revision of classical electrodynamics and mechanics in previously unimagined ways, especially at extremely high speeds, near the speed of light. The second was a study of microscopic particles in fluids that, when experimentally confirmed, removed any lingering doubts about the actual existence of atoms. The third was what Einstein called the "very revolutionary" insight that in some circumstances electromagnetic waves, such as light, may be considered to consist, not of continuous waves of energy, but of tiny individual bundles, or quantities—"quanta"—of light energy. Each tiny "light quantum" carried the same minute amount of energy, an amount equal to the frequency of the observed light multiplied by Planck's constant, a number Max Planck had introduced in 1900. The higher the frequency, the higher the energy contained in an individual light quantum, or what is today called a "photon."

In subsequent papers Einstein showed that the energies of tiny oscillators in matter that emit and absorb light, much like tiny antennae, are also quantized in that they possess specific, indivisible units, or quanta, of oscillation energy. This behavior, he showed, provided the only explanation for the thermal properties of crystal solids, made up of oscillating atoms, as the temperature decreased. But all of these results directly contradicted the widely accepted "classical" theories of matter, motion, and electromagnetic radiation, where energies are not at all broken up into discontinuous packets of energy but are, instead, smoothly continuous.

At the same time as Einstein was revolutionizing modern physics, efforts to comprehend the internal structure of atoms were reaching an impasse. Since we cannot actually see atoms, even with the most powerful optical microscopes, Ernest Rutherford at Cambridge, England, decided to smash the atoms of gold with high-speed charged particles to see what happened. He found that some of the high-speed particles sailed right through the atoms, but others seemed to bounce backward as if colliding with a hard "pit" within the atom. He had discovered that tiny atoms contain an even tinier positively charged ball, the nucleus, at their centers. He reasoned that the negatively charged electrons also existing inside atoms must make up the rest of the size of the atom. Rutherford suggested that the negative electrons are orbiting around the positive nucleus, to which they are attracted, much as the planets orbit the sun in our solar system. If they did not orbit, they would fall into the nucleus, and the much larger atom would not be any bigger than a minute nucleus. The main difficulty with Rutherford's model was that the circulating electron charges should also act as tiny antennas, since any accelerating electric charge, including one moving on an orbit, is required by classical electrodynamics to radiate electromagnetic waves. The orbiting electrons should radiate away all of their energy, with the result that the electrons should, again, spiral inward, coming to rest on the positive nucleus.

Danish physicist Niels Bohr, then a postdoctoral researcher in Rutherford's laboratory, put forth a surprising solution to the puzzle of his mentor's nuclear atoms: he incorporated Einstein's quantum hypothesis directly into Rutherford's planetary model.[2] In 1913, Bohr offered the bold hypothesis—actually a postulate, or assertion—that classical electrodynamics, which requires the radiation of electrons in

Sommerfeld and Bohr in 1919.

orbits, simply does not apply when electrons are moving in certain specific orbits corresponding to specific quantities of energy. As long as the electrons remain in these "stationary states," or "quantum states," they will not radiate away their energy. Furthermore, he stated as a postulate that if an electron happens to absorb from the outside a light quantum of energy exactly equal to the difference in energy between its current state and a higher quantum energy state, the electron will absorb the light quantum and make a "quantum leap" into that state. By the same token, if the electron jumps to an empty lower quantum state, the electron will emit a light quantum corresponding precisely to the energy difference. The only justification for Bohr's radical assertions in violation of classical physics was that they worked. Bohr showed that, in the case of the simplest atom, hydrogen (one electron orbiting a single proton), the calculated quantum states and the emitted and absorbed light quanta arising from the possible leaps between states accounted excellently for the lines of definite frequencies, the so-called Balmer series, appearing in the spectrum of hydrogen gas.

Bohr's work immediately captivated his Munich colleague, Arnold Sommerfeld. Further pursuing the analogy of the solar system and introducing the effects of relativity theory owing to the extremely high-speed motion of the electrons, Sommerfeld brought Bohr's quantum theory of the atom to its full potential by 1916.[3] In so doing, he provided an explanation for an array of observations regarding the emission and absorption of light by atoms and the energies required to strip, or ionize, electrons from atoms.

Nevertheless, despite the great success of the Bohr-Sommerfeld quantum theory of the atom, most physicists regarded it as a frustrating ad hoc combination of classical and quantum notions. The orbits of the electrons could be calculated using classical mechanics, like so many planets in the solar system, while the selected orbits and the jumps between orbits were strictly quantum effects. Theorists viewed the Bohr-Sommerfeld theory as a good intermediate step to a future theory. The future theory would be a "quantum mechanics" that would replace "classical mechanics" with a single, consistent, coherent theory of events at the atomic level.

Following World War I, new experimental techniques and more sophisticated quantitative analyses began to illuminate more and more areas in which the Bohr-Sommerfeld theory seemed to work less and less satisfactorily. Since an atom is so tiny, we can learn about its interior only by bombarding it with high-speed particles, or by observing what goes into it, what comes out of it, and how it interacts with other atoms and with electric and magnetic fields. Beginning with Heisenberg's entrance into the University of Munich in 1920, his work and studies brought him to three of the leading centers of quantum atomic research in the early 1920s, each of which focused on a different aspect of the effort to understand the interior of the atom. In Munich, Heisenberg worked with Sommerfeld on the puzzles of atomic spectroscopy, the attempt to create a model of the atom that could explain the complexities observed in the light emitted and absorbed by atoms. In Göttingen, Heisenberg worked with Max Born in an effort to push detailed planetary models of the atom to their limits and comparing the results with the observed stabilities and properties of simple atoms. The effort provided convincing evidence that the Bohr-Sommerfeld theory failed even for some of the simplest atoms. An entirely new theory was now needed. Moving to Copenhagen to work with Bohr on the interaction of light with atoms, Heisenberg began to perceive the contours of the new physics. The physics gradually emerged upon Heisenberg's return to Göttingen where, in 1925, in a fit of creative genius, he made the breakthrough to the long-sought quantum mechanics.

Bohr often cautioned that we must be prepared for the circumstance that the interior of a tiny atom may not behave the way that objects of our everyday world behave. Already the appearance of the quantum in that atomic world seemed to support Bohr's suspicions. Increasingly sophisticated studies in Munich and elsewhere of the electromagnetic spectra emitted by atoms were revealing ever greater puzzles at the atomic level.

When white light is sent through a glass prism or through a droplet of rain water in the atmosphere, it splits into a rainbow, or spectrum, of colors, each color corresponding to a different frequency of light. If the atoms of one element are stimulated by heat or high voltage and the light is sent through a prism and the emerging frequencies measured precisely by a spectroscope, the atoms are found to emit, not an entire spectrum of radiation, but only certain narrow lines of light at certain precise frequencies, or colors, characteristic of that element—a kind of fingerprint of the element. By the same token, when white light is shone on an unheated gas, these same lines are missing from the total spectrum that emerges from the gas. The atoms have absorbed light of the same frequencies that they have emitted. What made this behavior especially intriguing to Heisenberg and his colleagues was that the emitted and absorbed lines provided highly valuable clues to an understanding of the internal structure of the atoms that were emitting and absorbing them. According to Bohr, the observed lines of frequency arose from downward jumps of the electrons from high-energy quantum states to lower-energy states. By analyzing the observed frequencies emitted by an atom, or a gas of these atoms, one could re-create the internal structure of the quantum orbits within them.

Using his extension of Bohr's model in 1916, Sommerfeld, working with his assistant Peter Debye, took into account an important complication from classical electromagnetic theory. A charged electron orbiting around a nucleus should produce a magnetic field, much as a loop of wire carrying an electric current acts as an electromagnet. When an outside magnetic field is turned on, an interaction with the magnetic field of the orbiting electron occurs. Sommerfeld found that it should cause the electron's orbit to tilt at certain distinct quantized angles. The inevitable jumps between these quantum states and others would be observed as the splitting of an otherwise single observed line into three separate lines. This effect had, in fact, been observed by Pieter Zeeman and was known as the normal Zeeman effect. It now found a quantum explanation in the work of Sommerfeld and Debye.[4] This was the kind of success quantum theorists were seeking—an explanation of the observed properties of gases of identical atoms on the basis of a quantum model of the atoms making up the gas, a perfect match between the observed data and the unobserved interior workings of the atom.

Unfortunately, the perfect match for the normal Zeeman effect did not hold for heavier atoms likewise placed in an outside magnetic field. For these atoms, a single line was observed to split into many more lines than either classical or quantum theory could explain, an effect known as the anomalous Zeeman effect. What made the anomalous Zeeman effect so frustrating for physicists was that the unexplained splittings of a single emitted line in a magnetic field displayed such a wealth of numerical relationships among the observed frequencies that they knew there must be a very regular pattern of motions within the atom that produced these regularities. For instance, even before immersing an atom in a magnetic field, the individual quantum energy states seemed to divide themselves into doublets or triplets of energy states, according to whether the atom has one or two electrons in the outer orbit. (This was later shown to arise from the interaction of the electron's spin with the magnetic field it generated by its own orbital motion.) When a weak magnetic field was then turned on, these lines split into as many as six or eight components, which is the anomalous Zeeman effect. To make matters even more confusing, as the external magnetic field on the gas was increased in intensity, the many components gradually combined together into the three lines of the well-understood normal Zeeman effect.

Obviously a lot of complicated internal interactions were occurring within atoms, and Sommerfeld was determined to find out what they were. He began with the data, which were provided to him regularly, not by his Munich colleagues, but by experimentalists working at the University of Tübingen.

In 1919 Sommerfeld analyzed the highly regular Zeeman data of frequencies in search of empirical relationships and number harmonies that he hoped would provide clues to what he called a "model interpretation" of the data. By 1920 he had uncovered what he called a "number mystery" of complicated numerical relationships among the observed lines. The numbers and frequencies of the lines could be obtained from simple combinations of numbers for each quantum state. Some of these were the integer "quantum numbers," indicating the numbers of quanta of different types (energy or momentum) for each quantum state within the atom. In his textbook *Atomic Structure and Spectral Lines*, Sommerfeld seemed to echo Johannes Kepler in speaking of these numbers as "the language of spectra." This language, he wrote, is "an atomic music of the spheres, a harmonizing of whole number relationships."[5] Although Sommerfeld was writing in a era of German romantic mysticism, and perhaps delighting his readers with such talk, he was not really engaging in number mysticism but simply providing clues in the hope that they might help in revealing the underlying atomic model producing these numerical harmonies. Sommerfeld wrote, "The musical beauty of our number table will not hide the fact that it presently represents a number mystery. In fact I do not yet see any way to a model-based explanation either of the doublet-triplet data or of their magnetic influence."[6]

Soon after his arrival later that year, Heisenberg took up the search for the desired model-based explanation and was not long in finding an answer. Just a year after entering Sommerfeld's program, Heisenberg amazed his teacher by presenting a model of atoms that seemed to resolve every spectroscopic riddle at a stroke. But the model succeeded only because its daring inventor failed to follow the requirements of an acceptable quantum theory, as laid down by Bohr and Sommerfeld.

Atomic mechanics, abstruse line splittings, and mystical number harmonies constituted the highly rarefied atmosphere that Heisenberg inhaled from the moment he arrived in Sommerfeld's institute. The atmosphere was rarefied both in quality and in quantity. Students pursuing science studies were clearly in the minority at that time, even in Germany; among theorists, the overwhelming majority did not concern themselves at all with the abstractions of quantum theory or atomic spectroscopy. In 1920, only a smattering (8 percent) of all German students were studying science of any kind. Of the 337 doctorates awarded at the University of Munich at the end of Heisenberg's first semester, only 19 were in the sciences.[7] That quantum physics was a minority concern even among physicists is suggested by a study of those publishing in the major physics journals of the time. The study showed that of the German physicists born in Heisenberg's generation, from 1895 to 1909, only a little over a quarter devoted themselves to the quantum.[8] During the 1960s, the project Sources for History of Quantum Physics, sponsored by the American Philosophical Society, gathered interviews and archival materials from and about the main participants in the development of quantum physics through 1930. The project sought to preserve the historical record of quantum mechanics, one of the greatest intellectual achievements of the twentieth century. Worldwide, the project found that it could limit the information pool to a mere 200 individuals.[9]

While quantum atomic physics was statistically a rare and rarefied discipline, Heisenberg was statistically common within it. Perusal of the personal characteristics of the main contributors to the first breakthrough to quantum mechanics in 1925 indicates that nearly all stemmed from upper-middle-class academic families; most received their degrees from and were closely associated with the Munich-Göttingen-Copenhagen triad of research centers; all had worked in quantum spectroscopy; the overwhelming majority were German; and, excluding their mentors, their average age in 1925 was 24 years.[10]

Nevertheless, a quarter century earlier Germany was already leading other nations in number of theoreticians. This was the result both of international competition in cultural achievement and of the internal dynamics of the mathematical physics profession in Germany. Spurred by public fascination with science, technology, and the new discoveries in atomic science, by 1920 Germany had exploited its advantage in theoretical science by directing the efforts of its theoreticians into the abstruse yet internationally prestigious realm of atomic physics.[11]

In order to test and challenge the newcomer to his seminar, Sommerfeld early initiated Heisenberg into the mysteries of Zeeman spectroscopy. His sources of data in Tübingen (Friedrich Paschen and Ernst Back) had forwarded a new set of Zeeman data in the fall of 1920. Working backwards from the observed spectra in the Tübingen data to the quantum jumps giving rise to them, Sommerfeld was able to reduce the data to a new integer quantum number, the "inner quantum number," that he had discovered in earlier data.[12] It seemed to correspond to some unknown inner, hidden rotation taking place within the atom; hence its name. Four weeks after Heisenberg began attending his seminar, Sommerfeld suggested that he try his hand at analyzing the data. The eager novice promptly immersed himself in the intricacies of Zeeman spectroscopy, poring over Sommerfeld's book and latest papers on the subject. Toward the back of Heisenberg's only surviving gymnasium notebook is a neatly drawn scheme of the Zeeman effect for all stationary-state combinations, with intensities and polarizations of each line of the effect carefully indicated in then-standard fashion.[13]

The Tübingen data must have been for doublets, quantum states that split in two even before a magnetic field is applied. Sommerfeld's precocious student reported that the Zeeman lines could be easily obtained from stationary states by assigning not integers, but half-integer inner quantum numbers to each state: 1/2, 3/2, 5/2, and so on. Sommerfeld was shocked. "That is absolutely impossible," he retorted. "The only fact we know about quantum theory is that we have integral numbers, and not half numbers."[14]

Sommerfeld's seminar backed him up. The most striking feature of quantum theory was the existence of indivisible quanta of energy, each consisting of a single, identical parcel of energy that could not be further divided—much like an atom of energy. Such a notion was completely foreign to the continuous classical mechanics of Newton and the electrodynamics of Maxwell and others, for which no quanta existed at all. But on the atomic level quanta clearly manifested themselves in Einstein's hypothesis of light quanta, which had been experimentally confirmed; in the hypothesized stationary states of the Bohr-Sommerfeld atom; and in Sommerfeld's method of quantizing a continuous classical variable. Each of these required positive integer quantum numbers: 0, 1, 2, 3, and so on. Half-integer numbers simply had no physical meaning or place in quantum theory.[15]

To his credit, the co-author of quantum atomic theory tolerated his pupil's transgression. By early fall 1921, Sommerfeld had found a partial formula for the anomalous Zeeman effect by treating the atom as a simple contraption of electrons oscillating on springs and emitting the observed lines much like little antennae. Sommerfeld delayed publication until December.[16] The formula seemed to work, but he lacked an acceptable "model interpretation." He did not lack a capable student. The 19-year-old Heisenberg, unencumbered by integers, rewrote Sommerfeld's formula with half-integer numbers and thereby obtained all of the exact observed data. He now sought to derive this formula by working backwards from the data to a quantized model in which the orbiting electrons displayed half-integer orbital momenta—even if half integers found no place in quantum physics. The model he invented entailed one or two valence electrons orbiting outside an atomic core. This core consisted of the nucleus surrounded by the inner electrons orbiting in closed shells of orbits. The half integers arose from the circumstance that, for an unknown reason, the valence electrons each shared a half unit of their momentum with the core. This enabled Heisenberg to obtain the doublet and triplet energies observed for these atoms from the magnetic interactions of the electrons with the core. When the atom was then subjected to an outside magnetic field, Heisenberg simply modified the interactions of the core and the outer electrons in such as way as to produce the observed regularities of the anomalous Zeeman effect.

In the end Heisenberg's model worked, but it violated accepted methods of careful model building, along with nearly every basic principle in sight—the sharing of half units of quanta, the behavior of the core in magnetic fields, and even questions about the conservation of energy in this scheme. Yet, somehow, it worked. As Heisenberg built his model to yield Sommerfeld's formula and related data, the professor revised his manuscript at least twice to keep pace. The collaboration between professor and pupil proved vital to both: Heisenberg stimulated Sommerfeld to rethink and revise his theory; Sommerfeld tolerated his pupil's fracturing of accepted physics.[17]

Heisenberg brought his early ideas on his controversial new model to his first physics conference, the meeting of the German Physical Society held in Jena in September 1921. There he presented his new ideas to Pauli and to another researcher in this field, Alfred Landé. After one of the lecture sessions, the three physicists retired to consider the Zeeman effect. While Pauli and Landé argued over the inner quantum number, Pauli and Heisenberg disagreed over the acceptability of half-integer quanta. Pauli chided that once halves were introduced, then fourths, eighths, sixteenths, and so on would inevitably follow. To Heisenberg, the physical sense of half integers was less important than achieving success. Heisenberg took particular pride in reporting to his mother his apparent victory over his opponents: "That was now a three-way battle, in which each had to defend himself against the other two. Naturally we did not come to any conclusion. However, in the evening I got hold of the professor [Sommerfeld] and he had a letter from Paschen in which it turned out that once again I was completely in the right. Especially Pauli was completely defeated with that." Two days later Heisenberg even dared to challenge "the professor," telling him that a newly written section of the next edition of his textbook was all wrong. "Now that too has been gotten rid of," he wrote home.[18] Heisenberg's arguments were so persuasive that Sommerfeld requested his assistance in completely rewriting the chapter in question, while holding back his own publication to see what developed.

Heisenberg held back, too. As his model gradually took form amid an intense exchange of letters with Landé, Heisenberg grew ever bolder in breaking the rules. When Pauli wrote again to complain, Heisenberg responded with his now famous motto: "Success sanctifies the means."[19] Heisenberg was going to make his model succeed at all costs.

Sommerfeld finally conferred his blessing upon Heisenberg's model—he realized the young man was clearly onto something—and Heisenberg submitted a paper in December 1921 to the *Zeitschrift für Physik* (*Journal for Physics*), the preferred journal for quantum physics.[20] The paper contained what became known as Heisenberg's core model. The model displayed his incredible intuition, his ability to achieve a breakthrough when others could not, and his audacity in achieving success in physics even at the expense of accepted methods. Moreover, in hindsight, his model was, in fact, correct! The half-integer momenta are now understood to arise from the spin of the electron, and it is the valence electron alone that accounts for the Zeeman effect. Rather than the problematic interaction between the core and an outer electron, the spinning electron interacts with the magnetic field produced by its own orbiting motion. As the first—and for the next few years the only—theoretical atomic model that could reproduce the observed data of the Zeeman effect, it had to be taken seriously. Heisenberg was indeed on to something. Yet because it violated a host of accepted quantum principles and procedures of the day, most physicists reacted to it with caution.

Pauli and Sommerfeld, who had each encouraged Heisenberg, were both uncomfortable with what he had wrought. Sommerfeld described the situation in a letter to Einstein in January 1922. Informing Einstein of new and "wonderful numerical laws of line combinations," he wrote: "A pupil of mine (Heisenberg, third semester!) has even interpreted these laws and those of the anomalous Zeeman effect using a model (*Zeitschrift für Physik*, in press). Everything works out but remains however in the deepest sense unclear. I can only promote the technology of quanta; you must make your philosophy."[21] Apparently concluding that his pupil should fill the gaps in his knowledge of classical and quantum physics displayed in his model, Sommerfeld urged Heisenberg to write his doctoral thesis, not on quantum spectroscopy, but in the more traditional and less controversial field of hydrodynamics, a subject in which Heisenberg had demonstrated an ability in Sommerfeld's seminars.

Bohr made no secret of his displeasure with the model after Heisenberg sent him a copy of his manuscript in early 1922. Not only had Heisenberg publicly recognized only two of his deviations while supposedly relying on Bohr's work, but also, Bohr complained to Landé, "The entire mode of quantization (half-integral quantum numbers, etc.) does not appear reconcilable with the basic principles of the quantum theory, especially not in the form in which these principles are used in my work on atomic structure."[22] For Bohr, the source of the anomalous Zeeman effect lay, as it had since 1913, in a failure of classical electrodynamics, not in a failure of quantum physics. Bohr insisted that only a program of consistent applications of quantum rules and procedures, joined by well-recognized and well-supported deviations from the rules, offered "a hope in the future of a consistent theory."[23] Heisenberg's half-integer model contradicted that program on every score.

Bohr told Heisenberg so that summer. Denmark's leading physicist and a world authority on atomic physics, Niels Bohr, then 37 years old, would be regarded, together with Einstein, as one of the two leading physicists of the twentieth century (often followed, in third place, by Heisenberg tied with Dirac, Pauli, and a few others). In June 1922, Bohr, who would receive the Nobel Prize later that year for his work on the quantum atom, delivered a series of comprehensive lectures on quantum atomic physics to German theorists and their students assembled in Göttingen—an event known affectionately thereafter as the Bohr festival. The festival marked Heisenberg's first meeting with the master of quantum physics. It was the start of a lifelong, sometimes difficult, collaboration and friendship that was as important for Heisenberg as his relationship with Pauli.

Bohr's lecture festival was also something of a political statement. German scientists were still under an international cultural boycott.[24] They turned for information and stimulus to each other and to scientists from neutral countries, such as Denmark and the Netherlands. Bohr was unsympathetic to the boycott, impressed with German atomic theory, and grateful to Sommerfeld for supporting grant proposals for his institute.[25] He readily accepted an invitation to deliver the first postwar Wolfskehl lectures in Göttingen, with which his brother Harald, a mathematician, had a long-standing relationship. The lectures had to be postponed for over a year due to Bohr's heavy workload. By April 1922, he was relishing the prospect of lecturing to the Germans on quantum physics.[26] He had good reason. Although the third and latest edition of Sommerfeld's textbook seemed more favorable toward his latest contributions, the new papers coming out of Munich increasingly disregarded the content and methods of his research program.

Bohr delivered seven lectures over two weeks in June 1922 to packed audiences in the main lecture hall of the Göttingen physics institute. The smells of the garden roses and the flight of an occasional honeybee drifted through the open windows overlooking the rear of the institute. Nearly fluent in German, Bohr presented in his characteristically soft and convoluted speech a careful and systematic account of the quantum theory of atomic structure, its problems, and how they might be resolved.[27] For many in the audience, Bohr's festival lectures were their first systematic exposure to the subject, and they served as a basis for most of their research over the next several years.

Mindful of Bohr's complaints and himself convinced that Heisenberg should meet other theorists, Sommerfeld paid Heisenberg's way to the quaint university town of Göttingen in the northern state of Hannover. Inflation had forced the Heisenbergs to curtail Werner's travel and to rely more and more on "Gold Uncle" Karl for support. With Sommerfeld's help and his own initiative, Heisenberg easily gained access to the inner circles in Göttingen. During his stay, Heisenberg quartered on the couch of a local mathematician, probably Richard Courant. He delivered a private lecture on hydrodynamics to Ludwig Prandtl, the leading hydrodynamicist of the day, and eagerly joined the endless rounds of discussion in apartments and coffeehouses and on walking tours. "This afternoon everyone is meeting in a café," he wrote to his benefactors Uncle Karl and Aunt Helen. "Thus I must simply be there. I never get to bed before 1:00 AM."[28]

Unfortunately, no contemporary record remains of Heisenberg's oft-recalled first encounter with Bohr, which probably occurred on June 14. On that day Bohr presented to his audience a favorable account of a calculation by his assistant, H. A. Kramers, on the splitting of spectroscopic lines in an *electric* field.[29] Heisenberg had already carefully studied Kramers's paper and had criticized it in the Munich seminar, perhaps as early as his first semester. The audience listened approvingly to the speaker's summary of the paper and expected little comment from Bohr's peers during the discussion following the lecture. When Heisenberg, a mere student, rose from his seat, the astonished audience fell silent. It was an unspoken rule that students do not contradict their professors, especially in public. Seemingly without qualm, Heisenberg contradicted the master with a criticism of Kramers's calculation.[30] A shocked Bohr responded a little uneasily and afterward invited his critic for a walk to get a closer look at him.

Bohr already knew of Heisenberg and of his disturbing core-model paper, and he may already have made his acquaintance in Göttingen. As the tall, distinguished, well-dressed Professor Bohr walked alongside the slightly built, 20-year-old youth-movement veteran, their walk led them to a hill, the Hainberg, overlooking the town. Their discussions during the walk elevated with the terrain and ranged far beyond physics. As Heisenberg recalled years later (with much romantic admiration for his companion), they delved especially into some of the same issues he had discussed with his comrades—philosophical questions concerning atoms, the use of familiar conceptions, and the precise nature of a consistent "understanding" in physics.[31] Such philosophical and methodological issues were important elements throughout their collaboration. Bohr's interests beyond pure physics surely impressed young Heisenberg, who had previously known only Sommerfeld, the quantum technologist.

The inevitable confrontation over the core model occurred early the next morning, on June 15. There were no lectures that day, and after breakfast Bohr played host to Heisenberg and Sommerfeld in his sumptuous lodgings at a local guesthouse. The young man had now to answer to the co-authors of the dominant theory of the atom, which he had so easily set aside. Later that day, Heisenberg could brag once again to his family of his success. Always mindful of his family's high expectations, his letters redound with boyish pride. They show the same driven, almost reckless attitude toward his work that can be seen in his letters from Jena. He had to be successful—for his own obsessions and for his own survival in his new and competitive field. Heisenberg carried the burden of his background into his profession.

Sommerfeld opened the breakfast-table debate with a brief lecture on Munich physics, including descriptions of the core model and a new helium model possessing half-integer momenta. Bohr responded briefly, Heisenberg reported, "and then there developed a rather extensive discussion between Bohr, Sommerfeld, and me over my early paper. That was interesting. One can easily come to terms with Bohr."[32] The discussion was reportedly a vindication for Heisenberg: "In any case it was determined that until now a proof against my views is not to be found anywhere; at most only generalities and matters of taste speak against them."[33]

Bohr's reserved and diplomatic manner must have misled Heisenberg as to how easily they had "come to terms"; neither Bohr nor his first three systematic lectures apparently impressed on Heisenberg the profound nature of their differences. In his fifth lecture, five days later, Bohr inserted an explicit complaint about Heisenberg's "very interesting paper" in the strongest words he would ever use in criticism: "It is difficult to justify Heisenberg's assumptions."[34] For Bohr, physics was more than a mere technical achievement gained at any price, even though his own model had displayed similar problems. Understanding had to occur within the context of systematic theoretical study. Traditions and methods imposed by old or new authorities had to be respected, or at least subjected to careful deliberation before being ignored. Though always distrustful of authorities, intuitive, and unsystematic, Heisenberg gradually learned to appreciate these lessons during the next few years. Nevertheless, half integers and the related half quanta remained viable entities in quantum physics, until the discovery of spin four years later, and a principal point of contention between Bohr and the German physicists.

Bohr's diplomatic handling of Heisenberg obviously captivated the young man, even though he had spent several years rebelling against authority figures. Heisenberg wrote glowingly to his parents of the man who would come to exert a profound influence on him in every respect: "Bohr is the first scientist who also makes an impression as a human being. Always exercising only positive criticism . . . he is not just a physicist but much more. With me he was always especially nice. He always comes to me when he sees me anywhere, and he has invited me to see him once again next week."[35] Bohr would see much more of the young man in the months and years ahead.

CHAPTER 8

MODELING ATOMS

HEISENBERG'S TRANSFER TO GÖTTINGEN FOR THE COMING SCHOOL YEAR COINCIDED WITH the ascendance of that school of theoretical physics, becoming the third point of a quantum triangle with Munich and Copenhagen as the other two points. It also enabled his own ascendance onto the forefront of quantum research.

Heisenberg transferred to Göttingen while Sommerfeld accepted a guest professorship at the University of Wisconsin for the 1922–1923 school year.[1] During the Bohr festival, Sommerfeld arranged for his advanced pupils to study in Göttingen, and, on Pauli's recommendation, Max Born, Göttingen's professor of theoretical physics, openly considered Heisenberg as Pauli's successor as his privately funded assistant. Sommerfeld agreed to the plan but only on condition that Heisenberg return to Munich the following summer to complete his doctorate: Sommerfeld did not want to lose his favorite pupil so quickly. After the Bohr festival, Heisenberg returned with Sommerfeld to Munich, where they hurriedly co-wrote two papers before Sommerfeld left for Madison, Wisconsin, in August.[2]

The summer and fall of 1922 were busy seasons for Heisenberg. In addition to writing the last-minute papers with Sommerfeld, Heisenberg delivered his first invited talk in September to a conference on hydrodynamics in Innsbruck.[3] That same month he attended the Leipzig meeting of the Society of German Scientists and Physicians, or GDNA. Earlier that summer he had led his youth group on a month-long outing to South Tyrol.

The Tyrol trip served as the first of the "foreign policy" ventures instituted by the New Pathfinders. As did many of their elders during the Weimar era, the "apolitical" New Pathfinders, ignoring the Weimar government, developed their own foreign policy. As youth leader Franz Ludwig Habbel expressed it, their aim was "to work successfully against the subjugation of German culture in the world."[4] Whether Heisenberg knew of such an aim or not, it directly motivated his youth group's extended trips abroad. Heisenberg joined the one to South Tyrol in the summer of 1922 and a trip to visit German-speaking Finnish nationals following the conferral of his doctorate in 1923. In 1924 his group journeyed without him to the German-speaking regions of Hungary and Poland.

In a 1927 retrospective titled "The Foreign Policy of the German Pathfinder Movement," Habbel, second in command of the New Pathfinders, declared that the degradations of the Versailles Treaty, followed by the boycott of German Boy Scouts during a 1920 international jamboree in London, impelled them to turn inward, much as the boycott did for German physicists. The New Pathfinders concentrated on their own interests while taking personal responsibility for defending German culture abroad and in the "occupied territories"—those that Germany had lost to the Allies.

"The effects of the peace treaty," wrote Habbel, "forced us to defend and struggle against the suppression of German compatriots in the occupied territories, to support Germans in the separated boundary lands. The state as people, compared with the state as political accident and as a changing form of appearance, was the completely clear understanding and guide for all of our actions."[5]

For Heisenberg and the New Pathfinders, the current regime—beset as it was by transitory Social Democratic cabinets and plagued by a faltering economy—seemed little more than an unpleasant interlude between two strong and stable regimes. Heisenberg, for whom the war had brought an end to childhood comfort, wrote to his father in this vein in late 1922: "All of this was really only the fault of the war, which had destroyed what was earlier extraordinary and beautiful. Now we are at the point at which only one chapter is closed, only the end of the previous period is here, and the beginning of something new and 'solid' is not yet upon us."[6]

By 1922, the plight of the southern Austrian province of Tyrol had captured the attention of all Germans. As a reward for supporting the Allies in the war, Italy had been granted South Tyrol all the way north to the Brenner Pass, just south of Innsbruck. This territory included not only the Italian-speaking region of Trentino but also the overwhelmingly German-speaking Bozen province, whose people wanted to remain Austrian. When the Fascists came to power under Benito Mussolini in 1922, they began a systematic suppression of German culture and language in Bozen, now called Bolzano.

Germans rushed to defend the German-speaking province. Scientists employed one of their favorite devices: meeting in sensitive locations. Ludwig Prandtl held his September 1922 hydrodynamics conference at Innsbruck, the capital of Tyrol, and the GDNA scheduled its 1924 meeting there, too. In a secret directive to Bavarian leaders in early 1922, Martin Völkel, head of the New Pathfinders, ordered all units to travel abroad, especially south into Tyrol. "With this the separated Germans shall be greeted and at the same time a rigorous activity will be demanded of the groups."[7] In March, Bavarian Baron Karl Sonntag informed his subordinates and his own tribe of Völkel's orders, leaving no room for argument: "I expect from everyone unhesitating postponement of personal plans and wishes and faithful obedience."[8] Heisenberg and his boys had just joined Sonntag's tribe. They headed for Innsbruck with Sonntag on July 15. Traveling farther into South Tyrol, they demonstrated their support of German Bozen by establishing camp for several weeks in the beautiful mountainous region. To make their point obvious, they journeyed south all the way to Venice before returning to Munich in mid-August 1922.

During the following summer, the New Pathfinders exchanged visits with several hundred scouts in Hungary and Finland. In early August, the new Dr. Heisenberg led his group diagonally across Germany toward Finland. Stopping on the way in the Fichtel Mountains for a two-day youth festival celebrating the anniversary of their league, they eventually arrived at Stettin, near the northern Baltic coast of Germany.[9] From Stettin, Heisenberg accompanied ten of the older Munich New Pathfinders, among them Robert Honsell and Kurt Pflügel, on a visit with families among their German-speaking Finnish counterparts. Their hosts were descendants of Austrian immigrants who had recently helped drive Soviet Russians out of neutral Finland.

The Finland trip was probably the most successful of their foreign policy gestures; the cultural contacts and new friendships lasted long afterward. Yet despite this and the political implications of the trip, the naive Heisenberg and his followers regarded it as nothing more than an innocent adventure. They happily recounted their journey in a series of articles, published until as late as 1926 in *Die Spur in ein deutsches Jugendland*, the magazine for New Pathfinders followers. One of their first reports was an unsigned contribution, published in early 1924, titled "The Battle for the Crossing."[10] During Heisenberg's sixtieth-birthday celebration with his "boys," he revealed with fond memories that he had written the piece himself.[11] In it, Heisenberg recounted how he had applied charm, perseverance, and a bribe of some of Uncle Karl's dollars to convince the reluctant captain of a Finnish pleasure boat to take him and the older boys across the Baltic to Helsinki. The captain apparently had little use for inflated German marks. Only after a sympathetic Finnish passenger added 500 Finnish marks to the offer did the captain grudgingly consent. Heisenberg and his companions clambered aboard just as the ship glided from the dock.

In another story, Pflügel recounted how he, Heisenberg, and a friend named Wolfhard set out by boat on a hunting expedition in the Finnish lake region. They managed to shoot three ducks—but after getting wetter than the ducks, they recovered only one.[12] On their return to Munich, Heisenberg and his comrades wrote enthusiastically to their Austrian-Finnish hosts to express their thanks. In November 1923, Heisenberg traveled from Göttingen to Berlin to receive a delegation of Finnish scouts paying a reciprocal visit.[13]

The foreign policy initiative proved a smashing success. Yet, apolitical policy aside, the most striking feature of Heisenberg's foreign adventures is their juxtaposition with his other major pursuit during this time. He was a physicist of extraordinary abilities, already near the top of the profession that would soon produce quantum mechanics. He was immersed in research that was complex, sophisticated, and highly demanding. Yet he still engaged in adolescent romps and other uncritically naive activities as a boy scout. Apparently such immature behavior and the extended periods of outdoor fun provided a necessary counterbalance to and relief from the intense, technically abstruse physics he was creating.

Letters to his family and colleagues during and after long camping tours indicate that Heisenberg completely banished physics from his thoughts during these trips. For instance, after a month-long tour through Upper Bavaria in 1925 that occurred just after he had laid the foundations for the matrix form of quantum mechanics (an enormously exciting time for him professionally), the physicist wrote Bohr: "Obviously I have not thought at all about physics during the entire last month and I don't know if I still understand anything of it."[14]

Heisenberg came face to face with the hard reality of current events soon after returning from the Innsbruck hydrodynamics conference in September 1922, which had followed the Tyrol excursion. He was home less than a week before setting out again, this time for Leipzig, the site of the biennial meeting of the GDNA. It was the centennial of the founding of the prestigious society, and Max Planck, then chairman, decided to use the occasion to promote a new sense of unity among German and Austrian scientists. He scheduled a general lecture for September 18, the first day of the conference, to be given by Albert Einstein, Germany's most renowned scientist. He would speak on the theory of relativity, regarded as one of Germany's most famous scientific achievements. Before leaving for Wisconsin, Sommerfeld had encouraged Heisenberg to attend so that he might finally meet the great man. Heisenberg's father generously provided the round-trip train fare plus 2,000 marks, some of which Heisenberg planned to use for another visit with the Berlin New Pathfinders contingent after the meeting.[15]

Unfortunately, Planck's hoped-for demonstration of unity faltered in the face of mounting discord over relativity and mounting anti-Semitism aimed at Einstein. Still leading the attack on both fronts was the experimental physicist and Nobel Prize laureate Philipp Lenard. After a debate with Einstein on relativity at the previous GDNA meeting in 1920, Lenard was unconvinced of relativity and unswayed from his own alternative, a classical ether theory of electrodynamics. He was already nurturing paranoid anti-Semitism in response to the rejection of his ideas.

As the 1922 meeting approached, Lenard published "A Word of Warning to German Scientists" in his latest monograph on ether theory. In it, he dismissed relativity theory as a mere hypothesis and closed with an anti-Semitic diatribe against his critics.[16] Unbeknownst to Heisenberg and to Einstein's opponents, Einstein had withdrawn temporarily from public appearances after the shocking assassination in June 1922 of Walther Rathenau, the well-known Jewish foreign minister. Einstein was replaced as the featured speaker on relativity by the man who would support him a decade later—Max von Laue.

Heisenberg arrived in Leipzig on September 17 and checked in to a cheap youth hostel in the poor quarter of town to conserve his funds for travel and the endless rounds of coffeehouse conversations.[17] As he approached the lecture hall the following evening, one of Lenard's disciples pressed a handbill into his hand. The handbill had been signed by 19 scientists and physicians—with titles prominently displayed for effect—who proclaimed that they "not only regard the relativity theory as an unproved hypothesis, but even reject it as a basically failed and logically untenable fiction."[18]

Heisenberg was shocked. Unlike the earlier Munich episode, which involved fanatical local students—an incident that Heisenberg himself did not witness and of which he learned only a year later—this was a direct confrontation with anti-Semites supported by learned professors, including the eminent Lenard. The incident apparently shook him at last into brief recognition of the political implications (although moral issues were also at stake). "I felt as if my world were collapsing," he wrote years later in his memoirs. He had always thought that science was above politics; indeed that was one reason he had chosen physics as a career. "And now I made the sad discovery that men of weak or pathological character can inject their twisted political passions even into scientific life." Heisenberg said he wondered at the time whether physics was "really worth bothering with" after all.[19] Such doubts, however, did not deter him from his richly promising future.

Despite Heisenberg's sudden realization about the political susceptibility of physicists, it did not inspire him to take much action. If anything, the experience caused him to cling even more tightly to his apolitical illusions for science until forced to relax his hold a decade later. Nor did he display any increased interest in or concern for political affairs, even as a safeguard against their corrupting influence.

Heisenberg returned to Munich the very next day in a depressed state—not only because of politics. He had returned to his hostel after a lecture to find that he had been robbed of all his money and belongings. Faced with showing up at the conference unwashed and unshaved, he returned to Munich, where he put in a stint as a woodcutter in order to earn back his money and to buy new belongings.[20] He did not know that the man who spoke that night in Leipzig was not Einstein but Max von Laue.[21]

Heisenberg finally arrived in Göttingen in late October 1922 for the start of the winter semester. To Professor Born, the slightly built sometime woodcutter looked "like a simple farm boy, with short, fair hair, clear bright eyes, and a charming expression."[22] Despite the farm boy's excellent references and his own intention to hire him, Born decided to see first what he was getting. For his part, Heisenberg wanted to see what his new environment would offer him. What it offered was his first systematic introduction to mathematical atomic physics. Göttingen's mathematics tradition, Born's appointment, Bohr's lectures, and the generosity of various philanthropies had ensured Göttingen's place as one of the world's leading centers of atomic physics.

Göttingen mathematics boasted a long line of luminaries, among them Carl Friedrich Gauss, Georg Riemann, and Felix Klein. Klein, who arrived in 1886 to head the mathematical-physical seminar, later founded a series of institutes and research programs for pure and applied mathematics that made Göttingen the leader in such research.[23] By the time Heisenberg arrived in 1922, Göttingen could boast of a Mathematics Institute headed by such now-famous people as Richard Courant, David Hilbert, and L. D. Landau, an Institute for Applied Mathematics and Mechanics directed by Carl Runge and Ludwig Prandtl, and three separate institutes for physics. The last of these had been headed until 1920 by two full professors, Peter Debye and Woldemar Voigt, and an associate professor, Robert Pohl. When Debye left for Zurich in 1920, Pohl was promoted to full professor and head of the experimental physics section. Born, a highly regarded former student of Hilbert and a former assistant to the famed Hermann Minkowski, was called from Frankfurt to succeed Debye.

Born in Breslau of an academic Jewish family, Max Born had early devoted himself to mathematics and, like Sommerfeld, had turned to theoretical physics after encountering Klein in Göttingen. Briefly assisting Minkowski until Minkowski's untimely death in 1909, Born continued Minkowski's work on the mathematical formulation of relativity theory, then turned to the quantum theory of crystals and molecular structure under the influence of Einstein's quantum theory of solids. He was best known by 1920 for his book on the dynamics of crystal lattices and for his work on the chemical consequences of a theory of ionic crystals that he had developed with Alfred Landé.

During World War I, Born, a noncommissioned army officer, performed artillery research for the army in Berlin, while occupying an associate professorship at the university. While in Berlin, Born and his wife, Hedwig, a writer of romances, became close friends of Einstein (then between marriages), with whom they frequently corresponded thereafter.[24] In 1919, von Laue, then in Frankfurt, suggested to Born an exchange of job positions. Born readily agreed; it would mean promotion to full professor. Although he and Hedwig thoroughly enjoyed Frankfurt for its cultural offerings, they remained in the Goethe city barely two years. Born, then 38 years old, seemed an ideal candidate for the vacancy left by Debye in Göttingen.

Born hesitated. The shy and retiring theorist, plagued by hypochondria, was not attracted by big-science administration and had no desire to teach experimental physics in addition to theory. Meanwhile, the Frankfurt faculty was doing everything they could to keep him. Born went to the Prussian Culture Ministry in Berlin, which oversaw university appointments, to discuss the matter. He later recalled that, in reviewing the Göttingen files at the ministry, he had discovered a notational error that provided for an extra associate professorship at the institute. Born easily convinced the Prussian ministry to make the extra position a full professorship, to be held by an experimentalist. To clinch the deal, the ministry doubled its salary offer.[25]

With Born's arrival in 1921, Göttingen physics was reorganized in typical fashion for Germans at the time. Three completely independent institutes were created, each headed by a full professor and all housed in one building, the box-shaped Physics Institute at Bunsenstrasse 9. Since experimental physics still enjoyed more prestige and more direct connections with the original institute, Pohl directed the First Physics Institute, which was devoted to experimental physics. Born selected his good friend James Franck, who with Gustav Hertz had lent experimental support of Nobel-Prize quality to the Bohr atom, to head the newly created Second Physics Institute, also devoted to experiment and to the beginner's laboratory. Born himself directed the Institute for Theoretical Physics, consisting of one small room, one assistant, one "private" assistant, and a half-time secretary.[26]

The bashful Born seemed overwhelmed by the number of students flocking to Göttingen—only the Berlin Technical College had a greater enrollment. "There are students here like hay," he wrote a colleague.[27] During the semester in which Heisenberg studied in Göttingen, more than a third of the students enrolled at the university studied mathematics or science, by far the most popular subjects in Göttingen.[28] Each of Born's lectures, like Sommerfeld's, drew about 80 students from all majors and degree programs, but many more physics doctoral candidates attended than in Munich. Born originally had nine advanced students in the winter of 1922 to 1923; Sommerfeld sent four more from Munich during his absence, one of whom was Heisenberg.[29]

Heisenberg delighted at first in the picturesque walled town of Göttingen, with its "narrow alleys and strange dialect" and with a university only a third the size of Munich's.[30] A guide for English-speaking students described Göttingen as a small town where "life is comparatively quiet, and there are no noisy factories. In the outer town, peace and quiet are almost undisturbed."[31] The major industries produced instruments for the university laboratories, and town leaders were professors, army officers, and retired bureaucrats. Although it offered little public culture compared with Munich or Frankfurt, many of Göttingen's science professors were music devotees. They often had Heisenberg to their homes for musical evenings. "Heisenberg is at least as [scientifically] talented as Pauli," Born reported to his friend Einstein, "but personally more pleasant and delightful. He also plays the piano very well."[32]

But as winter set in Heisenberg sank into depressed loneliness. The many physics students around him were no replacement for his real friends, his Munich youth group. For the first time, he was away from both home and friends for an extended period. He soon felt trapped in the northern German town and penned numerous mournful letters to family and friends. "In general there just aren't any people here, or I can't find them," he complained to his brother.[33] Heisenberg tried to distract himself by cramming his weekdays with physics. On weekends, despite the cost, he traveled to Berlin to be with Heini Marwede and other expatriate Bavarians from his youth unit—but only rarely did he visit Erwin, still studying chemistry at the University of Berlin.

"You appear to give up your entire day only to physics," one of Heisenberg's comrades noted.[34] By Christmas vacation, which Heisenberg planned to spend skiing with his youth group in the Bavarian mountains, he had burned himself out. "If the [Christmas] holidays had started only 10 days later, I would go crazy with physics," he noted. "In Munich I will not speak one word of it. In personal terms Göttingen remains a completely desolate hole."[35] Periods of intense and lonely work, followed by long and relaxing outings with the group, became the pattern of Heisenberg's life.

Like Munich, Göttingen had experienced a postwar soviet upheaval, but it was quickly suppressed after a battalion of government troops took up permanent residence in the town. As typical of the times, Göttingen students tended to be right wing, nationalistic, and anti-Semitic. As everywhere in Germany, the economic plight of students had grown desperate in Göttingen, but because of the central place of the university in Göttingen civic life, town and university paid more attention to their needs. Although all the universities were publicly funded, private contributions helped to establish perhaps the earliest student Mensa, a cafeteria serving subsidized meals, and to acquire a building to serve as a dormitory.[36]

With so many students flocking to the university town from all over Germany, the housing shortage was acute. Heisenberg began the search for a room early and soon found one through the efforts of Erich Hückel, Born's university assistant. Heisenberg lodged in a spare second-floor bedroom of the large and stately Biedermeier home of the widowed Mrs. Ulrich at Walkemühlenweg 29, just a block from the Physics Institute.

Money was again a problem. When Annie Heisenberg stopped off in Göttingen on her way home from Osnabrück that November, shortly after the start of the winter semester, she gave her son 8,000 marks for his monthly expenses. Inflation, though still in its early stages, quickly ate it up. For 1,000 marks a month, Mrs. Ulrich supplied a bed, breakfast, afternoon tea, and evening potatoes. On top of this, Heisenberg had to pay for heat, which cost 2,600 marks; seminar fees for the semester were 718 marks; and once he had to buy Mrs. Ulrich a pound of butter for his fried potatoes, which cost him 750 marks. When in December Mrs. Ulrich gave him three eggs and a bouquet of flowers for his birthday, Heisenberg fretted over whether he could afford to reciprocate with a Christmas gift.[37] The only bargain was lunch. Instead of eating at the Mensa, for 50 marks Heisenberg and the other physics students attended a subsidized private table at a home across the street from the institute. This was the biggest meal of the day and even sometimes included a meat dish. The subsidy was probably arranged by Born, who, like most science professors in the period, could rely on the generosity of a befriended philanthropist.

By the end of November, August Heisenberg had to forward another 3,000 marks to his son and to raise his allowance for December to 10,000 marks. A thankful Heisenberg offered: "In case you should go bankrupt, I can make out in an emergency with 1,000 marks less per month."[38] By January his money problems were solved, at least temporarily. Professor Born had finally offered him the private assistant's position with a generous salary of 20,000 marks per month, supplied by the American financier and philanthropist Henry Goldman, the co-founder of Goldman Sachs.

Born's offer came within days of Heisenberg's first triumph in the Physics Colloquium, the high court of Göttingen physics. There local and guest speakers submitted their latest work to careful scrutiny and themselves to withering cross-examination. "It was customary to interrupt the speaker and to criticize ruthlessly," Born recalled.[39] The Physics Colloquium was an example of the emphasis placed on oral and personal interaction in Göttingen. Born, a former mathematics assistant who had handpicked one of his colleagues, worked much more closely with the local mathematicians and experimentalists than did Sommerfeld in Munich. Consequently, a variety of joint lectures, seminars, and colloquia were held as course work, in addition to each professor's individual seminars.

Heisenberg registered for all of these symposia. In each, he was expected to defend his position, which was not difficult for him, and to employ proper rhetorical and diplomatic skills, which was. "Thus I will no doubt learn how to lecture," he wrote in a letter home. And learn he did. In January he reported to Sommerfeld: "The result so far is that Born and Hilbert are of the opinion that I can lecture very well."[40]

Heisenberg's new skill, combined with his intellect and demeanor, enabled his early acceptance by the Göttingen critics. Within weeks of his arrival, Heisenberg reported to the Born-Hilbert seminar on the recent Sommerfeld-Heisenberg papers, and in December he was called before the Physics Colloquium to report on his first and still only individually written paper, his controversial core model of the atom for the anomalous Zeeman effect. Usually only physicists attended the colloquium, but on this occasion Hilbert and Courant of the Mathematics Institute were in the audience, and the aged Runge showed up to hear firsthand the author of the only atomic model that accounted for a set of rules he had found for the lines appearing in the Zeeman effect. No doubt these luminaries wanted also to see for themselves the lowly student who had dared to challenge the author of the Bohr model of the atom.

The audience in that colloquium was already well aware of Heisenberg's core model and equally aware of Bohr's criticism of it during the Bohr festival. As expected, the distinguished audience that gathered in the same lecture hall where Bohr had spoken the previous summer was already skeptical of the model and prepared for ruthless criticism of the upstart. Heisenberg was prepared for the challenge. He poured as much "verve and élan" as possible into his performance in an attempt to sway the experts.

"The result was resounding," he bragged to his brother. The audience did interrupt him several times—but only with applause. "So now all of Göttingen is convinced of the theory," Heisenberg proudly proclaimed.[41] Within a month the newly polished lecturer was working for Born, and Born admiringly confided to Sommerfeld Göttingen's infatuation with the young man: "I have become very fond of Heisenberg; he is very well liked and highly regarded by us all. His talent is unbelievable, but his nice, shy nature, his good temper, his eagerness and his enthusiasm are especially pleasing."[42]

If Heisenberg impressed Göttingen physicists, they at first did not impress him. He found them "very strange" compared with his Munich colleagues. The former mathematics disciple complained of too much concern for mathematics. "Even the physicists are actually interested much more in mathematics than in physics," he wrote to his father in November. "The result is that one has a somewhat bored impression of all the physics here; no one has the initiative to try something new; they pick out mathematically interesting topics that are in most cases exhausted as physics."[43] The quiet and subdued Born paled in comparison with the dynamic and forceful Sommerfeld, and Born's preference for rigid mathematical and physical consistency seemed much too tame for Heisenberg.

Between the Bohr festival in June and Heisenberg's arrival in October 1922, Born had settled on his own research program, which indeed contrasted sharply with Munich's: an even closer adherence to the elements of quantum atomic theory, both the quantum rules and the classical mechanics, than Bohr had just argued in Göttingen. "It was the time before the establishment of quantum mechanics," he later wrote, "and I was trying, with my collaborators, to find weak points and contradictions in Bohr's semi-classical theory of atoms."[44] It was "semi-classical" because the electrons moved in their orbits according to classical mechanics, but the orbits were selected and jumps between them occurred according to quantum rules imposed on the atom according to Bohr's original postulates.

While Munich physicists unraveled number harmonies in spectroscopy and constructed ad hoc models to explain them, Göttingen theorists, when studying atoms, had moved to the opposite extreme: they were now attempting to construct thoroughly orthodox planetary atomic models by adopting the mathematical methods and mechanical techniques of planetary astronomy to the orbits, then subjecting them to the rules of quantum theory. Absolute consistency would illuminate the weaknesses and contradictions in Bohr's quantum theory (and its extensions by Sommerfeld), they reasoned, and thereby, hopefully, point the way to a new and better theory. Despite his skepticism, Heisenberg quickly appreciated the approach: "For me personally Göttingen has the great advantage that for once I will learn correct mathematics and astronomy."[45]

By the time Heisenberg arrived, his two predecessors, Pauli and Ernst Brody, had helped to establish Born's program—and to encounter its difficulties. Bohr's quantum theory of atoms and molecules worked quite well— but only when applied to the simplest atoms consisting of just two particles. These included hydrogen (one electron orbiting a positively charged nucleus) and ionized helium (one electron orbiting a doubly charged nucleus). It failed when applied to anything more complicated, as Pauli had just demonstrated in his dissertation on the ionized hydrogen molecule (one electron orbiting two bound hydrogen nuclei).

Mathematically, however, Pauli's calculation was only a rough first approximation, since he assumed that the two nuclei were completely at rest. Allowing the nuclei to oscillate, as they do, as if bound together by a spring, required more detailed approximations and more sophisticated techniques. Born and his first assistant, Brody, developed such techniques by adapting to atoms of two or more orbiting electrons the sophisticated nineteenth-century methods for calculating the motions of planets.

Planets orbiting the sun are attracted to the sun by the gravitational force. Following Isaac Newton, the orbit of a planet can be calculated fairly easily on the basis of this force. But this is only a rough, first approximation to the actual motion, because the planet is also attracted, though with much smaller forces, to all of the other planets in the solar system. These additional attractions perturb the motion and produce "perturbations" of the basic orbit, as Newton himself had first shown. During the nineteenth century a sophisticated mathematical apparatus, "perturbation theory," was developed to handle these complicated motions. Born now applied these techniques to the quantum theory of orbiting electrons in atoms, viewed very much as planets orbiting the sun.

Unfortunately for Born and his assistants, the planetary calculations could not be transferred directly to electrons in atoms for a variety of reasons. Most importantly, the theory was limited because the negative electrons in atoms *repel* each other with a force that is almost equal to their electrical attraction to the positive nucleus. Planets only attract each other and with gravitational forces that are much weaker than their attraction to the sun. Finding an adaptation of planetary physics that would accommodate this and other complexities kept the Göttingen theorists occupied.

Born and Pauli did manage to develop more general techniques applicable to quantum atoms during the spring of 1922.[46] Since Bohr's original quantum theory already worked quite well for systems of two particles, such as the simple hydrogen atom, they and others focused on the second element of the periodic table, neutral helium, which consisted of three particles: two electrons orbiting a doubly charged nucleus. But even this seemingly simple problem was further complicated by the existence of two different forms of helium, apparently displaying two different possible model arrangements.

Along with the core model, Sommerfeld and Heisenberg had brought a model for one form of helium to the Bohr festival in which they had utilized half-integer quantum numbers. Bohr summarily rejected it, as he did the core model, and for the same reason. Half integers were simply not acceptable. He had already decided to blame another facet of quantum theory for helium's problems and thus retain the integers required by quantum theory. When his assistant Kramers carefully calculated a model for one form of helium (orthohelium) with integer numbers, he obtained a result that was unstable and did not correctly predict the energy required to ionize the atom. As he had done once before with his breakthrough to the quantum theory of hydrogen in 1913, Bohr simply declared that Kramers's model was, in fact, correct, but, he asserted, the use of classical mechanics to calculate the orbits was not. Kramers echoed Bohr's sentiment nearly word for word in his paper. "Mechanics is not valid in this simple case," he wrote in December 1922. "As recently emphasized by Bohr, one must generally expect that in the stationary states these laws are different from those of the usual mechanics."[47] Born and Heisenberg would soon test that assertion.

Although Bohr had rejected the helium model containing half-integers, Sommerfeld encouraged Heisenberg to analyze it anyway, when he found time in Göttingen. Heisenberg had plenty of other work to fill his lonely hours. Before leaving for Wisconsin, Sommerfeld had assigned each of his pupils a research problem to keep them busy. Heisenberg received the difficult task of calculating the conditions for the onset of turbulent flow in hydrodynamics—with the promise that he could submit the results, if satisfactory, as his doctoral dissertation the following spring. Heisenberg delivered a short report on his preliminary results to the Innsbruck hydrodynamics conference in September 1922, and he sought further advice on the problem from Prandtl in Göttingen. He wrote Sommerfeld and Landé that fall that he was too busy with hydrodynamics to worry about atoms.[48] But in his last letter to Sommerfeld before leaving Munich for Göttingen at the end of October 1922,

Heisenberg reported that he had been unable to contain his curiosity: he had already carefully studied the paper by Born and Pauli on an improved technique for applying planetary calculations to quantum atoms and had just applied it to Sommerfeld's half-integer helium model.[49] A rough calculation yielded precisely the measured energy required to ionize the atom (free an electron)!

Sommerfeld was ecstatic. After Heisenberg forwarded the detailed calculation, carried out with the help of a Göttingen student, Sommerfeld published the result in an American journal (with a thank-you to his German pupil).[50] Both Heisenberg and Sommerfeld were now convinced of the necessity of the errant half-integer quantum numbers and momenta—prominent components of the core model and now of their helium atom. Although Bohr and his Copenhagen colleagues emphasized, to the contrary, the failure of classical mechanics in atoms, fractional numbers and momenta soon became a permanent option in Göttingen and Munich. They were, of course, later justified as arising from the spin of the electron, a notion completely foreign to physics before 1926.

Nevertheless, Bohr's lectures on consistency during the Bohr festival at first inspired a new dedication in Göttingen to consistent adherence to quantum theory. Five days after Bohr's last lecture in June 1922, Born announced his new approach: "The time is perhaps past when the imagination of the investigator was given free rein to devise atomic and molecular models at will. Rather, we are now in a position to construct models with a certain, although still by no means complete, certainty through the application of quantum rules."[51]

Demonstrating his new devotion, Born freely constructed all imaginable models for the neutral hydrogen molecule (two hydrogen atoms bound together, thus containing two nuclei, each with an orbiting electron). He then eliminated all but one imagined model through strict adherence to the quantum rules, the one displaying only integer numbers. A student, Lothar Nordheim, received the problem for his dissertation, while Born turned to a similar but more complicated problem: excited neutral helium, a helium atom with one electron orbiting in a quantum state of high energy far away from the nucleus and the inner electron.

Excited helium would serve as an explicit test case of Bohr's original quantum theory of atoms. It would entail strict adherence both to the quantum rules and to the physics of classical planetary orbits. In this way, Born reasoned, consistency would show if, and exactly where, the current quantum atomic theory failed. But to handle this problem, the Born-Pauli apparatus for calculating electrons like planets had to be extended to even greater heights of mathematical sophistication. Hints on how to do this could be gleaned from the advanced textbook on planetary mechanics written earlier by the French mathematician and philosopher Henri Poincaré.

Shortly after Heisenberg arrived, Born started a Poincaré reading circle on Monday evenings in his home. Due to the housing shortage, the Born home, located near the institute on a street named for another Planck, was a rented ground-floor flat consisting of three enormous rooms and a kitchen. Born, his wife, and their two young children (one the future mother of the contemporary pop singer Olivia Newton-John) subdivided two of the rooms into bedrooms and designated the third a combination parlor, study, and music room. A first-rate Steinway grand, which Heisenberg often played, occupied one corner, the Poincaré reading circle another. Heisenberg attended along with several other advanced students and assistants, and in seminar fashion each prepared a talk on a section of the material, which the group then discussed with much enthusiasm. Continuing his hydrodynamics with Prandtl and his mathematics with Hilbert and Courant, Heisenberg studied his Poincaré with Born "with every ounce of energy."[52]

By late December 1922, Born and Heisenberg had obtained the needed extension. They quickly used it to test Bohr's explanation for the periodic arrangement of elements in the periodic table—the successive filling of electron orbits as one proceeds from one element to the next across the table. They confirmed Bohr's account of the periodic table, what he had called the "building-up principle," to their own satisfaction, but they still invoked the controversial half integers.[53]

At a time when problems in quantum theory were mounting, Göttingen had produced two positive results: the Sommerfeld-Heisenberg helium model and solid support for Bohr's building-up principle. But Bohr and Pauli (now in Copenhagen) both complained, rendering Heisenberg more than a little uneasy: "I am somewhat unhappy over the fact that with all of these papers I constantly contradict Bohr and Pauli."[54] Half-integer momenta, detailed mechanical planetary models, and strict adherence to stationary-state mechanics all contradicted the latest Copenhagen line. For Bohr and Pauli, the use of classical mechanics in the stationary states, not the integer quantum numbers, was to blame for the theoretical impasse.

Heisenberg was not attracted to the Copenhagen alternative. He responded with sharp criticism of Pauli's latest efforts to avoid the half integers of the core model. Pauli had tried to explain the anomalous Zeeman effect with little reliance on any model at all. Pauli responded in January 1923 with a two-day stopover in Göttingen to argue the Copenhagen point of view on his way back to Copenhagen from Vienna. Born, Heisenberg, and a handful of students tangled with the visitor. Although "he is besotted with Bohr," Heisenberg reported, "Pauli admitted that our standpoint is very consistent." But Pauli would make no concessions on Bohr's behalf: "Bohr in any case wants to allow mechanics to be no longer valid, i.e., only to a certain approximation. He does not yet believe in half quanta and helium."[55]

To resolve the debate over classical mechanics in quantum atoms, Born and Heisenberg turned at last to the excited helium atom as a careful, systematic test to prove or disprove the viability of classical mechanics and quantum rules in the reigning Bohr-Sommerfeld quantum atomic theory. A rigorously consistent application of the most sophisticated planetary mechanics to the quantum orbits, with strict adherence to integers, followed by a careful comparison of the results with available experimental data, could decide the matter better than any personal preferences. Heisenberg informed Bohr of the plan in early February 1923: "The other work of which I wanted to write you is a general investigation of all mechanically allowed orbits of excited helium. If in the end the experimentally found terms are not included, then one knows that the mechanics is wrong."[56]

Heisenberg, however inconsistent otherwise, kept his word, as did Born, in following to the letter the demands of quantum orthodoxy.[57] The advantage of excited helium was that it could be treated almost exactly like a hydrogen atom—a perfect setup for planetary "perturbation theory." In the highly excited atom, one of the two helium electrons orbited much farther out from the nucleus than did the other, more tightly bound inner electron. Since the negative charge of the inner electron can be seen as balancing one of the two positive charges of the nucleus, the effect on the outer electron could be treated as a small alteration, or perturbation, of the motion of a single electron orbiting a single positive charge, such as that found in the well-established model of the hydrogen atom. This simple arrangement, a slight variation of Bohr's original hydrogen model, should definitely prove whether or not the prevailing quantum theory of the atom was valid.

Bohr had earlier shown how the downward jumps between the quantum states of the orbiting electrons occurring in many hydrogen atoms at once would yield a series of individual spectroscopic lines, the so-called Balmer series. Bohr had also shown that a correction to the Balmer series could be obtained for heavier atoms if the effect of the nuclear charge were taken into account. Using their new quantum planetary mechanics, Born and Heisenberg derived an expression for the correction to the Balmer series for their hydrogen-like model of the excited helium atom. The exact value of the predicted correction depended on the orbit chosen out of all possible orbits for the outer electron. The physicists obtained four possible stable orbits. They dutifully discarded one of the four because it involved a non-integer quantum number. Then a careful comparison between the experimentally observed value of the correction for excited helium and the derived values for the remaining three possible orbits turned out, Born told Bohr, completely "catastrophic."[58] None of the values agreed!

As early as February 19, 1923, Heisenberg was able to inform Pauli of the result, and of his own conclusion: "This result appears to me to be . . . very bad for our present conceptions. One must probably introduce entirely new hypotheses—either new quantum conditions or new modification proposals for mechanics." A month later the Göttingen conviction had grown even more radical: "All present helium models are just as wrong as is the entire atomic physics."[59]

Meanwhile, in Copenhagen, Bohr and Pauli were coming to a similar conclusion regarding the atomic models of the Zeeman effect. Pauli had received a fellowship from the International Education Board, a branch of the Rockefeller Foundation, to work in Bohr's institute for a year beginning in the fall of 1922. While Born and Heisenberg scrutinized the properties of carefully calculated planetary atoms in Göttingen, Bohr and Pauli took the same approach as their Göttingen colleagues in examining a new model for the Zeeman effect. They constructed it in such a way that it adhered to every quantum rule required, including the one that permitted only integer numbers. Heisenberg expressed pleasure at the change of opinion in Copenhagen regarding the study of detailed models, but Pauli experienced only frustration. He wrote Sommerfeld that he had "tortured himself" for weeks with the anomalous Zeeman effect. "But it would not and would not agree! . . . For a while I was completely discouraged." By early March the Copenhagen model had failed as badly as did Göttingen's helium model. It had been, wrote Bohr, "a desperate attempt to remain true to the integer quantum numbers in that we hoped to perceive in the very paradoxes a clue to the ways by which we might search for the solution to the anomalous Zeeman effect."[60]

By the summer of 1923, the now dispirited Heisenberg, Born, Pauli, and Bohr, along with many of their colleagues, were willing to accept the inevitable. In July 1923 Born declared, in a review article on the state of quantum theory, that the need for "deliberate deviations" in spectroscopy, and the obvious failure of all quantum models for some of the simplest atoms and molecules clearly demonstrated "that not only new assumptions in the usual sense of physical hypotheses will be necessary, but the entire system of concepts of physics must be rebuilt from the ground up."[61] It was time, Born wrote, to begin the search for a new theory that he was now calling a "quantum mechanics."

The situation in quantum theory by the summer of 1923 was similar in many ways to what Thomas Kuhn has called in his analysis of scientific revolutions a "crisis." A period of scientific and psychological distress, he suggested, often precedes the onset of a new paradigm to replace one that has failed. In this case an entirely new theory, which seemed very much like a new paradigm, was not long in coming.

CHAPTER 9

CHANNELING RIVERS, QUESTIONING CAUSALITY

SOMMERFELD RETURNED TO MUNICH FROM THE STATES IN MAY 1923 FOR THE START OF summer semester. In the same month Heisenberg left Göttingen to complete his dissertation under Sommerfeld during his sixth and final semester as a student. It was a great relief to be back in Munich. Not only did living at home mitigate the effects of the accelerating inflation, but it also brought to an end Heisenberg's painful separation from his comrades. Even as he looked forward to his return to Munich, on the eve of his departure from Göttingen Heisenberg wrote a depressing letter to Eberhard Rüdel (Wolfgang's elder brother) in Erlangen, to which Eberhard replied: "From what you write in your last letter it appears to me that you were not in the best mood, that you did not know what you could do other than physics and music and nowhere really fit in, in other words that you missed your group."[1]

Heisenberg also returned eagerly to the circle of his Munich colleagues. Of those who were there a year earlier, Karl Herzfeld and Hans Kienle were still in Munich, Gregor Wentzel had habilitated and was now a lecturer, but Adolf Kratzer had gone to a chair in Münster. Since the mathematicians Alfred Pringsheim and Artur Rosenthal did not lecture that semester, and Oskar Perron and Aurel Voss offered only elementary topics, Heisenberg concentrated on his dissertation, Sommerfeld's lectures and seminars, and Wien's four-hour laboratory course. (Wien's eight-hour course conflicted with Sommerfeld's advanced lectures on spectroscopy.) Sommerfeld's main lecture that semester, on partial differential equations in physics, was probably most useful to Heisenberg in his work on his thesis problem: solving the horribly complicated equations for the stability and turbulence of flowing fluids.

Heisenberg had been working on the difficult problem for over a year. It derived from the work of the English physicist Osborne Reynolds in the 1880s and from early experiments at Sommerfeld's institute, conducted at the request of the Isar Company. Heisenberg's puzzle concerned the determination of the transition from laminar (smooth) flow to turbulent flow for channeled liquids, a phenomenon that occurred in the greenish Isar River as it flowed north through Munich from the foothills of the Alps. Sommerfeld's student Ludwig Hopf had experimentally examined the problem over a decade earlier, but no one had yet discovered how to predict the precise transition to turbulent flow.

Reynolds had treated the problem on the basis of energy conservation and found that a constant, Reynolds number, governed the transition to turbulence. Heisenberg set out to derive Reynolds's results from the fundamental equations of hydrodynamics.[2] In his 59-page thesis submitted to Philosophical Faculty II (science) on July 10, 1923, Heisenberg divided the problem into two parts. In the first he examined the conditions under which laminar flow becomes unstable, while in the second he investigated the role of Reynolds's number.[3] Having already studied these problems for Sommerfeld's second-semester seminar, he found little difficulty in solving either part for his thesis using various approximation and simplification techniques.[4] Professor Wilhelm Wien accepted the results for publication in his prestigious journal, *Annalen der Physik* (*Annals of Physics*). But after the mathematician Fritz Noether later challenged the results, they remained in doubt for nearly a quarter of a century before they were finally confirmed. Heisenberg did not publish again on hydrodynamics until 1946.

According to university regulations, a Munich dissertation was first submitted to the dean of the subfaculty, who passed it along to the student's advisor for critical assessment and a vote for or against acceptance. The work was then circulated among the entire subfaculty; if they accepted it, the candidate was admitted to the final orals.[5] Sommerfeld worried about possible objections to the fact that Heisenberg's solution of the problem was only approximate. While conceding in his two-page, typed report that "the work still leaves much to do with respect to the mathematics," Sommerfeld argued that the equations were so complicated that even approximate solutions were sufficient for a dissertation. But Sommerfeld's strongest argument in favor of acceptance was founded on the talent of the candidate even more than on the content of his work. "In the handling of the present problem," Sommerfeld concluded, "[Heisenberg] shows once again his extraordinary abilities: complete command of the mathematical apparatus and daring physical insight. I would not have proposed a topic of this difficulty as a dissertation to any of my other pupils. I therefore move for acceptance of this work."[6]

More at home with classical hydrodynamics than with contemporary quantum theory, Willy Wien seconded the motion, "even though doubts may be raised from the mathematical side against the considerations presented." The rest of the science faculty signed without reservation, and Heisenberg's oral *examen rigorosum* was set for 5:00 PM on Monday, July 23, 1923, in the seminar room of the Institute for Theoretical Physics.[7]

In Munich, a doctoral candidate's grades were based solely on his or her dissertation and performance on the final orals. Four passing grades were possible: I (summa cum laude), II (magna cum laude), III (cum laude), and IV (pass). At the completion of the orals, grades were given for the major subject, for each of two required minors, and for overall performance, the last being the most significant. The examining committee consisted of the professors in the candidate's two minor subjects—in Heisenberg's case, Perron for mathematics and Seeliger for astronomy—and the professor of his major subject. Since Munich physics was split between Wien and Sommerfeld, both attended the orals and both had to agree on a single grade.

Trouble was already brewing. Aside from the differences between Wien and Sommerfeld, Heisenberg and Pauli had made the mistake of working more on theory than on experiment when they first took Wien's laboratory course. Having taken a second lab course, Pauli managed to satisfy Wien in his orals and graduated with the overall grade of I.[8] Heisenberg did not fare so well in Wien's course that summer. Wien presented him with a particularly difficult problem: Heisenberg was to use the Fabry-Perot interferometer to measure the "hyperfine structure" in the anomalous Zeeman effect of mercury, an even finer splitting of lines in the spectroscopy of atoms subjected to magnetic fields that was later attributed to electrons' interactions with the nucleus.

Raging inflation made the acquisition and even the repair of equipment in Wien's laboratory nearly impossible. But Heisenberg did not know—or bother to find out—that he could use the institute workshop to construct his own equipment. Fresh from successes in the rarefied realms of Göttingen's mathematical physics, the overconfident theorist had little use for laboratory exercises. Nor did he trouble himself to consider the workings of his instruments, despite their obvious relevance to his own work on the Zeeman effect. For his experiments, Heisenberg simply threw together a slapdash contraption with cigar boxes and sealing wax. Such negligence only further incensed the already provoked professor, who proceeded to pounce on the offender in his orals.[9]

As the three professors joined Sommerfeld in the seminar room of his institute late that summer day, the 21-year-old doctoral candidate seemed confident. But that quickly changed. He easily handled Perron's mathematics questions and Sommerfeld's questions on theoretical physics, but he began to stumble over Seeliger's inquiries on astronomy—and fell flat on his face when confronted by Wien. Wien's ire mounted as Heisenberg proved unable to derive the resolving power of the Fabry-Perot interferometer or even that of the telescope or microscope, all of which Wien had discussed extensively in his lectures. Wien then asked Heisenberg to explain in detail how a storage battery works. The candidate was still lost, despite his earlier adolescent fascination with electrical gadgets. The vindictive Wien saw no reason to confer a degree on this ill-informed upstart, even if he were another in Sommerfeld's parade of prima donnas. A row promptly broke out over the relative importance of theory and experiment, resulting in Heisenberg's receiving the poor grade of III for physics, an average of Sommerfeld's I and Wien's V. Fortunately, he performed better in his other two subjects, receiving a I in mathematics and a II in astronomy. But he had to accept an overall grade of III—the equivalent of a C—for his doctorate.[10]

Sommerfeld was shocked, Heisenberg mortified. Accustomed to brilliant oral examinations, to unassailable defenses of his work, and to applauded lectures before the leaders in his field, Heisenberg found a III for his doctorate hard to take. Sommerfeld held a small dinner party for the new doctor at his home that evening. All of the institute assistants and students attended, but Dr. Heisenberg excused himself early, packed a bag, and took the night train to Göttingen, showing up in Born's office the next morning.

Born had earlier obtained Sommerfeld's approval for Heisenberg to habilitate in Göttingen, to qualify as a permanent university lecturer, after he had received his doctorate. Born had already offered to continue his assistantship until Heisenberg completed his habilitation treatise. In a process somewhat like that for tenure in the United States, the treatise would be a major piece of fundamental research and the candidate would be subject to further oral examination by his entire faculty. Because he had been so unhappy away from home, Heisenberg had obtained Born's approval to remain in Munich until the start of the winter semester. Born was now astonished to see him in his office in midsummer and looking so depressed. Informing Born of the debacle of his orals, the youth asked sheepishly, "I wonder if you still want to have me." Born pressed for more details before answering. Together they went over Wien's questions and, once satisfied that "they were certainly rather tricky," Born let his offer of employment stand for the coming winter.[11] Born, who also had his difficulties with experiment, would hardly let Wien's objections hinder his employment of the wunderkind. The wunderkind left Göttingen a few days later on the foreign-policy trip to Finland with his youth group.

But Heisenberg's father, knowing nothing of physics and determined to see his son succeed, was not so easily reassured. Nor was he so easily persuaded that physics was still the best route for his son's academic career. By chance, August Heisenberg and Willy Wien were chosen deans of their respective philosophical subfaculties for the coming academic year, and both were members of the faculty senate the following year.[12] When the semester began in November, Professor Heisenberg received a disturbing firsthand report on his son. Wien no doubt maintained more than once that the boy did not know enough physics to survive in academe, the only place with jobs for theorists. At the end of November, Heisenberg wrote to tell his father to stop worrying, since it didn't help either of them. He had now placed himself fully in Born's hands, and if his father wanted him to learn more experimental physics, he would have to take this up with Born. "For, as long as I am here in Göttingen I must do what Born wishes, just as in Munich I had to do what Sommerfeld wished."[13] The elder Heisenberg had long since lost any say in his son's affairs.

Even if Heisenberg hid behind Born, it did not deter his father. In January 1924, he wrote directly to both Born and Born's experimentalist colleague James Franck, asking them what his son's chances were in physics. He asked Franck also if he would mind teaching the young man some experimental physics. Born attempted to calm the professor with his own report (now lost) of young Heisenberg's extraordinary abilities, and Franck obligingly admitted Heisenberg to his laboratory course. Shortly thereafter, however, Werner Heisenberg and Franck agreed that Heisenberg should leave the lab; the bored young man could make better use of his time doing theory.[14] Heisenberg had made his choice: if he were to survive at all in academe, he would do so only as a member of the minority discipline of theoretical physics.

As Heisenberg was fumbling through his doctoral exams in July 1923 and Born was proclaiming the failure of quantum theory, the German economic and political order was undergoing a meltdown. The crises reached their peak in October and November 1923. By the end of the year, they had been confronted and at least temporarily defused.

The coincidence was extraordinary. Just as runaway inflation and a Nazi putsch gripped Germany in those months, Heisenberg and his Göttingen colleagues responded to the challenge of new spectroscopic data with a new quantum principle and approach to atomic physics that again seemed to remove every difficulty in a single stroke—but again it succeeded only by abandoning accepted procedures. A nationwide state of emergency was nearly two weeks old when the 21-year-old Heisenberg announced his new principle in Göttingen on October 9. As the authorities brought Hitler and the inflation under control in November, Heisenberg circulated a manuscript that tamed the Zeeman data with his new principle. The paper became his habilitation thesis presented eight months later to the Göttingen faculty.

Heisenberg completed his doctoral thesis in Munich just as the German monetary system imploded. Inflation had already begun to gallop during the World War, as Germany poured its wealth into its war machine. During the last war years, only deficit spending kept the machine rolling, largely at the expense of the middle class. By January 1920, the mark had dropped to one-fifteenth of its prewar value relative to the U.S. dollar, the international standard. By then the Berlin regime had established political control, and the mark had nearly stabilized. The stability weakened in 1922 after the assassination of pro-democracy Foreign Minister Walther Rathenau in June. As investments slowed to a trickle, the mark fell to less than one percent of its prewar value. At first unwilling to pay reparations to the war victors, Germany now found itself unable to do so. To cover demands for more money, the central bank merely cranked up the printing presses. Inflation brought windfall profits to German industries, and the working class was temporarily content with commensurate wage increases. The less influential middle classes maintained appearances of prosperity as their savings evaporated.

Support for the mark faltered further in January 1923 when the French, demanding reparations, occupied the Ruhr Valley, the heartland of German industry. With Germany again humiliated and its economy deprived of heavy industry, the mark deflated to practically nothing. By November 15, 1923, the first day of the stabilized Rentenmark, the old mark had shriveled to a shocking one-trillionth of its prewar value. Already traumatized by recent events, the German people approached panic.

The Heisenberg family did not escape difficulty, nor could Heisenberg totally ignore "the economic crisis and the misery all around me."[15] Still, Heisenberg and his family fared better than most. The state continued to raise the salaries of professors, all of whom were civil servants, in pace with inflation. August Heisenberg's monthly salary jumped every few months, reaching 2 million marks as his son received his doctorate in July 1923. August was also paid seminar fees, plus a local increment and a cost-of-living allowance of 1.2 million marks.[16] Yet without further emergency increases and Uncle Karl's dollars, the Heisenbergs would have been in trouble.

In the month Heisenberg graduated, the minimum monthly cost of living for a five-member working-class family in Munich was estimated at roughly equal to Professor Heisenberg's total salary, 3.2 million marks.[17] According to another estimate, the monthly expenses for an average "intellectual worker" with two children came to 9.4 million marks.[18] By the end of 1923, a Munich working-class family required 100 trillion marks. A kilogram of rye bread went for half a trillion marks.[19] The Heisenberg family managed somehow to weather the storm, as did Professor Heisenberg's colleague Willy Wien. Wien claimed that his salary, slightly greater than Professor Heisenberg's, was just enough to pay for essentials. Wien received his salary every two days and handed it to his wife, who promptly converted it into groceries before prices rose again.[20]

Werner Heisenberg and his postdoctoral colleagues did not fare as well as the older academic generation. While the government salaries for professors regularly increased, government funding for apparatus, literature, and assistants nearly dried up. Young scientists managed to survive the inflation and the difficult years ahead only because of family support and the prompt measures taken by established scientists. Many astutely gathered private funds from embarrassingly rich German industrialists. They also befriended philanthropic foreigners, mainly Americans. Munich chemist Richard Willstätter found financial support through a New York brewer. Sommerfeld could count on the generosity of a Berlin industrialist, and Max Born tapped every source he could find "to feed my students."[21] Fortunately, Felix Klein had already established good relations with German industrialists, one of whom, Carl Still, funneled funds into Born's institute. When one of Born's friends left for the United States to marry an American woman, Born asked him, half in jest, to find a rich American willing to provide research dollars. Shortly thereafter, Born made the contact with New York financier Henry Goldman of Goldman Sachs. Disturbed by the postwar mistreatment of Germany, Goldman generously supported Born's institute. Born paid his private assistants Brody, Pauli, and now Heisenberg, with these funds. His one university assistant, Friedrich Hund, received state funding.[22]

Other Americans, also dismayed at the plight of German culture, and eager to influence it to their own advantage, set up emergency committees. Their financial support (save for direct grants) was received and distributed by the Emergency Association of German Scholarship (Notgemeinschaft der Deutschen Wissenschaft). German academics and cultural administrators had established the association in 1920 in part to aid German research but also as a way of asserting their autonomy from the Weimar government.[23]

With its military and economy in disarray, only world-class scholarship, German scholars argued once again, could sustain the country's international standing. Culture could succeed where diplomacy had failed. Extraordinary measures were thus required to bolster Germany's place at the forefront of world research. Still, German academics were reluctant to permit the democratic regime control over research funds, a prerogative that the prewar monarchy had enjoyed. In order to skirt government involvement, they revived the notion of self-administration, and so invented the modern project grant system of support. Under this system, neither governments nor foundations determine who should be blessed with research funds. Rather, in an innovative departure from previous practice, independent committees of leading scientists and policy administrators in each field, backed by panels of specialist referees, evaluate the merits of each grant application and make the appropriate awards. This unique and powerful system of research support, not adopted in the United States until after World War II, helps to account for the paradox of Germany's lead in atomic physics and other sciences during this adverse period.[24]

Thanks to Einstein and other German physicists, physics had become one of the most prestigious fields of German research by the early twenties. Accordingly, in its first report, published in 1922, the Emergency Association announced that it had funneled the greatest support into physics. The physics disciplines most preferred were atomic physics, radiation, and the structure of matter—in particular, experimental research on the relativity and quantum theories. While these specialized areas seemed to offer few practical applications, the generous funding was intended to maintain Germany's already impressive lead. Unless German physicists received necessary support, the report declared, "financially better endowed physics in foreign nations will soon push us back."[25] Those who disagreed with the generous funding of relativity and quantum physics—notably Willy Wien and Johannes Stark—formed a separate agency to support their preferences: classical and technical physics.[26]

Coupled with its strategy of financing leading areas of research, the Emergency Association created another influential novelty: research stipends to promising postdoctoral researchers in order to keep them in the field. One of the first stipends went to Heisenberg. Many of the other recipients would be, like Heisenberg, primary contributors to quantum mechanics.

Beginning in July 1923, an Emergency Association subcommittee, the Electrophysics Committee, devoted itself almost entirely to atomic physics. Created with an annual sum of $12,500 from General Electric (and matched, in part, by German industries) ostensibly to help engineering physics, its members turned instead to the abstract physics of the atom. Of the 140 grant applications the committee received between 1923 and 1925, it approved 71, most of which were in atomic physics. Of the 71 approved, 56 supported research assistants. Of these, 3 had been submitted by Born on behalf of Heisenberg.[27]

Heisenberg received his first "electrophysics" stipend as soon as the mark stabilized in November 1923. It amounted to only 50 marks a month, certainly not enough to live on.[28] It soon increased to 100 marks, then to 150 marks in March 1924—exactly half of an assistant's salary. The other half of Heisenberg's income came from Born's private sources. Heisenberg's Emergency Association salary remained at this amount until the end of 1925.[29] Late in 1925, with Heisenberg and Born's breakthrough to quantum mechanics already in press, Born, supported by Courant, then dean of the science faculty, obtained for Heisenberg a two-year private lecturer's stipend—another invention of the Emergency Association—from the Prussian Cultural Ministry in the amount of 127.88 marks per month.[30]

These salaries, which supported Heisenberg throughout this highly creative period of his career, did not add up to much more than the cost of living for an unmarried young man in the provincial town of Göttingen. Yet they served their purpose well: they kept talented young people like Heisenberg in science and especially in quantum atomic physics, where they helped to lay the foundations of quantum mechanics as a predominantly German invention. In its 1926 report on the impact of its support during the previous five crisis years, the Electrophysics Committee congratulated itself for its perspicacity in financing Heisenberg and Born: "As is well known, quantum mechanics stands at the center of attention among physics circles of all nations. The work of Heisenberg and Born, which the Electrophysics Committee has supported and without which the work would very probably not have been done in Germany but elsewhere, has shown the usefulness of the Electrophysics Committee in the development of physics in Germany."[31]

Soon after the poor showing on his doctoral examination and the Finland trip with his youth group, Heisenberg returned to Göttingen in September. He announced his new Zeeman principle to Pauli on October 9. He had begun putting it on paper in October and was well into it several weeks later when he wrote to his Finland traveling companion Kurt Pflügel of a new interruption: the latest news in Bavarian politics had "sent the blood to my face."[32]

Insane inflation, Bavarian separatism, and Germany's unnecessary humiliation at the hands of the Allied victors had reached crisis proportions when Chancellor Gustav Stresemann decided to end resistance to the French occupation of the Ruhr on September 26, 1923, igniting unrest throughout the country. At the same time that German President Friedrich Ebert declared a nationwide state of emergency, Bavarian officials decided to declare their own state of emergency, aimed primarily against supposed left-wing agitation. They named Gustav von Kahr to the dictatorial post of General Commissar of Bavaria. Kahr, while close to militant Bavarian separatists, gained the temporary support of violent nationalists who sought to replace the democratic Weimar regime with a right-wing monarchical dictatorship. Among Kahr's nationalistic supporters were General Otto von Lossow, commander of Bavaria's division in the national army, and the German Battle League (Deutscher Kampfbund), an alliance among General Erich Ludendorff, Hitler's Nazis, and illegal secret militias established to stamp out any remaining Bolshevism in defiance of the Allied limitations on German military units.

Matters escalated on October 20 when Berlin authorities ordered Kahr and General von Lossow to close down the Nazi Party newspaper, *Völkischer Beobachter* (*Volkish Observer*), for libeling Berlin officials. Lossow refused and was summarily dismissed by his Berlin commanders. Kahr immediately seized control of the Bavarian army, placing it fully under Bavarian authority—with Lossow back in command. The next day Heisenberg, with family ties to northern Germany, wrote angrily to Kurt of the Bavarian behavior: "I simply don't understand any of this: Lossow is a soldier, isn't he? If he doesn't obey, then a soldier is supposed to be shot, so it's written somewhere."[33] As did his family earlier, Heisenberg openly sympathized with the Weimar regime over Bavarian zealots.

The more provincial Kurt, the son of a Bavarian officer, had meanwhile rejoined the Bavarian army after Kahr's action and now attempted to explain to Heisenberg the grounds for Bavaria's anti-Berlin attitude.[34] Stresemann had just moved militarily against soviet regimes in Saxony and Thuringia. The Bavarians demanded that he now move against socialists of any stripe everywhere, while at the same time resuming resistance, even if only passive, against the French occupation. In his long reply on October 31, the more democratically minded Heisenberg, though calling Bavarian motives noble, wondered how the government could fight socialism when, in his view, a third of Germany was democratic socialist and had an equal right to national recognition.

Heisenberg felt that direct opposition to the French would ultimately lead to the "hero's death" of Germany as a nation, given its present economic and military weaknesses. As during Napoleon's occupation, patience seemed the only practical alternative. "Out of all of this I conclude that it is more correct (as our forefathers [did] a hundred years ago!) to hold the enemy in check by cunning and deceit and fright until a hope exists for weapons. That is precisely what the national government appears to be doing."[35]

Hitler and his followers were not so patient. At a celebration commemorating the fifth anniversary of the Armistice on November 8, 1923, in the Bürgerbräu beer hall in Munich, the Bavarian government and leading members of the upper classes, including professors, were present. At the high point of celebration, just as Kahr unveiled his authoritarian plans, Hitler and a band of ruffians burst into the hall, fired several shots at the ceiling, and proclaimed the revolution: "Tomorrow we're all dead or we have a national government!"[36]

Inspired by Mussolini's march on Rome, Hitler planned to march on Berlin with the Bavarian army, together with the secret army units. Kahr, Lossow, and police chief Colonel Hans Ritter von Seisser agreed, at gunpoint, to Hitler's plan. But later that night the three slipped away, withdrew their support, and ordered the army to defend the government. The Hitler-Ludendorff Battle League scheduled a march through Munich the next day to force the issue over control of the army. Gottfried Simmerding, a longtime member of "apolitical" Group Heisenberg and now its new leader, also belonged to an illegal machine-gun company led by a Lieutenant Werner.

At 2:00 AM, Simmerding recalled, he was awakened at home by a friend from the secret company. He dressed quickly, stuck a pistol in his pocket, and went to join the unit. As part of the German disarmament, Bavaria was allowed only one machine-gun company, not enough in Bavarian eyes to protect against Bolshevism. During the predawn hours, after the Bavarian authorities had reversed their support of Hitler, they ordered Company Werner to the Feldherrnhalle in downtown Munich to help confine Hitler's march that day to the inner city. Apparently supportive of Hitler, Lieutenant Werner refused and his company disbanded without complaint.

As Simmerding made his way home later that day, after seeing to company finances, he heard machine-gun fire from the direction of the Feldherrnhalle. The

Augsburg machine-gun company under the command of Kurt's indomitable father, now-Colonel Pflügel, had arrived in time to support the police in confronting the Nazi march, led by Hitler, Hermann Göring, Heinrich Himmler, and General Luddendorff. They killed 16 of the marchers.[37] Hitler escaped with a dislocated shoulder, only to face prosecution and a brief stint in prison for his putsch.

Over the weekend, as loyal army units broke up repeated demonstrations by pro-Nazi students, Kurt, an army reservist and a student at the technical college, met with other New Pathfinders to decide their political position, an action they would have shunned just a few weeks earlier. Kurt informed Heisenberg in quiet Göttingen that they, like most Bavarians, had decided to support the Berlin government for now and had even decided to assume what they regarded as their responsibility for helping to alleviate social distress: "We ought not withdraw from the world and live our lives only in the woods and dream of a new Reich. If we are blessed enough to see the Reich, then it is our sin if we keep it for ourselves."[38]

Heisenberg did not respond to his friend for nearly two weeks. In the meantime, most Germans had come to tolerate the democratic regime (although the short-lived Stresemann cabinet had since fallen). Most still opposed dictatorial nationalists. Soon the economic basis for popular support was assured. On November 15, the government introduced the new Rentenmark (RM), or "mortgage mark," based on a mortgage against all German agricultural and industrial property. The Rentenmark, equal to a trillion paper marks, or one prewar mark, met with immediate acceptance.

On November 24, just as he received his first Rentenmark stipend from Born and was about to unveil his optimistic new quantum principle, Heisenberg wrote Kurt of his pessimistic view of current affairs. While resistance to the French was still hopeless and Hitler's "November-carnival" ludicrous, the Munich youth movement's turn to active social involvement was for Heisenberg, as it had been earlier for the radical New Pathfinders, unthinkable. "What role could we few little men play in a population of millions!"[39] Heisenberg had already settled on retreat to the pure worlds of physics, music, and youth. Youth-group follower Wolfi Rüdel wrote to Heisenberg soon after the upheavals dissipated in early 1924 that the "time is now coming once again where we are the most strictly limited by external circumstances. But I can already imagine what your answer will be: physics and ever more physics."[40]

If Heisenberg's characteristic optimism in matters of theoretical physics did not prevail in the political arena, it was because he saw little hope of immediate success. Both in physics and in current events, the momentary situation was for Heisenberg only transitory, an unpleasant interlude between two more permanent arrangements. But in the rarefied realms of theoretical physics, his abilities enabled him to achieve both influence and significant results of lasting consequence. Political entanglements were, to the contrary, merely superficial "money-business," a business in which he felt he would be just one of millions. It was a position that rendered him and others like him dangerously naive and unprepared for what was soon to follow. "Therefore, Kurt," he wrote, "I believe that we should stay a while longer in the woods; there the air is cleaner. . . . We cannot help the masses, and the few young people whom we can help will come to us all right 'in the woods'; a wonderful close-knit group that at every hour is ready to forgo its life, the highs and lows of which it has come to know like few others. For, whoever has been at the peaks of life no longer needs to fear the end."[41] Heisenberg held this view throughout the Weimar years.

Success in physics was one of the "peaks of life" for Heisenberg, and when he wrote that statement he had just climbed again to the top by rescuing his much-maligned core model, still his only independent contribution to quantum physics, with his new quantum principle.

Heisenberg had invented the core model of the atom earlier in Munich. It was an attempt to account for the puzzling division of certain individual spectroscopic lines emitted by atoms into doublets and triplets of lines, each of which divided further into lines under the influence of a magnetic field. All of these divisions seemed to display a host of empirical rules and regularities, the existence of which remained a theoretical puzzle. Only Heisenberg's core model offered an account of most of them. But by 1923 not only was this model still under heavy theoretical challenge, but new experimental evidence had revealed a new experimental complexity: a single line could split into as many as nine closely spaced lines, something that not even the core model could explain.[42] For theorists like Heisenberg, the task remained as before: to make a new attempt to account for the newly observed lines and splittings in the laboratory by means of an imagined model, a theory, of the underlying atom.

The ever-ingenious Heisenberg found a way to append his old core model to account for the new data. He did so by introducing what he called an ad hoc "Zeeman principle," a mathematical procedure (today we would call it an algorithm) that is, in fact, now forgotten.[43] Its only historical interest, aside from getting Heisenberg his habilitation to teach at the university level, is that it foreshadowed his reliance on observed data in his subsequent formulation of the breakthrough to quantum mechanics. The technical "Zeeman principle" provided a mathematical procedure for converting the continuous "classical" energies of observed spectroscopic lines into discontinuous quantum jumps between states according to the quantum rules.

Amazingly, the sleight of hand worked! "Now everything comes out of this," a gleeful Heisenberg wrote to Pauli.[44] Göttingen physicists were enthusiastic. Heisenberg received the imprimatur of the weekly Göttingen colloquium after delivering a well-received talk on the principle in November. Born urged immediate publication.[45]

But Heisenberg demurred under pressure from home. Having just learned about Werner's questionable physics from Willy Wien, Professor Heisenberg openly worried that the new theory, still based on the controversial core model, might confirm his not-yet-habilitated son's ignorance of physics—and at the very height of frightful economic and political chaos. Sommerfeld liked what he knew of the new theory but urged caution, which August conveyed in the strongest terms to his son at the end of November. Werner responded angrily that he was trying to be more consistent and self-critical than ever before and that this time he would seek advice and approval not

Wolfgang Pauli in 1924.
(COURTESY ESVA, AIP NIELS BOHR LIBRARY)

just from Born but from the master of consistency, caution, and physical insight, Niels Bohr: "I realize ever more that Bohr is the only person who, in the philosophical sense, understands something of physics."[46]

True to his word, Heisenberg sent a copy of his manuscript of the new principle to Pauli in Copenhagen in December 1923, asking his friend to forward it to Bohr for his blessing. Bohr, who was then fundraising in the United States, responded over a month later with an invitation to Copenhagen.[47] Heisenberg accepted for the semester break in March 1924. It would be his first visit to Copenhagen, and it would confirm and accelerate the already coalescing physics of the Göttingen and Copenhagen schools.

The critical and consistent Pauli, now a lecturer in Hamburg, didn't like anything about Heisenberg's new principle or any other works coming out of Göttingen. "I even regard it as ugly," he told Landé and others. "It is not the theory for which I am hoping."[48] Pauli no longer believed in any atomic models. The task of physics now was not to stamp out individual problems but to build them into a new theory. Pauli also complained to Bohr of his frustration with Heisenberg's inconsistency. His letter to Bohr dated February 11, 1924, provides probably the best intimate depiction of Heisenberg and his relationship with Pauli in this period. Pauli wrote: "I always feel very strange with him. . . . For he is very unphilosophical, he does not pay attention to the clear working out of the basic assumptions and their connection with the prevailing theories. However, if I speak with him, he pleases me very much, and I see that he has all sorts of new arguments—at least in his heart. Aside from the fact that he is also personally a very nice fellow, I regard him as very significant, even genial. . . . I was therefore very pleased that you have invited him to

Copenhagen. . . . Hopefully then Heisenberg, too, will return home with a philosophical orientation to his thinking."[49]

Heisenberg arrived in Copenhagen by train and ferry in March 1924, to an elaborate reception.[50] Danish hospitality aside, his reputation had obviously preceded him. He wrote home often during his two-week stay to tell of numerous luncheon invitations to sample Danish food; musical evenings at the home of Mrs. Maar, a widow who rented rooms in her multistoried house to the institute visitors; a weekend excursion to Bohr's country cottage in Tisvilde, on the northern coast not far from Hamlet's castle; a meeting with Wickliffe Rose, head of the Rockefeller Foundation's International Education Board (IEB), who happened to be in Copenhagen at the time; and a touring trip to Jutland with three visiting Americans. "I speak very well englisch," he assured his parents.[51]

But the main events were his conversations (in German) with Bohr. These increased in length to several hours each day. Bohr's three-story institute, a component of the University of Copenhagen, lay at that time on the outskirts of the small inner city near the large and secluded Fælledpark, with its stately trees, budding springtime flowers, and seemingly endless walking trails. Their conversations often occurred during strolls through the park or to the Copenhagen docks, where ships unloaded their wares not far from the statute of the Little Mermaid, and occasionally in the evenings in Bohr's residence on the third floor of the institute with "one (or more) glass of port wine." Within five days of his arrival, Heisenberg had an invitation to Copenhagen for a year, funded through a probable Rockefeller stipend provided by Rose. "Everything has really gone better than I could ever have expected," he wrote home.[52]

Heisenberg's conversations with Bohr during the first few days of his visit were taken up, not with the core model, but with the common interest they had discovered during their first meeting the previous summer—"philosophy," which meant to them practically everything nontechnical. "We have always talked about the most general questions and have picked apart their philosophical foundations (I see you scornfully smiling here, Papa)," he wrote.[53] Papa Heisenberg, the positivistic philologist, made little secret of his abhorrence of philosophical matters.

Nevertheless, as Pauli had hoped, toward the end of his stay Heisenberg had begun to appreciate Bohr's systematic, "philosophical" approach to physics: "His manner of doing physics is really very 'practical,' he always attempts at first only progress in the details."[54] As the two physicists turned, after nearly a week, to discussion of specific problems—the core model and Heisenberg's new principle—the well-versed Bohr probably made clear to the younger physicist that he required a deeper understanding of his subject. Heisenberg also reported to his father that he was spending his free time in the institute library reading physics textbooks, "in order to 'elevate' my general physics education."

One of the three Americans whom Heisenberg met that March in Copenhagen was the recent Harvard PhD John C. Slater.[55] During the outing to Jutland and in the institute with Bohr and Kramers, Heisenberg learned of the physical and philosophical implications of a radical new theory developed in Copenhagen at that time, the so-called Bohr-Kramers-Slater (BKS) theory.[56]

The BKS theory concerned the third of the three main areas of quantum atomic research in this period: the quantum theory of radiation, the study of the interaction of light with atoms. The other two areas, in which Heisenberg was already involved, were spectroscopy (in Munich) and planetary atomic models (in Göttingen). Together, each of these areas provided problems and clues for those seeking to create the new quantum mechanics.

The Bohr-Kramers-Slater theory was designed to resolve another of the growing fundamental mysteries of the period: the so-called wave-particle dualism. For over a century, light had been accepted to be a continuous wave. This interpretation had found decisive support in Maxwell's theory of electromagnetism during the 1860s, in which visible light is one form of electromagnetic wave. Only a wave could display such well studied effects as interference, diffraction, and dispersion, such as occur when light is bent through a glass prism or sunlight is split into the colors of a brilliant sunset. On the other hand, in 1905, Einstein had argued the revolutionary hypothesis that light also behaves as a stream of particles, or light quanta. In 1922, American experimentalist Arthur H. Compton offered support for this interpretation by observing the collisions of particle-like light quanta with free electrons, the Compton effect. He showed that the electrons and the light quanta collide and ricochet like billiard balls, rather than like a wave washing over a stone.

But didn't light also behave like a wave? Faced with two mutually exclusive interpretations of light—wave and particle—the cautious and "philosophical" Bohr chose the well-established wave interpretation. He attempted to account for the particle behavior of light without resorting to light quanta. The BKS theory was a product of this attempt. Unfortunately for him, experimental evidence soon disproved the theory, but, like the core model, it still served a purpose. In 1929, Heisenberg characterized this theory as a perspective that "contributed more than any other work at that time to a clarification of the situation in quantum theory."[57] It provided the background to his breakthrough to quantum mechanics a year later.

Quantum atomic physics was still based primarily upon calculations of the energies of electrons orbiting like planets in stationary (fixed) orbits. Even though this picture of the atom was shattered by the work of Born and Heisenberg, the frequencies of the light emitted or absorbed by atoms were still regarded as arising from the supposed quantum jumps of electrons among these calculated orbits. Quantum radiation theory concerned the scattering, dispersion, and absorption of light, viewed as particles, by an atom or groups of atoms, in which the likelihood for an electron to jump to another state must also be taken into account. The likelihood, or probability, combined with the number of atoms, provided the intensities of the observed radiation that was emitted or absorbed.

Rudolf Ladenburg showed in 1921 that, for radiation problems, the frequencies of the radiation could be treated more simply *as if* they arose, not from quantum jumps of electrons, but from the emission and absorption of wave radiation by tiny, imaginary oscillators within the atom—a set of tiny antennae, each consisting of a charged electron ball on a spring oscillating harmonically, that is, in simple back and forth motion.[58] The use of these tiny oscillators in atoms was comparable in effect to an explanation of the notes of an electric keyboard that are made to sound as if they are coming from vibrating piano wires. This made the solution of radiation problems much easier, since a great deal of classical physics could be applied to an atom treated *as if* it consisted of an array of tiny "virtual oscillators," each acting as an antenna for light waves of a specific frequency. Of course, the oscillators, like the piano wires, were only imaginary or "virtual." No one believed that atoms actually were made of balls on springs. But it was such a useful simile that it became the starting point for Heisenberg's work toward quantum mechanics a year later.

Slater came from Harvard to Copenhagen in 1923 with the remarkable idea of a "virtual radiation field" or "ghost field" that carried no energy or momentum (hence "virtual") but that was continually emitted and absorbed by all of the virtual oscillators of an atom.[59] Bohr and Kramers saw a way to use Slater's virtual field, together with the virtual oscillators, in their effort to preserve the wave theory of light and thus ignore both light quanta and the mechanics of stationary states (two elements that Bohr himself had introduced into atomic physics 11 years earlier). But to do so was no simple matter. They found that maintaining the wave theory required a severe trade-off, the abandonment of some of the most cherished principles of physics—the conservation of energy and momentum and the causal connection between events in distant atoms "so characteristic for the classical theories," as Bohr put it.[60]

Bohr and Kramers decided to accept the trade-off as the price to pay for moving beyond classical theories. For the first time in quantum physics, the Bohr-Kramers-Slater paper abandoned conservation laws and the causal connection between events. This arose because the emission and absorption of light were induced by Slater's virtual field, which did not carry energy and momentum. As a result, the three authors declared that causality and conservation laws were no longer precise, absolute assertions in physics, but were mere statistical concepts, arising from the average results of billions and billions of individual atomic events. But the trade-off had its benefits. Employing such radical notions, the otherwise conservative Bohr, Kramers, and Slater managed to account verbally for numerous radiation phenomena on the basis of the wave theory of light alone. Although the BKS theory was soon abandoned, the radical idea that individual events may not be absolutely determined but only predicted as statistical probabilities would resurface, most importantly, in another of Heisenberg's breakthrough works—this time in 1927 in his paper on the principle of indeterminacy, or uncertainty.

Returning to Munich after Copenhagen, Heisenberg headed to the hills with his boys for a brief respite before journeying back to Göttingen for an intense round of work. He was so busy that spring of 1924 that Born threatened to place a notice in the local newspaper: "lost assistant."[61] Added to the burden of reworking his Zeeman principle and preparing it for his imminent habilitation under Born were the demands of completing papers on ion polarization and half-integral momenta, helping Born prepare his lectures for publication, and laboring with Landé on a joint publication on yet another new Zeeman rule.

Bohr, visiting Germany on financial business, stopped off in Göttingen in early June 1924 to discuss developments. He had received preliminary approval of a Rockefeller stipend for Heisenberg and wanted to discuss Heisenberg's visit to Copenhagen with the two physicists.[62] During Bohr's visit, Heisenberg invited Bohr, Born, and the Danish scientist Svein Rosseland, Bohr's traveling companion, to his rooming house. While Heisenberg's landlady, the hospitable Mrs. Ulrich, provided tea in the parlor, Bohr approved of Born's latest work on the search for a new quantum mechanics. They also agreed that Heisenberg's latest Zeeman principle had now attained sufficient development and utility to constitute a habilitation thesis.[63]

Within days of Bohr's visit, Albert Einstein, on his way to northern Germany, probably to Kiel for a health cure, stopped off in Göttingen for a few days to visit friends and colleagues. Göttingen seemed overwhelmed by the succession of great men. Only Ernest Rutherford was missing, Mrs. Franck exclaimed.[64]

Heisenberg was elated. Einstein's visit would at last bring about his first meeting with the great physicist. Einstein, then 45 years old, already rotund and with slightly graying hair, a moustache, and a frequent pipe, walked from the train station to the physics institute for brief discussions with the assembled physicists and for longer discussions with Born and Heisenberg on the BKS theory. Einstein honored Heisenberg with a 15-minute stroll through the surrounding neighborhood, during which they further discussed the theory and Heisenberg's latest work. The next day, a disappointed Heisenberg reported to Pauli that "Einstein has a hundred objections [to BKS]."[65] Bohr's willingness to go as far as to relax the demands of such essential elements of physics as causality and conservation laws was simply unacceptable to Einstein. He had already written Born before his visit that, if these elements had to be given up, "then I would rather be a shoemaker or an employee in a gambling casino than a physicist."[66] It was a position he would hold to the end of his life.

Born acquiesced without complaint. In his new paper, which he titled "On Quantum Mechanics," Born announced that he would "make use of the intuitive ideas" of the BKS theory, especially the notion of virtual oscillators, but that "our line of reasoning will be independent of the critically important and still disputed conceptual framework of that theory, such as the statistical interpretation of energy and momentum transfer."[67] In it, Born introduced his own new principle (or algorithm), a rule for turning the continuous equations of classical mechanics into the corresponding discontinuous equations of quantum mechanics. Important for Heisenberg in the months ahead, the now-unified Göttingen physics promoted virtual oscillators and discontinuous equations, while ignoring the controversial (and soon disproved) BKS theory.

Heisenberg returned to Bohr's institute the following fall, where, combining the approaches of Göttingen and Copenhagen, he would soon produce the initial breakthrough to quantum mechanics. On July 6, 1924, Heisenberg wrote to Landé of Göttingen's optimism that the new quantum mechanics was finally coming into sight with the appearance of the recent work in both locations: "The beautiful thing about the new Bohr and Kramers dispersion theory is precisely that one now knows (or suspects), especially on the basis of Born's calculations, how the quantum mechanics will look."[68]

At the same time, Bohr, now back in Copenhagen, thanked Heisenberg for sending a copy of the final draft of his latest principle. He seemed pleased with the paper—it coincided with the tendencies of BKS and Kramers's dispersion theory—and he informed the young man that the IEB had officially granted him a generous stipend of $1,000 to work in Copenhagen for a year.[69] On July 28, 1924, after Heisenberg delivered a successful lecture on his new principle to the Göttingen science faculty, it voted to habilitate the author, conferring official certification of Heisenberg's suitability to occupy the position of full professor at any German university. The next day, the 22-year-old left with his youth group for a three-week vacation in the Bavarian hills.[70]

CHAPTER 10

ENTERING THE QUANTUM MATRIX

IN THE SEPTEMBER 1925 ISSUE OF THE *ZEITSCHRIFT FÜR PHYSIK* (*JOURNAL FOR PHYSICS*), the publication of choice for quantum theory, Heisenberg published a 15-page paper with the harmless-sounding title, "On a Quantum-theoretical Reinterpretation of Kinematic and Mechanical Relations." The paper was far from harmless. It aimed "to establish a basis for theoretical quantum mechanics, founded exclusively on relationships between quantities which, in principle, are observable."[1] It dealt exclusively, not with unseen atoms, but with the observed frequencies and intensities of emitted and absorbed light appearing in a laboratory. In so doing, Heisenberg's paper accomplished the long-sought breakthrough to quantum mechanics—the new physics of the atom. Heisenberg, Born, and their closest colleagues quickly brought the new physics to fruition during the months following Heisenberg's initial breakthrough. That advance precipitated the culmination of the quantum revolution of the first decades of the twentieth century, a revolution that reached its conclusion just two years later in Heisenberg's uncertainty principle and the Copenhagen interpretation of Bohr, Heisenberg, and Born.[2]

Heisenberg's path to quantum mechanics was neither direct nor his alone. In the introduction to his paper, Heisenberg cited in particular Born's 1924 rule for generating discontinuous equations and his own recent work with Kramers on the dispersion of light waves by matter. To these important first steps, one must add Pauli's destruction of Heisenberg's core model, Heisenberg's further work with Bohr on electromagnetic radiation, and the radical wave theory of Bohr, Kramers, and Slater.

Heisenberg's triumph came only after a winter spent in Copenhagen, where all of the elements that blossomed into new theory were in play within what Heisenberg called the hothouse "atmosphere of quantum theory" that pervaded Bohr's Copenhagen institute.[3] There Bohr and his exotics—Heisenberg, Pauli, and Kramers—struggled intensively and exhaustively with each other and with each other's idiosyncratic approaches to cultivate their achievements.

Bohr's Institute for Theoretical Physics was, like its inhabitants, on an upward trajectory to success in 1924 and 1925. During the last years of World War I, Bohr, professor of theoretical physics at the University of Copenhagen since 1916, had convinced the Danish authorities and the Carlsberg Brewery Foundation to replace his one-room office with a three-story institute.[4] The sons of a famous Copenhagen university professor, Niels Bohr and his brother, Harald, a mathematics professor, easily moved within the elevated circles of Copenhagen social and cultural life. As with most of the young physicists and mathematicians who would come to work and study with the Bohrs, culture and breeding made an unspoken commonality of interest and outlook—a commonality expressed in such joint endeavors as musical evenings, horseback riding, hiking tours, and frequent trips to the local movie house to view the latest silent films.

In 1921, Bohr had inaugurated his new building in the nearly rural outskirts of town. The institutional-looking, rectangular building, with its grey stucco facade, pitched, red-tiled roof, and gabled third-floor windows, stood behind a wire fence only a few yards from the sidewalk at Blegdamsvej 15. Within a few years, flowers had sprouted by the front gate to beckon visitors, and collegiate ivy had grown to cover the entire first floor of the outer walls, reaching almost to the large letters embedded in the wall above the entryway: "Universitets Institut for Teoretisk Fysik 1921."

With a permanent staff of eight and long-term visitors numbering at least nine, by 1924 the institute was overcrowded. Less than half the space was actually devoted to theoretical physics. The top two floors were residential and the basement was given over to experimental work, a typical feature of theoretical physics institutes at the time. Bohr, his wife, Margarethe, and their two growing boys lived on the second floor; the family maid, the lab demonstrator, and special guests occupied the third. With the upper floors and basement occupied, only the ground floor was left for the lecture hall, library, functionally furnished offices for Bohr and his close assistant, Dutch physicist H. A. Kramers, and a drab study hall of wooden desks for the many visitors. Bohr often invited visitors to his elegantly furnished, three-bedroom apartment, and during his first trip to Copenhagen, in March 1924, Heisenberg had shared more than one bottle of wine there with the professor. "Often, the next day, what we had discussed actually turned out to be correct," Heisenberg wrote his parents.[5] He occupied the guestroom then and on many occasions in the years to follow.

Bohr had returned from a visit to New York in November 1923 with a grant of $40,000 from the IEB to expand the building and living quarters. And, despite raging inflation, the Danish Culture Ministry simultaneously had increased the institute's funding to a remarkable 5 percent of the university's total budget—a clear indication of the esteem in which the Danes held the 1922 Nobel laureate and his institute.[6] New construction began soon after Heisenberg's arrival in 1924 for the start of his fellowship. The expansion was not completed until Heisenberg returned again in 1926. By then, Bohr had moved to a single-story house next to the institute.

During his visit to Göttingen in June 1924, Bohr had reached agreement with Born on Heisenberg's stay in Copenhagen. Born, who expected to be in the United States during the 1924–1925 winter semester, allowed his newly habilitated lecturer to absent himself from Göttingen during the summer and coming winter.[7] But Born wanted Heisenberg back in Göttingen by May 1, 1925, for the start of the summer semester. Under prodding from Bohr, the IEB allowed Heisenberg to complete the remainder of his one-year grant during semester breaks, an arrangement that, in the end, greatly benefitted the course of quantum mechanics.

Kramers arranged for Heisenberg's room and board in Copenhagen in the turn-of-the-century home of Mrs. Maar, the recent widow of a university professor, whom Heisenberg had met the previous spring. Mrs. Maar warmly befriended her young tenants, often inviting them for weekend outings at her family's country villa. Heisenberg occupied a "small but nice" bedroom on the second floor of the Maar home, next to a room occupied by an American visitor, a Dr. King, chemist and vio-linist. The room, overcrowded with "many chests," faced west, but because the house was so close to its neighbor, little sunlight found its way through the lace-curtained windows.

Meals were taken with Mrs. Maar and Dr. King in the dining room promptly at 8:30 AM, noon, and 6:00 PM. They turned into language labs. Dr. King spoke little German, Heisenberg little English, and Mrs. Maar all three languages. They agreed, to Heisenberg's advantage, to speak only English during meals. After breakfast, Mrs. Maar then helped improve their Danish by reading the newspaper aloud for a quarter hour before the two left for their respective institutes. During his stay in Copenhagen, Heisenberg apparently learned enough Danish to write and lecture in the language, a distinct advantage when a job opening later occurred at the institute.[8]

Soon after arriving in Copenhagen, Heisenberg sent word home about his daily routine. During the day, he worked alone at his desk in the study room, visited Bohr's large ground-floor office, and discussed "all sorts of questions" with Bohr during long walks in the quiet and verdant Fælledpark behind the institute. But in none of his reports home did Heisenberg mention discussions with Bohr's assistant Kramers, with whom relations were strained. By all accounts, the problem was a touch of envy. To the Göttingen wunderkind, Kramers seemed superior in every respect.[9]

The tall, broad-cheeked Kramers, with his receding hairline and smoldering pipe, seemed older than his 28 years. The Dutch physicist was fluent in several languages, an excellent musician, and far more knowledgeable in the Copenhagen specialty of quantum radiation theory than Heisenberg. Above all, he occupied the envied posi-tions of Bohr's personal assistant, confidant, and heir apparent. The moody and occa-sionally depressed Kramers exhibited a somewhat detached, condescending, and ironic attitude toward everyone, but his lack of deference toward Heisenberg and his near disdain for Heisenberg's core-model physics were worse. Kramers displayed obvious professional ambitions in an institute populated by a crowd of ambitious boy wonders, all of whom circled around the insistent, preeminent Bohr, the "Pope" of quantum physics, and his dutiful assistant, "His Eminence" the cardinal, as Heisen-berg and Pauli sneered.

Heisenberg, in second place behind an older competitor, worried again whether he would be accepted into the inner circle. As in Heisenberg's boyhood home, where he and his brother competed fiercely for their father's approval, the competitive relationship encouraged Heisenberg's urge to excel and to achieve the status of the master's favorite, no matter how entrenched the principal disciple. Even before he set foot in the institute, Heisenberg was already in competition with Kramers. And in the end, he succeeded. In May 1926 Kramers left Copenhagen for a professorship in Utrecht, whereupon Heisenberg stepped into his rival's shoes as Bohr's first assistant and confidant.

Heisenberg carefully timed his arrival in Copenhagen while Kramers was abroad at the 1924 Innsbruck meeting of the GDNA "so that I can then help you," he confided to Bohr.[10] After a stopover in Berlin to visit Erwin and his new fiancée, Marianne, Heisenberg arrived in Copenhagen on September 17, at the very peak of mounting difficulties with two essential features of Copenhagen physics: virtual oscillators and Bohr's "correspondence principle" for relating quantum properties of atoms to their classical counterparts.[11]

The complicated properties of spectroscopic lines, emitted by atoms only at definite frequencies, still served as the primary clues to the internal workings of the atom. In the Bohr-Sommerfeld theory of the atom, electrons orbited around a nucleus in certain fixed quantum orbits, or stationary states. A quantum jump between these states was accompanied by the emission or absorption of a light quantum—a particle of light energy bearing the frequency of one of the observed lines of light in a spectroscope. The frequency of the emitted or absorbed light quantum was proportional to the energy difference between the two stationary states. Since the concept of light particles contradicted the well-established electromagnetic wave theory of light, it could be avoided in practice only by the trick noted earlier: by treating each observed frequency and intensity of the emitted or absorbed radiation *as if* it arose from a tiny imaginary antenna, a virtual oscillator, within the atom—a charged electron ball on a spring oscillating with the observed frequency and amplitude (related to the intensity) of the emitted or absorbed light.

The BKS theory demonstrated the utility of the fictional virtual oscillators as a viable tool for handling the problem of the interaction of a light with atoms, just as the sound of an electric keyboard can be reproduced (even better) by the vibrating wires of a grand piano. Kramers argued in July 1924 that the virtual oscillators, together with Born's rule for creating discontinuous equations, could yield a new quantum theory of dispersion, the scattering of an incoming light beam by an atom into its different frequencies.[12] Kramers began work on the theory that summer, while Heisenberg toured the Bavarian hills with his boys. Upon his arrival in Copenhagen in the fall, Heisenberg joined in the effort, despite their differences.

Working in separate rooms—Kramers in his office and Heisenberg in the study hall—the two managed to produce a joint paper, written, however, mostly by Kramers. The Kramers-Heisenberg paper on the dispersion of light by atoms, submitted to the *Zeitschrift für Physik* on January 5, 1925, was a fundamental contribution to this field.[13] It served as the apex of the virtual-oscillator technique and, in retrospect, as the final touch needed for Heisenberg to "fabricate quantum mechanics," as he called it, six months later. The key ingredient of Heisenberg's breakthrough was his focus only on what can be observed in a laboratory, much as Pauli was already arguing at that time. Heisenberg later wrote of the Kramers-Heisenberg paper: "The necessity for detachment from the intuitive models was for the first time stated emphatically and declared to be the guiding principle in all future work."[14] Kramers and Heisenberg declared in their paper: "In particular, we shall obtain, quite natural-ly, formulas that contain only the frequencies and amplitudes characteristic for tran-sitions, while all the symbols referring to the mathematical theory of periodic systems [atoms] will have disappeared."[15]

Nevertheless, Heisenberg was not ready just yet to renounce the unobserved motions of electrons orbiting in atomic models. Especially, despite Pauli's rejection of these models, he would not abandon his now "symbolic" core model or his ulti-mate aim of finding what he called "the real Zeeman model."[16] The transition from mechanical models of the atom, however symbolic, to observed laboratory data as the foundation for quantum mechanics was an extremely difficult one—even for the audacious and inconsistent Heisenberg. Heisenberg's intensive struggles with atomic models, accompanied by his gradual retreat to mere observable quantities, became evident in his response to Pauli's newest assault on Copenhagen physics during the first half of 1925.

Early in the year, Pauli vehemently complained to Heisenberg of Copenhagen's "'virtualization' of physics," and he still grumbled about the core model.[17] At Einstein's suggestion, Pauli had begun a close examination of the core model's behav-ior on the basis of relativity theory. In several letters to Landé at the end of 1924, Pauli had presented his relativistic analysis. It decimated the core model. The core had to be rendered physically inert. This meant that all of the spectroscopic properties of the atom had to arise from the behavior of the electrons alone, in particular those in the outer orbits of the atom. Pauli had written to Sommerfeld: "One now has the strong impression with all models that we are speaking a language that is not sufficiently adequate to the simplicity and beauty of the quantum world."[18]

As the simplicity and beauty of the quantum world continued to elude the physi-cists, Bohr invited Pauli to Copenhagen in March 1925 to confer on the next steps to be taken. But Heisenberg's youth-movement buddy Wolfi Rüdel was at that moment also on an extended visit with Heisenberg. Not until after Rüdel's departure in late March did the physicists get down to business. With these four enormously gifted and energetic physicists together in one institute, the intensity of work, lasting daily from 9:00 AM to midnight, helped make up for lost time.[19] First there was the matter of recent experiments showing that light scattered by free electrons does indeed behave like tiny light particles, or quanta, rather than like waves. Energy and momentum are indeed conserved, and Bohr's cherished BKS theory had to be jettisoned. Second, before going back home on April 7, Heisenberg completed a paper in which he showed that two different symbolic atomic models seemed to work equally well for the same phenomena, even though they were incompatible.[20] Once again, no atomic model offered a clear path forward. All seemed to be very inadequate symbolic rep-

Heisenberg in Göttingen, c. 1924.

resentations of what is really going on inside the apparently simple and beautiful quantum world of atoms.

Heisenberg left his colleagues in Copenhagen in April for a two-week tour with his youth group through the romantic medieval villages of Württemberg. He returned home to find a depressing letter from Bohr. "There is much to report, for the most part negative, for I look at many things even more doubtfully than at the time when you were here." Still clinging to BKS, Bohr told of the "torture" that, with Pauli's help, he endured "in order to get used to the mysticism of nature. . . . I am attempting to prepare myself for all eventualities."[21]

Heisenberg returned to Göttingen on April 27, 1925, to begin his summer lectures and to follow what seemed to him might be a path to quantum mechanics: a careful analysis of the simplest atom of all, hydrogen, a single electron orbiting a singly charged nucleus. Without the devastating repulsions between electrons, Born's rule and the loose linkage between virtual oscillators and the orbital motions of the electron should enable him, he thought, to guess the correct quantum-mechanics model for hydrogen, just as Kramers had managed to do for the dispersion of light.[22]

A mathematical crutch existed for making the right guess. The planetary orbit of the electron as it revolved around the nucleus was periodic (it repeated itself). The French mathematician Jean Baptiste Joseph Fourier had shown that such a motion could be treated as an infinite summation of simple individual harmonic oscillators.

Bohr had shown that under certain conditions, the frequency of each oscillator should then correspond in the limit of high energies to one of the frequencies observed in the spectrum of hydrogen gas. The squared amplitude of the oscillator would give the corresponding intensity of the observed line. A careful derivation of these features for the light emitted by the hydrogen atom, Heisenberg hoped, would agree with the intensities of the observed hydrogen lines being obtained by Ralph Kronig, a German-American physicist who was then visiting Copenhagen.

The plan seemed promising, especially after an encouraging letter from Kronig.[23] But by the middle of May Heisenberg was bogged down in mathematical details. "The present conditions together are still not entirely sufficient to get the intensities uniquely," Heisenberg told Bohr, "but I still want to try to make progress."[24] He worked at it a while longer before finally giving up and trying a much simpler problem, the "anharmonic" oscillator, a simple ball oscillating on a spring but with a slight complication, such as a small amount of friction, introduced.

Pauli, having returned to Hamburg, had come to a similar impasse at about the same time. He, too, had become interested through Kronig in the problem of obtaining the observed line intensities for hydrogen by somehow modifying Born's rule to obtain the correct amplitudes of vibration of the corresponding virtual oscillators (which yield the intensities). But by the end of May 1925, Pauli, too, was enmeshed in difficulties and was nearly ready to give up. "Physics is at the moment once again very wrong," he exclaimed to Kronig. "For me in any case it is much too difficult, and I wish that I were a film comedian or something similar and had never heard of physics!"[25] Charlie Chaplin was then all the rage in Copenhagen.

Right up until Heisenberg's breakthrough in July, Heisenberg and Pauli—when not viewing Chaplin films—devoted all of their energies to the observed intensities rather than to finding the actual quantum mechanics. Remarkably, both continued to keep their positions, despite despair and rebellion. As late as June 24, Heisenberg explained their differences: "Our theoretical points of view differ insofar as you regard the line splitting as given and apply suitable oscillators to your representation; while I always attempt to hold on to the mechanics of the model."[26]

The more profound difference was that Heisenberg—ever pragmatic and audacious—could advance on two fronts at the same time and respond quickly on each. In letters to Pauli and Kronig on atomic intensities, his new approach to oscillators appeared almost as an afterthought. In the same letter to Pauli in which Heisenberg defended the use of mechanical models, he made a leap to Pauli's side, declaring that observed data, not models, would lead the way to quantum mechanics. Two weeks later, Heisenberg strayed further from the orbital motions of electrons. He even chided Pauli for writing of electron orbits that fall into the nucleus. "My entire meager efforts go toward killing off and suitably replacing the concept of the orbital paths that one cannot observe," Heisenberg declared.[27] Instead of the unobserved orbital paths, Heisenberg introduced a series of mathematical entities representing the observed radiation emitted or absorbed by an atom. Heisenberg, along with most other physicists, would soon learn that these entities were the elements of what was to them a little-known mathematical object, a *matrix*.

Heisenberg's breakthrough occurred because of his willingness to turn to a simpler problem (the anharmonic oscillator) and to try another approach (reliance only on the observed quantities, frequencies and intensities), even without being converted to it. It was Heisenberg at his best—simultaneously pursuing incompatible methods and employing inconsistent arguments intensely and brilliantly. Still, the work frustrated him. He could not advance beyond the simple problem to obtain a full quantum mechanics of even the simplest atom, hydrogen, and he dreaded his colleague's scathing criticism. "I have almost no desire to write about my own work," he told Pauli, "because everything appears to me still unclear and I can only surmise approximately how it will turn out."[28] Pauli held his tongue.

At this time, Born and his new private assistant, the 22-year-old German physicist Pascual Jordan, distinguished for his brilliance and an obvious lisp, were already at work independently of Heisenberg on a quantum theory of objects such as anharmonic oscillators. By May 1925, Born and Jordan sought "to reform radiation" by using only the frequencies and amplitudes of oscillation. They, too, raised the connection between these quantities and the observed data to a guiding principle of their new theory. In their published paper, completed by June 11, 1925, Born and Jordan announced "a fundamental postulate of great significance and fruitfulness. . . . Only such terms enter into the true natural laws that are in principle observable and determinable."[29] Physical quantities that can not be observed have no place in the laws of nature.

Heisenberg was fixated, but he thought the task his Göttingen colleagues had set for themselves was far too difficult. Irritated, he complained to his parents that his own work was at a standstill and that "everyone here is doing something different and no one anything worthwhile."[30] Leaving Born and Jordan to their calculations, Heisenberg tried to guess the quantum mechanics of an anharmonic oscillator from the myriad corresponding virtual oscillators that make up its motion. Even this simple problem required intensive thought and work, as revealed in his letters to Kronig and Pauli. But by relying only on observable quantities, he had most of the breakthrough to quantum mechanics by June 5, 1925, which he laid out in a letter to Kronig—not Pauli—on that date.[31] Without justification for the approach, and mindful of Pauli's past criticism, uncertainty bred restraint.

The observable quantities on which Heisenberg relied were the frequencies and the intensities of the individual lines of spectroscopic light observed in a laboratory. He considered the light to be emitted by electrons stuck onto "anharmonic" oscillators. In electromagnetism, an electric charge that is accelerated back and forth, as were these electrons, will emit electromagnetic waves, just as an antenna does. The complicated overall motion of an electron on the anharmonic spring could, as before, be treated as being composed of a summation of very simple (harmonic) oscillations, each with a definite frequency and amplitude. The problem Heisenberg set for himself was to work backwards from the observed frequencies and intensities of such a device to the simple individual amplitudes and frequencies of the oscillartors. Heisenberg first convinced himself that all of the observed data concerning the emitted radiation could be expressed by these imagined oscillators. One can assume that this basic idea also holds in quantum theory, he wrote Kronig. He reasoned that he could get to quantum mechanics by "reinterpreting" each amplitude and frequency as representing not an oscillation, but a quantum jump between two quantum energy states. In this way, the mathematical expressions for each classical oscillator could be reinterpreted as quantum expressions.

How to reinterpret the amplitudes? "The essential thing in this reinterpretation," he told Kronig, "appears to me [to be] that the . . . quantum amplitudes must be so chosen as to correspond to the connection of the [observed] frequencies."[32] It is here that the new quantum mechanics first appeared. When, for example, an electron makes a double jump—first to the next lower state, then to the state below that one—the two emitted frequencies must add together to produce the frequency that is actually observed. One can obtain this addition of frequencies by multiplying together the expressions for the virtual oscillators corresponding to each of the two jumps. Heisenberg found that, mathematically, if the two frequencies do add together, then the two amplitudes do not simply multiply together but are subjected to a new and strange multiplication rule involving all of the possible intermediate states—just in case the electron takes a circuitous route in getting from one place to another.

This simple rule constituted the central feature of Heisenberg's "quantum-theoretical reinterpretation of classical kinematic and mechanical relations," as the title of his breakthrough paper proclaimed. But Heisenberg had no idea what this multiplication rule actually represented. Not until two months later did he learn that it was identical to the rule for multiplying together two tables of numbers called matrices. In a quantum jump, all of the possible intermediate steps must also be arrayed in the table—a matrix. Then, when multiplying physical quantities together, such as the amplitudes of two motions, the corresponding matrices must be multiplied together according to the abstract rules of matrix mathematics.

However baffling, Heisenberg had his reinterpretation, but it was hardly a quantum mechanics. Nor was Heisenberg yet satisfied with this result. Even calculating the intensity of the radiation emitted by his anharmonic oscillator was still too complicated. He wrote Kronig, "The physical interpretation of the above scheme . . . results in a very peculiar point of view." Still uncertain, in a letter to Bohr three days later Heisenberg was silent about his work.[33]

At about this time, Heisenberg came down with a horrible attack of hay fever. Seeking relief, he left Göttingen for a barren rock in the North Sea, the tiny resort island of Helgoland off the German coast. He took a room in a guesthouse near the south shore, where he was alone with the boulders and the sea and a strange multiplication rule for oscillator amplitudes. What exactly happened on that barren, grassless island during the next ten days has been the subject of much speculation and no little romanticism. Years later, Heisenberg recalled suddenly realizing late one night that the total energy must be held constant, as recently reaffirmed with the defeat of the BKS theory. This apparently enabled him to derive the energies of the quantum stationary states by applying the new multiplication rule to the corresponding classical expressions. He hurriedly calculated the energies of the harmonic oscillator and the rigid rotator and obtained satisfactory agreement with his recollection of known observations in time to watch the morning sun dawn over the eastern sea.[34] Heisenberg had his breakthrough, but it was still only barely that.

On his return trip to Göttingen, Heisenberg stopped off in Hamburg to inform Pauli of his new physics. Heisenberg immediately began to draft a paper on his results. He managed to add another advance. The motions of electrons can be quantized by first reinterpreting the positions and momenta of the electrons, which are just numbers changing over time, as expressions that soon became quantum matrices. He then required that these expressions satisfy the condition that the multiplication of the momentum and position variables (summed over all possible values) must result in an integer (a quantum number) multiplied by constants, including Planck's constant. Heisenberg reinterpreted this expression as well and found once again a very strange result. As Born and Paul Dirac later showed, when the matrices for the quantum position and momentum of an electron are multiplied together, the order of multiplication makes a huge difference—an odd circumstance since this is not the case when multiplying two numbers together. But with quantum matrices, one does get a different result, depending on whether one multiplies position times momentum or, conversely, momentum times position. In the classical world there is no difference. In the quantum world there is.

Pauli was skeptical, if not openly critical. He still objected to Heisenberg's formalism and his introduction of half-integral numbers, this time for the quantum energies.[35] By June 24, 1925, Heisenberg had evidently seen the Born-Jordan paper on radiation, completed and submitted during his absence on Heligoland. Heisenberg met Pauli's skepticism by raising the observation of all quantities in his equations to a postulate, and in nearly the same words used by Born and Jordan: "The basic postulate is: In the calculation of whatever quantities, for example energy, frequency, and so on, only relationships between quantities that are controllable in principle [by observation] may appear."[36]

In the same letter to Pauli, Heisenberg continued to write of his own lack of joy about the paper and still wondered "what the equations of motion really signify if one conceives them as a relation between transition probabilities [amplitudes]."[37] By July 9, he was eager to abandon orbits entirely, turning instead to data and mathematical manipulation—after all, it worked! He sent a copy of his manuscript to Pauli and handed his only other copy to his mentor, Born, who apparently knew even less than Pauli about what his private lecturer was up to. Heisenberg had written his father shortly before, "My own works are at the moment not going especially well. I don't produce very much and don't know whether another [paper] will jump out of this at all in this semester."[38]

In early July 1925, a remarkably uncertain Heisenberg went to Born with two requests. First, he wanted permission to leave Göttingen before the end of the semester in order to accept an invitation to lecture in Cambridge, England. Second, he asked Born to look at his paper and to decide whether or not it was worth publishing. Born acceded to the first request and was fascinated by the second. But he too was at first puzzled by the strange multiplication rule for two amplitudes. Somehow it all looked vaguely familiar. Eventually, it reminded him of a rule he had encountered years earlier in a course on linear algebra that he had taken while a student in Göttingen. It was none other than the rule for the multiplication of two matrices.[39] At the end of the month, during Heisenberg's absence in England, Born forwarded Heisenberg's paper to the editor of the *Zeitschrift für Physik*. The breakthrough paper and its multiplication rule ignited the development of what became the matrix formulation of quantum mechanics.

CHAPTER 11

AWASH IN MATRICES, RESCUED BY WAVES

PAULI RESPONDED "WITH JUBILATION" TO HEISENBERG'S MANUSCRIPT OFFERING A QUANTUM reinterpretation of classical mechanics. The rapid evolution of this reinterpretation into the full-fledged physics of quantum mechanics over the following months gave the lately despairing Pauli "new hope, and a renewed enjoyment of life," as he wrote to Kronig in October. "Although it is not the solution to the riddle, I believe that it is now once again possible to move forward."[1]

Nearly two years later, Born maintained that he still remembered the exact day and hour when he realized that Heisenberg's paper involved the old but little-known and highly abstract mathematics of matrices (a branch of linear algebra).[2] While Heisenberg was lecturing on spectroscopy in Cambridge and finishing his Rockefeller fellowship in Copenhagen that summer and fall, Born and his assistant, Pascual Jordan, fashioned Heisenberg's matrix multiplication into what they called "a systematic theory of quantum mechanics."[3] Their paper, sent to the *Zeitschrift für Physik* in September 1925, introduced most physicists, including Heisenberg himself, to the erudite methods of matrix calculation.

Applying these methods to quantum problems, Born and Jordan laid the foundations for the new quantum mechanics of matrices, often called "matrix mechanics."[4] The new mechanics was (and still is) nearly incomprehensible to the technically uninitiated. Again, it is not necessary to comprehend the details of this theory in order to appreciate the overall tendencies of this approach. In fact, its abstract character is one of its most important tendencies, and one that encouraged other physicists to accept with great relief the more accessible alternative of "wave mechanics" soon offered by Austrian physicist Erwin Schrödinger.

In Göttingen's "matrix mechanics," a matrix entailed an ordered array, a table or a spreadsheet, of quantities that is then mathematically manipulated and multiplied with other matrices representing physical quantities, according to the rules of matrix algebra, in order to obtain physical results. In this scheme, nearly every variable and function of classical mechanics was reinterpreted as a corresponding quantum matrix and subjected to quantum rules. The joining of matrix mathematics with quantum principles so that actual, observed quantities could be derived from it involved, at that time, training in sophisticated matrix methods. Today a semester course at the advanced level would usually be required. The mathematics were made even more difficult by the circumstance that there are, in principle, an infinite number of possible stationary states in any quantized system. Fundamental variables, such as position and momentum, thus could easily turn into infinitely large matrices containing infinite numbers of rows and columns.

Because of this, Born and Jordan devoted most of their paper to explaining the technical methods of matrix manipulation and adapting them to quantum physics. They discovered a matrix analogue to nearly every prior classical and quantum equation. Their efforts focused on the equally sophisticated mathematics of classical planetary physics, which Born and his assistants had earlier adapted to the study of planetary atomic models. Reinterpreting these equations as abstract matrix expressions, they managed to obtain the energies of the stationary states in an atom. This, together with the conservation of energy, seemed "strong grounds" to Born and Jordan "to hope that this theory embraces truly deep-seated physical laws."[5]

Working independently of Göttingen, Paul Adrien Maurice Dirac, a 23-year-old Cambridge physics student, confirmed Göttingen's hopes. The shy and retiring son of a secondary school French teacher from Switzerland, Dirac was just completing his bachelor's degree in electrical engineering and applied mathematics at Bristol University several years earlier when he learned of the British confirmation of Einstein's prediction of the bending of starlight in relativity theory. Fascinated by relativity through the popular works of cosmologist Arthur Eddington, Dirac headed for Cambridge to pursue a doctorate in physics. As a student of the noted theorist R. H. Fowler, his interests in relativity expanded to include atomic physics, and in the summer of 1925 Dirac attended a lecture on atomic spectroscopy in Cambridge by one of Fowler's famous foreign guests—Werner Heisenberg.

During his visit to Cambridge, Heisenberg spoke publicly only about the Zeeman effect. He was still too uncertain about his quantum paper. But he did speak of it privately with Fowler, who soon obtained the page proofs from Bohr. When Dirac learned the full details of Heisenberg's reinterpretation from the galleys, he obtained valuable analogies between Heisenberg's strange multiplication rule and a specialized group of classical equations (Poisson brackets) in which the order of multiplication is important.[6] At the same time, Göttingen further confirmed its own hopes by extending the Born-Jordan formalism to systems of any number of particles and motions. The result was the famous and fundamental "three-man paper" titled "On Quantum Mechanics II, " submitted by Born, Heisenberg, and Jordan (in alphabetical order) in November 1925. It became the fundamental treatise on the new physics for those who could withstand the mathematics.[7]

Heisenberg contributed to the effort by letter from Copenhagen until Born ordered his return to Göttingen so that the three men could finish the paper. Also, Born was planning to leave for a visiting professorship at the Massachusetts Institute of Technology that winter.[8] Heisenberg was to substitute for Born in his absence, and Born wanted to familiarize him with institute affairs and teaching assignments.[9] Born,

Heisenberg, and Jordan, experts on adaptations of planetary theory to quantum atoms, had little difficulty developing the matrix-mechanical analogue of their earlier work. Their method, however, raised the theory to new heights of abstraction.

Nevertheless, the three men proudly announced that their new quantum mechanics did achieve the long-sought aim. It contained the basic postulates of quantum physics in its very foundations—the existence of distinct stationary energy states in atoms and jumps between states accompanied by the emission or absorption of light. And it allowed calculation in principle of any system displaying periodic motions, such as atoms, and in close analogy with classical mechanics. The previously puzzling properties of atoms could now be derived from the new quantum matrix mechanics. Pauli and Dirac immediately used the new mechanics—and their own ingenuity—to derive the well-known Balmer series of the hydrogen atom.[10] And two recently arrived postdoctoral researchers applied the theory to the so-called band spectra of molecules. One was Lucy Mensing in Göttingen; the other was J. Robert Oppenheimer, who utilized Dirac's version of the theory in Cambridge.[11]

Still, the new physics was not yet fully applicable to atomic phenomena until the advent of the additional concept of electron spin—the hypothesis that an electron in an atom rotates on its axis like a spinning top, or a miniature earth, with a half unit of rotational momentum. The half-integer value of the electron's spin suddenly resolved at a stroke all of the many riddles regarding half-integer numbers that had plagued atomic spectroscopy for the past five years! The success solidified attitudes among the matrix mechanicians regarding their physics and regarding the characteristics of nature that they believed the new mechanics represented. Matrix mechanics and spinning electrons provided the culmination and conclusion of five years of intensive work in atomic physics.

Heisenberg and Pauli were at first not very receptive to the idea of spinning electrons. The idea seemed a throwback to pictorial models of atoms and electrons. Even worse, it required a point on the whirling equator of the electron to move faster than the speed of light, which, according to relativity theory, is not permitted. But the utility of the idea finally won out, and Pauli "capitulated" with "heavy heart" on March 12, 1926.[12] Just four days later, Heisenberg and Jordan submitted a manuscript from Göttingen applying electron spin to the quantum matrix mechanics of the Zeeman effect, resolving over a decade of headaches regarding this phenomenon.[13]

The new Zeeman theory also constituted fulfillment of the first of three "program points" agreed upon by Bohr, Pauli, and probably Heisenberg in March 1926.[14] These points called for the cleanup of all old business by means of the new quantum mechanics of matrices and spinning electrons—much like what Thomas Kuhn referred to as the "mopping-up operation" that follows the establishment of a revolutionary new paradigm. With the hydrogen atom already absorbed into the new physics, the physicists' agenda called for derivations of all of the observed behavior of spectroscopic lines emitted by atoms; a derivation of the observed behavior of the two types of helium atoms; and a complete explanation for the structure of the periodic table of elements. The three physicists expected this backlog of accumulated problems and puzzles to find its natural resolution in the new quantum mechanics. The remarkable fulfillment of two-thirds of this program by late July 1926—except for the closing of the electron orbits in the periodic table—preceded and paralleled a surprising new challenge to their approach coming from a complete outsider. Meanwhile, the closing of the shells was handled simply by a new principle, Pauli's exclusion principle.[15]

Beginning in early 1926, Schrödinger, a rakish 38-year-old professor of theoretical physics at the University of Zurich, published a series of papers that laid the foundations for a completely different approach to quantum mechanics, one that did not rely on electrons and matrix algebra. This new quantum mechanics became known as wave mechanics.[16] Schrödinger approached the puzzles of quantum atomic physics from a very different direction and with very different aims from those defended by the inventors of quantum matrices. In his series of papers, working alone, he incorporated the opposite side of the wave-particle duality into the foundations of his formulation of quantum mechanics. While matrix physicists focused on quantum jumps between energy states and distinct spinning electron balls of charged matter, Schrödinger drew on the hypothesis of matter waves propounded at the time by the French doctoral candidate Louis de Broglie and recently accorded Einstein's favorable notice.

De Broglie, the son of a noble member of the French Assemblée Nationale, had set out for a civil service career in Paris. But after reading the works of Henri Poincaré he was converted to physics by his elder brother, Maurice, a physicist. He learned of atomic physics firsthand in his brother's laboratory and published a series of papers leading up to his 1924 dissertation on matter waves at the Sorbonne.

If electromagnetic waves can behave like particles—light quanta—then, de Broglie reasoned, under certain conditions, particles should behave like waves of matter—matter waves.[17] If so, Schrödinger now suggested, there should be a "wave equation" representing the evolution, or propagation, of the matter waves through space, just as there is one for the propagation of electromagnetic waves. There should also be a "wave function" that represents the waves themselves.

In his first paper, Schrödinger made use of an old analogy between mechanics and wave optics and applied it to the matter waves, together with the mathematical principle that small variations of a function should yield only vanishingly small effects. He eventually arrived at a wave equation that looked very similar to other wave equations. The solutions of this equation each yielded a distinct energy of the wave, and each energy corresponded to a standing wave in a distinct mode of vibration. Applying this equation to electrons orbiting in atoms, Schrödinger viewed the electron matter waves as vibrating in orbits around the nucleus. He interpreted each distinct vibration mode as a stationary state. In other words, electrons orbiting in atoms exhibit a series of distinct quantum states, each with a distinct energy, because each state represents a different possible standing oscillation of the electron wave, each with a unique energy. The orbits are distinct, as they were in Bohr's original quantum theory of the atom, because only certain orbits at certain energies can

The Nobel Prize winners arrive in Stockholm, 1933.
RIGHT TO LEFT: *Schrödinger, Heisenberg, Dirac, Dirac's mother,*
Schrödinger's partner, Heisenberg's mother.

accommodate standing waves. It was a very novel idea, quite unlike anything imagined in Göttingen, Copenhagen, or Munich.

Schrödinger easily applied his new wave mechanics to the derivation of the well-known Balmer series of lines observed in the spectrum of hydrogen from his calculated values of the energies of the quantum states. Introducing time as a variable in a subsequent publication, Schrödinger obtained the more general equation for the propagation of these waves, known as the Schrödinger equation and found today in nearly every textbook on quantum physics. It emerged as the centerpiece of the new quantum wave mechanics, fully applicable to the same sorts of dynamic problems as Göttingen's erudite quantum mechanics of abstract matrices.

The power of Schrödinger's wave mechanics was awesome, its advantages obvious, its profound importance loudly proclaimed. Most physicists were already well familiar from optics with the wave equations of the sort contained in wave mechanics, and many now preferred the more familiar notions on which this mechanics rested. Recently presented with the abstractions of matrix algebra, these physicists, even matrix mechanicians, gladly welcomed Schrödinger's more familiar task of solving a mere wave equation, however difficult that might prove. Sommerfeld, for instance, who at first considered Schrödinger's approach "totally crazy," declared during a talk in Hamburg some months later that "although the truth of matrix mechanics is indubitable, its handling is extremely intricate and frighteningly abstract. Schrödinger has now come to our rescue."[18]

Schrödinger's facile derivation of the Balmer series in his very first communication obviously impressed the ever-critical Pauli, who had just completed a tortuous derivation of the same result using matrices. He wrote Jordan in early April 1926, "I believe that this paper numbers among the most significant that have been written lately. Read it carefully and thoughtfully."[19] The possibilities afforded by Schrödinger's physics also encouraged Max Born, who was then working with Norbert Wiener at the Massachusetts Institute of Technology on an extension of matrix mechanics to colliding particles. Impressed that wave mechanics enabled a facile treatment of collisions, Born voiced enthusiasm for it upon his return to Germany: "I would regard [wave mechanics] as the deepest form of the quantum laws."[20]

Heisenberg exhibited far less enthusiasm for Schrödinger than did his colleagues, and was "not very pleased" with Born's apparent defection to wave mechanics.[21] His skepticism derived as much from acquired conviction as it did from personal competitiveness. He called Schrödinger's paper "incredibly interesting," especially for its mathematical simplicity, but unlike Born, he refused to acknowledge that, as physics, Schrödinger's work had any advantage over his. Born wrote to Schrödinger a year later, "Heisenberg from the very beginning did not share my opinion that your wave mechanics is physically more significant than our quantum mechanics."[22]

During the formulation of matrix mechanics and electron spin, which occurred simultaneously with Schrödinger's work, Heisenberg had acquired "very great confidence" in Göttingen's matrix-mechanics approach.[23] But he already worried about the reception of the Born-Heisenberg-Jordan point of view regarding a more fundamental point: the characteristics of nature itself, which he believed required the existence of particles, quanta, and discontinuities. But it was a point of view that seemed obscured by the abstractions of matrix algebra. He had complained earlier to Pauli that Göttingen scientists, other than the three men, had fallen again into two camps: those who welcomed the unparalleled successes of matrix physics, regardless of its abstraction, and those who had given up hope of ever understanding matrix physics.[24] Previously so motivated by success at any price, Heisenberg himself at first joined neither camp. He was not unsympathetic toward Schrödinger's work, but in the end, he argued, one could not replace jumps and quantum states with waves and vibration modes.

Schrödinger took the opposite position. While the founders of matrix mechanics stressed the existence of quantum jumps and the elements of discontinuity, lack of pictorial representations of atomic motions, and the use of matrices for continuous variables, he wrote, wave mechanics entailed just the opposite: "a step from classical point mechanics towards a *continuum theory*."[25] He argued that his theory was based on a continuous field, although he admitted that his field existed not in our real, everyday space, but in an abstract multidimensional space. Nevertheless, he believed that his theory returned physics almost entirely to the classical pictures and atomic models of the physical processes at work.

"All philosophizing about 'principle observability' only glosses over our inability to guess the right pictures," Schrödinger told his colleague and supporter Willy Wien. The future development of quantum physics would be best served, practically and intellectually, by adherence to visualizable wave mechanics, he wrote, instead of "considering ourselves bound in atomic dynamics to suppress intuition and to operate only with abstract concepts such as transition probabilities, energy levels, and the like."[26]

In a complete surprise to both sides, in the end the two alternative versions of quantum mechanics—wave mechanics and matrix mechanics—existed independently for less than two months. In May 1926, Schrödinger published a proof that the two proposed formalisms—wave equations and matrix algebra—seemingly so different in form and content, were in fact mathematically equivalent! With that, the two sides began to deride each other more openly, and at times the ensuing debate became emotional and even personal.

Schrödinger himself was no help. In his paper proving the equivalence of wave mechanics and matrix mechanics, he argued not for the equal consideration of the two interpretative contexts of the equivalent formalisms but "perhaps not wholly impartially" for the superiority of his own. Moreover, in a famous footnote to this paper, Schrödinger declared, "My theory was inspired by L. de Broglie . . . and by short but incomplete remarks by A. Einstein. . . . No genetic relationship whatsoever with Heisenberg is known to me. I knew of his theory, of course, but felt discouraged, not to say repelled, by the methods of transcendental algebra, which appeared difficult to me, and by the lack of visualizability."[27]

Heisenberg expressed similar sentiments toward Schrödinger's theory in a letter to Pauli written shortly after the publication of Schrödinger's equivalence paper: "The more I think about the physical portion of the Schrödinger theory, the more repulsive I find it. . . . What Schrödinger writes about the visualizability of his theory 'is probably not quite right' [an echo of Bohr], in other words it's crap."[28] The only advantage to Schrödinger's method, he declared publicly, was that it enabled a simple calculation of the atomic transition probabilities for plugging into the matrices of matrix mechanics.[29]

It is important to note that it was not the mathematical equivalence itself that evoked such passion—Pauli had, in fact, proved that earlier and without much ado—but rather what each side made of it. Nor did the alternative yet equivalent formalisms actually cause much debate. Although matrix mechanics continued to be the most applicable to some problems, especially those involving spin, Schrödinger's wave equation soon won out as the formalism of choice by most physicists for solving most problems. Rather, the differences had to do with interpretation, with the two wholly different understandings of nature and atoms underlying each formalism, each of which was now found to be mathematically equivalent to the other formalism. With mathematical equivalence established in May 1926, attention focused on the physical concepts that underlying each formalism—and neither side was willing to concede.

The conflict between the respective defenders of matrix and wave physics formed the immediate background of the interpretation that eventually was accepted

—Heisenberg's uncertainty principle in early 1927 and Bohr's complementarity principle in the same period, the foundations of the Copenhagen interpretation of quantum mechanics.

Historians and others have made much of this conflict and its consequences.[30] One view is that the fight was actually not about physics at all but over who would dominate the future of quantum mechanics. Some have pointed to Heisenberg's envy, ambition, and snobbery as the cause. Heisenberg was, of course, no stranger to these attributes, and neither were some of his close colleagues. But such traits seem less significant in accounting for the matrix mechanicians' attitudes toward the wave-mechanical challenge. Judged in a broader context, it is important to recognize that Heisenberg and his close colleagues had devoted their entire academic careers to struggling with the new and strange characteristics of nature contained in quantum theory—discontinuities, quantum jumps, distinct particles, and particle-like light quanta. They believed profoundly that such characteristics actually existed in nature and that such phenomena were successfully incorporated into their matrix mechanics. Schrödinger seemed to deny all of this.

But something less elevated was also vying for their attention: the sudden availability of academic jobs. Just as Schrödinger published his equivalence paper in 1926, several German teaching chairs in theoretical physics became vacant. The lure of top jobs captured the attention of all of those young geniuses who, like Heisenberg, had recently habilitated and who now looked to the defense of matrices as a defense of their professional status, and thus a requirement for landing a decent job.

Late in 1925, as Born, Heisenberg, and Jordan put the finishing touches on matrix mechanics, Bohr's Copenhagen assistant, H. A. Kramers, accepted a teaching chair (a full professorship) in Utrecht. This gave Bohr the chance to offer Heisenberg Kramers's dual position as university lecturer and institute assistant, a spot that Heisenberg had long coveted. Heisenberg also had the advantage that he was now nearly fluent in Danish. He tentatively accepted Bohr's offer, but made no secret of yet another goal. As Bohr wrote Born, "No one in Germany need fear that [Heisenberg] will stay here very long."[31] Born, still in the United States, agreed to allow Heisenberg to go to Copenhagen, provided Heisenberg would stay in Göttingen until Born's return in April 1926.

Heisenberg's desire to settle in his homeland paralleled the extraordinary popularity of theoretical physics in Germany and abroad, especially following the recent breakthroughs to quantum mechanics. German work in relativity and quantum physics gained international recognition just as Germany itself suffered crises in other areas. Nearly every faculty in Europe and abroad sought to have a leading German representative of the new physics at its school. As Heisenberg's usual luck with timing would have it, the deaths and retirements of several senior professors at German-speaking universities opened up room at the top.

In early 1926 the theoretician George Cecil Jaffe left an associate professorship in Leipzig for a full professorship in Giessen. On October 1, 1926, Max Planck retired from the University of Berlin (but remained active for another 20 years). A week later, Theodor Des Coudres, Leipzig professor of theoretical physics, suffered a fatal heart attack. In the same month, Karl Schmidt, theorist in nearby Halle, submitted his request for retirement. Three months later, in January 1927, Otto Wiener, Leipzig professor of experimental physics, also died suddenly of heart disease. "This damned Leipzig is a constant source of unrest!"[32] Pauli exclaimed to Heisenberg (not to mention a constant source of heart problems).

In the ensuing complicated "appointment chatter [*Klatsch*]," as Heisenberg called it, Heisenberg and Pauli found themselves at the top of nearly everyone's list of preferred candidates. In April 1926, Augustus Trowbridge, IEB European representative, inquired of Bohr what it would take to lure Heisenberg across the Atlantic.[33] At the same time, Heisenberg, Pauli, and Gregor Wentzel learned that they were all (in that order) on Leipzig's list to succeed Jaffe.

Heisenberg had already accepted Bohr's offer to replace Kramers and was scheduled to be in Copenhagen on May 1, 1926. In April he had sent a startlingly late inquiry to Bohr about financial arrangements, which occasioned a worried telegram from Bohr announcing a salary increase for Heisenberg. Heisenberg was then in Dresden (the seat of the Saxon cultural ministry), where he was engaged in negotiations regarding the Leipzig position.[34] German academic tradition required a young person to accept the first offer of a professorial appointment, or he would be considered "impossible" for some time afterward. Heisenberg's father, aware of this formality, strongly urged his son to accept the Leipzig offer, even if it meant backing out of the Copenhagen post.

An uncertain Heisenberg turned to Göttingen for advice. Born, newly returned from the States, and Richard Courant, then dean of the science faculty, strongly urged Heisenberg not to go to Leipzig, but rather to take the splendid opportunity to work with Bohr in Copenhagen. Courant informed Bohr the next day: "I urged him to go to you under all circumstances and not to sacrifice the scientific and human advantages of his stay in Copenhagen to the superficial advantages of the call from Leipzig. It is my opinion that Heisenberg can calmly forgo this first opportunity [of a position] in Germany."[35]

The advice was repeated in Berlin. Earlier in April 1926, Max von Laue, in charge that year of the famous physics colloquium at the University of Berlin, had invited Heisenberg to lecture on the new matrix mechanics. Heisenberg presented a grueling, comprehensive two-hour lecture on April 28.[36] He wrote to his parents from Berlin the next day that all of the "bosses of physics"—Einstein, Laue, Walther Nernst, Rudolf Ladenburg, and Lise Meitner—had assembled to hear him speak. During their private conversations, the Berlin bosses, despite obvious doubts about Heisenberg's physics, unanimously urged him to take Bohr's post and to reject Leipzig's offer. They argued that Heisenberg would certainly learn much more from Bohr; perhaps they even hoped that the Copenhagen connection might improve Heisenberg's physics. They were certain that another chair would open for him later.

"If I continue to produce good papers," he wrote, "I will always receive another call; otherwise, I don't deserve it."[37]

Heisenberg accepted the unanimous advice and went to Copenhagen, but only on a temporary leave from Göttingen. He wanted to keep the option open to return to Germany if necessary. In early May 1926, he arrived in Copenhagen to begin his full-time duties as lecturer and assistant to Niels Bohr.

Pauli, for his part, now promoted to full professor, refused to leave Hamburg. Wentzel, third on the list, accepted the Leipzig associate professorship, just as Schrödinger's equivalence paper appeared in print. If Heisenberg were to receive another call to a German chair, the quality of his papers would now be judged in terms of the new situation brought about by the equivalence of his less-popular theory with Schrödinger's. This became clear when, just two weeks after Heisenberg's arrival in Copenhagen, Grandfather Wecklein showed up, accompanied by his unmarried daughter, Heisenberg's aunt.[38] The aged Wecklein never ceased to regret that he had settled for anything less than a university chair. His grandson's habilitation and recent substitution for Born had been the highlight of his last years. Now Heisenberg's refusal to accept a German chair in favor of an assistantship abroad must have seemed incomprehensible to the ill and aged Greek scholar—as it did to his son-in-law, August Heisenberg. Three months later, Bohr received a worried letter from Heisenberg's father about his son's future. Bohr responded with strong assurances that Heisenberg was indeed on the right track toward a chair.[39] Before the year was out, Heisenberg received news of his grandfather's death, made doubly sad by the fact that neither of the two grandsons—Erwin had not even received his doctorate yet—seemed close to achieving a chair. Heisenberg would nevertheless realize his grandfather's dream within a year. But the pressure was now on.

Finally arriving in Copenhagen, Heisenberg intended to room again with Mrs. Maar, but when he arrived, she was on holiday. Until her return at the end of May, he occupied a cramped but nicely furnished third-floor service room beneath the roof of Bohr's institute. His window looked out on a splendid view of the Fælledpark, where he and Bohr took their frequent walks. The Bohrs, who had recently moved to the new house next to the institute, treated Heisenberg as a family member, providing regular meals and access to their piano until Mrs. Maar's return. "I am already half at home with the Bohrs," he told his parents.[40]

Music and recreation, never neglected by those upper-class academics, consumed Heisenberg's leisure time and that of his Copenhagen colleagues throughout this period. Apart from his usual daily walks and piano playing, Heisenberg took numerous weekend tours to the countryside alone or with other institute members. He went on long sailing excursions and a hiking tour in Norway with Bohr, with whom he also took weekly horseback riding lessons. Such activities were never undertaken frivolously but "in order not to degenerate in physics."[41] Still, as strenuous as these activities may have been, they could not counterbalance the intensity of his work.

Heisenberg's duties in Copenhagen consisted of helping Bohr direct the work of their many visitors. He corrected the research papers of numerous students and visiting fellows and delivered a weekly one-hour lecture in Danish. Heisenberg seems to have managed well with Danish, as he informed his parents, and as can be judged from his lecture notes from that period, written in Danish.[42] But Heisenberg's most demanding job was his work with Bohr to elucidate the interpretive situation in quantum mechanics now that Schrödinger was challenging the very foundations of the matrix version of quantum mechanics.

Schrödinger had publicly championed his continuum theory of visualized waves over the discontinuity and quantum jumps of spinning particles forming abstract matrices. Heisenberg and his colleagues, while they accepted Schrödinger's wave method, rejected his claims. During the months following Schrödinger's publication of the equivalence and Heisenberg's move to Copenhagen, they reaffirmed their belief in the existence of discontinuity and their desire to resolve the difficulties it presented.

Soon after reading Schrödinger's paper in May, Pauli wrote to Schrödinger that a discontinuous element must be introduced into an understanding of quantum phenomena.[43] Continuum physics alone would not do, he repeated in November: "Do not think, however, that this conviction makes life easy; I have already tortured myself a great deal because of it and will probably have to do so again!"[44] After a face-to-face encounter with Schrödinger, Heisenberg complained to Pauli that Schrödinger's theory "did not fit the facts"; that is, it did not solve problems that seemed to require discontinuity. "Schrödinger simply throws overboard all 'quantum-theoretical effects,' such as photoelectric effect, Franck collisions, Stern-Gerlach effect etc.; then it is not hard to construct a theory."[45]

Heisenberg's encounter with Schrödinger occurred in Munich at the end of July 1926. Having heard Heisenberg's lectures on matrix mechanics, Berlin physicists were eager to hear Schrödinger on wave mechanics and invited him to Berlin in July. On his return trip to Zurich, Schrödinger's avid supporter, Willy Wien, now rector of the University of Munich, and his less-than-avid supporter, Arnold Sommerfeld, invited him to stop off in Munich to deliver two lectures on his new physics at the university —one to the Bavarian section of the German Physical Society on Friday, July 23, the other to quantum specialists the next day.[46] Heisenberg returned to Munich for the occasion—his first appearance in Germany after moving to Denmark—and for a holiday with the young men of his youth group, whom, as usual, he missed terribly.

Heisenberg did not deliver a formal talk on this occasion, but he attended both of Schrödinger's lectures presented in the crowded lecture hall of the physics institute. Heisenberg withheld his objections to Schrödinger's claims until the end of Schrödinger's second lecture, entitled "New Results of Wave Mechanics." As the bespectacled Professor Schrödinger completed his discussion of applications that Saturday morning, the 24-year-old Heisenberg rose from his seat, as he had four years earlier in Göttingen, to argue that the new theory could not explain even basic quantum phenomena such as light quanta and the Compton effect (scattering of quanta by electrons), which seemed to require discontinuity and quantum jumps. The audience clearly disagreed, and the aged Willy Wien, obviously annoyed, vehemently motioned to Heisenberg to sit down and shut up. Later, Wien tried to console Heisenberg, whose physics "and with it all such nonsense as quantum jumps," was, in Wien's view, "finished."[47] Having failed to convince anyone of his views, even Sommerfeld, the despondent Heisenberg headed for the hills with his youth group. But after the meeting and subsequent discussions with Bohr and Born during a conference in England, Wien wrote a worried letter to Schrödinger asking how indeed one could account for light quanta and the Compton effect without discontinuities. Schrödinger was, for the moment, equally uncertain.[48]

Heisenberg was still eager to take on his opponent. "A few days ago I heard two lectures here by Schrödinger," he told Jordan after the Munich encounter, "and I am rock-solid convinced of the incorrectness of the physical interpretation of Q. M. presented by Schrödinger. That Schr.'s mathematics signifies a great progress, however, is clear."[49] Yet several months later, with most physicists now enamored of wave methods and mechanics, as he had observed in Munich, Heisenberg seemed even more defensive. In September 1926, he delivered a lecture in Düsseldorf to a joint meeting of the German mathematical and physical societies held during the large biennial conference of the GDNA. His address to the audience, representing the most powerful bodies in German physics and mathematics, was published in the widely circulated *Naturwissenschaften* (*Sciences*). In it, he argued neither for a rejection of the rival wave conception (which would have been unlikely) nor for the superiority of the particle conception, but for a fuller appreciation of the particle conception on an equal footing with waves, much as Bohr preferred. This meant a return to a type of wave-particle dualism: "There exists in our intuitive interpretation of the physical phenomena and the mathematical formulas a dualism between wave theory and corpuscular theory of such a type that many phenomena can be explained most naturally by a wave theory . . . while other phenomena can only be explained on the basis of a corpuscular theory."[50]

Heisenberg's physical arguments in support of his views touched off a major debate at the Düsseldorf meeting. In response, Born, Heisenberg, and Jordan contemplated a new three-man paper to set the record straight. In a letter sent to Jordan, Born, and an opponent (Adolf Smekal), Heisenberg claimed he was withdrawing from polemics, but Schrödinger's anticipated rejoinder to Heisenberg's remarks soon changed his mind.[51] Although Schrödinger actually had no intention of responding to Heisenberg's assertions, Heisenberg submitted to the *Zeitschrift für Physik*, a journal usually reserved for noncontroversial research reports, a paper harmlessly titled "Fluctuation Phenomena and Quantum Mechanics."[52] In it he attempted to prove "that the fact of discontinuity is contained in a natural way in the system of quantum mechanics." But the real aim of the paper was indicated by its tone and by Heisenberg's admission to Pauli that it was "only a pedagogical paper for the gentlemen of the continuum theory"—a polemical response to a perceived polemical situation.[53]

Heisenberg's pedagogical paper was written in the wake of an unsettling visit by Schrödinger to Copenhagen in early October 1926, shortly after the Düsseldorf meeting.[54] Bohr had invited Schrödinger to Copenhagen after receiving a letter (now lost) from Heisenberg following the Munich encounter the previous summer. Bohr, too, was determined to resolve the conflict over waves and particles in a way that at least accounted for, if it did not preserve, quantum jumps, discontinuous energy quanta, and rotating particles.

Schrödinger arrived in early October for the quantum summit meeting and was housed alone in an institute guest room. Discussions began almost immediately. The usually gentle and congenial Bohr could also, wrote Heisenberg, "insist fanatically and with almost terrifying relentlessness on complete clarity" when it came to accounting for quantum phenomena.[55] According to Heisenberg's account of the meeting, Bohr's fanatical insistence soon rendered his unassuming visitor physically ill. Bohr pursued Schrödinger even to his sickbed—but without achieving the desired resolution. Heisenberg, still Bohr's lecturer and assistant, witnessed the struggle and no doubt joined in, but he recollected only its general features years later. A subsequent flurry of correspondence indicates that the main issue was, as expected, the existence of jumps and discontinuity—specifically, the existence of quantum jumps between distinct stationary states as revealed by various quantum phenomena, and as recently supported in a new interpretation published by Born.[56]

In the second of his papers treating the collisions of two particles, received by the *Zeitschrift für Physik* in July 1926, Born had reinterpreted one of Schrödinger's early assumptions concerning his wave function.[57] When squared (multiplied by itself or by its complex conjugate), it acted like a density—but the density of what? Schrödinger had assumed that, when multiplied by the electron charge, this density function represented the density of the electric charge.[58] But the assumption fell apart in Born's analysis. In order to derive such scattering phenomena as the collision of a matter-wave electron with an atom, Born interpreted the squared wave function, not as the density of the charged particles, but as the density of the *probability* for scattering, that is, as the probability that the waving electron will scatter into a certain quantum state after it hits the atom.

Probability and the closely related notion of only statistical (rather than exact) predictions for the outcomes of experiments had entered physics most recently in the failed theory of Bohr, Kramers, and Slater. Einstein had shown in 1916 that the jumps of electrons from one quantum state to another require the introduction of a probability for each possible jump. This feature had been incorporated into the BKS theory and the notion of virtual oscillators: the greater the probability of a specific jump, the greater the intensity of the radiation emitted by a gas of atoms at that frequency.

Born's interpretation of Schrödinger's waves went even further than prior assertions, providing one of the essential features of the forthcoming Copenhagen interpretation, and an important challenge to Schrödinger's claims. The square of Schrödinger's wave function had *nothing* to do with the density of matter or charge in space. It referred to the probability of finding a given matter wave in a given state, either within an atom or after a scattering process. For numerous particles, the probability thus represented the number of particles found in each state.

For Born, Schrödinger's equation did not describe the propagation of matter waves through space and time, as Schrödinger would have it. Instead, it described the propagation of the *probabilities* of finding particles at certain locations in space and time. "We free forces of their classical duty of determining directly the motion of particles," he told a British Association meeting at Oxford in August 1926, "and allow them instead to determine the probability of states. Whereas before it was our purpose to make these two definitions of force equivalent, this problem has now no longer, strictly speaking, any sense."[59]

The implications of Born's probability interpretation of the wave function would soon grow even more radical and profound in the context of Heisenberg's indeterminacy, or uncertainty, principle as it developed in the months following Schrödinger's October 1926 visit.

During that visit, neither Schrödinger nor Bohr would budge from his position. The upshot for Bohr was the realization that no available interpretation of the formalism was entirely adequate, that is, no consistent correlation between the equations of either version of quantum mechanics and the observed data in the physicist's laboratory had yet been established. Bohr concluded that anyone's "wishes for a future physics" would be realized only after such an adequate interpretation had been found.[60] Bohr's 24-year-old assistant, Werner Heisenberg, now concentrated all his energies on the search to realize his wishes.

CHAPTER 12

DETERMINING UNCERTAINTY

ON MARCH 22, 1927, WERNER HEISENBERG SUBMITTED ANOTHER FUNDAMENTAL BREAK-through paper to *Zeitschrift für Physik,* this one titled "On the Perceptual Content of Quantum-Theoretical Kinematics and Mechanics."[1] The 27-page paper, forwarded from Copenhagen, outlined one of Heisenberg's most famous and far-ranging achievements in physics: his formulation of the uncertainty, or indeterminacy, principle in quantum mechanics. Bohr's subsequent principle of complementarity, Born's statistical interpre-tation of Schrödinger's waves, and Heisenberg's uncertainty principle together would form the Copenhagen interpretation of quantum mechanics, an explication of the uses and limitations of the mathematical apparatus of a now-unified quantum mechanics that fundamentally altered our understanding of nature and of our relation to it. Uncertainty and the Copenhagen interpretation marked the end of a profound transformation in physics that has not been equaled since; nor has an alternative theory proved as success-ful or as widely applicable to the phenomena of the atomic scale as has quantum mechanics since its completion in the Copenhagen interpretation of 1927.

Just two weeks after he submitted his paper enunciating what became the uncer-tainty principle, Heisenberg published the first of his non-technical summaries of the nature and significance of his work for nonphysicists.[2] In his summary, titled "Fundamental Principles of Quantum Mechanics" and published in a popular German science periodical, Heisenberg followed his uncertainty paper in arguing that the con-tent of a physical theory may be easily recognized, not by its mathematical equations, but by the new concepts to which it gives rise. Fundamental concepts and their mean-ings, rather than equations, had been the point of contention between the proponents of the matrix and wave versions of quantum mechanics.

Until the turn of the century, Heisenberg wrote, Newtonian mechanics was regarded as the foundation of all of physics. This theory involved the concepts of space and time, force and mass, along with the fundamental assumptions of a direct connection between cause and effect and an objective nature existing more or less independently of the observer. Relativity theory changed our notions of space, time, and mass and showed that under certain conditions—at extremely high speeds approaching the speed of light and under intense gravitation—Newtonian mechanics had to be replaced by a new relativistic mechanics.

A similar transformation, Heisenberg continued, is now required in the realm of small masses moving within extremely short distances, such as the behavior of electrons inside of atoms. But here physicists encountered a difficulty. They could not actually observe the internal workings of a minute atom; they could observe only the gross properties of large numbers of atoms in their laboratories. Among these properties, the frequencies of the light emitted and absorbed by atoms and observed in the laboratory using spectroscopic equipment provided some of the best clues to an understanding of the behavior of the electrons inside the atoms. Intensive analyses of this spectroscopic light associated with atoms, carried out over the previous years, indicated the need to replace Newton's mechanics of atomic electrons with a new quantum mechanics. The new mechanics incorporated the appearance of fundamental discontinuities in an otherwise seemingly continuous world. These discontinuities entailed the existence of discrete quanta, or bundles, of energy and momentum, and quantum leaps between distinct states of energy. Because of their extreme minuteness, the discontinuities of quanta and quantum leaps, surmised on the submicroscopic level of atoms, are not recognizable on the macroscopic level of the everyday laboratory, where the continuous classical mechanics and electrodynamics of Newton and Maxwell still held, as it does today.

The same division between the atom and the laboratory occurred for other basic concepts. The equations of the new mechanics indicated that the electrons within atoms cannot be described using everyday pictorial descriptions and concepts, such as position, velocity, and orbit, wrote Heisenberg in this uncertainty paper. Unlike billiard balls on a pool table or planets orbiting the sun, the motions of the electrons seemed to escape precise mental pictures or textbook illustrations. Such illustrations were now regarded as only symbolic or approximate to the actual motion. In the language of the day, the actual motions of the electrons were simply not *anschaulich*— literally, not look-at-able. Again the physicist had to find a prescription—an interpretation—of the symbols and equations of the quantum world that would enable him or her to proceed with the research program of science: to explain the laboratory data gathered from studying the properties of atoms in the everyday world on the basis of the strange behavior and concepts and quantum laws governing the interior of the atomic world.

These were highly philosophical issues about the very nature of science itself and of the work of scientists. What Heisenberg was doing in the uncertainty paper was not the derivation of a new mathematical theory. Instead, he was setting forth an argument in support of the matrix mechanics view of the limitations on the measurements and meanings of fundamental atomic concepts, such as the position and momentum of an electron. The arguments, and their origins, that he, and later Bohr, presented on the basis of their physics have been researched and debated ever since, within and far beyond the confines of physics and its history.[3]

Previously one could always easily describe the motion of an electron, as one would a billiard ball on a table, by noting its position and velocity at any given instant. Inspired by Pauli, Einstein's earlier work, and his own struggles over the past few years, Heisenberg now made a profound philosophical claim in his uncertainty paper, and in the essay "Fundamental Principles" explaining the argument for a broader audience. He stated that such basic concepts as position and velocity of an atomic particle are meaningful only when they are referred to, or defined by, the actual experimental procedures, or operations, used to measure them. A property of electrons or atoms that cannot be observed or measured has no place in the theory. In other words, a physicist cannot know any more than what he or she actually measures; and the physicist can know it only to the extent that the measurements allow. It is here that the uncertainty, or indeterminacy, principle makes its bold appearance.

Since making a measurement is the condition for the physical meaning of a concept, Heisenberg considered the following imaginary "thought experiment." Let's say a researcher wants to measure the exact position and the precise momentum (mass times velocity) of an electron on a bench in a laboratory. In order to see the location of the very tiny electron, the researcher would need a microscope of very high resolution. Since a microscope works by reflecting light off the object into the objective magnifying lens of the microscope, light of extremely short wavelength, such as gamma rays, would be needed to illuminate the electron. But, according to quantum theory, the shorter the wavelength, the greater the energy of the light quanta (or the pressure of the light wave) hitting the electron, and thus the greater the recoil momentum of the electron. The precision of the measurement of the position is thus fairly good, but at the expense of the measurement of momentum. On the other hand, one can make the momentum measurement more precise by using longer wavelengths of light, thus reducing the recoil of the electron, but then the position measurement becomes more imprecise—that is, more inexact or uncertain. Because of this, Heisenberg noted, there seems to be a reciprocal relationship between the imprecisions—the uncertainties—of the results of these two measurements. He showed in his uncertainty paper that the equations of quantum mechanics strongly suggest this interpretation, although it was not a rigorous proof. That proof came only later, in a treatise published by Hermann Weyl in 1928, in which Weyl thanked Pauli for suggesting to him the derivation.[4]

Nevertheless, Heisenberg argued that the only conclusion consistent with the quantum mathematics and the experimental procedures is that one cannot simultaneously measure with absolute precision both the position and the momentum of an electron at any given instant. As he put it:

The more precisely we determine the position, the more imprecise is the determination of momentum in this instant, and vice versa.

The reciprocal relationship between the imprecisions, or uncertainties, of these two measurements also holds for other pairs of variables, such as energy and time. These relations have become known as the uncertainty relations, and this behavior has become known as Heisenberg's uncertainty principle.[5]

The implications of this simple statement about the limitations of actual measurements of atomic properties were profound and far-reaching, and still are. Heisenberg proudly enunciated one especially significant implication in his original paper and in his nontechnical summary. Again, in his words:

> *The above-mentioned boundary of precision, as determined by nature, has the impor-*
> *tant consequence that in a certain sense the law of causality becomes invalid.*[6]

This was a revolutionary assertion that went far beyond the technical minutiae of quantum equations. The law of causality, so fundamental to physical science, requires the association of a cause with every effect, things don't just happen without a cause. Moreover, the cause must precede the effect. This seemingly commonsense law had been challenged philosophically in the past, but it became an underpinning of physical science following Immanuel Kant's defense of the notion early in the nineteenth century. The law of causality is an implicit or explicit assumption in practically every form of rational research.[7]

Pierre-Simon Laplace is credited with one of the simplest and most widely assumed definitions of causality in mechanics, the physics of motion: if one knows the exact position and momentum of a particle at a given instant, along with all of the forces acting on the particle, then one can use Newton's laws of motion to calculate the particle's exact position and momentum at any given future instant. For example, the motion of the particle, say an electron moving through an electric force, is fully determined by the mechanical laws of motion for all future time. This is because all of the "effects"—the positions and velocities of the electron at all future moments—are linked directly and uniquely to their causes—the electron's position and momentum at the start, together with any forces acting on the electron. The uncertainty principle denies this, Heisenberg declared:

> *In the strict formulation of the causal law—if we know the present, we can calculate*
> *the future—it is not the conclusion that is wrong but the premise.*[8]

Because of the reciprocal uncertainties, or imprecisions, in the measurements of position and momentum, we cannot know both the present position and the present momentum of an electron with absolute precision. And because of this reciprocal imprecision, we cannot determine with absolute precision exactly where the electron will be at any future moment. The best we can do is to calculate a range of possibilities for the position and the momentum of the electron at a future time, one of which will be the result of an observation made by an experimenter at that future time. Because of the uncertainty of the beginning condition of the electron, the future motion of the individual electron cannot be determined exactly. As a result, declared Heisenberg, the laws and predictions of quantum mechanics "are in general only of a statistical type." This has become known as Heisenberg's principle of indeterminacy, although he did not call it this at the time. Initially, people spoke only of this principle, while referring to the uncertainty relations. Today, we often speak of the uncertainty principle, while referring to the indeterminacy of quantum events.

Combined with Born's probability interpretation of the wave function, the indeterminacy of quantum events means that one can never determine exactly the outcome of a single observation of any atomic process; the scientist can predict only the *probability* of each outcome among a wide range of possibilities. However, for a large number of observations, the probabilities do lead to precise statistical predictions, which the experimental results will display. For example, if an electron has a one in three chance of being at a certain location at a certain time, then out of billions of identically prepared electrons on average one third will be observed in that position at that time.

Heisenberg had arrived at the astounding assertions of the uncertainty/indeterminacy principle and its implications during the months following his move to Copenhagen as Bohr's assistant in May 1926. Also during those months, Erwin Schrödinger had completed his formulation of wave mechanics, had published the mathematical equivalence of the differing wave and matrix formulations of quantum mechanics, and had called in print and in person (in Copenhagen and Munich) for a return to continuity and the direct visualization of moving electrons. Heisenberg, for his part, employed less visualizable electron spin and matrix mechanics to account for atomic spectra and the behavior of systems containing many electrons, and he used the behavior of minute oscillation (fluctuation) phenomena to argue the existence of fundamental discontinuities in nature.[9] In the wake of the Bohr-Schrödinger encounter in Copenhagen in October 1926, the search for a proper physical interpretation of the quantum formalism became paramount in Bohr's institute. The hoped-for interpretation would, in contradiction to Schrödinger, incorporate discontinuity and quantum jumps into a prescription for linking the worlds of the atom and the laboratory.

Two wider issues lent a sense of urgency to this task during the fall and winter of 1926–1927. They were the overwhelming popularity of Schrödinger's wave mechanics and a second flurry of job openings in central Europe. Following Schrödinger's proof of the equivalence of wave and matrix mechanics, Heisenberg did not fail to notice the sudden drop in the number of publications employing matrix methods, along with the simultaneous increase in publications using the wave-mechanics formalism.[10] He complained to Pauli nearly a year later, in May 1927, of a particularly threatening feature of such defections: some physicists had the audacity to rework old matrix mechanics papers, such as the Heisenberg-Jordan Zeeman work, in wave-mechanical terms. "I am always annoyed when even now physicists are writing a 'conjugate' wave paper to every matrix paper. I think that they should better learn both [wave and matrix mechanics]."[11]

By November 1926, the full extent of the new "appointment chatter" had also become evident. Although Heisenberg strove to hold his research above such mundane matters as a job, he could not ignore family pressures, or his own ambition. Of course, there was no question that he would eventually land a prestigious job, but he wanted one *now* and in Germany, where many of the existing chairs for theoretical physics were suddenly up for grabs.

Because of his stature and his close familiarity with nearly every quantum theorist, Sommerfeld played a pivotal role in landing the new generation of quantum theorists in university professorships. Soon after the death of Leipzig theorist Theodor Des Coudres in October 1926, the head of the Physics Institute and holder of the experimental physics chair, Otto Wiener, asked Sommerfeld to suggest a successor.[12] Sommerfeld replied that Born and Schrödinger were probably not to be had, that Heisenberg and Pauli were still the best prospects, and that several other physicists would be deserving.[13] On December 18, 1926, the Leipzig Philosophical Faculty unanimously accepted the search committee's recommendations for Des Coudres's chair: Peter Debye, Erwin Schrödinger, and Max Born. But within a month Wiener had died, and the search expanded for replacements for both Des Coudres's theoretical physics chair and Wiener's experimental physics chair. Wiener's favorite, Debye, more experimentalist than theorist, became the leading candidate for Wiener's chair. He accepted well before the July 27, 1927, faculty meeting. At that meeting the search committee announced a new list of candidates for Des Coudres's chair, as approved by Debye, the new head of Leipzig physics: Heisenberg, Wentzel, and Pauli, in that order.[14] Two months earlier, Heisenberg, still in Copenhagen, had published his uncertainty-principle paper and the nontechnical exposition.

As the juggling of jobs played itself out, quantum physicists pondered their sudden surfeit of theories: two very different yet equivalent mathematical schemes for quantum mechanics—wave mechanics and matrix mechanics—and two very different interpretations of the fundamental properties of nature and the fundamental meanings of the symbols used in the two theories. Unlike most similar situations, in which scientists must make do without a theory, they now had one too many. Everyone recognized that a "fusion" of wave and matrix mechanics, as Born called it, was now required, along with a fusion of their interpretations.[15] The search culminated for most physicists within a matter of months in a newly unified mathematical formalism produced by Dirac and Jordan, and called transformation theory.[16] Within a year of the new mathematics, the Copenhagen interpretation, with its uncertainty principle, was in hand.

Heisenberg's intellectual route to the uncertainty principle lay through the work of his closest colleagues—Born, Jordan, Pauli, Dirac, and Bohr. As each struggled during the last months of 1926 with finding a suitable interpretation of the newly unified quantum equations, each informed Schrödinger of his opposition to Schrödinger's assertions that a theory involving continuous waves alone would suffice to account for atomic phenomena. To Heisenberg and his colleagues, the particle side of the dualism of waves and particles seemed paramount. Pauli and Jordan even tried to throw the weight of majority opinion against Schrödinger. Pauli wrote, "But I am convinced now as before (together with many other physicists) that the quantum phenomena cannot be encompassed with the conceptual resources of the continuum physics alone."[17] A continuum theory could not account for phenomena that seemed to require jumping, rotating, orbiting balls of charged matter—electrons.

Following his work on electron spin and his encounters with Schrödinger, Heisenberg himself had grown even more committed to proving the existence of particles, jumps, and discontinuities. In this context, a prime impetus toward uncertainty came, as usual, from Pauli. Born had earlier interpreted Schrödinger's wave function, not as a matter wave as Schrödinger would have it, but as a probability wave. The square of the wave function represented the probability for a transition to occur to a certain quantum state in a collision between two electrons.

Writing to Heisenberg from Hamburg on October 19, 1926, Pauli reinterpreted Born's interpretation as representing not only the probability for the results of a collision of electrons but also the probability of finding an electron at any given time in a given quantum state within an atom. Looking again at Born's work in which two electrons, treated as waves of matter, collide with one another, Pauli noted that when the electrons are far away from each other, the precise position and momentum of each electron may be chosen without difficulty. But when the electrons approach each other in a collision or in an atom, their quantum behavior comes into play. In so doing, they manifest a "dark point": if the momenta are measured precisely then the positions take on a range of values, and vice versa. "Thus, one cannot speak of a definite 'path' of the particle," nor, wrote Pauli, "may one inquire simultaneously about momentum and position."[18]

Heisenberg was "*very* enthusiastic" about Pauli's letter, which "continually made the rounds" in Copenhagen, and especially about Pauli's "dark point," to which Heisenberg often returned over the following months.[19] His enthusiasm culminated in a 14-page letter to Pauli on February 23, 1927, in which Heisenberg used the new mathematics of Dirac and Jordan to derive the reciprocal relationship between the uncertainties, or imprecisions, of measurements of position and momentum, followed by a discussion of nearly all the features of the paper that he would submit a month later—his paper on the uncertainty principle.[20]

As Schrödinger had done in his papers the previous year, Heisenberg, eager to undercut Schrödinger's alternative views, included in his uncertainty paper a number of bold and profound assertions about physics and scientific inquiry. In particular, Heisenberg declared it "fruitless and senseless" to inquire any further about the actual motions of electrons beyond the reach of our observations. The momentum and path of a particle really have no meaning at all. He told Pauli,

> *The solution can now, I believe, be expressed pregnantly by the statement: the path only comes into existence through this, that we observe it.*"[21]

In essence, we can never know nature as it really is; we can know it only as it appears to us through our experimental data. As Bohr would soon emphasize, the experimenter, by choosing the data to be collected, in effect becomes part of the experiment.

This was philosophical empiricism of the most radical sort, even more radical than that in Heisenberg's original matrix mechanics paper or in his remarks about statistical laws. What he was now saying is that not only is it fruitless and senseless to inquire about reality beyond our observations, but as far as he was concerned, there really is no more to reality than what we can observe of it. Moreover he declared in the concluding paragraph of his paper: "Because all experiments are subject to the laws of quantum mechanics, and thereby to the [uncertainty relations], the invalidity of the causal law is definitively established through quantum mechanics."[22] How did Heisenberg arrive at such radical ideas?

One answer may be the influence of Pascual Jordan. In February 1927, eighteen days before Heisenberg came to his radical conclusion about the path of a particle, Jordan had published his Göttingen habilitation lecture, titled "Causality and Statistics in Modern Physics."[23] In it, Jordan asserted that causality is not a given, universal law, but an experimentally defined concept, and this seems to be the source of Heisenberg's interest in causality and statistics in his uncertainty paper. In addition, Jordan's paper helped to induce Heisenberg's shift from abstract definitions of concepts such as position and momentum to data-based definitions and assertions, from which emerged his notions about reality and causality.[24]

But Heisenberg himself, while citing Jordan in his paper at this point, years later recalled only Einstein's influence.[25] Before taking up his post in Copenhagen as Bohr's assistant in May 1926, Heisenberg had lectured on matrix mechanics before the Berlin physics colloquium. Following the lecture and a long discussion with the many skeptics in the audience, an intrigued (though skeptical) Einstein had invited young Heisenberg to accompany him on the walk home to his apartment. Heisenberg had accepted gladly, and during the half-hour walk along the tree-lined streets of Berlin to his apartment on Haberlandstrasse, Einstein had gotten to know the brilliant young man a little better. Heisenberg had first met the great physicist two years earlier in Göttingen. But it had been only a brief encounter and had concentrated on Einstein's objections to the Bohr-Kramers-Slater theory. This time Heisenberg was a principal author of a revolutionary, yet baffling, new mechanics, and the two had exchanged several letters on the subject during the previous months.[26] In Berlin, Einstein, then 47 years old, had wanted first to know more about Heisenberg's background, education, and research; Heisenberg, half Einstein's age, had wanted Einstein's opinion on whether or not he should refuse the Leipzig job offer in favor of working with Bohr. Einstein had urged the young man to work with Bohr.

When the two men finally had arrived at Einstein's elegantly furnished apartment —with its heavy oak furniture, glass-enclosed breakfront, overstuffed leather sofas, and built-in bookcases containing the complete works of Goethe, Schiller, and Humboldt—the conversation turned to the issue at hand: quantum mechanics.[27] In a sense the conversation reflected Einstein's own role in quantum physics. From the very beginning of the quantum revolution at the turn of the century, Einstein had been a principal player, but never a principal contributor, to an encompassing quantum theory. His work, more than that of any other physicist, had indicated the very existence of quanta of energy and the necessity of radical revisions in physics to encompass them, but not the solutions to the puzzle.

When presented with the radical notions of matrix mechanics, Einstein had not been supportive. He had preferred instead the approach of wave mechanics. He had shared with Schrödinger the conviction that the quantum had to be understood in traditional terms, not merely accepted or assumed. Thus, while Heisenberg, Bohr, and others struggled to obtain a new atomic theory that would somehow encompass the nonclassical notions of quanta, jumps, and discontinuities in a consistent fashion, Einstein would not be satisfied until all appearances of the quantum could be properly explained away, more or less, on existing principles. Schrödinger's approach, based on the continuous wave nature of matter, which Einstein had encouraged, had coincided with his own aims and seemed to hold the promise of an understanding of quantum phenomena without relying on quanta, discontinuity, problematic particles, or nonvisualizable motions. Just two days before Heisenberg's visit, Einstein had written Schrödinger that he was convinced that Schrödinger's work represented "a decisive step forward . . . just as I am convinced that the Heisenberg-Born approach is off the track."[28]

Heisenberg's much-later recollection of the meeting focused on Einstein's objections to the observation-based elements of the Heisenberg-Born approach. Heisenberg had built his multiplication rule on equations that, he argued, involved only quantities that could be observed in the laboratory—primarily the frequencies and the intensities of the emitted radiation—and, in his enthusiasm, he had elevated this approach to a prescription for the formulation of any cogent quantum theory, including, later, the uncertainty principle.

"But you don't seriously believe," Einstein had objected, "that none but observable magnitudes must go into a physical theory?" In his defense, Heisenberg attempted to raise Einstein's formulation of the special theory of relativity, in which Einstein had excluded such notions as absolute space and time because they could not be observed. Muttering that a "good trick should not be tried twice," Heisenberg's recollected Einstein had called all such empirical reasoning nonsense. "In reality the very opposite happens," he had declared. "It is the theory that decides what we can observe."[29]

Confronted ten months later with the unified formalism of Dirac and Jordan, but without a satisfactory interpretation of its symbols, Heisenberg recalled suddenly remembering Einstein's statement just before writing his uncertainty paper, thus probably just after reading Jordan's paper on causality. Definitions of fundamental concepts based only on observations and subject to quantum mechanics and the uncertainty principle quickly followed. The theory did indeed decide what could or could not be observed.

Heisenberg's paper nevertheless contained an error that even his inconsistent pragmatism could not surmount. He first argued that the reciprocal relation between the uncertainties in position and momentum, energy and time, arose from the Dirac-Jordan formalism.[30] Then he tried to show their consistency with various experimental examples, as well as with his "thought experiment" involving the gamma-ray microscope outlined earlier. Since theory and experiment agreed, the uncertainty relations provided, in Heisenberg's opinion, a satisfactory and sufficient interpretation of the quantum mechanics.

Bohr begged to differ. He had been discussing with his assistant and visitors the problems of interpretation and measurement ever since Schrödinger's unsettling visit to Copenhagen in October 1926. While Heisenberg grew ever more attached to particles and discontinuity, the judicious Bohr became ever more convinced, following the refutation of the BKS theory, of the need to accommodate both waves and particles equally in any interpretation of the quantum formalism. Heisenberg saw no reason for such generosity—particles would suffice! By mid-February 1927, Bohr had had enough of arguing and left for a skiing trip in Norway, without Heisenberg.[31]

In order to clarify his own thoughts, several days after Bohr's departure Heisenberg wrote his 14-page letter to Pauli to work up toward the uncertainty principle paper. He relied exclusively on particles, discontinuity, and the Dirac-Jordan theory. Bohr returned around the time that Heisenberg's final manuscript espousing the uncertainty principle was received by the *Zeitschrift für Physik* (March 23, 1927) —but without the customary prior approval of the institute's director. Two days later, Bohr sent out a cry for help to Pauli, offering to pay his way to Copenhagen.[32] He and Heisenberg were locked in a deepening dispute over the origins of uncertainty.

Bohr insisted, and later demonstrated, that Heisenberg's microscope experiment was, as Heisenberg put it, "not quite right." Earlier, this instrument had nearly cost him his doctorate, when he had been unable to derive its resolving power during his final oral exams. Now it threatened to cost him his physics. Bohr informed the author of what became the uncertainty principle that the principle arose, not from the recoil of the electron under bombardment by a light quantum, but from the scattering of the waves making up the light quantum into the aperture of the microscope's objective lens—an essential limitation on the resolving power of any microscope. Not only was the finite aperture of this lens essential to the analysis, Bohr argued, but most important, the analysis required a wave interpretation of the scattered light quantum.[33]

Heisenberg at first refused to accept Bohr's argument, and especially his suggestion that Heisenberg should withdraw his paper, which was now in press. Not only did the paper argue nearly every aspect of nature and physical inquiry to which Heisenberg had become committed during the past year, but the uncertainty principle seemed to him a fully consistent consequence of the unified quantum formalism, which contained those commitments in its very foundations. The basis, the innovations, the radically sweeping claims, the demonstration of Heisenberg's abilities that this paper contained could not be withheld because of a mere error in a supporting thought experiment—or because of the mere neglect of one side of the wave-particle duality.

Bohr strongly insisted, Heisenberg stubbornly refused, and Pauli greatly regretted that he could not come to Copenhagen as mediator. Heisenberg soon learned what had driven Schrödinger to his sickbed. As the controversy raged through the offices and halls of Bohr's institute, Swedish theorist Oskar Klein, then on a visiting fellowship, entered the fray and, as Heisenberg reported it, supported Bohr out of friendship

Heisenberg and Bohr in Copenhagen, 1930s.
(COURTESY ESVA, AIP NIELS BOHR LIBRARY)

to Bohr and out of opposition to his assistant, who had apparently offended Klein with overweening criticism of his work. The controversy soon degenerated into what Heisenberg called "gross personal misunderstandings."[34] In the heat of the battle, Heisenberg, who at one point burst into tears, managed to wound Bohr with his sharp tongue. Obviously much was at stake for Heisenberg—his past commitments, his alternative insights, and, especially, his future career in Germany.

Apparently the prime cause for tears was neither the enormous pressure exerted by Bohr and Klein, nor the microscope error, but frustration at Bohr's regarding Heisenberg's argument as "too special a case of the general rule" of what Bohr was now calling "complementarity," which, for Bohr, encompassed and superseded Heisenberg's uncertainty principle.[35]

Complementarity, this additional component of the Copenhagen interpretation, entailed Bohr's preference for both waves *and* particles. Bohr argued that wave and particle notions appear simultaneously in any given experiment. In fact, objects in nature encompass both wave and particle properties until they are observed by experimenters, whereupon they suddenly become one or the other. Because of this, both wave and particle notions, although mutually exclusive and incompatible, are nevertheless both essential for a complete account of the behavior of nature. They are, in fact, *complementary*. The experimenter's necessary choice of either the wave or the particle picture for the experiment introduces a disturbance into the experiment that reduces the wealth of alternatives that nature offered before the observation, and this is manifested in the limitations imposed by the uncertainty principle. The disturbance enters through the experiment, but in a way that differed from Heisenberg's notions: not in the act of fixing a variable in a measurement, but in the necessary act of choosing one side or the other of the wave-particle duality in the design and performance of the measurement. The experimenter becomes, in effect, a part of the experiment.

In Bohr's view, complementarity provided the general framework for the existence of uncertainty, which in turn provided the grounding for the unified formalism of quantum mechanics. Together, uncertainty and complementarity, along with the probability interpretation of the wave function, represented the interpretive culmination of quantum mechanics, the basis for comprehending the dualities of waves and particles, wave functions and matrices, atoms and laboratories, continuity and discontinuity, causal and acausal descriptions, researchers and their experiments—the foundations of the Copenhagen interpretation of quantum mechanics. Heisenberg's paper, resting solely on particles and discontinuity, was, Bohr suggested, both too narrow and too premature.

The Copenhagen duality now flourishing between Heisenberg and Bohr remained unresolved through Easter. Not until May 16, 1927, after further clashes, was Heisenberg willing to capitulate in writing. The scattering of light waves into the aperture of the microscope lens was, he conceded, the basis of the uncertainty principle in that thought experiment. At Klein's urging, Heisenberg agreed to add a postscript to the galleys of his otherwise unaltered paper.

The postscript concerned several "essential points that I had overlooked" and that, he wrote, Bohr had since brought to his attention.[36] Foremost among them was the realization that the "uncertainty in the observation" arises not exclusively from discontinuous particles or continuous waves but also from the attempt to encompass simultaneously phenomena that arise from both wave and particle origins.

Bohr, already at work on a paper under the rubric "there are waves and corpuscles,"[37] attempted to clarify the situation even further at the Lake Como celebration of the hundredth birthday of Alessandro Volta, the inventor of the voltaic cell, the battery, in September 1927. Nestled in a villa overlooking the deep-blue mountain lake in northern Italy, the celebrants listened as the soft-spoken Bohr raised the existence of discontinuity in nature to the status of a quantum postulate, an assertion that no doubt pleased one member of his audience, Heisenberg.

In the course of his somewhat confusing paper, perhaps a reflection of his own imprecision at that point regarding his ideas, Bohr offered the perfect compromise between the opposite positions espoused so forcefully by Heisenberg and Schrödinger.[38] Wave physics and particle physics were not, as hitherto supposed and debated, antithetical, either wholly right or wholly wrong; they were complementary—mutually exclusive yet jointly essential. Rather than explaining away the wave-particle duality in favor of one or the other extreme, Bohr incorporated it into the very interpretation of quantum mechanics. Observation and measurement by an experimenter force a choice between one or the other set of complementary physical concepts and thus, by reducing the choices to one or the other side of the wave-particle dualism, disturb the system. Because of this disturbance, precise, simultaneous observations of complementary pictures or measurements of complementary variables are limited according to the precise demands contained in Heisenberg's uncertainty principle. Heisenberg rose from his seat in the audience that day to confer his public approval on Bohr's interpretation of their physics.[39] The Copenhagen interpretation was born.

Heisenberg's battle with Bohr had reached its climax and diffusion in the spring of 1927 just as several faculty committees were deciding on candidates for vacant teaching chairs. Heisenberg's academic ambitions seemed close to realization. He had produced another good paper, and faculties were meeting to award appointments. The changing situation on the job front is reflected in part in his changing relationship to Bohr after the publication of his uncertainty paper—but only in part. Jobs aside, Heisenberg always looked up to Bohr as a father figure and could not let a disagreement over waves and particles come between them for long.

Probably during his Easter vacation, but certainly by June 1927, Heisenberg learned from Sommerfeld of Debye's intention to have him called to Leipzig as successor to Des Coudres. (In the German tradition, a candidate, even one who had not applied for the job, was not appointed but "called" to the position; the professing of knowledge was a calling, not a mere job.) Heisenberg received the official call at the end of July. By then he had received two other calls to full professorships. "All good things come in threes," he gloated.[40] In June 1927, a search committee in Halle had recommended Heisenberg, Gregor Wentzel, or Friedrich Hund as a replacement for Karl Schmidt. A month later, the Saxon Education Ministry, which had officially issued the call, requested Sommerfeld's confidential judgment of the three. But Heisenberg was already carefully weighing the alternatives. The call to Halle, coming at a time when he felt he had not yet made up for his injurious remarks to Bohr, caused Heisenberg considerable concern. In a personal letter to Bohr written in a very shaky hand, Heisenberg informed his mentor in apologetic terms of the Halle appointment. Diplomatically, Heisenberg asked for advice: should he stay in Copenhagen or "seek my own future"?[41]

But Sommerfeld offered another alternative. Both Sommerfeld and Schrödinger were on the list to succeed Planck in Berlin. In gratitude for his turning down the Berlin appointment, the Bavarian Education Ministry increased Sommerfeld's salary and his institute budget significantly. They also promised to establish a long-requested associate professorship in Sommerfeld's institute.[42] Sommerfeld hoped that Heisenberg would fill that post. Having heard from his father that Heisenberg might want to remain in Copenhagen a while longer, Sommerfeld wrote to Heisenberg in June offering to reserve the new position for him. The post itself was actually secondary to its most important features: it was in Heisenberg's beloved Munich, and it would give Heisenberg "simultaneously the right after further years to become my successor in the full professorship." The implication was clear. With his sixty-fifth birthday approaching in 1934, Sommerfeld had decided on Heisenberg as his heir apparent. In the event that Heisenberg decided to leave Copenhagen, Sommerfeld also told him unofficially of Debye's intention to recommend Heisenberg rather than Schrödinger to succeed Des Coudres in Leipzig. Sommerfeld counseled acceptance: "I thus advise you with heavy heart, but at Debye's wish, to accept the Leipzig professorship, which is indeed especially attractive because of Debye and Wentzel and colleagues."[43]

Having turned down one call already to a German chair, Heisenberg could not afford to reject another. He decided to accept either the Halle or the Leipzig offer. Bohr reluctantly conferred his blessing. Heisenberg's only real preference was for Munich. If he could not be called immediately to a full professorship there, he told his parents, it did not matter to him where in Germany he resided until he could move to Munich. He duly thanked Sommerfeld for his kind offer, which he wanted to reconsider later.[44] Expressed less diplomatically, Heisenberg had accepted the mantle as Sommerfeld's heir apparent, but he would occupy a chair elsewhere until Sommerfeld retired.

Saxony's official call to Heisenberg to occupy the Leipzig chair occasioned another shaky letter to Bohr.[45] Heisenberg wanted to reassure Bohr that he was not courting callers because of their earlier differences. Two more offers followed: one from Augustus Trowbridge of the IEB, on behalf of American universities eager for a German quantum theoretician, and another from the esteemed Zurich Polytechnic as Debye's successor. The Polytechnic experimentalists wanted a full-time quantum theoretician as Debye's replacement in order to keep the institute abreast of "the enormous development of the-oretical physics."[46] Neither of these offers had a chance with Heisenberg, although he did dangle the Polytechnic offer in his negotiations with the Saxon ministry. Negotiations finally concluded in late October 1927 with Heisenberg's acceptance of the Leipzig professorship for theoretical physics, retroactive to October 1. The position came with a lucrative salary, an opportunity to work with Debye, and, equally impor-tant, a place in Germany—none of which the Swiss could match.

A week later, having recently returned from Lake Como, Heisenberg headed for Brussels, where he and Born were to lecture on the latest developments in quantum mechanics to the prestigious Solvay Congress of selected quantum physicists. There Heisenberg also tried to smooth over any lingering personal and scientific misunder-standings with Bohr. The undertaking did not quite succeed, however, apparently in part because Heisenberg somehow overslept on the morning of Bohr's departure.[47]

On delivery of his inaugural lecture before the Leipzig faculty on February 1, 1928, Heisenberg was celebrated as Germany's youngest full professor.[48] He had just turned 26. With Heisenberg and Debye finally settled in Leipzig and Sommerfeld firmly established in Munich, the rest of the central European appointments quickly fell into place. Schrödinger succeeded Planck in Berlin; Wentzel followed Schrödinger at the University of Zurich; Hund took Wentzel's place in Leipzig; Pauli, second on the Zurich Polytechnic list, succeeded Debye; Jordan followed Pauli in Hamburg; and Bohr's ally Oskar Klein took over Heisenberg's post as Copenhagen lecturer and assis-tant. Within a year of the Como and Solvay meetings, a new generation of quantum theorists had come of age. With their new status as tenured professors, institute direc-tors, and established leaders of their field, Professors Pauli, Heisenberg, and Bohr at last began to address each other with the familiar German pronoun *du*.

CHAPTER 13

REACHING THE TOP

HEISENBERG'S APPOINTMENT TO THE LEIPZIG CHAIR FOR THEORETICAL PHYSICS IN October 1927 and his attendance at the prestigious Solvay Congress in Brussels in the same month represented milestones in the completion and acceptance of quantum mechanics. During the six-day conference—the fifth of occasional meetings of leading physicists to consider fundamental developments—Heisenberg and other former matrix mechanicians rallied behind Bohr and the Copenhagen interpretation of quantum mechanics, refuting objections raised by Einstein and others more closely associated with Schrödinger's alternative of waves and continuities. The debate was not over the physics equations but over the interpretation of the symbols and the results.

The Copenhagen faction managed to rebuff its opponents, but it could not convince them for long. Despite this failure, acceptance and exploitation of the Copenhagen interpretation spread rapidly across the international physics community. This was facilitated by two factors: first, the almost missionary zeal displayed by proponents carrying the "Copenhagen spirit," as Heisenberg called it, around the globe; and, second, the establishment of the new science in the seats of academic power, the university teaching chairs.[1] The settling of the European teaching appointments redefined the aim and structure of the physics institute to accommodate the new science. To the few already established centers of quantum physics were added new centers that employed the science to train future generations wholly within the new spirit. Heisenberg's Leipzig institute became a prime destination for many of the new generation of students.

Heisenberg had closed ranks behind Bohr's joining of uncertainty with complementarity at the Solvay Congress and earlier at Lake Como. The reconciliation had rescued the doctrines of the matrix and Copenhagen adherents from the challenges of their opponents—at least for the moment—and for that Heisenberg was clearly grateful. But he had other motives, as well, for lauding Bohr. Heisenberg had always admired and respected Bohr, and he wished to make up for his earlier adolescent behavior in defending uncertainty. During the five years following the Como and Solvay meetings, the personal and professional relationships between the two men, as judged by the extraordinary number of mutual visits, personal letters, and joint outings, grew as close as they ever would.

Beyond the healing of old wounds lay a professional motive. Just a month before the Como conference, the Saxon Education Ministry had informed Heisenberg of his call to Leipzig, and within weeks the new professor was already lecturing.[2] Heisenberg's need to underscore his unique contributions to quantum mechanics suddenly evaporated. It was replaced by a new ambition that now required Bohr's help to realize. Heisenberg was determined that he would build a permanent, first-class research program in Leipzig on the basis of a new, solid, and successful physics. Bohr's views not only buttressed Heisenberg's debatable arguments for uncertainty, they also provided a rallying point for Bohr's followers. Bohr's disciples were, like Heisenberg, eager for a completed physics that they could propound from their chairs and podiums and exploit in their papers. Even the ever critical Pauli declared himself "very much in agreement" with Bohr's Como paper, "both with the overall tendency and with most of the details."[3] After Como, Heisenberg and other Copenhagenites no longer gave their allegiance to individual programs and discoveries, matrix mechanics or uncertainty, but to the "Copenhagen spirit"—a spirit epitomized, they believed, by Bohr's Como presentations, personified by Bohr himself, and propagated by his ordained ministers, namely themselves.

Max Born promptly announced the new consensus in his reply from the audience to Bohr's Como lecture. "Herr Professor Bohr has presented the views that we have formed about the basic concepts of quantum theory in such an appropriate manner that there is nothing left for me to do but to add a few remarks."[4] A month later Born and Heisenberg, his former assistant, surveyed those basic concepts for the fifth Solvay Congress. Should any doubt remain, as it did for Schrödinger, discontinuity topped the list of basic concepts—not continuous waves but discrete, discontinuous particles, energy quanta, and quantum jumps were fundamental to understanding nature. Their opening sentence: "Quantum mechanics is founded on this idea, that atomic physics is distinguished essentially from classical physics by the existence of discontinuities."[5]

After first hailing Bohr for grounding uncertainty in complementarity, Born and Heisenberg in their joint paper declared the completeness and sufficiency of quantum mechanics in even bolder and more provocative terms than Heisenberg had used in his uncertainty paper. "We regard quantum mechanics as a complete theory for which the fundamental physical and mathematical hypotheses are no longer susceptible of modification." Nor, echoing Heisenberg, would future developments ever alter any of the fundamental features of this theory. Its implications for indeterminacy in particular were fixed forever: "Our fundamental hypothesis of essential indeterminism is in accord with experiment. The subsequent development of the theory of radiation will change nothing in this state of affairs."[6] Like it or not, quantum mechanics and the Copenhagen interpretation of it were, in their view, now etched in stone. Some historians and philosophers looking back on this episode argue that this was by no means the case, and that the triumph of the Copenhagen interpretation was by no means inevitable. Only circumstance and promotion by its supporters enabled this interpretation to achieve dominance for decades thereafter.[7]

The Copenhagen circle could hardly have been surprised to learn of objections raised by those outside the "community of true believers," as Pauli once called the

Solvay Congress of 1927. FRONT ROW: *Einstein and Marie Curie, Planck to her right.* SECOND ROW, RIGHT TO LEFT: *Bohr, Born, de Broglie, A. H. Compton, Dirac, Kramers.* THIRD ROW, RIGHT TO LEFT: *Heisenberg (3rd),Pauli (4th), Schrödinger (6th).*

(COURTESY ESVA, AIP NIELS BOHR LIBRARY)

Copenhagenites.[8] Some physicists, notably those who preferred Schrödinger's early interpretation, were still unconvinced of the Copenhagen doctrine and unwilling to condone what seemed to them a fundamental failure of physicists to do their job. Physics should enable one to comprehend nature so completely that what occurs in any given experiment can be precisely explained and predicted. Settling for anything short of that age-old goal was seen by some as resignation, or even despair.

With such a divergence of attitudes and opinions, the many formal and informal discussions during the Solvay Congress became a running debate between imbibers of the Copenhagen spirit and their more sober opponents. Ever since physicists Walther Nernst and H. A. Lorentz convinced the wealthy Belgian industrialist Ernest Solvay in 1910 that his generosity could best be used in support of occasional conferences at which a select group of leading physicists would present and discuss their latest research on a basic problem of physics, the Solvay Congress had come to mark both the culmination of a phase of research and the stimulus toward new directions.[9] Lorentz, the organizer of the fifth Solvay Congress, settled on the topic of electrons and photons. Among the 29 invited participants, nearly every major contributor to quantum physics—Heisenberg, Bohr, Schrödinger, Einstein, Born, Planck, de Broglie, and many others—was in attendance at this quantum summit. As they assembled in the sumptuous Hotel Métropole in downtown Brussels, their papers and discussions in the meeting hall, dining room, and hallways revolved around the interpretation now being promoted by the Copenhagen faction.

Both supporters and opponents of the Copenhagen doctrine had already reaffirmed their positions during the month since the Como conference. Their debate, which resumed with even greater force at the next Solvay meeting in 1930, ultimately pitted Einstein against Bohr.[10] The opposing arguments often revolved around a favorite device of the era: imaginary thought experiments involving various ideal arrangements of photons and electrons in boxes or passing through slits in infinite, impermeable screens. During the scheduled sessions and even more during the informal evening discussions, Einstein, a master of the thought experiment, contrived numerous such arrangements to demonstrate the inadequacy of quantum mechanics as a physical theory. By the following morning, Bohr, the master of Copenhagen dialectic, managed with the help of his supporters to refute each one.

What particularly bothered Einstein was the statistical nature of Copenhagen's understanding of quantum mechanics. Because of the uncertainty principle and the disturbance introduced into the experiment by the observing physicist, the present and future results of any measurement of an individual system, such as a single electron, cannot be predicted with absolute precision—the theory is essentially indeterministic, or acausal. The Copenhagen doctrine maintained that only the probabilities for a range of possibilities, or the statistical number of systems with a given property out of a large number of systems, can be obtained from the quantum equations. Although Einstein himself had introduced probability into quantum physics as early as 1916, such a notion was, for him, intended as only temporary. Every individual event in nature is governed by precise physical laws, and any physics that permanently settles for only probabilistic or statistical predictions about the outcome of a given experiment was, in his view, incomplete. Einstein did concede that quantum mechanics and the Copenhagen interpretation together formed a closed, logically consistent, and complete theory of statistical events, but the theory was nevertheless incomplete when applied to individual events. Choosing to rely on probabilities rather than searching for an even more revolutionary theory that would provide precise results was intellectual resignation to Einstein. He had already expressed himself on this point in his now-famous theological response to Born's probabilistic interpretation of Schrödinger's wave function: "The quantum mechanics is very worthy of regard. But an inner voice tells me that this is not the true Jacob. The theory yields a lot, but it hardly brings us any closer to the secret of the Old One. In any case I am convinced that *He* does not throw dice."[11] Einstein wrote to Sommerfeld after returning from the Solvay meeting: "The quantum theory may be a correct theory of statistical laws, but it is an inadequate conception of individual elementary processes." In Pauli's view, Einstein had taken a "reactionary" position.[12]

The demonstrated self-consistency of the Copenhagen interpretation reinforced the Copenhagen penchant for sweeping statements. Convinced by the last day of the 1927 Solvay meeting that his opponents had been refuted, if not silenced, Heisenberg wrote home: "I am satisfied in every respect with the scientific result. Bohr's and my views have been generally accepted; at least serious objections are no longer being made, not even by Einstein and Schrödinger."[13] Nothing, they now believed, could stop the spread of their conveniently closed and unified quantum mechanics.

Having triumphed within their own profession, the Copenhagenites now moved outside it, attempting to take their doctrine to other fields. Yet when they announced the doctrine and even attempted to apply it outside their own discipline, they made claims for it that went far beyond its actual meaning and significance, encouraging others to stretch the interpretation even beyond the breaking point.[14] Bohr, for instance, used his contribution to the 1929 Planck issue of *Die Naturwissenschaften* (*The Sciences*) to inform philosophers of the new doctrine. He then proceeded, during the early 1930s, to apply complementarity to practically everything: biology, law, ethics, religion, even life itself. Some quantum physicists began exploiting the Copenhagen doctrine for vitalist, neo-romantic, even mystical philosophies. Their fundamentally antiscientific and antirationalist attempts continue to this day.

Yet while the philosophical implications of the Copenhagen doctrine were enormous, most philosophers later acknowledged complete surprise at what the physicists had wrought and were slow at first to respond. Like his mentor Bohr, the otherwise unphilosophical Heisenberg lost no time in waking them up, and he did so in the most direct way possible: he began lecturing to assembled audiences of philosophers on how his new doctrine now applied to *their* field.

Heisenberg's philosophical works require caution regarding what they reveal about their author. For physicist Heisenberg, philosophical issues were never of such moment as to merit the devotion he lavished on the issues of quantum formalism. Systematic philosophical positions always were of lesser importance to him. Some readers have not fully appreciated that Heisenberg's philosophical pronouncements were always tailored for public consumption and most were informed and motivated by his specific aims in addressing each particular audience. Only rarely did any of Heisenberg's philosophical writings fail to derive from one of his public addresses. Significantly, these writings began to appear only after the formulation of the Copenhagen interpretation and after his appointment to the Leipzig professorship.[15] The new and revolutionary physics provided a convenient basis from which to pursue broader horizons opened by his new position within his own profession.

The earliest of Heisenberg's prescriptive addresses, a lecture to Leipzig philosophers entitled "Epistemological Problems in modern Physics," served as a practice run for the most influential and widely read of his early philosophical pronouncements, a paper entitled "Causal Law and Quantum Mechanics." Heisenberg delivered the paper in 1930 to an early gathering of the influential Vienna Circle of "exact epistemologists," the philosophical school of logical positivism that has struggled to construct a logically coherent model of an empirically grounded theory. The gathering took place at the Königsberg meeting of the Society of German Scientists and Physicians, GDNA (epistemology being the study of how we know something). The paper was published in the circle's widely read journal, *Erkenntnis* (*Knowledge*). It was followed by similar talks delivered at the University of Vienna in 1930 and 1935, the

Saxon Academy of Sciences in 1933, the universities of Munich and Göttingen in 1933 and 1934, the 1934 meeting of the GDNA held in Hannover, and for readers of the *Berliner Tageblatt* (*Berlin Daily Paper*) in 1931.[16]

As theoretical physics came under increasing suspicion and disrepute following Hitler's ascension to power in Germany in 1933, Heisenberg's lectures became noticeably more defensive of the profession. Nevertheless, Heisenberg's philosophical positions had been set as early as his first unpublished lecture to his Leipzig colleagues. His uncertainty principle was a challenge to the notion of causality in atomic processes, and causality, or the lack of it, became his major public theme. His claims, like Bohr's for complementarity, went far beyond their narrow quantum-mechanical significance.[17]

Heisenberg and Bohr were not alone. Born had earlier called quantum mechanics a deterministic theory of probabilities. Quantum mechanics yields only probabilities—not exact predictions—for the outcomes of atomic events, but the probabilities themselves evolve in a precise, determined fashion in accordance with the Schrödinger equation.[18] Thus, acausal laws in the microscopic world did not imply for Born indeterminism in the macroscopic world. But by the time of Bohr's Lake Como address, Born's remarks during the discussion contained no hint of his earlier restraint: "It seems to me important to emphasize that the new quantum mechanics gives up determinism, which has dominated natural research until now."[19] Lay audiences throughout Germany, apparently critical of physics for its supposed subservience to "mechanistic materialism," would have welcomed such a statement. Moreover, a well-known thesis argued by Paul Forman asserts that quantum theorists were responding to public pressure in Weimar Germany by introducing and emphasizing the emergence of acausality and indeterminism in fundamental physics.[20] Certainly the people represented within Heisenberg's youth movement longed for such results; nor was it a coincidence that, within days of submitting his uncertainty paper, Heisenberg, eager for a job in Germany, immediately informed the German public that "the meaninglessness of the causal law is definitely proved."[21]

By 1928 Heisenberg was ready to take on nearly all of Kantian epistemology, an essential underpinning of classical physics. He told his first philosophical audience that the relativity and quantum theories challenged Kant's claims that space, time, and now causality are suitable bases for establishing a rational, objective science ("synthetic a priori judgments"). The challenge called, he declared, for the "very difficult task of rolling out the Kantian basic problem of epistemology once again and, so to speak, starting all over again. . . . But this is your task, not that of the scientist, who should only deliver material with which you can work further."[22] Physicists, if criticized by the public and non-physicists for their mechanistic determinism, could at least render a service by pointing the way for others to undercut philosophical determinism, while at the same time they could establish their crucial significance, and that of the Copenhagen doctrine, by feeding other professions material from their work that could not be easily rejected.

Heisenberg delivered the promised material to the assembled epistemologists in Königsberg and Vienna in 1930. Many in both audiences, such as Philipp Frank, Hans Reichenbach, and John von Neumann, were associated with the Vienna Circle. Relying more explicitly on Bohr's analysis of quantum experiments, Heisenberg shifted the philosophers' task from a reevaluation of Kant's entire philosophy to a reevaluation of the concept of causality—the assignment of a specific physical cause to each individual phenomenon. He argued that the impossibility of an objective world of perception was the "source of indeterminacy," and it rendered the traditional definition of causality not invalid but simply "empty of content." "I hope," he told the exact epistemologists, "to have made clear . . . to you that the situation created by atomic physics really does make a renewed discussion of the concept of causality necessary." A year later, he told newspaper readers, "Now it is the task of philosophy to come to terms with this new situation."[23]

Numerous epistemologists heeded Heisenberg's call for a rethinking of causality, a task that was already of concern to academics at that time. Not surprisingly, the philosophers drew heavily on the Copenhagen interpretation. One noted member of the Vienna Circle, Moritz Schlick, even sought extensive direction from Heisenberg in his analysis of quantum causality.[24] As philosophers finally took note of the implications of the Copenhagen doctrine, Heisenberg, his purpose achieved and his priorities set (he would pursue physics rather than philosophy), began to portray his own field as having little more to add to the discussion. "The researches of the physicist that have affected the field of philosophy came to a certain conclusion two years ago," he told readers of the *Berliner Tageblatt* in 1931.[25] Bohr's many discourses on complementarity through 1929, which manifested his own strenuous efforts to refine his originally vague definition of the principle, now served as the standard source.[26]

Heisenberg's subsequent philosophical views on the Copenhagen doctrine evolved primarily out of conversations with his Leipzig student and close personal friend, Carl Friedrich von Weizsäcker, and with Grete Hermann, a former philosophy student in Göttingen, who came to Leipzig in 1932 to study quantum acausality at its source. After Hermann's arrival, the conversations grew increasingly intense, yet Heisenberg's own participation grew equally diminished. As Weizsäcker recalls and as Heisenberg's correspondence indicates, Heisenberg did not concern himself with such matters beyond what he would need for his public addresses.[27]

With both physicists and philosophers now scrutinizing his claims, Heisenberg, while removing himself from discussion, decided to protect his position by checking the weakest point of his own argument, the microscope thought experiment. By 1930, quantum mechanics had advanced to include relativistic quantum fields, which altered the details of the argument. Heisenberg assigned his student, Weizsäcker, the task of thoroughly analyzing the microscope experiment—an analysis that he himself had twice flubbed in the past—using the Copenhagen doctrine and the most recent formalism of quantum field theory. Weizsäcker's analysis, completed in 1931, yielded no reason to question the irrefutability and universal sufficiency of the Copenhagen doctrine.[28]

With their confidence reinforced and the acceptance of their views spreading rapidly, Heisenberg, Bohr, and their followers had little reason to fear any challenge. But complacency turned into shock and confusion in the wake of a powerful and subtle attempt to refute their doctrine in 1935 in the pages of the leading American physics journal, *Physical Review*. As Léon Rosenfeld, Bohr's collaborator, put it: "This onslaught came down on us as a bolt from the blue. Its effect on Bohr was remarkable."[29]

The bolt from the blue bore the title "Can Quantum-Mechanical Description of Physical Reality Be Considered Complete?" Its authors were none other than their old adversary, Einstein, then at the Institute for Advanced Study in Princeton, and two collaborators: Boris Podolsky, a Russian physicist on a fellowship to the institute who apparently actually wrote the paper, and Nathan Rosen, a recent graduate of the Massachusetts Institute of Technology (MIT) who was now a research fellow with Einstein.[30] Despite the sudden shock, the now famous Einstein-Podolsky-Rosen (EPR) argument threatened less to undermine Copenhagen orthodoxy among its true believers than to hinder efforts to maintain a loyal following among the less devoted. "A certain danger . . . exists," Pauli wrote to Heisenberg, "of a confusion of public opinion—namely in America."[31] By 1935, the physics community in the United States was the largest in the world and it was producing first-rate work. American physicists constituted an audience that the Copenhagenites did not want to lose.

Published in May 1935, the EPR paper, consisting of a mere four pages, is to this day the subject of considerable scientific, philosophical, and historical study and debate.[32] Eight years after the Como and Solvay meetings of 1927, the paper and the responses to it revealed the persistence of deep-seated differences over fundamentals that the Copenhagenites, despite their pronouncements, could neither resolve nor silence. The new arguments of 1935 echo the earlier Bohr-Einstein debates, and in a sense the EPR paper represents the distillation of those debates. But again there was no universal resolution or agreement.

The authors of the EPR argument answered its title question with a definite "no." Despite Copenhagen's pronouncements about the completion and closing of quantum mechanics, the EPR authors argued that quantum mechanics must still be considered incomplete, because it predicts only probabilities for the outcomes of atomic experiments, not the exact results they expect from a theory that is complete in all respects. Their argument devolved not from formalism or description but from certain preconceptions and prior criteria fundamental to their position. The paper proceeds, as did Einstein's Solvay arguments, from a so-called realist point of view: the assumption of a knowable, precisely defined physical reality to the world, existing independently of the physicist's observations. (Bohr and Heisenberg had made the experimenter a part of that reality.) The opening sentence reads: "Any serious consideration of a physical theory must take into account the distinction between the objective reality, which is independent of any theory, and the physical concepts with which the theory operates." Acceptance of this distinction underlay the validity of their criterion for completeness two paragraphs later: "Every element of the physical reality must have a counterpart in the physical theory." This criterion required, however, a definition of an "element of the physical reality," hence the EPR reality criterion: "If, without in any way disturbing a system, we can predict with certainty (i.e., with probability equal to unity) the value of a physical quantity, then there exists an element of physical reality corresponding to this physical quantity." Bohr, on the contrary, maintained that the disturbance they exclude cannot be avoided. It is, in fact, so important that it limits our knowledge of reality. With their definitions and criteria, Einstein, Podolsky, and Rosen claimed, on the basis of a thought experiment involving the entanglement of two quantum particles, that the disturbance arising from the observation can be avoided, and that because quantum mechanics does not take this into account it is thus an incomplete theory.

The EPR argument provoked immediate public and scientific reaction. The *New York Times*, already attuned to public fascination with Einstein, carried a report under the headline "Einstein Attacks Quantum Theory." But in that article, Princeton physicist E. U. Condon pointed out the weak link in the EPR chain of reasoning: "Of course, a great deal of the argument hinges on just what meaning is to be attached to the word 'reality' in connection with physics." Pauli was even less generous: "If a student in the early semesters had made such objections to me, I would have regarded him as very intelligent and hopeful."[33]

Perhaps regarding Americans as somewhat less advanced than his imagined pupil, Pauli urged Heisenberg to respond immediately in the pages of the *Physical Review*. Heisenberg did draft a response (in German) titled "Is a Deterministic Completion of Quantum Mechanics Possible?" He sent copies to Pauli and to Einstein himself, but he deferred to Bohr, who had already submitted a response to the *Physical Review*, and withheld the paper.[34] When Heisenberg learned several months later, in August 1935, of attempts by Schrödinger and Max von Laue to build on the EPR argument, he revived his manuscript, but Bohr found fault with its logical coherence, which killed it for good.[35] (It is now among his manuscripts.) Bohr apparently did not like Heisenberg's use of Hermann's recent argument refuting the possibility that additional, unknown, and undetected (so-called hidden) variables could yield a deterministic theory after all.[36] Heisenberg dropped her argument when he incorporated the manuscript into one of his philosophical addresses at the University of Vienna at the end of the year.[37]

Despite—or because of—the profound arguments on both sides of the EPR debate, neither side ever managed to convert the other. For Heisenberg, Bohr, and their closest colleagues, EPR was more a nuisance than a serious objection, a threat only to maintaining the commitment of their followers. The tone of Heisenberg's unpublished manuscript is not defensive but condescending toward his critics and skeptics. The essentially statistical character of quantum mechanics had already been fully explored by Bohr, Heisenberg declared, and "the essential content of the following train of thought is already available in the first papers on the fundamental interpretation of quantum mechanics."[38] For him, everything had already been discussed and settled years earlier.

Yet everything was far from settled for those outside the circle of Copenhagen devotees. Einstein and his followers, to the ends of their lives, insisted upon various versions of the EPR argument. In his response to the contributors to a two-volume work in honor of his seventieth birthday in 1949, Einstein reiterated the argument nearly word for word in asserting the incompleteness of quantum mechanics. He expressed his general views as follows: "Above all . . . the reader should be convinced that I fully recognize the very important progress that the statistical quantum theory has brought to theoretical physics. . . . This theory and the (testable) relations, which are contained in it, are, within the natural limits of the indeterminacy relation, complete. . . . What does not satisfy me in that theory, from the standpoint of principle, is its attitude towards that which appears to me to be the programmatic aim of all physics: the complete description of any (individual) real situation (as it supposedly exists irrespective of any act of observation or substantiation)."[39]

Heisenberg was equally adamant to the end of his life. His most extensive rebuttal of the attitudes espoused by the critics of the Copenhagen doctrine appeared in a series of lectures on the intellectual history of physics delivered at the University of St. Andrews, Scotland, during the winter of 1955–1956. These lectures were later published as *Physics and Philosophy: The Revolution in Modern Science*. The Copenhagen interpretation found its natural place in his sweeping account of physical notions since the Greeks. These notions entailed an ever-increasing renunciation of classical realism, a belief in the existence of a real world behaving in precise ways completely independently of an observer. Instead, the old separation of the observer from the observed had become increasingly blurred. Every attempt, including the EPR paper, to return to classical realism contradicted this historical trend and, Heisenberg now asserted, such efforts to reintroduce realism represented nothing more than a desire to return to an outmoded philosophy of materialism. "It cannot be our task to formulate wishes as to how the atomic phenomena should actually be," he declared. "Our task can only be to understand them as they are."[40] The implications of the Copenhagen doctrine were, as before, true, irrevocable, and nonmaterialistic.

Pauli's worries notwithstanding, the American physics community hardly wavered in its preference for Copenhagen. Although Einstein spent the rest of his life in the United States, his objections did not persuade many of his American colleagues. The majority had already pledged their allegiance to Copenhagen by 1935. This came about not only because the Copenhagen interpretation always seemed to work when needed (regardless of philosophical disputes), while its opponents offered no viable alternative, but also because Americans were already receptive to the proselytizing influence of European quantum mechanicians during the late 1920s and early 1930s. American physics had come of age by the early 1930s, and American physicists were eager to participate in the new fields of research.[41]

Like Americans, audiences the world over eagerly welcomed the new European atomic physics, especially when received directly from its practitioners. Backed by well-endowed philanthropic foundations, Americans mounted aggressive and successful efforts to import the new physics and to establish it at American universities. Contrary to popular belief, American physics, though heavily oriented toward classical and practical fields, did not lag far behind European physics. Although the United States still lagged behind Europe in such exotic contemporary fields as relativity and quantum physics, the country was already building momentum in those areas by the early 1930s and actually led in such classical fields as acoustics and electro-technology.[42]

A fertile institutional setting, already in place by the end of World War I, provided favorable conditions for the flourishing of imported European specialties. According to one study, the democratic, practical, and collaborative American university physics department proved more conducive to the further development of quantum physics than did the hierarchical German professorial institute, once American physicists had mastered the new discipline.[43] They achieved their mastery of it through a consciously conceived twofold strategy: dispatching bright students to Europe to learn the latest science at the source, and inviting guest lecturers to teach it to those who remained at home. Karl T. Compton, one of many Americans sent to Max Born's institute, reported: "In the winter of 1926 I found more than 20 Americans in Göttingen, at this fount of quantum wisdom."[44]

Americans continued to pursue European advances after the creation of quantum mechanics and especially after most efforts to lure leading quantum theorists to American professorships had failed. Nearly every European atomic physicist of any renown lectured in the United States during the late twenties and early thirties. One report notes that by late 1929 American universities had invited at least eight of the leading European contributors—Heisenberg, Born, Dirac, Debye, Friedrich Hund, Sommerfeld, William L. Bragg, and Léon Brillouin—to lecture to their physicists on quantum mechanics over periods ranging from six weeks to a full semester.[45] The benefits were mutual and lasting. By rapidly assimilating quantum mechanics, American physicists could, by the early thirties, work independently of, or in equal partnership with, their foreign colleagues. The annual Ann Arbor summer school sessions, sponsored by the University of Michigan and funded by foundations, helped to maintain the momentum until the onset of the Second World War. Heisenberg and other prominent Europeans were invited, often repeatedly, to deliver series of summer lectures to American physicists and their students on their latest work. The arrival of refugee scientists from Europe during the 1930s accelerated the already rapid pace of American science, helping to ensure its lead by the time the world entered its second global war.

The benefits to visiting Europeans were of equal significance (even aside from their handsome salaries).[46] The contacts they established with foreign colleagues were essential to the further development of their field. Those colleagues also provided important assistance during the massive relocation of European scientists beginning in 1933. Lecture tours required the quantum mechanicians to develop (many for the first time) coherent and structured presentations of the new science for both physicists and the general public. Textbooks by the masters proliferated, another means of spreading the new doctrine (and the author's fame). Sommerfeld's lectures in Pasadena served as the basis for a volume on wave mechanics—supplementary to his oft-revised classic text *Atomic Structure and Spectral Lines*—which enabled him to organize his thoughts on the subject.[47]

Heisenberg had a similar experience. As a condition for accepting the appointment to Leipzig in 1927, he had negotiated an eight-month leave of absence in order to accept a series of lecture invitations from abroad. With bookings at MIT and the University of Chicago, and then in Japan and India, Heisenberg boarded a ship at Bremerhaven in early March 1929 for his first transatlantic trip. He eagerly anticipated the journey, which would ultimately take him around the world; and he looked forward to meeting fellow visitors Sommerfeld, Dirac, and Hund "over there among the wild Americans."[48] But less than a day out to sea, the ship was trapped for several days by an early spring fog and a solid sheet of ice. Adrift in the unending fog prison, Heisenberg began to entertain second thoughts. He wrote to Bohr of how much he would rather spend his holiday in his usual fashion, with his youth group at their Bavarian ski hut "instead of traveling all the way to America—but one has to try everything."[49]

The trip must have been exhausting for the 27-year-old. After delivering the MIT and Chicago lectures, Heisenberg traveled to Washington for a few days to lead a session on atomic structure and spectral lines at the American Physical Society meeting. From there he went west to climb in the Rocky Mountains, writing excitedly of the unanticipated beauty of the mountain landscape, which reminded him of home; then through Colorado and Arizona to the Grand Canyon; on to Pasadena for a week of lectures at the California Institute of Technology and touring in the Sierras; then back to Chicago by June. From Chicago he wrote to Bohr that in this "unsettled life" he traveled 1,000 kilometers (620 miles) per week—and in those days one went only by train.[50]

In Chicago Heisenberg stayed with German-American physicist Carl Eckart. When not lecturing and working, he sailed and swam in Lake Michigan, went on a fishing expedition to the northern Wisconsin lakes with Barton Hoag, a Chicago experimentalist, stopped in Madison to see Dirac, and engaged in tennis and music with numerous other physicists. On July 20, he described for Pauli the rest of his exhausting schedule: he would meet Dirac in Yellowstone National Park in mid-August and leave for Hawaii and Japan, where they would stay and lecture until mid-September. While Dirac headed from Japan through Siberia to Moscow, Heisenberg would travel through China to India to lecture, tour, and visit the Himalayas, finally arriving back in Leipzig in time to begin the winter semester in November 1929. "Then I hope to be able to really do physics once again."[51]

Aside from mountaintops, the series of ten lectures that Heisenberg delivered to the University of Chicago in March and April 1929 were the high point of his tour. The lectures were the basis of his textbook, *The Physical Principles of the Quantum*

At the University of Chicago, 1929.
FRONT ROW, LEFT TO RIGHT : *Heisenberg, Dirac, R. A. Millikan, Hund.*

Theory, which was probably the most influential and widely read early account of the Copenhagen doctrine and the inspirations behind it. In Eckart's translation from the German, the text also marked the transition of the name of his famous principle in English usage from "indeterminacy" to "uncertainty."[52]

Heisenberg's textbook is remarkable both for its clarity and for its overreliance on Bohr. While he frequently cites various versions of Bohr's complementarity paper, Heisenberg does not mention his own name anywhere in the text. Even though he portrayed the uncertainty principle as a consequence and clarification of quantum phenomena, citations of his works, particularly his uncertainty principle paper, are conspicuously absent from the bibliography. Bohr alone is cited, even for uncertainty. One reads in the (awkwardly translated) foreword: "The purpose of this book seems to me to be fulfilled if it contributes somewhat to the diffusion of that 'Kopenhagener Geist der Quantentheorie' [Copenhagen spirit of quantum theory] . . . which has directed the entire development of modern atomic physics."[53] For Heisenberg, Bohr as noted earlier, obviously embodied that "Geist."

Other itinerant lecturers outside Bohr's inner circle were less bullish on Bohr and complementarity. While Born and Jordan fondly dedicated their new textbook to Bohr, Sommerfeld neglected to include complementarity in the supplement on wave mechanics to his own textbook. Pauli published an encyclopedic account of quantum mechanics in 1933 that opened with the uncertainty principle and complementarity as the formal basis for constructing all of the new physics. But Enrico Fermi founded his influential Ann Arbor lectures on the uncertainty relations alone. Numerous textbook writers have since followed Fermi.[54]

The uncertainty principle and its mathematical expression lay closer to the derivation of actual, experimental applications of the quantum mechanics than did the philosophy of complementarity. This circumstance, combined with the easier comprehension of uncertainty, facilitated the ready acceptance of uncertainty and other useful exotica of quantum physics by the more practical-minded Americans as the foundation of the new physics. At the same time, they tended to leave the more refined aspects of complementarity and similar rarefied issues of the Copenhagen doctrine to the "philosophers."[55] The EPR debate of the mid-1930s came too late to convert the already entrenched Americans, or to undermine practical allegiance to the doctrine.

With a full professorship, a successful and accepted physics, and worldwide recognition as a leader of the quantum revolution, Heisenberg arrived back in Leipzig from his world tour having fulfilled most of his life's ambitions. In an exquisite recollection of their journey together to Japan that year, Dirac recounted how, while strolling one day with Heisenberg in the vicinity of a beautifully constructed Japanese pagoda, he watched in astonished amazement as young Heisenberg, without uttering a word, gingerly climbed to the top of the pagoda and, in triumphant defiance of certain injury and even death, balanced himself precariously on one foot on the very pinnacle of the building as a fierce wind swept wildly about him.[56] As he balanced in the wind beneath the open sky, the daring and ambitious physicist could be satisfied in knowing that he had attained at last the pinnacle of his life's work and ambitions.

CHAPTER 14

NEW FRONTIERS

HEISENBERG ARRIVED IN LEIPZIG AS HEAD OF THE INSTITUTE FOR THEORETICAL PHYSICS within days of the Solvay Congress of October 1927. Heisenberg's research and his new institute propelled the rapid dissemination and acceptance of the Copenhagen interpretation of quantum mechanics. Leipzig physics epitomized the new quantum physics centers that would produce a new generation of physicists educated into quantum mechanics and nourished on the Copenhagen spirit.

Physicists everywhere looked to the enormous possibilities of applying and expanding the new science to numerous areas of physics and to other sciences. For many scientists, especially those in the United States, the obvious utility of quantum mechanics spoke louder than any lecture on philosophical principles. Heisenberg and the Leipzig institute took an early lead in demonstrating its uses. "The gate to that entirely new field—the quantum mechanics of the atomic shell—stood wide open," Heisenberg recalled, "and fresh fruits seemed ready for plucking." After nearly a decade of puzzles, anomalies, and failure, the five years following the 1927 Solvay Congress seemed to many physicists, including Heisenberg, "so wonderful that we often spoke of them as the golden age of atomic physics."[1] Even so, difficulties encountered in expanding the new physics to high-energy research and upheavals among the German public and students multiplied in that same period. The euphoria of those wonderful years often turned as well into equally deep despair.

Heisenberg's institute was a subdivision of the university's Institute for Physics, headed by Dutch physicist Peter Debye. Heisenberg and Debye both arrived in 1927 and together they immediately transformed the institute and its physics. During the last years of their predecessors, Otto Wiener and Theodor Des Coudres, the once-powerful Leipzig physics had atrophied to near nonexistence. Although the grand old men had been at the forefront during the first decade of the century, neither had fully assimilated Einsteinian physics. Des Coudres concentrated on classical and technical subjects; Wiener, until the last, on a discredited kinetic theory of electromagnetism and the search for its experimental confirmation.[2]

Gregor Wentzel, appointed to a theory position in 1926, summed up the sorry state of Leipzig's pre-Heisenbergian physics in a letter to Sommerfeld in the same year: "Meanwhile the old routine continues on in even worse form than before. Not much more can be made of the older student generation. What I have experienced in knowledge of physics on state final examinations is indescribable. Only one candidate has registered for the doctorate, and he appears to be hopelessly untalented. He is, by the way, a businessman, which is often the case here."[3] But by 1932, Leipzig physics had revived to such an extent that a civil servant in the Saxon Culture Ministry proclaimed, "In fact, the extraordinary representation of physics in Leipzig (Debye, Heisenberg, Hund) is one of the greatest internationally active posts in the university."[4]

Debye inherited from Wiener his dual position as professor of experimental physics and director of the Physics Institute. The turn-of-the-century institute building was just outside the city center at Linéstrasse 5; nestled uncomfortably between a cemetery and a mental institution. The horseshoe-shaped building housed lecture rooms, laboratories, offices, instrument collections, and staff apartments. Heisenberg's Institute for Theoretical Physics occupied the northern wing of the two-story building.[5]

Debye also inherited Wiener's three aging, loyal assistants, along with the three engineering sections that Wiener had staffed with tenured faculty—applied mechanics and thermodynamics, radio physics, and applied electrical science. Although the electrical section enjoyed some repute as the creator of the first successful television receiver, Debye ignored the technical sections and placed Wiener's assistants in charge of the basic beginners' laboratories.

As part of its deal with Debye, the Culture Ministry, located in the Saxon capital, Dresden, agreed to hire Debye's Zurich assistant, Heinrich Sack; it recommended an Emergency Association grant for his second assistant, H. Falkenhagen; it provided 3,000 marks to fund an annual conference—the world-famous Leipzig Lecture Week held each spring—and it empowered Debye to negotiate with Heisenberg to bring him to Leipzig.[6] Upon the success of the last of these, Debye presented the new professor with his first student, Debye's former Zurich pupil Felix Bloch.

Heisenberg started practically from scratch, with one student, no assistant, and no place to stay. Des Coudres, a lifelong bachelor, had occupied a small, sparsely furnished service apartment under the roof of the institute, which now fell to Heisenberg. During his last years, Des Coudres had hired a caretaker to look after him, Frau Gretschmer, who lived in an assistant's apartment in the institute. In early November, Heisenberg arrived to find Frau Gretschmer in Des Coudres's apartment, Wentzel in the assistant's apartment, and the institute reeking of a dusty odor. He ordered the doors and windows thrown open before going off to find bed linen. By December Heisenberg had moved from a hotel to the institute, had bartered the offer of a job at Columbia University for free rent, and had hired a full-time cleaning woman to spruce up the institute, make his meals, and tidy his apartment.[7] Heisenberg also had his apartment refurbished, installed a piano in his rooms, and ordered a Ping-Pong table for the basement. But, he regretted to his mother, the refurbishing had left the toilet outside his apartment, across the hall.

As his first professorial acts, Heisenberg tried to steal an assistant from Sommerfeld and began lecturing without formal announcement. Apparently he took over

Wentzel's course on the theory of electricity, for which, surprisingly, 40 students showed up the first day. "That's indeed a good beginning," he wrote to his proud parents. But his letter to Sommerfeld was less well received. "Your letter reeks of your bad conscience from beginning to end. You want to steal assistants, and of course only the best ones!"[8] He did without an assistant until the summer semester, one of the two semesters at most German universities, the other being the winter semester.

In the summer of 1928, the first semester in which Heisenberg's name appeared in the course catalogue, a record-setting 150 students enrolled in his lectures on classical mechanics, 80 students attended his advanced course on atomic physics, and his advanced research seminar with Wentzel grew to 12 participants, thanks in part to 3 students whom Sommerfeld had urged to go.[9] Sommerfeld also furnished Guido Beck, who received a four-year appointment as Heisenberg's first assistant. Bloch, Heisenberg's second assistant, graduated that semester as his first doctoral student. Like most neophyte professors, Heisenberg was nearly overwhelmed by the sudden strain of academic duties, prompting a complaint to his parents: "I have doctoral papers to correct, galley proofs of my own papers are here, and the lectures require a lot of preparation. Sometimes everything is almost too much for me; also I have unfortunately very little time for music."[10]

With the sudden revival of Leipzig physics, the number of physics students exploded. Within one semester alone, from winter 1927–1928 to summer 1928, their number nearly doubled, while the total number of all Leipzig students increased by only 28 percent. Between the winter semesters of 1927–1928 and 1929–1930, the number of physics students shot up 166 percent (from 41 to 109). Physics classrooms and laboratories overflowed, causing an irate student delegation to demand more laboratory instructors. Debye managed to obtain Rockefeller Foundation funding for a new laboratory, only to have the Culture Ministry refuse to fund the needed new assistants.[11]

The growth of Leipzig physics exceeded even the tremendous general increase (71 percent) of physics students throughout Germany. The latest developments in atomic physics had obviously captured the imaginations of young people. In addition, between 1927 and 1933 an extraordinary number of future notables flocked from abroad to work and study with Heisenberg. Among them were Edward Teller, Laslo Tisza, Lev Landau, George Placzek, Isidor I. Rabi, John Slater, and W. V. Houston. Even J. Robert Oppenheimer showed up for a lecture series on his way to work with Pauli in Zurich. The flood continued nearly unabated until the start of World War II.[12]

Heisenberg built his institute on the Munich and Göttingen models. He transferred the experimental laboratory to Debye, keeping one technician for lecture demonstrations, and he elevated the director of the section for mathematical physics—Wentzel, followed in 1929 by Friedrich Hund—to the rank of full professor (but with authority and salary below his own). Heisenberg gave his three-hour basic theory lecture each semester successively on classical mechanics, thermodynamics, electrodynamics, and optics. He also offered a weekly one-hour special seminar on an advanced topic of current interest, a research seminar with Wentzel or Hund on the structure of matter, and a weekly physics colloquium with Debye in which local and invited speakers lectured on their latest research.

During the early 1930s, Heisenberg began formal lectures on quantum theory and atomic physics, while Hund offered courses on basic theory or on his own research areas, primarily the properties of solid bodies. A lively exchange of assistants and postdoctoral students with Pauli's institute in Zurich and Bohr's in Copenhagen established a new European network of quantum research centers.[13] After his habilitation in 1932 and a stint as Pauli's assistant, Bloch joined the Leipzig professors with courses on general relativity, the quantum theory of magnetism, and the absorption of high-speed particles in matter. Bloch, of Jewish descent, returned to Zurich in 1933, and eventually emigrated from there to Stanford University.

Although Friedrich Hund, the assistant director of the theory institute, was five years older than Heisenberg, the two men had been good friends since the Göttingen days when they had worked and wandered together as Born's students and assistants. They had used the familiar *"du"* long before it was common among colleagues. Until the birth of his second child, Hund's family lived in the small assistant's apartment above Heisenberg's. But Hund was never happy with his role as Heisenberg's shadow and with often being assigned the second-rate students. He was also unamused by a standing joke in Leipzig: the course catalogue always listed the heads of the research seminar as *"Heisenberg mit Hund"*—Heisenberg with dog. Furthermore, Hund never got over being ordered by Heisenberg to return to Leipzig from a visit to Chicago in 1929, before he had taken up his post, so that Heisenberg could continue his own travels.[14] But after Hitler's rise to power in 1933, Hund became one of Heisenberg's closest political allies and confidants.

Hund's even more strained relationship with Debye reflected the wide divergence between theory and experiment in Leipzig. One American visitor to the institute wrote to a colleague back home in early 1933, "I see Bloch occasionally, but being an experimental physicist, I have very little to talk with him about. The categories of theoretical and experimental physicists are more sharply defined here in Leipzig than anywhere I have ever seen before."[15] Perhaps comparing Hund unfairly with Heisenberg, Debye always regarded Hund as a second-rate theorist. Hund, in turn, regarded Debye as "clever but lazy." Heisenberg reported that he often saw the cigar-chomping Debye in the institute garden, watering the roses during duty hours. "Debye had a certain tendency to take things easy," he once noted.[16] Hund stayed on in Leipzig nevertheless and eventually inherited the directorship of the institute after Heisenberg followed Debye to Berlin in 1942.

Because of their youth and their perpetually young lifestyles, Heisenberg and Hund shared an easy rapport with their students. Hund recalls that Heisenberg often ran their joint seminars like a youth-movement *Thing* (criticism session), to which Hund, a former Wandervogel, had little objection. For his advanced special seminars, Heisenberg often bought pastries at the corner bakery and prepared tea on his chem-

Heisenberg with his seminar studens, 1931. FRONT, LEFT TO RIGHT: *R. Peierls, Heisenberg.*
REAR, LEFT TO RIGHT: *G. Placzek, G. Gentile, G. Wick, F. Bloch, V. Weisskopf, F. Sauter.*

ical stove.[17] Until the formation of political student groups after 1933, the two professors often hiked in the hills with their students on weekends (though never with
Debye and his pupils), and assistants often accompanied Heisenberg to his Bavarian
ski hut. During the summer months, Heisenberg took daily horseback riding lessons
at 7:00 AM and frequently attended large swimming parties, once bragging to Bohr
that he could still perform difficult dives.[18] Competition never lagged: student chal-
lengers found in Heisenberg a ready opponent on the university tennis courts. The
basement Ping-Pong table became a special focus of institute fun. Following his trip
to the Far East and his hours of practice with Asian experts aboard ship, Heisenberg
remained the undefeated champion.

As his twenty-seventh birthday approached in 1928, the youthful professor occupied his thoughts during an important faculty meeting with a letter to his mother on
what he wanted for his birthday: a new pillowcase or a bread basket. He thanked her
profusely for sending both. "I am so incredibly happy when such a token of my home
penetrates my all-too-physics-oriented Leipzig existence." The familiar longing for his
Munich home and the companionship of his youth-group comrades continued unabated,
and, as usual, he threw himself into his work as an antidote to such feelings.[19]

The earlier plaint to his parents having ameliorated, music soon occupied much
of Heisenberg's time. He practiced for hours alone in the evening in his apartment.
Music was again a way of escaping his daily life and a way of relaxing from the strain
of work, of reestablishing his equilibrium. Even though he had many friends and col-

leagues in Leipzig, none had shared the intense experiences of his youth and few could accompany him to the mountaintops of scientific reflection and physical exertion to which he drove himself. Those who could were not in Leipzig. An intensive correspondence continued unabated with Bohr and Pauli.

Still, not even music could cure him of his relentless drive. Heisenberg would not let up, even while he was relaxing. Bloch once recalled listening one evening from his own rooms as Heisenberg practiced over and over again a few bars from the Schumann concerto. After finally getting it right, Heisenberg came down to Bloch's apartment for a chat before retiring. Bloch recognized that behind Heisenberg's often adolescent behavior was an extraordinary seriousness, an intense concentration of purpose that went far beyond mere adolescent fun or the suppression of unwanted emotions. "We all knew the dreamy expression on his face," Bloch remembered, "even in his complete attention to other matters and in his fullest enjoyment of jokes or play, which indicated that in the inner recesses of the brain he continued his all-important thoughts on physics."[20]

Located in central Germany, south of Berlin between Halle and Dresden, Leipzig enjoyed special status as a trading center for goods, culture, and people from the East. Leipzig's book trade was the largest in Europe, its library organization dominated the German library system, and the annual Leipzig Trade Fair served as an international industrial showplace. One eighteenth-century visitor wrote, "The Leipzig inhabitants are to be regarded as a small cultural republic. Everyone produces his little original contribution. Wealth, knowledge, talents, possessions of all kinds lend the location its fullness."[21] Goethe chose Auerbachs Keller in Leipzig as the location where the obsessively learned Dr. Faust and a group of students met under the influence of the evil Mephistopheles.

Culture thrived in Leipzig, as it did elsewhere in Germany, during the late Weimar period. The university, one of the oldest in the world (founded in 1409), still housed the four traditional faculties of medicine, law, theology (Lutheran), and philosophy (including science). The original legal and economic independence of the university within the free city of Leipzig had long since been replaced by the authority of the Saxon Culture Ministry, but culture and learning enjoyed unhindered reign. Heisenberg's friend and former student Carl Friedrich von Weizsäcker judged the cultural offerings the best in Germany, while Berlin offered more of everything.[22] Music was especially available; there were free concerts by the renowned Bach choir every Friday evening in the Thomas Church and weekly Mendelssohn concerts by the equally renowned orchestra in the Gewandhaus auditorium. Heisenberg often wrote home of his visits to both. After one letter in which he told them of dancing until 1:00 AM following a Gewandhaus concert, his parents chided him for leading a life of pleasure in Leipzig. They had obviously missed the point. Work and career were his main preoccupations, he insisted. "Many things are much more serious to me than you imagine."[23]

Musical pleasures were indeed more serious than mere relaxation and escape: music also provided Heisenberg a direct entrée into the elite social circles of Leipzig.

During that period, Leipzig social life was dominated by a plutocracy, as Hund called it, composed of book publishers, university professors, and judges and attorneys at the Imperial Court (Reichsgericht).[24] Culture united them all. "Social life blossomed very nicely during those years in Leipzig, although in simpler style than before the war," the Leipzig philosophy professor Hans Driesch recalled. "We circulated with colleagues from all of the faculties, but in addition also with many families on the Imperial Court."[25]

Politicians circulated in such elevated circles only on an individual basis. Political maneuvering and intrigue remained beneath the supposed dignity and objectivity of the cultured social elite. The main exception was Dr. Carl Goerdeler, the monarchically inclined conservative mayor of Leipzig (and later a participant in the 1944 assassination plot against Hitler), who frequented the same circles as did Heisenberg.[26]

Music served those circles as a common denominator, and Heisenberg attended many musical soirees in the homes of Leipzig plutocrats. One of his most frequent musical hosts was the distinguished Otto Mittelstaedt, an attorney at the Imperial Court whose grandson, Peter Mittelstaedt, later studied physics under Heisenberg. Heisenberg and other physicists regularly visited the Mittelstaedt home and that of their in-laws, the Bückings, an old publishing family. Although it is difficult to imagine a genteel Heisenberg in semiformal evening attire, chatting about the latest gossip over tea and cookies while waiting to join the evening's chamber ensemble, these circles of cultural acquaintances served him well, not only by ensuring his membership in the Leipzig social elite, but also by providing personal support and professional influence during the difficult years after 1932. It was at one such musical evening that Heisenberg met his future wife.

During the early 1930s, Professor Heisenberg also circulated in two other influential professional and cultural groups, which had overlapping memberships. In June 1930, the mathematical-physical section of the Saxon Academy of Sciences in Leipzig elected Heisenberg to full membership.[27] At first, such academies were intended to stimulate scientific discussion, to promote science and learning, and to counsel ruling authorities, but with the revival of universities and the establishment of professional journals, their activities mostly resembled an academic men's club. Six months after his election, Heisenberg thanked his supporters by delivering a paper on energy fluctuations in an electromagnetic field.[28]

Heisenberg and a few other members of the Saxon Academy also belonged to a professors' club called the Coronella. This informal group of eight to ten younger male members of the philosophical faculty met regularly for evenings of music and academic chatter. Both types of cultural clubs grew in popularity as the self-styled "bearers" of German culture (*Kulturträger*) felt themselves increasingly challenged and politically isolated in the last years of the Weimar era. Heisenberg never attained an administrative position in the Leipzig philosophical faculty, but several of his close colleagues did. These connections gave him access to a wide circle of influential people, access that would prove crucial after Hitler's rise to power.

Heisenberg's most immediate circle of acquaintances consisted, of course, of his professional colleagues, assistants, and students. By 1932, this circle encompassed mainly Hund; Bloch; B. L. van der Waerden, professor of mathematics; Karl-Friedrich Bonhoeffer, professor of biochemistry and brother of anti-Hitler theologian Dietrich Bonhoeffer; and von Weizsäcker, who completed his dissertation in 1933. Heisenberg was not especially close to Debye or his staff.

Heisenberg had met young Weizsäcker in Copenhagen when, as Bohr's assistant in the winter of 1926–1927, Heisenberg had been invited to a musical soiree at the home of the German ambassador to Denmark, Ernst von Weizsäcker, Carl Friedrich's father. When Heisenberg was called to Leipzig, Carl Friedrich, still a gymnasium pupil, decided to follow him there to the university. But when he graduated in 1929, Heisenberg was abroad. Weizsäcker enrolled for a year at the University of Berlin to await his future teacher's return. Although ten years younger than his teacher, Weizsäcker became one of Heisenberg's closest friends and companions until the former transferred to Berlin in 1936. He often traveled with Heisenberg on weekend outings or joined him with other physics students and youth comrades at the youth group's Bavarian ski hut. By all accounts, their discussion centered on physics and philosophy, to which Weizsäcker was well attuned. When asked if they ever discussed politics or economics during the early thirties, Weizsäcker replied: "Not much, actually no." Hund, however, responded to the same question: "Yes, daily."[29] If Weizsäcker, a student and assistant, was Heisenberg's intellectual colleague and companion, Hund, a full professor, was his academic and political confidant. But for all of them quantum physics captured the greatest attention.

There was plenty to do. As the political situation fell into disarray during the last five years of the Weimar republic, so did post-Solvay efforts to expand the successful quantum mechanics to new realms. Although quantum mechanics provided the foundation for comprehending atomic events, and the Copenhagen interpretation provided the connection between the atom and the everyday world, an important element of the connection was still missing. Electromagnetic radiation, emitted or absorbed as atomic electrons jumped between stationary states, provided the main source of information on the internal workings of atoms. But what was the precise connection between the electrons and the light waves? Since the light existed neither before it was emitted nor after it was absorbed, how can we understand what seemed to be the very creation and annihilation of electromagnetic energy by charged particles? The answers to such questions lay beyond the capabilities of quantum mechanics in 1927.

Nor did any strong connection yet exist with the other great advance in physics during the twentieth century, the theory of relativity. Einstein had shown that the seemingly universal concepts of mass, time, and space are, in fact, relative concepts. The results of measurements of these variables depend upon the speed of the object relative to the observer. The higher the speed, the more the relative aspect plays a role. This result of relativity theory should apply in particular to electrons moving rapidly through space or orbiting at high speeds within atoms. Electrons, of course, were already subject to the laws of quantum mechanics. The obvious next step was to find a combined theory of electrons and electromagnetic radiation that also brought together quantum mechanics and relativity theory. The new theory would be what we now call, in technical terms, a relativistic quantum field theory.

Heisenberg, together with Dirac, Pauli, Pascual Jordan, and their colleagues during the late 1920s and early 1930s, laid the foundations of this extremely technical and mathematical branch of quantum physics.[30] The efforts to understand the fundamental nature of matter, fields, and their interactions with each other, and the search for the proper way to wed the lessons of quantum mechanics with the fundamentals of relativity theory, are still topics of intensive research today. Over the years they have spawned some of the most exotic and expensive research in physics, both experimental and theoretical. Heisenberg was also a leading member of the small band of abstract theorists who laid the foundations of relativistic quantum field theory as it has been pursued ever since.

Even before Heisenberg left Copenhagen for Leipzig, he and his colleagues had already set themselves the tasks of reformulating the newly acquired equations of quantum mechanics in such a way as to render them compatible with relativity theory, and of expanding the equations to include the emission and absorption of light. As Heisenberg began lecturing in Leipzig, he and Pauli settled by letter on a new research program: to build upon work already available in an effort to find a way to join matter in the form of individual charged electron particles with the continuous electromagnetic fields emitted and absorbed by the electrons. Pauli and Jordan had already managed to develop a procedure for handling electromagnetic waves consisting of light quanta.[31]

The difficulty now lay with the electrons, which were to satisfy both quantum mechanics and relativity theory. Their ingenious Cambridge colleague Dirac provided a crucial impetus. On January 2, 1928, as Heisenberg prepared to deliver his formal inaugural address to the Leipzig faculty, Dirac submitted to the Royal Society of London perhaps his best known achievement: "the Dirac equation," a new equation that replaced Schrödinger's non-relativistic wave equation for electrons where the requirements of relativity theory come into play.[32] The "Dirac equation" also contained the half-integer spin of the electron in a natural way. With the new Dirac equation describing relativistic, spinning electrons, the Jordan-Pauli theory for quantum electromagnetic fields, and Dirac's further work showing how to treat the interaction between the two, a full-fledged relativistic quantum field theory suddenly seemed ripe for harvest.

The fruit remained out of reach a good deal longer. One problem in particular seemed so fundamental that it shook the confidence of even the usually optimistic Heisenberg. The problem arose from the puzzle of the electron's size. As indicated earlier, the electron carries mass and charge, but it is usually considered to have no size at all—that is, it is assumed to exist as a mathematical point, with zero radius. In classical theory, an electron sitting by itself is surrounded by an electric field. The electric field contains energy, and a calculation of the amount of energy, often called the "self energy" of the electron, is related to the electron's radius. If the electron has no radius, then the energy becomes infinite, an obvious impossibility. In Heisenberg's and Pauli's new theory, the infinite energy reappeared, this time from the interaction of the electron with an infinite number of light quanta in its own field.[33]

There seemed no way around this infinite result. Ironically, at the very height of the outward success of quantum mechanics, despair grew audible in Leipzig and Zurich, to which Pauli had recently moved as Debye's successor. When Dirac lectured on his new equation during Debye's first annual Leipzig Lecture Week in June 1928, the results were disheartening. As he had done nearly a decade earlier, Pauli resigned from quantum physics to await "a fundamentally new idea."[34] He spent the rest of 1928 working on thermodynamics and writing a utopian novel—not necessarily in that order. Heisenberg likewise turned from "the more important problems" to applications of nonrelativistic quantum mechanics, "in order not to frustrate myself continually with Dirac."[35] Fruitful uses of nonrelativistic quantum mechanics suddenly blossomed from the despair in Leipzig and Zurich after Dirac's depressing lecture.

Inspired by Sommerfeld's recent electron theory of metals, in which the electrons were treated like a gas within the metal, Leipzig's despairing theoreticians turned to the study of solids.[36] Immediately following the formulation of quantum mechanics and the discovery of electron spin, Pauli and Heisenberg had earlier considered the quantum properties of atoms and electrons in metals. A more captivating topic, the Copenhagen interpretation, had drawn them away. But now frustration and the arrival of Bloch in Leipzig brought a sudden revival of the earlier ideas. In his 1928 doctoral thesis—the first under Heisenberg—Bloch presented a quantum-mechanical electron theory of metals.[37] Most metals are conductors of electric current, which is a stream of electrons. One puzzle involved the alacrity with which the current is produced, even in very long wires. A current flows almost immediately after a wire is plugged in to an electrical socket, even though the floating electrons in the metal should be slowed by collisions with the metal atoms in their long trek down the wire.

Bloch solved the problem by considering forces of attraction formed in the wire by the metal ions. If the ions are assumed to be arranged in a crystal lattice, then the overall force on the electrons appears as a kind of wave of attraction all along the wire. Solving Schrödinger's equation for a quantum electron moving in such a field, Bloch managed to explain how electrons are able to propagate over long distances through conductors in short time intervals. The chance of an electron colliding with an ion in a nearly perfect crystal lattice is very low, even at room temperature.

One of the students sent by Sommerfeld to study with Heisenberg was Rudolf Peierls. Heisenberg presented Peierls with a problem: to consider the effect on Bloch's theory of an insight that had proved crucial to Heisenberg's earlier explanation of helium spectra (and that had enabled Heisenberg's Nobel Prize-winning work). This insight was the notion of a what he called an "exchange interaction," a force generated simply by the circumstance that two identical quantum particles, electrons, can exchange positions without altering anything else about the physical situa-tion.[38] Heisenberg suggested that Peierls examine the effects of this quantum force on Bloch's theory of metals. Heisenberg himself, avoiding quantum fields, examined the same effect as a possible explanation of one of the most puzzling phenomena exhib-ited by metals—the magnetism of iron magnets, or ferromagnetism.

Materials such as iron can be magnetized when they are rubbed by a magnet or placed, in a molten state, in a magnetic field. They remain magnetized even after the iron has cooled and the field is removed. One possible explanation of this effect derived from the spinning valence electrons of the iron atoms. Any rotating charge acts like a little closed current loop, and a closed current loop behaves like a little magnet. A molecular magnetic field inside the iron—the so-called Weiss field (named for Zurich physicist Pierre Weiss)—maintains the alignment of all the spinning valence electrons in the same direction, and this renders the magnetism of an iron magnet. But Heisenberg could not prove this until new mathematical methods became available for handling huge numbers of electron loops. Once he applied the new methods (group theory), the long-time puzzle of ferromagnetism readily succumbed to the completed quantum mechanics of the Copenhagen school.[39]

Heisenberg's publication of the solution to the puzzle of magnetism in 1928 became at once another triumph of the new quantum mechanics and an argument for the practical utility of the new physics for solving long-standing problems. He went out of his way a year later to emphasize the latter in an address on the subject to the German Metallurgical Society, which appeared in the society's organ, *Metallwirt-schaft (Metal Economy)*.[40]

But Heisenberg never stopped thinking about how to join relativity theory with the quantum theory of electrons and fields. By the time of his lecture to the metallurgists, he had hit upon a mathematical trick for dealing with some of the other technical diffi-culties standing in the way of the new theory. Pauli traveled from Zurich to Leipzig in January 1929 for a district meeting of the German Physical Society. By the end of the meeting he and Heisenberg had settled on a joint paper to be completed by the time Heisenberg departed for the United States in March.[41] The paper, entitled "On the Quantum Dynamics of Wave Fields," constituted the first full-fledged relativistic quan-tum field theory. It laid the foundations for field theories to follow. That paper and its sequel later that year were the only two papers that Heisenberg and Pauli, who contin-ued to work closely together until Pauli's death in 1958, would ever publish jointly.[42]

As the Heisenberg-Pauli papers went to press, the man who would become Heisenberg's American doppelgänger, J. Robert Oppenheimer, was at that moment visiting Pauli's Zurich institute on a Rockefeller stipend. Three years younger than Heisenberg, the American-born Oppenheimer had emerged, like Heisenberg, from the cultured upper-class elite. He had attended the best American universities, and he had recently studied with Max Born, receiving his doctorate from Göttingen in 1927.

Working for several months with Pauli before returning to professorships in California, Oppenheimer carefully examined the energy of an electron in the Heisenberg-

Pauli field theory. He proved beyond a doubt that the self-energy of the electron once again went to infinity, and that this impossible infinite result was inescapable in any quantum theory of electromagnetic fields then available. Oppenheimer demonstrated that this behavior destroyed even such simple, yet essential, calculations as the spectrum of the hydrogen atom, a single electron orbiting a proton. The infinite energy caused an infinite displacement of the quantum atomic states, rendering any application of the Heisenberg-Pauli theory to atoms impossible.[43]

Endowing the electron with a size, a finite radius, so that it looked like a ball of mass and charge seemed to be the only way out of this impasse. Yet there still seemed to be no way to fit a finite size into the theory. Not only would the electron blow itself up in electrical repulsion, but relativity theory made such a notion nearly unacceptable. Just three years after the completion of quantum mechanics, Bohr, while eagerly applying complementarity to nearly every dilemma of life, began to concede the possibility of a limit to the validity of the new mechanics. In his prestigious 1930 Faraday lecture to the British Association in London, Bohr suggested an analogy with the earlier development of quantum theory. In penetrating to ever-smaller sizes, down to the structure of atoms, scientists moved ever further away from the familiar world of the laboratory, and the usual classical concepts found their limit at the surface of the atom, where quantum concepts come into play. So, too, he argued, quantum concepts may fail at the next-smallest level of distances, those about the size of the nucleus and elementary particles, such as protons. (A proton is the nucleus of a hydrogen atom.)[44]

Quantum mechanics, Bohr noted, had been constructed in close correspondence with classical theory, which assumes only infinitesimal point particles. Perhaps there is a minimum fundamental length, about the size of the proton or electron, below which quantum mechanics simply breaks down. Bohr even began to speculate in private that a new quantum revolution might be necessary to handle the nucleus and elementary particles. He wrote Dirac several months after his lecture: "I . . . believe firmly that the solution of the present troubles will not be reached without a revision of our general physical ideas still deeper than that contemplated in the present quantum mechanics."[45]

While Pauli was correcting the galleys of their joint paper on quantum field theory, his warnings to Heisenberg, then touring the United States, over the infinite electron energy grew shriller. But Heisenberg had apparently already imbibed his share of American pragmatism. Even after a long discussion at the California Institute of Technology with Oppenheimer, lately returned from Europe, Heisenberg wrote Pauli: "The catastrophal interaction of the electron with itself does not concern me very much, despite your warnings. You are of course correct that this interaction makes the theory temporarily unusable; but it is indeed that already because of the Dirac jumps [to apparently negative energies]."[46]

Heisenberg's flippancy regarding infinite energy may also have arisen from an inkling of what he might do about it. Within months of returning to Leipzig late in 1929, Heisenberg presented to Bohr another piece of extraordinary ingenuity. The contrast in Leipzig between failing field theories and successful lattice theories—those of Bloch, Peierls, and Heisenberg—may have prompted Heisenberg's newest flash of genius: Why not turn the whole world into one big lattice—a "lattice world"? Space itself might be construed as a honeycomb of minuscule cubic cells of the size of an elementary particle. If such cells existed, they would constitute an absolute minimum distance, below which the size of any elementary particle, or any size whatsoever, could not shrink. Such a minimum distance would make infinitesimal particles impossible—thus rendering their energies finite—and, as Bohr declared in London, the lattice cells would supply a lower limit to all theorizing using current quantum concepts.[47]

Reconnoitering this brave new lattice world in letters and a paper published in 1930, Heisenberg managed to derive the correct electron mass in the case of such a world of one dimension. But when he expanded the theory to the real world of three dimensions, he encountered "thoroughly radical changes of our present quantum theoretical concepts."[48] At the very least, the theory required abandonment once again of the hallowed conservation law of energy, as well as those of mass and charge. Bohr, now convinced of the need for radicalism, would once again tolerate the loss of energy and mass conservation, if necessary, but not of charge. Nor would he support such a crass proposal as a world composed of cubic cells. After much debate in Copenhagen, Heisenberg quietly dissolved his lattice world, and the infinite energies of electrons loomed larger than ever.

Renewed frustration echoed through Heisenberg's letters home: "My work is still not going as well as I wish; nature is apparently constructed in remarkable subtlety."[49] The untimely death of Heisenberg's father compounded his slump. In early November 1930, Heisenberg's mother called to tell him that his father was suffering from typhoid fever, the disease the younger Heisenberg had contracted a decade earlier.[50] August had drunk some tainted water while attending an international philology conference in Greece. He died a short while later, on November 22, 1930, two weeks short of Heisenberg's twenty-ninth birthday.

Despite his father's obvious support of Heisenberg's career and their camaraderie over chess games whenever Heisenberg was at home, their relationship had remained rather distant and strained since the close of the World War. August constantly fretted over his son's career and constantly worried how far the youth-movement business might take him from his work. Heisenberg, for his part, though reveling at times in rebellion, had known early what he wanted in his career, and by 1930 he had already achieved most of it. He did not always appreciate his father's meddling as he made his rapid climb to the top of his profession.

In his first letter from Leipzig to his mother after the funeral in Munich, Heisenberg seemed moved less by sorrow at the tragic loss of his father than by the reminder it brought of his own mortality and waning youth. A month before his father's death, the 28-year-old had attended a youth meeting from which he came away feeling "very old." Looking back on his young life in his letter, he mused, "A saying occurs to me that ends thus—*aber ging es leuchtend nieder, leuchtet's lange noch zurück* [but if it sinks in shining splendor, still it shines a long way back]. I believe as long as we ourselves are in this world, we must be satisfied with feeling this shining-back. . . . I remember the time when I myself was at my liveliest, you know, about ten years ago; that was also the most beautiful time of my life, such that my happiness also transferred itself to others."[51]

Each year, on the anniversary of his father's death, near his own approaching birthday on December 5, Heisenberg offered his mother reflections on his life, his boyhood, and his father. During the difficult thirties, Heisenberg felt a heightened sense of mortality, nostalgia for the happier days of his bygone youth, and a twinkling frustration with his professional urge "to build up something definitive that will last as long as I can work all work"—quantum mechanics and uncertainty apparently did not suffice.[52]

Thwarted in 1930 by such impossibilities as infinite-energy electrons jumping to negative energies, the "aging" professor turned his attention to his students and to his teaching program. He also appeared before public audiences, speaking to the philosophers in Königsberg, to his colleagues at the Saxon Academy, and to British scientists assembled in a follow-up to Bohr at the British Association in 1931. For research, he concentrated on further applications of quantum mechanics to "less important problems," those derived mostly from experiments performed in Debye's section of the Leipzig institute. These included theories of magnetization, the incoherent scattering of X-rays by nuclei, and a so-called formula collection for phenomena induced by the impacts of high-energy upper atmospheric particles, the so-called cosmic rays.[53]

"On the other hand," he told Bohr in 1931, "I have given up concerning myself with fundamental questions, which are too difficult for me."[54] In the Christmas issue of the *Berliner Tageblatt*, he wrote that progress on such fundamental questions as the quantum mechanics of small distances, such as in the nucleus, would have to await further probing of that minuscule ball of matter at the center of the atom: "Whether indeed the year 1932 will lead us to such knowledge is quite doubtful."[55]

Just six months later, Heisenberg would stun his colleagues with the first contemporary nuclear theory—a quantum mechanics of the nucleus that laid the foundations of nuclear physics as it has been practiced ever since. Heisenberg's theory would come in the wake of a new discovery. In March 1932, British physicist James Chadwick announced the discovery of the neutron. Before that, only three elementary (noncompound) particles were known: the electron, the proton (hydrogen nucleus), and the light quantum, or photon. Now there was a fourth particle, the neutron, an elementary particle of about the same mass as the electrically positive proton, but possessing no electric charge. Even though electrons came out of the nucleus as one form of radioactivity, so-called beta rays, they could not exist in the nucleus because the uncertainty principle prevented their confinement to such a small space. That left only protons as the constituents of nuclei, since light quanta carried no mass. But protons alone could not make up the entire mass of the nucleus since that would result in too

*Bohr and Heisenberg during a
Bavarian ski trip in 1932, shortly
after discovery of the neutron.*

much positive charge. Protons carry a positive charge of the size of the negative electron charge. Moreover, the protons would electrically repel each other, preventing the
nucleus from holding together. In addition, it appeared from previous work that quantum mechanics failed at the small size of the nucleus. The neutron suddenly offered
the ever-imaginative Heisenberg a way out of this thicket of conundra at last.[56]

Just before the discovery of the neutron, Bohr and Heisenberg appeared to be in
quite different states of mind. In February 1932, Bohr was deeply immersed in completing the manuscript of an overview of the sorry state of nuclear physics. The manuscript was to be a more carefully constructed exposition of some informal remarks
he had made the previous October to a conference on nuclear physics in Rome.[57] But
Bohr was still having trouble expressing his views on this difficult subject. Bloch volunteered to meet Bohr in Salzburg in February to help him complete the manuscript
before going off together to Heisenberg's Bavarian ski hut, where Heisenberg and
Weizsäcker awaited their arrival for a skiing vacation.[58]

Bohr's nuclear pessimism initially infected Heisenberg, who engaged Bohr in an
incessant inquiry into the difficulties of nuclear physics. When Bohr finally returned
to Copenhagen in March, he found Chadwick's letter announcing the discovery of the
neutron.[59] Within a month, Heisenberg was back in Copenhagen, mentioning some
vague new ideas on nuclear physics. He then fell strangely silent for two months. The
Copenhagen visit apparently stimulated Heisenberg's thoughts in a new direction.

Gradually, he came to recognize the full potential of a new point of attack centering on the neutron. On June 20, 1932, he broke his silence with a letter to Bohr. It was accompanied by the first installment of what would become a classic three-part paper, "On the Constitution of Atomic Nuclei."[60]

"The basic idea," Heisenberg told Bohr, "is to shove all fundamental difficulties onto the neutron and to do quantum mechanics in the nucleus."[61] While Bohr awaited a revolutionary new physics to handle such small dimensions as those of the nucleus, Heisenberg had been poised for an attack that would enable him to advance into the nucleus by sidestepping the usual problems, such as infinities. Rather than attempting to imagine an all-encompassing theory—as Bohr might have done—Heisenberg exploited the neutron as a scapegoat that would shoulder a multitude of sins.

Among its sins, in Heisenberg's theory, was the circumstance that the neutron was at once both a fundamental particle and a composite of two fundamental particles. In the opening paragraphs of "On the Constitution of Atomic Nuclei," Heisenberg treated the neutron first as an "independent fundamental constituent" of nuclei, together with the proton; the nuclei would consist just of neutrons and protons, the two together making up the mass of the nucleus, the number of protons alone making up its positive charge.

But then Heisenberg also treated the neutron as a composite of a proton and an electron. Such a composite of two particles to form one was incomprehensible in any theory, and even violated his own uncertainty principle. Nevertheless, radioactive beta decay, which entailed the ejection of a high-speed electron from the nucleus with the transmutation of a neutron into a proton, seemed to suggest such a structure. In other words, Heisenberg's neutron was simultaneously indivisible (fundamental), compound, and a contradiction of both quantum mechanics and conservation laws! Nevertheless, as with the atomic core model a decade earlier, the ends richly justified the means. They enabled no less than the beginnings of contemporary nuclear theory.

Thanking Pauli for "valuable discussions" (presumably by letter), Heisenberg developed his theory of the binding of neutrons to protons and to other neutrons to form nuclei. Since protons repel each other electrically, there must be an even stronger force that overcomes the electrical repulsion and attaches protons to each other. It must also attract protons to neutrons and neutrons to neutrons. This "nuclear force" must be very strong, and it must be very short-ranged, otherwise nuclei would be much bigger than the tiny balls they are at the centers of atoms. Heisenberg did not obtain this force, but he could account for the binding of protons and neutrons to form nuclei in close analogy with the chemical bonding of atoms to form molecules. Thanks to Linus Pauling and others, chemical bonding had become one of the greatest successes of quantum mechanics. Heisenberg had also contributed with his invention of the ever adaptable idea of the "exchange force"—a quantum-mechanical force arising from the circumstance that two identical particles can exchange their positions without affecting the physical properties of their environment. Heisenberg reasoned that if the neutron were a composite of a proton and an electron, then the neutron-proton force could arise from exchange of the electron between the two, much as an ionized hydrogen molecule is held together by the "exchange force" of a single electron being shared equally by two protons. In a sense, the neutron and proton play a wild game of catch with the one available electron, the proton turning into a neutron when it catches the electron, the neutron turning into a proton when it releases the electron—then back again. When one side occasionally misses the ball, the previously "virtual" electron escapes from the nucleus as a real electron. It appears in the laboratory as the radioactive radiation of beta decay.

Such a daring proposal required the intuition of a Heisenberg. With it the entire apparatus of nonrelativistic quantum mechanics lay at Heisenberg's disposal, enabling him to create the modern neutron-proton model of nuclei. With this theory, he was also able to account for the stability of helium nuclei (the alpha rays of radioactivity) and the binding of a neutron to a proton to form the deuteron, the version (isotope) of hydrogen that, when combined with oxygen, forms heavy water. Heavy water, as would nuclear theory, soon played a crucial role in the search for the controlled and uncontrolled release of nuclear energy.

Heisenberg had opened the door to the entire nucleus. His proton-neutron model set in motion the field of contemporary nuclear structure studies, and it stimulated the branch of quantum theory that focused on two new forces of nature within the nucleus—the strong, or nuclear, force that holds nuclei together, and the weak force that holds together protons and neutrons. Decades later, these two forces, together with that of electromagnetism, have been joined to form the so-called standard model, which appears to have been confirmed by high-energy accelerators. It is the search for a unification of the standard model forces with gravitation that has driven a good deal of contemporary research.

Save for a brief paper on lightweight nuclei, summary reports on his theory and its applications during the seventh Solvay Congress, held in Brussels in October 1933, and to a celebration of Zeeman's fiftieth birthday in The Hague in 1935, Heisenberg did not write again on nuclear physics until the outbreak of World War II, seven years later.[62] He had achieved all that he wanted to do with it. Although the nucleus continued to attract Heisenberg's students and visitors throughout the next decade, the discoveries of the positive electron (positron) and the appearance of showers of cosmic-ray particles in the laboratory, both also occurring in 1932, returned Heisenberg's attention to what he called the "more important problem" of finding a suitable fundamental quantum field theory without infinite energies—a program and an undertaking in which Bohr had little interest or faith.

No correspondence between Heisenberg and Pauli survives from 1932, when Heisenberg worked most closely with Bohr on nuclear structure, but their collaboration and correspondence rapidly escalated again thereafter, largely replacing those between Bohr and Heisenberg on the "more important" matters. While Bohr continued to explore nuclear physics, developing the widely used liquid drop model of heavy nuclei, Heisenberg turned to high-energy physics and to the insights it provided into the fundamentals of field theory. The higher the energy with which two particles smash into each other, the closer their centers approach each other and the more information they provide about the behavior of matter and fields at distances even smaller than the sizes of elementary particles.

Besides the different direction in research that Heisenberg pursued after 1932—research that brought him away from Bohr's line of work and closer to Pauli's—Bohr and Heisenberg also became intensely involved in the ominous events unfolding in Germany. Both men were soon drawn into activities far removed from the rarefied physics of nuclei and quanta.

Less than six weeks after Heisenberg completed his third paper on nuclear physics, Adolf Hitler attained the position of German Chancellor. On the evening when Hitler came to power, Heisenberg happened to be visiting the Weizsäcker family's home in Berlin. As Heisenberg and his student, Carl Friedrich, looked out from the Weizsäcker home over the darkened Berlin streets that evening, wondering aloud to each other what the future might hold, a torchlight parade of Hitler's brown-shirted storm troopers marched below, row on row, in celebration of the Führer's ominous triumph.[63]

CHAPTER 15

INTO THE ABYSS

IT WAS AT THE END OF JANUARY 1933 THAT GERMANY'S PRESIDENT, FIELD MARSHAL Paul von Hindenburg, appointed Adolf Hitler, then chairman of the influential National Socialist German Workers Party, chancellor of Germany and head of a new cabinet in Berlin. Most Germans were relieved. A cabinet crisis was finally resolved, and with nationalist conservatives in control. The *Leipziger Neueste Nachrichten* (*Leipzig Latest News*), a conservative newspaper not allied with the Nazis (as National Socialists were disparagingly called), extolled the event: "The first day of the Hitler cabinet has closed in the brightest glitter. The day was dominated by a feeling of widespread joy at the unification on the [political] right. It cannot be better expressed than in Hitler's own words at his first cabinet meeting: 'Faith and trust shall not be disappointed!'"[1]

From the moment Hitler gained control of the chancellery, he and his party held the "nation of poets and thinkers" in an ever-tightening grip. Within a day, the Reichstag (parliament) was dissolved; within a month, the constitution was suspended. By the summer, thousands of Jews and political opponents had lost their jobs, and many were leaving the country. The first concentration camps—intended to concentrate opponents, criminals, and other undesirables in a common prison—were already in operation. Political efforts to halt the National Socialist takeover were thwarted by the imposition of one-party rule. A year later, by the end of August 1934, Hitler had created for himself the position of national Führer, the unquestioned leader of Germany; the Nazi dictatorship had stamped out Germany's first democracy.

The frightening rapidity and apparent ease with which Hitler and his henchmen seized the German state seems to have arisen from a combination of unique demonical genius and the particular susceptibility of the populace to demagoguery. Although politically the National Socialists gained their greatest support from the unemployed and the economically threatened lower-middle class, most observers agree that they could not have taken over so rapidly and completely after January 1933 had they not received the crucial support of the army and the initial acquiescence of the upper-middle class—civil servants, industrialists, and such opinion makers as professors and news media.[2]

Through manipulation, propaganda, violence, and intimidation, the National Socialists outmaneuvered and overwhelmed great blocks of the all-too-willing German people. Uninspired by the impotence of the Weimar state in foreign affairs, disillusioned by heavy political infighting that caused a perpetual cabinet crisis in Berlin, and anxious for the future as Germany and the world sank into economic depression, many Germans—and academics most particularly—longed for the return of a powerful, unified empire (Reich), free of so-called party chaos. Even more so than before, German academics shunned overt involvement in political intrigue as antithetical and detrimental to scholarly objectivity. Politics were beneath the dignity of an academic aristocrat. Thus, for many people, openly opposing the new regime meant descending into the dirty world of politics, while openly supporting or even tacitly acquiescing to one-party rule was seen as somehow apolitical and objective— as it had been in the days of the Kaiser. Heisenberg's earlier views that politics were a "money-business" and that the Weimar democracy seemed merely a transitory phase preceding a more fundamental system suggest that he too embraced, at least in part, such dangerous political notions.

Such notions were demonstrated by the numerous widely publicized written declarations by academics, many of whom were not party members, in support of Hitler and his evolving dictatorship.[3] Because of German reverence for professors, the public impact of these manifestos must have been significant; some were signed by as many as 68 professors at a time. Anti-Nazi academics could have issued their own equally powerful declarations, at least before Joseph Goebbels's Propaganda Ministry seized control of the public media late in 1933. Heisenberg's and Debye's names appear on an appeal, published late in 1932, for more financial support of science, which was signed by a total of 141 academics.[4] But not a single counter-manifesto appeared from them or others in opposition to Nazi measures. Older academics, such as the influential Max Planck, were still smarting from the ludicrous manifestos of World War I, and they apparently discouraged any new appeal. Political manifestos smacked of party politics, while quiet diplomacy and judicious compromise seemed more certain of protecting what really mattered to them most— German science and scientists.[5] But another reason for reticence was also at work— the use of violence.

During the early days of the Weimar era, some splinter parties advanced their programs by literally beating the opposition: they unleashed gangs of armed thugs organized into private street armies. Hitler's party employed its own paramilitary organization, the brown-shirted storm troopers (*Sturmabteilung*), or SA. Although the storm troopers' street violence had waned in the last years of the republic because of official proscription and an effort to give the Nazi Party an appearance of decency, violence suddenly intensified after the party came to power. The left, abandoned by voters and in disarray, resorted in some cities to a desperate urban guerrilla warfare that was quickly suppressed by superior numbers of Nazi troops and by the depletion of leftist ranks through constant arrest. In Leipzig, where the majority had voted consistently for the Socialist and Communist parties, newspapers were filled with reports of street battles, followed by power displays by marching brigades of Nazi troops waving swastika banners.[6] The establishment of the Dachau concentration camp just outside Munich with 43 leftist opponents, among them members of the clergy, received wide publicity.[7] (It was not yet the death camp it would become.)

The new regime turned more openly in the spring of 1933 to the persecution and expulsion of Jews following the suppression of overt political opposition and the suspension of civil rights after the mysterious burning of the Reichstag building on February 27, 1933—blamed, naturally, on a Communist. This was followed by the appointment of Reich commissars and governors in each state to ensure "order and security." To the political onslaught was now added a moral affront—the imposition of anti-Semitic laws. As Nazis employed their usual coercive tactics to achieve their aims and stifle opposition, April 1933 opened with a "day of boycott" of Jewish stores. Then followed the enactment of the infamous anti-Semitic laws for the "restitution" of the civil service (April 7) and against "overcrowding" of German schools and universities (April 25).[8] The bloodletting had begun.

During the day of boycott, which actually took place over a four-day period, storm troopers harassed Jewish store owners and their customers and rampaged through laboratories, libraries, classrooms, and courtrooms, forcibly expelling Jews from their work. Not all Germans sympathized with the action.[9] Heisenberg's associate, the upright nationalist mayor of Leipzig, Carl Goerdeler, defended a Jewish shopkeeper from storm-trooper harassment.[10] Critics of the regime tended to keep silent about these and other dictatorial measures for fear of provoking more, but again there were exceptions. One particularly vociferous exception was Albert Einstein.

On a visiting professorship to the United States when Hitler came to power, Einstein made known his decision not to return to Germany and declared in an interview, "As long as I have any choice in the matter, I shall live only in a country where civil liberty, tolerance, and equality before the law prevail. . . . These conditions do not exist in Germany at the present time. Men, among them leading artists, who have made a particularly great contribution to the cause of international understanding, are being persecuted there."[11]

Shortly thereafter, Einstein resigned his nonteaching professorship at the Prussian Academy of Sciences and called on all democratic nations to unite in opposing National Socialist Germany.[12] His appeal was in vain. Nazis responded by accusing Einstein of participating in a foreign defamation crusade against Germany and raided his summer home on the pretext of seeking terrorist weapons. Master propagandist Goebbels declared that the boycott and harassment of Jews would continue until such anti-Nazi "propaganda" ceased.[13] Physicists Max Planck and Max von Laue, attempting to ameliorate the situation through quiet diplomacy, informed their colleague that he was making matters difficult by violating the unspoken rule against political involvement: "Here they are making nearly the entirety of German academics responsible when you do something political."[14]

Einstein, of course, would not keep silent. His response to Laue is refreshing in its candor: "I do not share your view that the scientist should observe silence in political matters, i.e., human affairs in the broader sense. . . . Does not such restraint signify a lack of responsibility? Where would we be had men like Giordano Bruno, Spinoza, Voltaire, and Humboldt thought and behaved in such a fashion? I do not regret one word of what I have said and am of the belief that my actions have served mankind."[15]

Despite Einstein's stature, on April 1 the Prussian Academy announced, while Max Planck was away on his annual Sicilian holiday, that it had no reason to regret Einstein's resignation.[16] Refusing appeals to terminate his vacation, Planck returned at the end of April to an appalling situation—friends and colleagues removed from their jobs and preparing to leave the country. Worried about posterity's judgment of the academy's handling of Einstein, Planck inserted a positive statement on Einstein into the academy record when he became its secretary a month later. But by its action, the most prestigious of German academic institutions had indirectly endorsed the expulsion of Jews from academic positions by expelling Germany's most famous scientist.[17]

The law for the "restitution" of the professional civil service, following on the heels of the boycott, confronted academics even more directly with the grim issues at stake—teachers and professors, as civil servants, were now personally affected. Paragraph 3 of the new law stated: "Civil servants who are not of Aryan ancestry are to be placed in retirement." The only exceptions, in deference to Field Marshal Hindenburg, were civil servants who had served at the front in World War I or had become civil servants before August 1, 1914. But the deference was withdrawn in practice by Paragraph 4, which allowed for no exceptions: "Civil servants who by their previous political activity do not offer the guarantee that they will stand up at any time for the national state without reservation can be dismissed from service."[18]

The Reich commissar and cultural ministry in each state decided who should be dismissed and passed the decision along to the appropriate faculty for prosecution. At first there was some confusion regarding the enforcement of the new law and a faint hope that it would not be applied too harshly. German physicist Hans Kopfermann, asked by Niels Bohr to report on the situation in Germany, wrote in May 1933 that state bureaucrats were defining "non-Aryan" rather narrowly. Many dismissals were actually temporary leaves of absence until the authorities could decide each case individually. Although the party had unleashed its Nazi students in a campaign against the "non-German spirit," there was at first only little support for the violent crusade at most universities.[19] The ambiguous status of the affected faculty became clearer in the fall of 1933—but for the worse. The application of anti-Semitic laws grew ever more oppressive until the end of the "battle phase" (*Kampfphase*) of Nazi control in 1935.

Like the Leipzig mayor and other nationalist-oriented non-Jewish Germans, Heisenberg was at first appalled at the crudity of the new leaders and the "excesses" of their new regime, but he greatly sympathized with the long-term national revival promised by the National Socialists. "Much that is good is now also being tried," he wrote as late as October 1933, "and one should recognize good intentions."[20] He and others expected that the regime, like its immediate predecessors, would hardly last out the year. An urgent political response, had they with their apolitical attitudes even considered one, would have seemed to them unnecessary. Because of this, Heisenberg and most other non-Jewish academics still in Germany did not make overt protestations like Einstein's. Nor did the obvious moral issue of anti-Semitism move Heisenberg and his colleagues to action. Although some no doubt sympathized with anti-Semitic policies, there is no direct indication that Heisenberg did so in the scientific sphere. Rather, as noted earlier, he and his colleagues tended to regard anti-Semitism as a mere political issue—thus an issue to be avoided entirely. Nevertheless, the plights of their closest Jewish colleagues and the damage to the German physics profession demanded a response.

Many Jewish academics, reading the signs, had left Germany as soon as the law was enacted. Nobel Prize-winning scientists Fritz Haber and James Franck resigned outright, refusing to take advantage of the "leniency" the law provided for front veterans. Max Born, barred from the classroom and placed on permanent leave, left for his summer home in northern Italy, intending never to return. In June 1933, he informed Einstein and Heisenberg that he did not want his children to live as second-class citizens in the country of their birth. Nor could he excuse the parting anti-Semitic accusations and epithets hurled at him in Göttingen.[21] Göttingen, one of the world's premier science centers, the town that had nurtured Heisenberg to the top of his profession and had witnessed the birth of quantum mechanics, was soon stripped of many of its world leaders in science.[22] In this crisis, uncertain and confused, Heisenberg turned to Planck for counsel and advice.

The match seems unlikely. Planck was then 75 years of age, Heisenberg 31. Planck, though a founder of quantum theory, could hardly be called a radical thinker or a political sophisticate. He had opposed the Copenhagen interpretation so dear to Heisenberg. Like Einstein, he had called for the revival of determinism and objectivity in atomic physics. Heisenberg lived in a world that glorified youth and rejected what was seen as the ossified ideas of the older generation. He embraced the radical elements of Copenhagen physics and excelled in a profession that richly rewarded new and successful ideas. At the same time, Heisenberg's world hated hypocrisy and revered integrity and devotion to duty above all else.

Planck, it was universally agreed, represented the epitome of ethical fortitude and consistency and was considered a paragon untainted even by the usual human frailties. As Prussian Academy secretary and president of the prestigious Kaiser Wilhelm Society, Planck, a wiry looking physicist kept spry by his long devotion to strenuous mountain-wandering tours, had presided for years as unofficial dean of Heisenberg's profession. He had chaired committees that provided Heisenberg's postdoctoral grants and had recently lessened his criticism of Copenhagen physics. Heisenberg's support of Planck's quiet efforts to protect their profession in the face of the worsening political situation was evident to readers of Heisenberg's reviews of Planck's collected essays in this period.[23]

The April dismissals, accompanied by the uproar over Franck's public resignation in protest, prompted an unsettled Heisenberg to seek out Planck even before Planck's return to Berlin from his Sicilian holiday.[24] Throughout the remainder of 1933 and into 1934, Heisenberg met and corresponded often with Planck and occasionally with Planck's Berlin colleague Laue, the anti-Nazi president of the German Physical Society, as they worked together to defend and protect their profession against regime policies. But direct, outspoken, broad-scale opposition to the regime was not a consideration.

Since the main professional concerns in the spring of 1933 were the dismissal of leading Jewish physicists from teaching positions and their emigration from Germany, Heisenberg and his co-strategists concentrated on two goals: persuading the dismissed physicists to remain in Germany, and working behind the scenes to rescind the dismissals. Evaluating such a strategy—and the situation in which the strategists found themselves when it ultimately failed—requires the advantages of historical hindsight. At that time, the Western world had no experience with a regime such as Hitler's and no conception that such a regime could lead its people inexorably into the evil nightmare that the Nazi dictatorship became. Nor did German academics have much experience with moral action or political commitment.

With the regime in control of the legal apparatus and the SA in control of the streets, perhaps the most effective opposition Heisenberg and other academics could have launched in those early months would have been the political mobilization of the middle and upper-middle classes, and especially their students, against the regime and its policies. But given the initial popularity of the regime, opposition would have been possible only if it had been mobilized well before 1933. That would have required a political sensitivity and commitment to democracy, and an education of students in democratic virtues, that did not then exist. Individual professors who did attempt to spark a democratic opposition were regarded as examples of the futility of individual protest—precisely because they were alone. A year earlier, Gerhard Kessler, a non-Jewish professor of economics in Leipzig, had sought to reach students and the public with pamphlets and lectures against National Socialist doctrine. His efforts were rewarded with mass demonstrations against him and the violent disruption of his lectures by rampaging students.[25] He was dismissed from his post in April 1933, was arrested by the Gestapo in July, and emigrated to the United States under the threat of ending up in a concentration camp. The several other attempts in Leipzig to speak out to students against the Nazis were similarly silenced.[26]

It could be argued that a mobilization of mass opposition could have been achieved in early 1933 had more non-Jewish academics followed the examples of Einstein, Franck, and Haber: that is, had there been a simultaneous resignation of professors in moral indignation at the dismissal of their colleagues and the treatment of Jews in general. With their international reputations, they could easily have found jobs elsewhere. The upright Otto Hahn, in fact, suggested just that to Planck. Such blatant persecution and anti-democratic measures certainly warranted the action.

Planck, however, refused; his view of politics led him to believe that their protest would go unreported and that their jobs would be filled by unworthy individuals—an even greater damage to their profession and to their beloved homeland.[27] A noted Leipzig geophysicist made the same argument to an American visitor: many lower-ranking academics had joined the Nazi Party and were only waiting for an opportunity "to force dismissals so they can get jobs."[28]

Actions often bespeak self-image. Since the new regime had come to power and was pursuing its policies under the guise of legality, and since professors did not recognize the utter contempt in which Hitler and his cronies held intellectuals and the law, they must have seen themselves and their situation not much differently than they had before 1933. Heisenberg, Planck, and other elite German physics professors were conscious of their positions as the prestigious creators of quantum physics, heads of major scientific institutions, and leading representatives of German culture. They felt that their importance to their nation transcended their personal preferences. They felt a special responsibility to defend their profession—and by extension their culture—against the intrusive excesses of a regime whose overall goals seemed otherwise admirable, even when defense required such personal sacrifices as engaging in political intrigue. For them, resigning in protest did not square with that responsibility. Instead, during the first year of Hitler's reign, Planck, Laue, and now Heisenberg responded as German professors had done for decades, relying on their high social standing and personal diplomatic skills to influence the new leaders in the right direction.

One retrospectively bizarre expression of the German academic position was the upright Max Planck's audience with the new chancellor, Hitler, just two weeks before Heisenberg's first meeting with Planck. Planck, as president of the Kaiser Wilhelm Society, a powerful network of government-sponsored research institutes, had returned to Berlin from his Sicilian holiday on April 28 to find official seventy-fifth birthday greetings from Hitler. He also learned that many of his closest Jewish colleagues had lost their positions and that the society was being forced to impose the new laws on Haber's Kaiser Wilhelm Institute for Physical Chemistry. Haber having resigned and left the country, Planck used his thank-you to Hitler and his status as society president to approach the new chancellor, hoping to convince Hitler that his policies needed correction.

According to chancellery records, Planck paid his visit on May 16, 1933.[29] Hitler, still dressing in the respectable suit and tie of a German politician, received the balding and mustached physicist in the chancellor's office. Reports differ about what transpired. After the war, Planck recalled that he had argued that Jewish scientists could be good Germans too and that some should be spared for the sake of German science. Haber, for instance, had shown his dedication to Germany by introducing gas warfare during World War I. Hitler responded that he had nothing against Jews—it was the communists he was against—and then flew into such a rage that Planck could do nothing but leave. Einstein heard in 1934 that Hitler had threatened the old man with imprisonment in a concentration camp.[30] But Heisenberg, who visited Planck at his home in the elegant suburb of Berlin-Grünewald at the end of May, transmitted only a positive report.[31] In a June 1933 letter to Born soon after the latter had left Göttingen for northern Italy, Heisenberg told of a pledge that Planck had supposedly received from Hitler: "Planck has spoken—I think I can pass this on to you—with the head of the government and obtained the assurance that nothing will be undertaken beyond the new civil service law that will impede our science."[32]

Perhaps Planck had not told Heisenberg everything that had happened at the meeting; perhaps Heisenberg did not convey everything to Born. Whatever actually transpired at the Hitler meeting, Planck, Laue, and now Heisenberg seemed far from discouraged. They began intensive efforts at working through bureaucratic channels to prevent, delay, and cancel job dismissal orders.[33] Optimism and faith in reasoned diplomacy abounded. Kopfermann recounted the mood and strategy to Bohr: "Laue, for example, who is probably the most optimistic, is attempting to delay all decisions for as long as possible [to provide time to argue against dismissals]. In this way he hopes to be able to save a large portion of those threatened."[34]

Although appeals by scientists and academics to public opinion remained unthinkable—both too crass and too confrontational—private petitions to state bureaucrats multiplied. In June 1933, Planck and Heisenberg circulated a petition in support of the Göttingen mathematician Richard Courant, who, unlike Franck and Born, had decided to stay and fight his dismissal.[35] The same month, Heisenberg issued a personal appeal to Born to return to Göttingen to await improved conditions: "Since only a very few will be affected by the law—you and Franck certainly not, Courant probably not either—the political transformation could take place by itself without any sort of damage to Göttingen physics. . . . Certainly in the course of time the ugly will separate itself from the beautiful. . . . Therefore, I would like to ask you not to make any decisions yet, but to wait and see how our country looks in the fall."[36] The few could be sacrificed for Göttingen and the hope of a better future. Born sent a typed transcript of this portion of the letter to Paul Ehrenfest as an example of the position of "our well-meaning German colleagues."[37]

Born found it difficult to respond to Heisenberg. He seemed genuinely touched by Planck's and Heisenberg's efforts on his behalf; he was willing, he said, to postpone until autumn his decision to leave Germany for good. He was aware that resettling in a foreign country would be difficult for him and his wife, given their age. But as one who had been unjustly placed on forced leave from his tenured position, he felt a responsibility to support others who found themselves in a similar situation. He felt especially allied with Franck, who had publicly resigned to protest the pernicious new law. Perhaps Heisenberg could also understand how hurt Born felt at being rejected by his own countrymen because of his religious heritage and how worried he was about his children's future: "What shall I do now? I will see if I can delay my decision."[38]

Heisenberg showed Born's letter to Planck, who expressed relief at Born's willingness to postpone his decision until autumn. "Perhaps then the situation will be more reasonable," Planck declared.[39] Encouraged by the effect that the Courant petition seemed to be having, Heisenberg initiated a petition in support of Born, which Arnold Eucken circulated in Göttingen.[40] Heisenberg also visited former Leipzig professors in the Prussian Culture Ministry in Berlin at least twice in support of his Göttingen colleagues.[41]

But Born could not wait after all. Pauli, in Zurich, headed one of the international refugee organizations that had sprung up to facilitate the emigration of German scientists to positions abroad. These organizations quickly developed into sophisticated conduits to freedom for persecuted scientists and science students. With generous foundation support, they maintained files on the plights of threatened individuals and provided stipends, at first primarily to British universities, to create three-year positions. At the end of the period, the committees helped the universities create more permanent arrangements. Soon, however, Britain and other European countries could not, or would not, absorb any more refugee scientists, many of whom moved (though also with difficulty) to the United States.[42]

Born had earlier worked in Great Britain, and through Pauli's emergency committee, in early July 1933 he accepted a three-year appointment in Cambridge. Born originally intended to remain on leave from his Göttingen post, still hoping to return one day.[43] But after his oldest children, left behind in Göttingen, experienced further indignities, he gave up that dream. On the eve of his departure for Cambridge, Born resigned. "One cannot serve a state that treats one as a second-class citizen and that treats children even worse," Born told Sommerfeld. Planck, who was vacationing at a nearby Italian villa, was, as might be expected, "very depressed."[44]

The autumn of 1933 did not bring the hoped-for return of reason—the amelioration of Nazi policies, if not the collapse of the Hitler government—but it did bring news of interest to Heisenberg: he was to receive the prestigious Max Planck Medal at the September meeting of the German Physical Society in Würzburg. Heisenberg chose not to attend the meeting and went instead to a concurrent Copenhagen physics conference, held annually at Bohr's institute.[45] Perhaps he wanted to reaffirm the Copenhagen connection for refugee physicists. During his travels to and from Copenhagen in September 1933, Heisenberg stopped in Berlin for daylong consultations with Planck and Laue on the state of their profession.[46] Laue, then head of the German Physical Society, had for months been gathering lists of dismissed physicists of all ranks and passing them to Bohr, who was a member of and fundraiser for a refugee organization and was preparing to establish a Danish committee to facilitate emigration to Denmark.[47] Heisenberg acted as an occasional conduit of information to Bohr.[48]

Planck and Laue were also fighting hard to ward off new attacks on their profession. A decree had been issued revoking the habilitation of non-Aryans. At the same time, the Nobel Prize-winning Nazi physicist Johannes Stark attempted to gain control of the German Physical Society and the Emergency Association of German Scholarship (*Notgemeinschaft*) and to join the powerful Prussian Academy as the new führer of German physics. Laue publicly and courageously opposed Stark at every turn. Although he succeeded in thwarting many of Stark's ambitions, he soon found himself defending the very mention of Einstein's name and the teaching of relativity theory in Germany. His actions prompted praise from Einstein and fears for his safety.[49]

Planck, soon to display the courage of his convictions by holding a forbidden public funeral service for Fritz Haber, who had died of heart disease in exile, submitted a memo to Heisenberg's former Leipzig colleagues at the Prussian Culture Ministry on behalf of non-Aryan habilitants in general and Lise Meitner in particular.[50] Although women scientists had been excluded from German laboratories until after World War I—they would distract the men from their more important work, it was thought—Meitner, an Austrian-Jewish physicist, had worked for decades under Otto Hahn's protection in the Kaiser Wilhelm Institute for Chemistry in Berlin. With Planck's support, she had become one of the first female physicists to habilitate in Germany and even received Einstein's adulation as "our Madame Curie." Now, Planck's renewed efforts on her behalf failed. Meitner lost her habilitation and was banned from her classroom. She remained isolated in Hahn's semi-private laboratories until she was forced to flee when Nazi Germany annexed Austria in 1938, just weeks before the most momentous discovery to come out of Hahn's laboratory—the discovery of nuclear fission. To formalize policy, a Reich statute, first promulgated in 1934, restricted habilitation, and hence future professors, to Aryans; and then, in a 1938 revision, only to those who could prove their political reliability by attending a Nazi indoctrination camp for university lecturers.[51]

Besides working to combat these affronts, Heisenberg, Planck, and Laue had to confront a new challenge to their professional strategy of holding dismissed colleagues in Germany while attempting to have the dismissal rescinded: the possibility that some professors would break ranks and resign of their own accord, even without threats from above. Their fears were realized when non-Jews Hermann Weyl in Göttingen and Erwin Schrödinger in Berlin left their posts in early September, citing health and working conditions. Heisenberg, still preoccupied with his profession and believing that others should show similar dedication, was especially angry with Schrödinger, "since he was neither Jewish nor otherwise endangered."[52] A shaken Planck wrote to Laue with a stiff upper lip: "I regard Schrödinger's resignation as a new deep wound to our Berlin physics, which we must endure with all of the energy available to us."[53]

On his return from Copenhagen in early October 1933, Heisenberg met with Planck and Laue to discuss the situation. Both of the older men despaired of their ability to halt the exodus of physicists, including those who were apparently not even in danger.[54] They tried on Schrödinger their previous strategy, even though it had failed in every case—buying time by turning the resignation into a temporary leave, then making a strong personal appeal to the individual to return to his post while maneuvering with bureaucrats behind the scenes. Heisenberg took the lead. He followed up successful visits to the former Leipzig professors in the Prussian ministry with a long exchange with Schrödinger, which he again shared with his mentor, Planck.[55] But again the strategy failed. Schrödinger accepted a temporary position at Oxford, arranged through a refugee organization. He then held a position in Graz, Austria, until the German army invaded in 1938. He ended up in Dublin, Ireland, where, as with Einstein in Princeton, an institute for advanced study was built around him.

Having failed in their efforts to keep their colleagues in Germany, Planck, Laue, and Heisenberg turned to what seemed to them the most positive response under the circumstances—finding worthy replacements to fill the teaching chairs left empty by dismissed or departing scientists. Heisenberg again played a leading role. While Planck attempted to have Max von Laue appointed to Einstein's former position at the Prussian Academy, Heisenberg made inquiries of non-Jewish foreign physicists, especially among the Dutch, concerning their interest in a German chair.[56] Planck considered Heisenberg himself a candidate for a key position. Heisenberg thought that Planck would have him replace his archrival, Schrödinger, in Berlin. "Among the blind, the one-eyed man is king!" he exclaimed.[57] But the decimation of Göttingen physics and mathematics and the ultimate failure of the petition on behalf of Courant (who, after a brief interlude in Cambridge, founded a now-famous mathematics institute at New York University) apparently convinced the dean of German physics that Heisenberg should fill Born's now-vacant position.[58] The optimism of the previous spring gave way to grim determination. Recalling the height of Göttingen physics only a decade earlier in a letter to Franck (then in Copenhagen) in early 1934, Heisenberg declared: "I fear that a long time will pass before such a time of scientific enthusiasm will be possible once again in Germany. But I want to hold out here. That I will do everything in my power for our Göttingen, you may be sure."[59]

The advantages of historical hindsight are again apparent. Although the exodus of leading scientists enhanced the influence of those who remained behind, there is no reason to think that Heisenberg's conscious intentions and those of his advisors were less than upstanding. Planck's sense of duty and ingrained commitment to upright behavior would not permit otherwise.[60] Heisenberg, Planck, and Laue, after all, regarded themselves as representing German's highest cultural ideals. Rather, it is in their focus on profession instead of individuals and instead of broader principles, and in their failure to take a moral stand (if not a political one), that Heisenberg and his advisors went astray. Their new response to regime anti-Semitism—filling vacated positions—may appear reasonable when viewed in terms of preserving German physics. Yet there is no indication that they ever reflected on the broader implication of this tactic, that the preservation of decent science under the Nazi regime would support the arguments that National Socialism was not so bad after all and that it was not fundamentally incompatible with the ideals of scientific inquiry. Nor did they appear to consider that the preservation of decent physics in such places as Göttingen might play into the hands of those who said that the Jewish professors were not needed anyway. Most disturbing is the ethical implication that was ignored: to participate in finding a replacement for a man who had been unethically dismissed or who had resigned in protest could be seen as tacit acceptance of the grounds for dismissal and denial of the legitimacy of the protest. The Nazi regime confronted Germans with extremely difficult moral and political decisions for which even the most upright among them were thoroughly unprepared.

Heisenberg traveled to Göttingen in early 1934 in pursuit of the new policy. After delivering a lecture entitled "Atomic Theory and Knowledge of Nature"—scheduled for a time when some younger faculty and students did not have storm-trooper drill— Heisenberg met in the home of the head of the University League (*Universitätsbund*) with local representatives of industry, science, and the Emergency Association.[61] In view of the "crisis situation which has surprisingly developed in the mathematical-physical fields at the University of Göttingen," they decided to revive an old academic-industrial alliance set up at the turn of the century by Felix Klein for the support of mathematics and physics and, in addition, to offer Heisenberg an immediate appointment as Born's successor.[62]

Although Heisenberg's colleague Debye did not believe that he would leave Leipzig, Heisenberg readily accepted the position.[63] But his appointment at Göttingen first required approval by higher authorities—the local Nazi University Teachers League and the new Reich Education Ministry (REM). Both caused trouble. Nazi opponents of the plan had little use for modern theoretical physics. They were alerted when, at Heisenberg's suggestion, Sommerfeld's recent student Fritz Sauter applied for and received faculty approval to serve as Heisenberg's assistant in Göttingen.[64] After bureaucratic maneuvers and the recovery of (deliberately?) mislaid files, the REM finally approved Sauter's appointment. He began lecturing on theoretical physics in the summer of 1934, giving the first course on the subject given at Göttingen in nearly a year. "The only thing missing is you as institute chief," he wrote to Heisenberg.[65]

But such great opposition to the appointment of a leading theorist had arisen in the wake of Sauter's approval that the REM, preferring to avoid a confrontation, simply refused to act further. Robert Pohl, the only remaining Göttingen physics professor, reported that he, the dean, and even the rector had pressured the REM, but to no avail: "I told you, the difficulties lie in Berlin."[66] Even Heisenberg's invitations to prestigious professorships in the United States could not be turned to his favor.

A year later, a frustrated Heisenberg finally agreed to drop the Göttingen offer, which was still pending. By then there was plenty of consolation in another opportunity to occupy a key position: Sommerfeld had officially selected Heisenberg as his successor in Munich. At the same time, ominous signs were appearing on the horizon. Heisenberg's calls to Göttingen and Munich had so incensed Nazi physicists that they mounted an increasingly nasty, increasingly public campaign against theoretical physics that temporarily prevented Heisenberg's call to any German physics chair. The Nazis' transition from surreptitious meddling with science education to public denunciation of theoretical physics marked the end of the Planck-Laue-Heisenberg replacement strategy and the beginning of a new phase of the Nazi challenge to academic physics.[67]

But at the end of 1933, the replacement strategy had seemed a good one, and Planck was extraordinarily pleased at the dedication shown by his younger colleague.[68] During the plenary session of the German Physical Society meeting on November 3, Planck conferred on his younger colleague the highest award a German physicist could receive, the Max Planck Medal. The aged Planck had knighted a new member of the German scientific elite. Less than a week later—just a month short of his thirty-second birthday—Heisenberg received a telegram from the Royal Swedish Academy. He had been awarded the Nobel Prize in Physics.

CHAPTER 16

SOCIAL ATOMS

ANYONE FORTUNATE ENOUGH NOT TO HAVE LIVED UNDER A DICTATORSHIP SUCH AS Hitler's would find it difficult to imagine life in Nazi Germany. Nazi control extended not only horizontally—over social, political, and legal institutions—but vertically, from professional and regional strata down to the daily lives of individual citizens. It was never static. Widespread sympathy with national revival and weak response to early repressive measures encouraged tighter control, thereby creating an even deeper sense of helplessness. Because of this vicious circle, the social and political horizons of the individual German began to shrink from the global expanse of nation and profession to the private sphere of one's closest friends and associates. This privatization of public political experience entailed what Hannah Arendt has called the paradoxical "atomization" of the new "mass society" into isolated individuals, existing in a world of twisted reality, inverted ethics, and continual fear. For us of the post-Nazi and post-Soviet era, it should be easy to recognize where such a world will ultimately lead, and hopefully we are inspired to encourage measures to counter it. But for those who lived within this first modern Western encounter with what would culminate in the incarnate evil of the death camps, the potentialities of unchecked demagoguery, character defamation, and intrusion into private lives could not be imagined during the early years of that regime.

Heisenberg's experiences and reactions were typical of those of many educated Germans. Like most, he was at first sympathetic toward the regime's nationalistic aims, if not all its individual policies.[1] At the same time, his letters to his mother display a narrowing sphere, as his professional efforts were thwarted and his personal life fell under increasing scrutiny and control. In 1933 and 1934, he maneuvered to counter the effects of the dismissal policy—the ethnic "cleansing" of the universities, as the Nazis called it, a word that has sadly come again into usage. But by 1938 he referred to a political assault against the whole of modern physics and equally against himself as merely "our small, private political problems."[2]

State intrusion into private life intensified with the dismissal policy, but again without any concerted resistance. In June 1933, Heisenberg, like other public servants, succumbed to the indignity of submitting his parents' birth and marriage certificates to state authorities to determine his ethnic origin. German documents invariably listed religion. (Similarly, American documents betray a special interest in race.) "I *must* turn them in," he lamented to his mother.[3] Travel abroad now required state approval;

spies operated in lecture halls and laboratories; party members informed on friends and associates; attendance was required at official ceremonies and marches and occasionally at a political indoctrination camp. Heisenberg, like most, did not openly resist such requirements. He nonchalantly informed Mrs. Bohr of his required attendance at a weekend indoctrination camp in February 1935. A month later, universal military service was introduced for all men to age 45, again without objection. A year earlier, Heisenberg had already planned to volunteer for an "army sports camp," in order, he wrote, "to acquaint [myself] a little more with this politics."[4]

Psychologically, the most devastating form of control was the solemn oath of personal allegiance to Hitler required of all civil servants and soldiers. The pledge, instituted by law in August 1934, was a final step in the imposition of dictatorial rule: control over the personal ethics of civilians and soldiers alike.[5] Such a matter was not taken lightly, even though many who swore the oath found they could not abide by it. According to university records, Heisenberg finally signed his civil-servant oath by January 1935.[6] Later, as an army reservist, he probably also swore the soldier's allegiance to lay down his life for führer and fatherland.[7]

Like most other Germans in his situation, Heisenberg rationalized his compliance with the more obvious measures of control by regarding them as the least compromise required to achieve his higher aims of preservation of what he held dear. Yet compliance, for whatever reason, did not hinder the even more devastating measures that inevitably followed, confusing as well as stunning most Germans. Heisenberg's future wife, then a new gymnasium graduate, later wrote of the mood in 1933, "Most people were insecure and frightened and were not really sure what things were leading to, or what to believe."[8] Young people, many of whom were in the vanguard of the Nazi movement, looked to their elders for guidance in such matters. They often found them either supportive of the regime or bewildered by it. Many turned to those who did offer answers—National Socialist demagogues.

Heisenberg's unit of the youth movement, the New Pathfinders, had dreamed for over a decade of an approaching third Reich, headed by a savior-like führer who would lead Germany out of Weimar decadence, greedy capitalism, and national disgrace. By the late 1920s Hitler was winning over a new generation of students and youth-movement members with similar phrases as the National Socialist Party experienced a steady upsurge in support. Heisenberg's generation of youth-movement alumni (*Altmannen*) warned against such seductions.

At a 1931 Easter camp of the *Jungmannschaft* (young men's group), one of Heisenberg's comrades presented a fireside lecture, "The National Movement and Our Political Situation," in which he tried to make the distinction between the youth movement's apolitical ideals and "Hitler's empty political phrases."[9] Receiving such apolitical encouragement and holding such ominously idealistic notions as Reich and führer, it is little wonder that when independent youth groups were outlawed in 1934, nearly all younger New Pathfinders joined the Hitler Youth, while many older members joined the SA (storm troopers) or the SS. Most founding elders quietly retired.

Altmann Heisenberg met with his aging group one last time that summer before dissolving their formal assembly forever.[10]

The bewilderment of the older generation is evident in the diversity of their modes of personal response. Heisenberg's future wife, Elisabeth, recalled that many "now felt called upon to add themselves to the equation by joining either the party or the SA."[11] Party records and postwar affidavits indicate that, like many middle-class professionals in those years, a number of individuals close to Heisenberg chose this route.[12] Some joined the party. Several key associates in Heisenberg's professors' club became party members, as did Heisenberg's close colleague and friend, Pascual Jordan. Some family friends joined for fear of losing their jobs. One colleague of Heisenberg's father found himself a member of the SA when, in 1934, the SA absorbed the Steel Helmet (Stahlhelm), a paramilitary veterans' organization to which he belonged. He quit six months later but still had to face postwar denazification.

Despite much that has been written by and about Heisenberg during those difficult years, his private position remains difficult to reconstruct in detail. The written record, the principal stuff of history, loses its reliability when viewed against the backdrop of dictatorial politics. In contemporary accounts, critical statements may be absent because they could have been used against their author. Positive remarks may have been made to win the favor of a bureaucrat. Third-party reports must be read through the opportunist lens of the informant. Postwar accounts of events may be clouded by the terrible suffering brought on by the regime and the war. Actions may speak louder than words, but only when they are reported in their own time and viewed within their own context.

While captured documents show that Heisenberg never joined the Nazi Party, he did hold high scientific positions throughout the Third Reich and performed required duties without complaint. He never openly broke with the regime and continued to pursue his science. Still, as indicated previously, on a personal level he disapproved of what he regarded as Nazi excesses and defended his profession as conditions changed. To us, from the vantage point of the present, many of Heisenberg's responses are frustratingly weak, insensitive, even repugnant. Of course, there are limits to what any individual or group of individuals can do, and can be expected to do, under such difficult circumstances. Not everyone is a hero or a martyr. Once recognizing those limits, however, why didn't Heisenberg refuse to cooperate? Why didn't he eventually emigrate in disgust at the new regime? How did his own responses appear to him at that time?

As with many Germans of that period, whether directly persecuted or not, Heisenberg's lifelong attachment to his homeland could not be broken except under the most extreme circumstances. His entire life and career were bound to the culture and people and soil of Germany. After the war, Heisenberg attempted several times to describe the situation as he and others perceived it. The most extensive of his statements, of which the others seem derivative, is an unpublished manuscript dated November 12, 1947.[13] It was written, apparently in support of Ernst von Weizsäcker, who was accused and convicted at Nuremberg in the so-called foreign ministries case. The manuscript displays the expected surfeit of postwar apologetics and special pleading. Its purpose was, after all, to exonerate the (directly and indirectly) accused. Moreover, much of what Heisenberg wrote applies to the war years, for which Weizsäcker had been brought to trial and during which home-front dictatorial control was at its zenith. Nevertheless, a sense of the world in which Heisenberg lived, and a vestige of the difficulties he endured even before the war, seems to rise through the special pleading and over-rationalizations.

An intensifying dilemma characterized the situation of those who, like Heisenberg, sympathized with the nationalistic aims of the regime but did not condone the repressive assaults against their profession and their friends. With most Germans supporting Hitler, those non-Jewish Germans who opposed Nazi measures were forced, wrote Heisenberg rather simplistically, to choose between two alternatives: passive or active opposition. For Heisenberg, passive opposition meant emigration, either literal emigration "to enjoy security from persecution in a foreign country" or "inner emigration," withdrawal "from all responsibility." As the wording suggests, passive opposition was, for Heisenberg, tantamount to desertion. After the war, he told one critic, "I have never been able to have the least sympathy for those people who withdraw from all responsibility and then tell you during an innocuous table conversation: 'Well, you see that Germany and Europe are going to pieces; I always told you so!'"[14]

In Heisenberg's reconstruction, active opposition offered only black and white alternatives. One side entailed direct opposition with armed resistance and public protest. This response, which Leipzig socialists and the coup d'état conspirators of 1944 had chosen, was for Heisenberg admirable, but at best hopeless and at worst irresponsible. You would only end on the gallows or in a concentration camp, he reasoned, and your sacrifice would go unreported and unknown. Consider the tragic example of the White Rose pacifist student group whose leaders were all executed during the war for the heinous crime of distributing antiwar leaflets at the University of Munich. Nevertheless, Heisenberg's argument exaggerates the extent of Nazi control over the news media and daily life during the early years of the regime. When pressed on this point, Heisenberg's Leipzig colleague, Friedrich Hund, simply shrugged in exasperation: "But what could a physicist do?"[15]

Heisenberg wrote in his 1947 statement that Ernst von Weizsäcker and others like him, including himself, followed another form of active opposition—the acquisition of a certain amount of influence that from the outside appeared to be collaboration. "It is important," he wrote, "to be clear that this was actually the only way really to change anything."

From the vantage point of history, Heisenberg's chosen mode of opposition manifests the dilemma and the debilitation experienced by the subjects of the Third Reich. While one could indeed help to ameliorate conditions somewhat, and could do so better the higher the office one attained, this type of opposition also implied full accept-

ance of the situation. The external appearance of collaboration was of less consequence than the internal intentions; what little one could change was more valuable
than how much one could not. Such thoughts reveal the political ambivalence and
misjudged perception of effectiveness in which Heisenberg and others indulged. Yet
because of this position, many non-Jewish Germans later felt that their own courage
and suffering were never fully appreciated by those who had never experienced a dictatorship. Theirs was not so much the fear of persecution and murder—to which there
can be no comparison—but of discovery, the constant fear that the appearance of collaboration would be suddenly stripped away to reveal their inner opposition, an opposition that only their courage prevented from breaking out into open resistance. The
failure of the regime to strip away the appearance preserved the larger illusion of
opposition that made collaboration a way of life for so long.

Theirs was a life confronted daily with painful contradictions and compromises,
all endured for the sake of appearance in order "really to change anything." In his
1939 novel of the Jewish exile, *Exile*, Lion Feuchtwanger captured the feeling in the
complaint of his fictional character Riemann, a highly regarded German musician,
who visits his expatriate Jewish friend Sepp Trautwein in Paris. In one scene
Riemann, after tying to explain why he will not leave Germany, conveys to Trautwein
what he must endure for his decision: "Do you think that it's pleasant? . . . When you
have to stay there day by day, and everything is much worse than you read in the
newspapers, and you have to keep your mouth shut and on top of that play music,
mein Lieber, that's no small thing. When, after you've played Beethoven, you have to
play the Horst-Wessel-Lied, try that sometime, Sepp, and see how it is." To which
Sepp replies: "But who put you up to it?"[16]

Within days of the appointment of a Reich commissar for Saxony, the culture
ministry appointed a National Committee for the Renewal of the University of
Leipzig. The committee consisted of Nazi students and faculty, charged with oversee-
ing the transformation of the university into a "leaders' school of the German Volk."
Students formed the vanguard at Leipzig, as at most universities, in carrying it out.[17]

Simultaneously with the April 1933 anti-Semitic laws, the National Committee
abolished all student organizations except for its own Nazi Students League, which
already controlled the student government. On April 12, the national office of the
Students League unleashed its infamous campaign against the "non-German spirit" at
universities, which in Leipzig took the form of open intimidation of Jewish and democratic professors, one of whom suffered a fatal heart attack.[18]

An Italian visitor to Heisenberg's institute reported home his observations soon
after arriving in Leipzig in 1933: "Leipzig, which has a Social Democratic majority,
accepted the revolution without difficulty. . . . Brown uniforms are not much in evidence, but swastikas can be seen everywhere. The Jewish persecutions delight most
Aryans. The number of people who will find jobs in public and private establishments
as a result of the liquidation of the Jews is considerable."[19] By the end of May, two
American visitors to Leipzig reported the situation at the university in relatively

favorable terms: "It is clear that conditions in Saxony are not so fanatical as in Prussia, and the professors dare to speak out much more. One sees comparatively few flags on the street."[20] Because of poor press, the Saxon authorities even ordered the student leader, SS Sturmführer Friedrich, to rein in his student minions.[21]

During the first semesters under the new regime, the National Committee acted as liaison between the Saxon authorities and the university in the matter of dismissals. According to one count, in 1933 the university lost 21 professors to dismissal, resignation, forced retirement, and death.[22] The Physics Institute lost one professor with the unfortunate name of Marx, who headed the radio division, two assistants—Felix Bloch and Heinrich Sack—and one technician, Heisenberg's lecture assistant, Herr Dornfeld.[23] The minutes of meetings of the Leipzig philosophical faculty, to which the Physics Institute belonged, yield little evidence of the behind-the-scenes diplomacy that resulted. An entry for the meeting of May 17, 1933, attended by Heisenberg, reads simply: "The dean [geophysicist Ludwig Weickmann] reports on the steps he has undertaken with the government as a result of the new professional civil-servants law. A discussion about it is not desired." The dean's reports of dismissals and ministerial orders likewise appeared in the official record without comment.[24] Nazi professors and state and party officials naturally enjoyed full access to such records.

Other less public (though written) sources yield a somewhat fuller picture. News of agitation against Jews and foreigners in the Physics Institute reached Dresden by the end of May. On May 30, 1933, a student member of the National Committee audaciously wrote the interior minister himself, demanding the investigation of Bloch, Sack, Marx, and mathematician B. L. van der Waerden on the basis of religious and political rumors about them.[25] Both Bloch and Sack, whom Debye had brought from Zurich, had gone home for the spring break, and both had already decided not to return to Leipzig. Sack, Debye's assistant, had asked for a year's leave to wait out further developments. Debye managed to obtain a fellowship for him in Brussels, but the interior minister recommended dismissal anyway. A protest by the Swiss consul in Dresden and a threat of retaliation against Germans in Zurich were without avail.[26] Sack was replaced in October.

Heisenberg's assistant, Bloch, had, like Sack, realized that little future remained for him in Germany.[27] Bloch had already applied for a Rockefeller stipend to work in the United States before he learned of the anti-Semitic laws. According to the rules of the Education Board, which administered the stipend, a fellow had to specify where he or she would work after the end of the grant. At Heisenberg's urging, Bloch had indicated Leipzig. As the new law became public, Bloch wrote a worried letter to Bohr, then in the United States, asking Bohr to intervene with the Education Board to change his future location to Copenhagen. Bohr agreed, prompting a thank-you from Heisenberg "for . . . your efforts on behalf of our young physicists, whose well-being lies on all our hearts," and an apology "for all of that which is now happening in this country." A year later Bloch, then in Rome, accepted an invitation from Stanford.[28]

Little remains to illuminate the extent of Heisenberg's role as conduit to Bohr and Pauli, both of whom were active in international refugee organizations. Most of the correspondence on such matters was destroyed when Pauli and Bohr themselves joined the exodus from Europe during the war. Records do remain regarding Heisenberg's reactions to the plights of three young physicists: Gerhard Herzberg, Herta Sponer, and Guido Beck. Because his wife was Jewish, Herzberg received notice in 1934 of dismissal from his assistantship in Darmstadt. The physicist-philosopher Michael Polanyi, then in Manchester, England, appealed to Heisenberg, who made inquiries in Darmstadt. Herzberg should not have been dismissed under the civil-service laws, but because of his state-defined position, local bureaucrats could do as they pleased. Heisenberg could do little more for Herzberg in Germany and suggested that Polanyi might do better for him in England.[29] Herzberg and his wife soon settled in Canada.

Heisenberg was more directly successful in helping Sponer and Beck—but again only by looking abroad. Though not Jewish, Sponer, a member of Franck's Göttingen institute, realized that her future as a woman physicist in Nazi Germany was bleak. "The German woman belongs in the kitchen!" as Hans Kopfermann facetiously put it.[30] Sponer lost her job, and Heisenberg learned of her decision to emigrate during his attempts to salvage Göttingen physics. After several inquiries abroad, he managed to find her a position in Madrid.[31] She later moved to Chicago, where she and Franck were married.

As an Austrian Jew who was unemployed when Hitler came to power, Beck was harder to place. Beck's assistantship with Heisenberg had already ended in early 1932 and could not be continued. With no job offer and no position in Leipzig, Beck's future was dark. Bohr had secured funds for him to go to Copenhagen until Easter 1934, but it was only a temporary remedy. Throughout most of 1933, Bohr and Heisenberg sought other funding for him. By the end of the year, they finally created a unique position for him at the German University in Prague with the combined support of Dutch, Danish, and Swedish foundations. Beck remained in Prague only a few years, traveled to the Soviet Union, then went to France, from which he escaped to Latin America during the German occupation.[32]

Erich Marx's dismissal, announced to the faculty on September 27, 1933, came in the midst of a power struggle in the Physics Institute and at a time when the university itself was undergoing internal battles.[33] More than a month before Marx's removal, and probably in anticipation of it, heads of the two other technical divisions in the institute petitioned the Saxon ministry for independent institutes of their own. With technical physics in the ascendant in the new state and Marx in trouble, the technical physicists made their move. Debye, apparently regarding defense of Marx as hopeless, countered with a replacement and a redefinition of his field from radio to radiation physics. With the help of Dean Weickmann, Fritz Kirchner, a nuclear physicist, was quickly appointed. The independence of the technical divisions was prevented, but not without the festering resentment of the division chiefs. A year later they aimed their hostility at Heisenberg.

While Heisenberg substituted as institute head during Debye's absence in Brussels in 1934 and 1935, Kirchner was invited to assume a full professorship with his own institute in Cologne.[34] The offer was too tempting to refuse. Heisenberg was now faced with thwarting the renewed pretensions of the technical physicists. He managed to hold them in check, but their close connections with Debye's longtime party-affiliated assistant may ultimately have worked against the institute heads. Heisenberg's Nazi opponents later enjoyed inside information on institute affairs.[35]

During the first full school year under the new regime, intensified university reorganization went far beyond the exclusion of unwanted faculty. Characteristically, a new university leader proclaimed: "The university in the new Germany will be *political*, an educational institution for political persons who place their knowledge and abilities in service to the nation."[36] Politics, not in the democratic sense of party politics but in the Fascist sense of "service to the nation," or Volk, meant subservience to the new representatives of the nation, the National Socialist German Workers' Party. To realize this aim efficiently, Nazi student leaders organized their followers into close-knit work and study teams. A "teachers' corps" was formed for middle- and upper-level lecturers to control senior faculty through the approval of new habilitations and faculty appointments.[37] Partly unsuccessful attempts were made to introduce the "Führer principle"—dictatorial one-man rule. The new führer of the university, Arthur Golf, joined with the student SS Sturmführer to proclaim an end to the "conflict" between the apolitical professors and their "political" students. Professors and students now joined hands as "comrades under Hitler."[38]

Throughout all this, Heisenberg maintained his academic position and offices but withdrew privately into his physics, his music, and his circle of friends and colleagues. In his memoirs, he recalled a challenging encounter early in the regime with a leader of the Leipzig Hitler Youth.[39] The session must have taken place before June 1933, when Heisenberg moved from the institute into a small house.[40] Heisenberg remembered meeting the student after he had finished practicing a Schumann concerto on the piano of his institute abode. He invited the student, who had been listening in a window well in the hallway, into his living room. The student, dressed in the starched brown uniform of the Hitler Youth, was puzzled. Why did Heisenberg, still a young man and a close associate of the youth movement, now hold himself so aloof from the younger generation, "like one of those conservative old professors who live completely in the past"? Heisenberg's sympathies seemed to lie with the youth leader, who claimed that "Germany needs political liberation from her state of inner corruption." In his recounting of the meeting in his memoirs, Heisenberg even appears to admire the youth's passion for bringing about change through direct action. But Heisenberg recalls drawing the line at mixing means and ends. "You must realize that I cannot help you when Germany is being ruined," he declared. "It's as simple as that."

But, of course, things were not quite that simple, for shortly afterward Heisenberg turned to Planck for advice. Events a few months later also challenged his preferred aloofness from National Socialists and from those outside his immediate circle of friends, associates, and institute members, learned that he would receive the prestigious Max Planck Medal, and on November 9 received the exciting news of having been awarded the 1932 Nobel Prize.[41] At the same time, Heisenberg refused to participate in a highly publicized national rally held on November 11 in Leipzig under the auspices of the National Socialist Teachers League. The rally, a widely publicized "demonstration of German scholarship," supported the nation's withdrawal from the League of Nations, to be decided ostensibly by a referendum and an election on November 12. Heisenberg informed the rally organizer, physicist Johannes Stark, that he would not attend.[42]

Numerous teachers and students, four university rectors, and six professors did attend, among them the noted philosopher Martin Heidegger (also a rector). A vindictive Stark informed Leipzig students of Heisenberg's refusal to join the "acknowledgment by professors to Adolf Hitler." Students, delighted by Heisenberg's prestigious prizes but angered by his failure openly to support the cause, were thrown into confusion. "How vehemently the debates swirled about you in those days," one lecture student recalled, "when at the beginning of the winter semester 1933 your refusal to participate in the election rally resulted in a small scandal in the institute! And how much support for you among the students finally outweighed everything else!"[43]

Elisabeth Heisenberg recounts an episode she heard only later about a befriended Nazi youth leader—perhaps the same young man Heisenberg had met earlier—who warned Heisenberg that a band of students planned to disrupt his main lecture the day after the announcement of his Nobel Prize. The student managed to defuse the demonstration by deftly leading the protesters out of the lecture hall. The remaining students broke into a standing ovation.[44] Only a few days later, on the foggy evening before the scheduled teachers' rally, nearly the entire Leipzig student body honored the new Nobel laureate with a torchlight procession—the admiring students marching row on row to his home in storm-trooper formation.[45] A week later, Heisenberg invited the district leader of the Nazi Students League to his home to defuse any lingering animosity. The young man officially exonerated the professor of any fundamental opposition to Führer and state.[46]

Following the November uproar and a mysterious trip to Zurich on behalf of refugee physicists, Heisenberg traveled by train to Stockholm with his mother to receive his Nobel Prize from the hand of the reigning king of Sweden. After stopping in Copenhagen on the way to thank Bohr personally for collaboration that had led to the prize (and for recommending him), Heisenberg and his mother arrived at the Stockholm train station where they were greeted by two other physicists, Paul Dirac and Erwin Schrödinger, who were there to share the 1933 Nobel Prize in Physics.

The conferral of prestigious prizes and the attribution of discoveries to individuals are always crucial matters for scientists. Recognition by colleagues and posterity are more important for scientists than for most other professionals. A special aura of prestige surrounds the Nobel Prize, and the Nobel committee, which selected the recipients, was acutely aware of this. The physics prizes conferred in 1933 were especially noteworthy in this regard. Reflecting the domination of physics by experimental work and the committee's preferences for tangible results, most physics prizes had previously gone to experimentalists or to those whose work greatly influenced experimental research. For the first time, three theoretical physicists were chosen primarily for their contributions to theoretical physics. The official citations, however, indicated the peculiar manner in which the committee viewed the nature and relative contributions of these three men. Heisenberg was awarded the reserved 1932 Nobel Prize "for the creation of quantum mechanics, the application of which has led, among other things, to the discovery of the allotropic forms of hydrogen"—at least a partly tangible result. Dirac and Schrödinger shared the 1933 prize "for the discovery of new and fruitful forms of atomic theory."

Bad feelings abounded. Max Born's contributions to quantum mechanics had been slighted, as were those of Jordan, Pauli, Bohr, and others; Schrödinger must have winced at the designation of Heisenberg as the creator of quantum mechanics, while he and Dirac seemed only half worthy of a full prize for "new and fruitful forms" of the theory.[47] Although the awkward situation may have been made worse by the professional situation—Born and Schrödinger had both recently left Germany, despite objections from Heisenberg, and were now without permanent jobs—their level of recognition and the apparent over-recognition of Heisenberg seemed the biggest difficulty. Anecdotal claims to the contrary, Heisenberg appears to have maintained cordial relationships with Born and Schrödinger to the end of their lives.[48] He expressed his own unease with the entire affair on several occasions in 1933. After writing an apologetic note to Born from Zurich in November, he wrote to Bohr on November 27, 1933: "Concerning the Nobel prize, I have a bad conscience regarding Schrödinger, Dirac, and Born. Schrödinger and Dirac both deserved an entire prize at least as much as I do, and I would have gladly shared with Born, since we have also worked together."[49] The Nobel committee eventually made up for its oversights, awarding prizes to Pauli in 1945 and Born in 1954. Bohr had received his for the Bohr atom in 1922.

The distress and excitement of Stockholm behind him, Heisenberg returned to his work and to his small circle in Leipzig. With Bloch now gone, Heisenberg's recent doctoral graduate, Carl Friedrich von Weizsäcker, emerged at the center of his closest associates. During this period, Weizsäcker met with Heisenberg daily for long discussions and soon became his teacher's closest confidant. Their relationship was so close at times that Heisenberg would often feel abandoned when Carl Friedrich was away.

The son of a long line of aristocratic German diplomats, Weizsäcker, who maintained his noble title, Freiherr von Weizsäcker (Free Lord from Weizsäcker), could always turn a diplomatic phrase when needed.[50] While his brother, Richard von Weizsäcker, studied law (he eventually rose to the presidency of postwar West Germany), Carl Friedrich intended to study philosophy. He added the field of physics to his interests after meeting Heisenberg in Copenhagen during the twenties. Although

Carl Friedrich von Weizsäcker, 1933.

11 years younger than Heisenberg, Weizsäcker, with his high forehead, narrow chin, and aristocratic demeanor, was more involved than Heisenberg with philosophical matters. They discussed philosophical issues endlessly in Leipzig and on their frequent hiking and skiing expeditions, often with former youth-movement members, in the Bavarian hills. As the political situation deteriorated following the National Socialists' rise to power, Heisenberg came to rely on Carl Friedrich and his intellectual understanding of worldly matters. "Only the friendship with Carl Friedrich, who struggles in his own serious way with the world around us, leaves open to me a small entry into that otherwise foreign territory," he wrote to his mother in October 1934.[51] Hitler had consolidated his hold on higher office just two months earlier.

Weizsäcker remained in Leipzig as Heisenberg's friend and assistant until the younger man habilitated and transferred to Berlin in 1936. During that period, Heisenberg's retreat into his personal circle hastened in the wake of the Leipzig events of 1935. Armed with the anti-Semitic Nuremberg laws and no longer restrained by a now-deceased Hindenburg—the president who had appointed Hitler and demanded special exemptions for Jewish front veterans—the regime introduced its second civil-service purge to prepare for its first four-year plan: economic and military mobilization in anticipation of the coming war. The new purge provoked a crisis in the Leipzig philosophical faculty. On May 1, 1935, at the request of the Saxon Education Ministry, Martin Mutschmann, the Reich potentate for Saxony, ordered the immediate dismissals of Professor Joachim Wach (theology) and Assistant Friedrich Levi (mathematics) and the forced retirements of Professor Benno Landsberger and

Professor Weigert.[52] Being only part Jewish or World War I front veterans, the four had been spared earlier.

Heisenberg reacted with shock and dismay. Not only did he believe that the worst was already over, but he was also well acquainted with the victims, Wach being a member of the professors' club. The order now made it obvious that there was no protection for Jews (or anyone else, for that matter) in public service, whatever their accomplishments or status. Heisenberg also realized that the university would never be free from bureaucratic meddling.

Heisenberg reviewed the situation, which would prove to be a turning point in his relationship to the state, with his closest colleagues, Weizsäcker, Hund, B. L. van der Waerden, and Karl Friedrich Bonhoeffer. He then met with another member of his professors' club, Dean Helmut Berve, to formulate a response. Berve, history professor and party member, had served as dean of the entire philosophical faculty since late 1933. They all agreed to express their disapproval at the next faculty meeting, knowing full well that a report of their protest would eventually reach Dresden, the capital of Saxony. A more direct confrontation with the authorities would probably not succeed, they reasoned, and might result in even more dismissals, including their own.

The somber professors assembled on May 8 in their faculty's oak-paneled branch library for the carefully planned and orchestrated meeting. Dean Berve opened the session with a formal announcement of the dismissals. Under the steady gaze of the portraits of their predecessors lining the library walls, Heisenberg, Hund, van der Waerden, Bernhard Schweitzer, and Konstantin Reichardt—all closely associated with Berve and the club—voiced their prepared objections. Reichardt inquired about the grounds for the dismissals; Rector Felix Krueger responded that his own inquiries in Berlin had been fruitless. Berve complained that the faculty had been kept completely in the dark. Van der Waerden, who wondered if there had been other grounds for the dismissals, was followed by Heisenberg. "Herr Heisenberg doubts that the measures now being taken are consistent with the intention of the law, according to which front veterans also belong to the Volk community, and he holds it a duty to his friends to help those affected in any way he can." Hund noted that nonveterans should be ashamed at the treatment of the veterans. Finally, Berve interjected—for the record—that he was permitting the discussion only to report to Dresden the faculty's mood.[53]

The meeting adjourned to await Dresden's response. It was not long in coming. Van der Waerden recalls that within a day, even before Berve submitted his report, the education ministry had been informed. As might be expected, the protests had little effect. The dismissals and retirements remained in force. Several days later, van der Waerden and probably Heisenberg received formal reprimands via the rector from Mutschmann himself.[54] On May 12, Heisenberg wrote his mother that the rector was pressuring him to enter the German Army as a reserve officer in order to demonstrate his loyalty to the Reich, despite his voiced objection to this application of regime policy. His hope of becoming dean of the science section of the philosophical faculty seemed doomed.[55]

Berve was also punished, having allowed his faculty to express its displeasure. The Reich commissar himself ordered Berve's replacement as dean at the end of the semester. Berve recalled after the war that the local head of the Nazi University Teachers League subjected him to considerable abuse for merely meeting regularly with his professors' club friend Wach in a Leipzig restaurant. On the day on which the club made its last outing with Wach before his departure from Leipzig, Berve received the order from Dresden to step down as dean.[56]

Perhaps because Heisenberg and the others were so shocked by the dismissals—which seemed so arbitrary, yet so symptomatic of life under Hitler—this episode emerged as a crucial moment in their prewar lives. Heisenberg, Hund, and van der Waerden remembered these events as having occurred two years earlier, when the dismissal policy was first introduced. Apparently it was not until 1935 that they fully understood how arbitrary the Nazis could be and how powerless they were to object by traditional means. Van der Waerden recalls that Heisenberg had expected at that time an even worse punishment than a reprimand—dismissal.[57] But in view of his Nobel Prize, international fame, and Max von Laue's example of survival in office despite opposition to Nazi measures, he was unlikely to suffer that fate at that time.

Years later, Heisenberg recalled that with the faculty protest failure, he and his colleagues at last contemplated resigning from the university.[58] Since they had been unable to turn back the dismissal policy or to halt the dictatorial transformation of the university during the preceding two years, resignation had become by then an appropriate—probably the only appropriate—political and moral alternative. Even as lower-ranking incompetents waited in the wings, the Nobel laureate's actions would send a signal far beyond the Leipzig faculty. A fundamental decision was now necessary. Torn and uncertain, Heisenberg turned again to the aged Max Planck for direction.

The now 77-year-old Planck received the 33-year-old physicist in the dark, old-fashioned living room of his home in the upper-class Berlin suburb of Grünewald. All that seemed to be missing, Heisenberg wrote, was an antique oil lamp on the table in the center of the room.[59] The tired and somber Planck seemed somehow years older since their last meeting. Despite their differences in age, background, and opinion regarding quantum mechanics, the two physicists had grown closer in outlook during the last few years under the mounting repression of academic physics. As in 1933, Planck saw in Heisenberg a hope for the future of German physics, and Heisenberg saw in Planck a man to respect, a symbol of the esteemed tradition, achievement, and integrity of German science. During the first phase of dismissals, Heisenberg had traveled often to Berlin to seek political advice from Planck and Laue, both of whom were well known for their rectitude. But this time even Planck seemed bewildered.

Resignation in protest was never an option for Planck. In 1933, he had talked his Berlin colleague Otto Hahn out of such a move in favor of fighting from within. At that time, he still hoped to influence government policy through personal diplomacy. But in 1935, as he and Heisenberg sank into the overstuffed chairs in the gathering darkness of Planck's dimly lit living room, Planck's only counsel was one of political acquiescence. Planck saw events unfolding as an avalanche, out of control, even out of the government's control. For Planck, further political or diplomatic maneuvering was now futile. Resigning a position in protest seemed to Planck equally hopeless, and even a form a desertion. Heisenberg's resignation would surely go unreported and unknown.

Instead, according to Heisenberg, Planck argued that "it is to the future that all of us must now look." They should withdraw from the present climate, but not from their posts. Planck believed they were now charged with a new responsibility—to preserve enclaves of decent German culture and science as "crystallization points" for a better world.[60] It was a return to the well-practiced politics of Weimar, to the very position that Heisenberg himself had settled on long ago amidst the throes of revolution and coup d'état: the dissociation of good German culture from the ephemeral world of bad political rulers, the separation of nation from state, profession from administration, intentions from appearances. In the end, such distinctions played right into the hands of their enemies.

Planck's advice was the kind that Heisenberg understood and welcomed, if indeed he had not mixed his own views with his memory of the meeting. It gave Heisenberg new impetus to fulfill his rationalized role of active collaboration—refusal to desert his high post to accept one of the many tempting offers he was receiving from abroad. A captain could not abandon his ship, no matter how violent the storm—especially not as the storm worsened. Planck may have prepared Heisenberg with similar advice in 1933. In his June 1933 letter to Born, just after having visited Planck in Berlin, Heisenberg wrote of his desire to work amidst "a circle of people who understand that we live for them and for the science entrusted to us"—much as he had lived for his youth group amid the chaos of post–World War Munich.[61] By late 1935, Heisenberg's horizons had shrunk to the circle of his science, his students, his friends, and his closest colleagues.

In the fall of 1935, as a new semester started and new troubles mounted, Heisenberg wrote his mother of his new task, as he perceived it: "But I must be satisfied to oversee in the small field of science the values that must become important for the future. That is in this general chaos the only clear thing that is left for me to do. The world out there is really ugly, but the work is beautiful."[62] Heisenberg had made his fundamental decision. If possible, he would remain in his homeland and at his post no matter how fiercely the storm raged in the years ahead.

CHAPTER 17

OF PARTICLES AND POLITICS

HEISENBERG'S RETREAT AT THE END OF 1935 INTO THE CIRCLE OF HIS STUDENTS, HIS and his music reaffirmed his previous position. He was a physicist, and a good one. The beauty of physics, like the beauty of classical music, offered an escape from the "ugliness" and "general chaos" of the "world out there," and his contributions to this field would last for years. The nature and intensity of his correspondence throughout the 1930s, especially with his longtime colleague Wolfgang Pauli, indicate his fervent devotion to those ideals. Only occasionally do hints of personal and political conflicts filter through the abstract discussions of technical minutiae. Physics and physicists, the scientists told themselves, were meant to exist in a world of their own.

Yet events "out there" continually intruded. Science possesses both a form and a content—personnel, workplaces, funding, and communications, as well as the theories, experimental data, and accepted criteria and procedures of advance. All were under siege in those years. Preservation of the form—the professional dimension of science—a duty that no German physicist could refuse, placed Heisenberg and his colleagues on a collision course with Nazi demagogues and in a position that required enormous compromises for the sake of their science. Before settling on what he called his "island of existence" in Leipzig, Heisenberg was forced to contend with these real-world conflicts and compromises—but not before his science received a new impetus.

In September of the first year of Hitler's regime, soon after the news of the Schrödinger and Weyl resignations, the discoveries of the positron and of cosmic-ray showers turned Heisenberg's physics in a new direction, toward high-energy physics and back to the problems he and Pauli had encountered earlier in their efforts to combine relativity and quantum theory in an overall theory of electrons and electromagnetic fields. Those efforts had resulted in what was known as "relativistic quantum field theory." Applied to electrons and electric fields, it was called "relativistic quantum electrodynamics." Although the basic approach of this theory seemed correct, it stumbled badly over the appearance of infinities. The worst infinity arose from the theory's prediction that the energy of the electron should be infinite, which it obviously is not, owing to it zero size. This becomes especially important when electrons and other elementary particles smash into each other at such high energies that they approach closer than their supposed sizes, even almost to zero distance.

The smashing of particles together at such high speeds that they collapse into each other is the basic process occurring in high-energy accelerator experiments even today. Before the advent of accelerators of such high energies during the late 1940s, research in this branch of physics focused on the observed behavior of cosmic rays smashing into matter. Cosmic rays are extremely high-energy light quanta and sub-atomic particles, such as protons and nuclei, entering the atmosphere from outer space. Their origins are still somewhat of a puzzle, but most seem to come from our galaxy and even beyond, and arise from exploding stars, including, in the case of a gamma-ray burst, from giant stars collapsing suddenly into black holes. These are among the highest energy events occurring anywhere in the universe. They form part of the natural background radiation to which we and everything on the surface of the earth are exposed.

In 1933 Patrick Blackett and Giuseppe Occhialini, working in Cambridge, England, had used the newly invented cloud chamber to observe cosmic rays reaching sea level and colliding with a thin lead plate placed within the cloud chamber. The cloud chamber consists of a vapor of gas that condenses into droplets when a particle passes through it. The track of droplets formed in the gas by the passing particle can then be photographed and studied. If magnetic and electric fields are placed on the chamber from outside, scientists can determine the charge and mass of the particle from the curvature of the track. Blackett and Occhialini, who later received the Nobel Prize for their work, made the surprising discovery that when a high-speed cosmic-ray particle hit the top of the thin lead plate in their cloud chamber, a "shower" of particles was ejected from the bottom of the plate, apparently created from the collision of the cosmic ray with a lead nucleus in the plate.[1] Shortly before this discovery, Carl D. Anderson, working at Caltech, had announced, on the basis of his studies of cloud-chamber photographs of cosmic-ray events, the discovery of a new subatomic particle. He had found a new "positive electron," or positron—a particle having the same mass as the electron but a positive charge, instead of the usual negative charge.[2] Previously, only negative electrons, along with protons, neutrons, and light quanta were the known elementary particles.

The only explanation for a positively charged electron was found in Dirac's relativistic quantum theory of the electron, published in 1928, in which he had predicted the creation of positive particles from collisions of high-energy light quanta with heavy nuclei.[3] Dirac thought that these positive particles might be the familiar protons. But when their predicted mass was found to be that of the much smaller electron, Dirac suggested that they might be a new type of electron, an "antielectron."[4] An antielectron would have the same mass as but the opposite charge of a normal electron. Most importantly, when an antielectron (or positron) meets an electron, the two particles would annihilate each other, leaving behind their energy in the form of two light quanta radiating away in opposite directions. By the same token, when a high-energy light quantum hits a nucleus, it could disappear upon the creation of an electron and a positron, a process known as "pair creation."

Anderson's new positron looked identical to Dirac's antielectron, inspiring renewed interest in Dirac's theory and in the creation and annihilation of particles, both by

*Dirac and Heisenberg in
Cambridge, England, 1930s.*

Heisenberg and Pauli in Europe, and by their American counterparts, J. Robert
Oppenheimer and his group of California theorists at Berkeley and Caltech.

Dirac's Cambridge colleagues, Blackett and Occhialini, suggested that the creation of electron-positron pairs might be the origin of the cosmic-ray showers they
had just discovered.[5] Upon hitting the lead plate, a high-speed cosmic-ray electron
would be deflected and absorbed by the lead, and in the process radiate away its
energy in the form of a light quantum, a photon. The photon would then produce an
electron-positron pair, which might then go on to produce more photons, and so on,
resulting in a small shower of particles that could then be observed emerging from the
bottom of the lead plate.

Other theoreticians working with the Cambridge experimentalists produced a
series of calculations on the processes involved in the stopping of high-speed particles in matter. Heisenberg, however, was preoccupied in 1933 with his research on the
structure of atomic nuclei, with the preparation of his report on the subject for the
Solvay Congress of October 1933, and with his efforts to respond to the dismissals
and resignations of his colleagues. He did not learn firsthand of the new experimental results and of Dirac's latest theorizing until the annual Copenhagen gathering of
physicists in September of that year.[6]

The theoretical results on the stopping of particles in lead plates, published soon
after that Copenhagen meeting by Oppenheimer and his student Wendell Furry in the
United States and by refugees Hans Bethe and Walter Heitler in Britain, were only partially encouraging.[7] Pair creation could account for the stopping of particles in matter, but only for energies well below those observed for showers. As it had earlier, the

theory contained infinities at high energies or small distances of approach, indicating that the electrons should be readily absorbed in the lead. This, however, directly conflicted with the long distances that high-speed cosmic rays are observed to survive in large blocks of lead. Because of this, it seemed to most theorists that their theory broke down at some upper limit of energy. Above that upper limit, the theory simply stopped working, at which point the electrons could travel as far as they wanted without interacting with matter, even though they should interact if the theory were correct. The theorists set the upper limit, as before, at an energy corresponding to the size of an electron in classical theory. Or, using the uncertainty principle, this energy corresponded to the energy contained within the mass of an electron at rest.

Returning to Leipzig from Copenhagen, with a Berlin stopover to meet on political matters with Planck and Laue, Heisenberg was both optimistic about reversing Schrödinger's resignation and excited about the positron. "What do you say about the positive electrons?" he asked Sommerfeld in October 1933. "It appears that once again Dirac was much more right than we had previously thought. . . . Hopefully you will find enough rest from politics during the winter semester to enjoy the positive electrons."[8]

Heisenberg himself enjoyed the positive electrons for several months, before political maneuvering again demanded his attention. During the negotiations in Göttingen with Dean Brandi on the filling of vacant teaching chairs, Heisenberg and Brandi had sought new ways to preserve theoretical physics in its former stronghold and throughout Germany. One way, they decided, would be to convince engineers and industrialists of the intrinsic value of theoretical physics and of the advantages of hiring young theoreticians. With the regime emphasizing research for immediate, practical ends, mainly in preparation for war—nearly the antithesis of Heisenberg's work at the time—seemingly impractical theoretical physics required bolstering.

Heisenberg arranged in Göttingen to present his case for theory personally to the chairman of the powerful Association of German Mining Engineers. Later he offered a lecture called "Science and Technical Progress" at the annual convention of the association in Düsseldorf. The talk appeared in the association's periodical, *Stahl und Eisen* (*Steel and Iron*). His efforts paid off. He managed to convince the influential chairman and many of his followers "that the existence of a good theoretical physics in Germany is of decisive significance for the collaboration of technology and science."[9]

But Pauli, working quietly in Zurich in neutral Switzerland, was less convinced of the positron physics that Heisenberg was also producing in this same period. By the following June, Pauli proclaimed his disgust with their joint positron research program, complaining of "far-ranging negative results from the physical point of view," in particular of infinite energies and even the existence of negative energies, which normally have no physical meaning—energy either exists or it does not; it can be either positive or zero.[10] Two days after Pauli's resignation, Heisenberg informed him that at least one promising physical phenomenon did result: the scattering of photons by photons.[11] Debye suggested that this effect might be the origin of the sun's corona. While Pauli and his assistant, Viktor Weisskopf, attempted to formulate an "anti-

Dirac theory," Heisenberg turned the scattering problem over to two of his brightest students at that time, Hans Euler and Bernhard Kockel.[12] Although he kept abreast of their work, Heisenberg was soon embroiled once again in other matters: "There are so many other urgent and remarkable things here in Germany with which one must concern oneself for the time being," he told Pauli.[13]

Among those other things was the need for an even more direct and public defense of theoretical physics as well as a personal response to Johannes Stark. Stark still held a grudge against Heisenberg for his refusal to rally for Hitler the previous November, and he could now call upon an even more powerful Hitler for support.[14] Hitler had gained new political strength through his bloody purge of the Storm Troopers (SA) during the infamous "night of the Long Knives" in late June. Within days of Hindenburg's death in early August 1934, Hitler announced the uniting of the offices of chancellor and president in himself. This gave Hitler, as Führer, control over both the state and the political apparatus of the Reich. Hitler's officials scheduled a plebiscite on the plan for August 19, and the Nazi Party launched a major propaganda campaign to ensure a favorable vote—which could hardly have been otherwise. Nazi physicist Johannes Stark sent telegrams to his fellow German Nobel laureates, inviting them to join in a public declaration of support for Hitler. Physicists Heisenberg, Laue, Planck, and Walther Nernst all refused, once again invoking their old argument that science and politics should not mix.[15]

Stark responded angrily in letters to his colleagues that public support for Hitler was not a political act at all, but merely an "avowal of the German Volk to its Führer." At the same time he claimed that the Nobel laureates were already taking a political position by speaking publicly in favor of Einstein.[16] It was the kind of perversion of science and politics that so characterized Stark's thinking. Laue told Stark not to write him any more personal letters.

A month after the Stark episode and Hitler's reported overwhelming victory as Führer, Heisenberg joined other leading scientists in presenting again the case for modern theory, including Einstein's theories, to the broader scientific public. This time, the prestigious Society of German Scientists and Physicians (GDNA), meeting in Hannover in September 1934, served as a ready forum. Heisenberg delivered the keynote address, entitled "Transformations in the Foundations of Natural Science in Recent Time," to the plenary session of the society. In his survey of recent developments, Heisenberg informed his audience that the newest theories, particularly the relativity and quantum theories, were not the products of purely speculative, "revolutionary ideas" foreign to science—as Stark would have it—but were rather the necessary results of years of careful research based upon cogent empirical methods and cherished elements of classical physics. Relativity and quantum theory "have been forced upon research by Nature in the attempt to carry the program of classical physics consistently to its end."[17]

Indeed, Heisenberg declared, the latest transformations in science had occurred step by step in concert with Stark's field of research, experimental studies. For instance, a revision in the classical notions of space and time had become necessary only after the results of Albert A. Michelson's experiments to detect the electromagnetic ether; and Philipp Lenard's investigations on the photoelectric effect had been adequately interpreted only through Albert Einstein's light-quantum theory. By referring to Jewish physicists by name and in the same sentence with the name of Nazi physicist Lenard, Heisenberg was defending the very mention of Jewish physicists and their work in German lecture halls. The professional audience in this particular lecture hall did not seem to object. In fact, two other speakers seconded his remarks—experimental physicist Walther Gerlach and the famous surgeon Ferdinand von Sauerbruch, then president of the GDNA. As a further show of support, Heisenberg's lecture, with names, was published in near record time in the pages of the widely read *Naturwissenschaften* (*Sciences*) and avidly reprinted elsewhere. Heisenberg seemed pleased with the result. The lecture "went decently," he reported to his mother, "and I was very satisfied with the behavior of the public; if it helps at all is another question."[18]

Einstein's name and his theories also appeared in Heisenberg's Leipzig course lectures that winter.[19] But by then the mention of Einstein and the explicit teaching of his theories were more dangerous acts than Heisenberg may have realized. After the Hannover lecture, one band of aged Nazis in Hamburg threatened to provoke students to disrupt his classes unless he publicly retracted the lecture.[20] A Nazi functionary complained directly to party ideologue Alfred Rosenberg, suggesting that "the concentration camp is doubtless the suitable place for Herr Heisenberg." Rosenberg agreed but regretted that because of Heisenberg's international stature, he could only issue a reprimand.[21]

After the inevitable tension of the Hannover meeting, Heisenberg headed for a brief respite in Bohr's Copenhagen institute. He returned to Leipzig in October 1934 to find his pupils, Euler and Kockel, well into their study of the scattering of light on light using the latest quantum electrodynamics.[22] The theory was beset by the same infinite results as before, and therefore subject to the same limitations at high energies and short distances. But now these problems were for Heisenberg no longer ends in themselves but means for obtaining clues to the much-needed new quantum theory. He wrote to Pauli that the situation seemed similar to an earlier one: "With regard to quantum electrodynamics, we are still at the stage at which we were in 1922 with regard to quantum mechanics. We know that everything is wrong. But in order to find the direction in which we should depart from what prevails, we must know the consequences of the prevailing formalism better than we do."[23]

Euler nevertheless pursued the calculation of the scattering of one photon on another for his dissertation, which he completed in November 1935 to the glowing approval of his mentor, Heisenberg, who was working along similar lines. But the results, despite Heisenberg's approval, were hardly to be welcomed. The Leipzig physicists had again discovered that the theory yielded infinities at high energy or small distances of approach on the order of the size of an electron in classical theory.

In a subsequent joint paper, Heisenberg and Euler seemed to relish this result, for, they declared, it provided "an important clue for the further development of the theory." Hard as it may seem following the great triumph of quantum mechanics during the 1920s, Heisenberg and Pauli were now, a decade later, again on the verge of despair and eager for a new revolution in quantum physics.[24]

Reacting as he had during the early 1920s, Pauli responded with a demand for new physical insights, not new mathematics. In November 1935, Pauli nearly repeated his famous remark to Kronig of ten years earlier: "It is very bad at the moment for theoretical physics, and I would grasp at any straw that offers itself to me, who is drowning in the Heisenberg-Dirac formalism."[25] For his part, Heisenberg responded as he had a decade earlier. He began searching the limits of the available theory for clues to the way forward.

Heisenberg now readily conceded that they had reached the limits of their abstract and mathematical theories. He and Euler concluded in their joint paper, submitted in December 1935, that further advance required, not just new mathematical techniques, but new knowledge of experimental results on events occurring at tiny distances shorter than the electron radius. In particular, an analysis of new experimental data on cosmic-ray showers, which still seemed to indicate a breakdown of positron theory at an energy corresponding to the size of the electron, "is probably indispensable."[26] Leipzig physics had now come full circle, from experiment to theory and now back to experiment.

Yet while circling through the formalism, Heisenberg and Euler persisted in their refusal to acknowledge, at least publicly, any nonscientific influence on their work. The names of Jewish scientists continued to appear without apology in their German publications (publications that most Nazis, save perhaps for Stark, probably did not read anyway). But the "ugly" outside world intruded nonetheless into their Leipzig "island of existence." As early as 1934, the authorities had forced Kockel to resign as Heisenberg's deputy assistant because he had once belonged to a socialist student group.[27] Although he was permitted to stay in Leipzig, his dissertation and career were delayed by the official disapprobation.

Euler experienced similar indignities. After he received his doctorate in 1936, Heisenberg wanted him to replace Carl Friedrich von Weizsäcker as his assistant. Weizsäcker had recently habilitated and could thus no longer serve as assistant. Also, as Heisenberg wrote to Debye, now director of the new state-sponsored Kaiser Wilhelm Institute for Physics in Berlin, Euler's talents corresponded more closely to Heisenberg's current interest, high-energy physics, than did Weizsäcker's.[28] But more importantly, Heisenberg hoped to provide political cover for Euler through the assistant's position. But Euler, who privately sympathized with Soviet communism, publicly refused to cooperate with Nazi authorities. He attracted regime attention when he refused to join any party organizations or to attend an indoctrination camp for prospective teachers. When Weizsäcker, with Debye's help, moved to Berlin as assistant to Lise Meitner, the Saxon Culture Ministry used Euler's nonparticipation to deny him Weizsäcker's post.[29]

After the ministry's refusal, Heisenberg managed to have Euler named to the temporary post of "extraordinary assistant," while working to have the position upgraded. Lengthy negotiations a year later between the ministry and the newly named prorector of the university, Heisenberg's professors' club colleague Helmut Berve, finally resulted in Euler's promotion to full assistant.[30] But a plan to habilitate him in 1938 caused further problems. After demanding repeated statements from Euler on his political and religious affiliations, the authorities again insisted that he attend a teachers' camp per regulation, and Euler still refused per political scruples. Heisenberg finally advised submission to what seemed to him a mere formality (and he reported for his own required military training that summer).[31] The sensitive young man followed his teacher's advice, only to return to Leipzig thoroughly unsettled by the experience. The weekend of paramilitary drill and frightening lectures by Nazi demagogues on National Socialist science and education ideology must have been unbearable. The treatment accorded Heisenberg's disciples is indicative of the kinds of pressures that the Nazi regime exerted on young academics—the future teachers, scholars, and scientists of the new Reich. It also points toward one of the party's main education goals during the middle years of its reign: the control of teaching personnel and their curricula.

Having lost the battle over dismissals by the end of 1935, physicists now turned their attention to the major remaining issue, other than the actual content of scientific research: the battle for control of the physics profession through the control of academic appointments, curricula, publications, and professional organizations. By protecting the teaching of science—who taught what to whom from what ideological standpoint—the physicists believed they could control the future of German science.

A number of state and party authorities, and individual pretenders to authority, also vied for control of the physics profession. The main institutional contender was the Reich Education Ministry (REM) under Bernhard Rust. Hitler established this cabinet-level ministry, an expansion of the Prussian Culture Ministry, in May 1934 with the initial consent of those with similar dictatorial interests in the area of education: Rudolf Hess (Hitler's deputy party boss), Alfred Rosenberg (party ideologue), and Heinrich Himmler (SS). The REM functioned ostensibly as the highest authority for the naming of university rectors, faculty, and student leaders and for the promulgation of degree and curriculum requirements. But under the weak and vacillating Minister Rust, it also served as a battleground for rival factions and their representatives.[32]

As usual, ideology served the political aims of the various protagonists. As the pretender to authority in all matters academic, the REM attempted to kill the protective notion of the apolitical professor—the renewal of the German Volk demanded submission of the individual to the aims of the Volk. Professors must be leaders and molders of young people, Rust declared in a public address, and would therefore be chosen for their ability to guide young people to their future roles as citizens of the Third Reich.[33] In practice, however, ability counted as much with Rust as ideology.

Nevertheless, Rust's homage to Nazi ideology placed academics in a difficult position, which their enemies could easily exploit. The academics' traditional response to ideological demagoguery—that the apolitical and objective nature of science and scientists necessitated neutrality in research and faculty selection criteria—could be considered not merely annoyingly uncooperative but overtly subversive of Nazi aims. For Nazi ideologues, Nazism was not a political movement for power and influence but a cultural and racial movement for the supremacy of the German Volk. Science, along with all other forms of learning, was thus to be perverted into a racially defined *völkisch*, *deutsch* (German), "Aryan" science; and in every field, those seeking personal or political gain began to propound their own Aryan doctrines.[34] One of the most pernicious of these doctrines was the Aryan physics movement (*Deutsche Physik*) propagated by the Nobel Prize-winning experimentalists Lenard and Stark.

These individuals and their ideology have been discussed and analyzed at some length.[35] They typified to an extreme the predicament faced by many late nineteenth-century physicists transported into the brave new world of twentieth-century relativity theory and quantum mechanics. Both had enjoyed honor and recognition as experimental physicists at a time when experimental physics was preeminent and the German empire was in order. Stark had even been one of the precious few early supporters of Einstein's light-quantum hypothesis; he had performed experiments confirming the hypothesis; and he had helped Einstein publish an early review article on light quanta. But as the monarchy collapsed in democratic socialist revolution and the new field of theoretical physics suddenly gained preeminence, Stark and especially Lenard reacted with bewilderment and anger. That their field was no longer dominant was made painfully clear to them by the world fame accorded Einstein for a new and strange relativity theory that Lenard, at least, despite his Nobel Prize, probably never fully comprehended. Both Einstein and Heisenberg later surmised that Stark's chagrin exacerbated existing pathological elements in his personality, rendering him a paranoid anti-Semite and, like his friend and cohort Lenard, an early follower of Hitler.[36]

Stark's paranoia was confirmed for him when, after he succeeded Willy Wien in Würzburg in 1920, he repeatedly failed to secure another teaching chair anywhere in Germany, despite his Nobel Prize. The difficulty was his own doing. In 1922, he had left his Würzburg chair and used his Nobel Prize money to buy a nearby porcelain factory, where he set up a private research laboratory—all in violation of Nobel Foundation rules. While railing against relativity theory and the Bohr atom from his porcelain works, Stark accused theoreticians of sabotaging his efforts to return to an academic appointment.

For his part, Stark's cohort, the black-bearded Lenard, refused in the same year, 1922, to follow a government decree to close his Heidelberg laboratory and lower the institute flag in mourning after the assassination of the liberal Jewish cabinet minister Walther Rathenau. Leftist students decided to persuade the crusty old physicist by dragging him from his laboratory and dunking him in the nearby Neckar River—an act that fired his paranoid hatred for democrats and Jews even more.

Stark made his final pre-Reich attempt at an academic post in 1928, this time in Munich. There he hoped again to succeed the now lately deceased Willy Wien. But a faculty committee headed by Sommerfeld refused to place him on the list of candidates. Nobel laureate Stark seemed to Sommerfeld unqualified for the job: he had renounced his teaching position, violated Nobel Foundation rules, and produced little of scientific value from his private lab. Nor could Sommerfeld imagine working with a man whose scientific and political views were so far from his own.[37] The Munich faculty settled on the conservative Nobel Prize-winning experimentalist Walther Gerlach, who had worked with Otto Stern on a famous 1922 atomic-beam experiment that lent overwhelming support to the existence of half-integer quanta in atoms. From then on, Stark regarded Sommerfeld as his enemy.

By the late 1920s, Stark and Lenard were turning out an increasing number of speeches and pamphlets on their anti-Semitic dogma of Aryan physics.[38] This dogma, though never formally articulated, had at its center the Nazi belief that human creations, such as science, are the sum of individual contributions, which are in turn expressions of each individual's race and ethnic origin. It was the antithesis of the Marxist concept of science, according to which science is the product of the material conditions of the class that produces it, and different material conditions would result in a different science. In counterpoint to then-current Soviet notions of capitalistic versus proletarian science, Lenard and Stark propounded "*völkisch*" notions of an Aryan versus a Jewish physics. In the foreword to his 1936 textbook on classical mechanics, titled *Deutsche Physik*, Lenard proclaimed a racist foundation of physics: "'Deutsche Physik?' one will ask.—I could have also said Aryan physics or physics of the Nordic-natured man, physics of the searchers of reality, of the seekers after truth, physics of those who have founded natural research.— 'Science is and remains international!' someone will object. He is however fundamentally in error. In reality science, like everything else that mankind produces, is conditioned by race, by blood."[39]

For Lenard and Stark, "Jewish physics" found its ultimate expression in the so-called formalistic relativity and quantum theories, which, they claimed, were not only intellectually obscure but in fact contradicted nature—that is, their naive realist conception of it. Aryan physics, on the other hand, was based on the objective and easily understood truth of Newtonian mechanics. This distinction, however, became explicit in their writings only as conditions changed after 1933.

As Nazi "old warriors," those who had joined the party in its early days, Stark, then 59, and Lenard, then 71, suddenly found themselves close to the seats of power in 1933. "At last the time has come when we can bring our conception of science and of scientists into play," Stark wrote to Lenard, and Lenard, though in retirement, boldly suggested to Hitler that in all university personnel matters Hitler should "obtain my counsel before making a decision."[40] Hitler did not need any advice.

Like regime policies as a whole, Lenard's and Stark's ideological writings through 1935 reflected less concern for academic dogma than for self-serving political manipulation—the freeing of physics from "Jewish-Marxist domination" through the appointment of themselves to leading positions. In a collection of articles published in 1934, a copy of which Heisenberg kept for evidence in his private files, Stark railed against the "Jewification of German science" during the Weimar years. This had supposedly resulted in "the holding down or exclusion of individual, German-conscious scientists"—presumably referring to himself—and in the propagation of "great dogmatic theories . . . Einstein's relativity theories, Heisenberg's matrix theory, and Schrödinger's wave mechanics."[41]

While Lenard remained in official retirement and served as unofficial patron and advisor to Nazi physicists, Stark was well rewarded for his polemics. In 1933, an acquaintance of Stark, Hitler's new interior minister Wilhelm Frick, appointed Stark head of the prestigious Physical-Technical Reich Institute (PTR) in Berlin, the German bureau of standards—even though Stark's colleagues had opposed and still opposed the appointment. At the Würzburg meeting of the German Physical Society that fall, Stark unveiled his fantastic plans for the introduction of the "führer principle" into German science, the reorganization of all scientific research and publications under the general direction of the president of the PTR. He then submitted his candidacy for president of the physical society and for membership in the Prussian Academy of Sciences as Einstein's successor.[42] With significant segments of the physics community supporting the new regime and the rest optimistically preoccupied with quiet diplomacy, the physics profession as a whole failed to mount a broad-ranging, public opposition to a physicist who could count on backing by the regime. As president of the German Physical Society, Max von Laue's single-handed but only partially successful opposition to Stark's dictatorial pretensions drew Stark's ire and earned Laue the praise and gratitude of physicists everywhere.[43] But a year later, neither Laue nor awakened scientists from numerous fields could prevent Stark's appointment as president of the Emergency Association of German Scholarship (Notgemeinschaft der Deutschen Wissenchaft), the highly successful funding agency of the Weimar era.

Stark's appointment ultimately led to his undoing. He soon found himself caught between competing regime power blocs within the Reich Education Ministry. As described by others from varying perspectives, hostilities began when Stark's personal plans to administer German science clashed with the government's decision to establish the REM for the same purpose.[44] One author argues that Theodor Vahlen, then head of REM office W-I for university affairs (W stood for "Wissenschaft," scholarship), had appointed Stark to head the Emergency Association in order to co-opt the vociferous physicist.[45] But plans by Erich Schumann and Rudolf Mentzel in REM office W-II (military research) to reorganize scientific research along military lines using the Emergency Association—renamed the German Research Association —met with violent opposition from their intended subordinate, Stark. Hardly envisioning himself a mere pawn of the German army, Stark summarily threw the two men out of his office. This alienated not only REM office W-II but also the army and the SS, of which Schumann and Mentzel, respectively, were members. In his efforts to become the new führer of physics, Stark now realized that he would need the active support and protection of a power center within the regime. Stark turned to Hitler's caretaker of party ideology, Alfred Rosenberg, whom Stark had earlier befriended and whom he now named a patron of the German Research Association.

Late in 1935, as Heisenberg's attention turned again to cosmic-ray showers and high-energy physics, Stark's attention turned to the control of REM appointment policy and to the control of Heisenberg in particular. The Munich faculty had begun to seek a successor for Stark's old nemesis, Arnold Sommerfeld, and Sommerfeld made no secret of his top choice for the position—his former pupil Werner Heisenberg.

CHAPTER 18

HEIR APPARENT

IF PLANCK WAS CONSIDERED THE DEAN OF GERMAN THEORETICAL PHYSICS, THEN Sommerfeld was the professor. Ever since his appointment as Ludwig Boltzmann's successor at the University of Munich in 1906, Arnold Sommerfeld had produced a steady stream of world-class theoretical atomic physicists. His textbook *Atomic Structure and Spectral Lines* had educated a generation of physicists, and even after Hitler's rise to power his institute remained a center for theoretical research. But two years into the Third Reich, his tenure in Munich neared its end. On January 21, 1935, Reich Minister Rust promulgated a new law, under Hitler's signature, mandating the retirement of university teachers over the age of 65 at the end of the semester. Sommerfeld was 66.

Professor Sommerfeld must have learned privately of the new law before its enactment: in early January he repeated his choice of Heisenberg, which he had made back in 1928. Heisenberg, barely half Sommerfeld's age, again could think of no better place to settle than his hometown, where his friends, his mother, the blue skies and Alpine hills of Bavaria, and the Sommerfeld tradition all beckoned his return. Leipzig had been only a temporary way station until he could realize his dream of returning to the place of his formative youth. Once again, he gratefully accepted the honor as Sommerfeld's chosen heir. "If fate should place me in this position," he wrote his former mentor, "I will make every effort to hold up the tradition of the 'Sommerfeld school.'"[1] Anticipating difficulties with the party, he outlined for Sommerfeld his political background in case the information should be needed—as indeed it was. But the initial hurdles to Heisenberg's appointment were bureaucratic, not political.

Heisenberg was still Göttingen's choice as intended successor to Max Born. In addition, Debye was in line to succeed Walther Nernst in Berlin and had just accepted the directorship of the soon-to-be-built Kaiser Wilhelm Institute for Physics; Friedrich Hund was under consideration for a call to Königsberg; and Otto Scherzer, Sommerfeld's assistant, was likely to move to Darmstadt. Despite Stark's ravings against theoretical physics, both the Saxon and Bavarian ministries worried that such a reshuffle would leave Munich or Leipzig without a theoretical physicist. (Their concerns, however, did not deter a new round of dismissals that spring.)

Cause for optimism abounded, nonetheless. With Rust's approval, the Bavarian Culture Ministry accepted an urgent request from the Munich faculty to allow Sommerfeld "to substitute for himself" until a successor could be found.[2] The law permitted such a bureaucratic circumvention. More important, the local Nazi teacher führer—appropriately named Dr. Führer—who had assigned himself the task of evaluating and approving all personnel matters, offered no objection.[3] After Debye assured Sommerfeld that he would not take Heisenberg with him to Berlin and that Heisenberg truly wanted to go to Munich, the Munich faculty submitted its list of candidates to the Bavarian ministry in early summer 1935. According to one source, the list consisted of a single name: Heisenberg. "We would like to help arrange for Heisenberg to work in the place of his choosing," it proclaimed.[4]

Neither the Munich rector nor the Bavarian ministry objected to the prospect of employing one of the most famous physicists remaining in Germany. Heisenberg's candidacy was duly forwarded to the man who now controlled all university appointments in the new Reich—the chemist Franz Bachér, mathematician Theodor Vahlen's deputy chief in office W-I of the Reich Education Ministry (REM).

Anyone who remained in Nazi Germany after the first wave of dismissals was well aware that compromises were required. This was especially so for professors, who were civil servants of the National Socialist state. Anyone who remained in a teaching position or attempted to advance his career by obtaining a better post had chosen to accept the obvious compromises that would be demanded of him. The Hitler salute was required before all public lectures, official correspondence had to be signed "Heil Hitler," and participation in faculty marches, outings, and indoctrination camps could not be avoided. One of the more significant compromises involved professional association with influential government bureaucrats, most of whom had attained their positions solely through their party connections.

By late September 1935 no decision had been made regarding Heisenberg's call to Munich. Debye visited Werner Studentkowski, the Dresden bureaucrat in charge of Saxon university affairs, on Heisenberg's behalf. Studentkowski had earlier demonstrated his "concern" for objective scholarship by founding the Nazi Students League in Leipzig. He had served as Germany's first professor of Nazi indoctrination, also in Leipzig.[5] Debye found in Studentkowski a surprisingly ready supporter of his former colleagues, the famous scientists under his administration. After discussing Heisenberg's transfer to Munich at length, Debye and Studentkowski decided to visit Bachér in Berlin. Both discovered in their separate visits that, with encouragement from Göttingen, Bachér had finally settled on Heisenberg as Born's successor in Göttingen. Because of this, Bachér had sent Munich's request for Heisenberg back to Munich with a demand for the usual list of three or more names from which to choose. Debye and Studentkowski could only object to the bureaucrat "that Heisenberg himself does not want to go [to Göttingen] and prefers Munich. . . . Another acceptable candidate will have to be found for Göttingen."[6]

On November 4, 1935, just as Heisenberg and Euler were completing their latest contribution to positron theory, the Munich faculty duly submitted to the Bavarian Culture Ministry (and ultimately to Bachér) the requested full list of candidates for Sommerfeld's chair. At the top stood Heisenberg once again. There followed "far behind Heisenberg" a list of practically all of Sommerfeld's available former pupils and colleagues: Hund, Wentzel, Kronig, Stückelberg, Fues, Sauter, Unsöld, and Jordan.[7]

Faced with renewed pressure from Debye and Studentkowski and the additional support of Leopold Kölbl, the dean of the Munich philosophical faculty, in which Sommerfeld's institute was located, Bachér and the REM were now inclined to approve Munich's choice. Chemist Dr. Rudolf Mentzel, the SS officer in charge of REM office W-II (office for military research), had already found another candidate for Göttingen—Richard Becker, a hapless theorist at the highly regarded Berlin Technical College who had run afoul of the local ballistics professor, an army general. Upon complaints from the general, Mentzel dispatched Becker to Göttingen without appeal. This move left the Technical College with the largest number of physics students in Germany but without a lecturer in theoretical physics—a circumstance that clearly demonstrated to German physicists the willful attitude of their masters.[8] But it also left an undeterred Heisenberg free at last to move to his beloved Munich.

The decision was postponed again. Old warriors Stark and Lenard, wielding their sharpest battle-ax—Aryan physics—suddenly unleashed a public campaign to thwart Heisenberg's appointment and to strengthen their own influence on REM policies. As Munich again settled on Heisenberg, and as Heisenberg and Euler sent their latest theory to press in December 1935, Heisenberg was forced again to deal with the world of "ugly" things.

The Aryan physics campaign for the control of university appointments, including Sommerfeld's chair, has been described in varying detail and perspective.[9] It was closely related to several significant changes in the Third Reich at about the same time. The first was a major shift from internal politics to police-state tactics, as exemplified in June 1936 by the shift of domestic power from Hitler's political interior minister, Wilhelm Frick, to his new Reich chief of police, Reichsführer-SS Heinrich Himmler. The second change concerned the beginning of overt preparations for war in the first four-year plan, which was aimed at achieving economic and military self-sufficiency by 1939. In addition, concurrent with these events, and following the second wave of dismissals in 1935, the Nazis escalated their suppression of any remaining opposition among non-Jewish Germans by practicing a new form of mental tyranny: accusing critics and opponents of simply thinking or acting "Jewish," thus threatening them with the same persecutions that actual Jews faced. Himmler's deputy, Reinhard Heydrich, spoke in 1935 of a "spiritualization" of the Nazi struggle for domestic control now that many of the Jews and overt political opponents had been ostracized—of a "struggle of the spirits" for the hearts and minds of those who remained.[10] Stark and Lenard, though not allied with the Himmler-Heydrich SS power bloc within the regime, eagerly joined the battle.

Stark and Lenard opened their ideological assault in December 1935 at the dedication of the new Philipp Lenard Institute for Physics in Heidelberg. In his laudation for the now retired Lenard, Stark—with his high-pitched voice and meanly contoured face—raved not only at Einstein and other Jewish physicists, who, in his view, had produced the "Jewish formalism" of the relativity and quantum theories, but at "their non-Jewish pupils and imitators" who were teaching and using such physics. Lenard's long struggle against Einstein and "Jewish" physics had, said Stark, only partially succeeded: "Now Einstein has disappeared from Germany; but unfortunately his German friends and supporters continue to act in his spirit." Planck was still head of the Kaiser Wilhelm Society; Laue was still a physics referee in the Berlin Academy; "and the theoretical formalist Heisenberg, spirit of Einstein's spirit, is now even to be rewarded with a call to a chair." Such a situation could not be tolerated in Nazi Germany. "May Lenard's struggle against Einsteinism be a warning," Stark bellowed. "And it is to be hoped that the responsible referees in the Culture Ministry allow themselves to be guided by Lenard in the filling of physics teaching chairs, including those for theoretical physics."[11]

These were ominous threats and serious accusations and demands. Of course, Stark and Lenard had raved earlier against "Jewish" physics and had even included Heisenberg's name in their ravings, but now the situation was much more serious. With the "spiritualization" of the struggle, the SS in ascendance, and war on the distant horizon, the political orientation of non-Jews was of paramount concern to the regime, and to call a non-Jew "the spirit of Einstein's spirit" was to call him an enemy of the Reich. To extract as much political gain as they could from their attack, Stark and Lenard turned to party ideologue Alfred Rosenberg, their ostensible patron in Hitler's cabinet.

Among the many ruthless competitors for power in Hitler's Third Reich, Rosenberg was no match for Himmler, Hess, Hermann Göring, or Joseph Goebbels. His primary claim to the title of party ideologue was his rambling anti-Semitic diatribe, *The Saga of the Twentieth Century*, which, like Hitler's *Mein Kampf*, few had ever bothered to read. Although Hitler continued to keep him in his cabinet, by 1936 his star was already in decline as others rose in power and stature. But he still held one point of light in his crown—he was editor-in-chief of the official Nazi party newspaper, the *Völkischer Beobachter* (*Volkish Observer*), or *VB*. Like official party organs in other totalitarian regimes, the *VB* exerted an influence beyond its estimated circulation in 1936 of 500,000. All party officials were required to subscribe to it; rank-and-file party members showed their support by buying the paper; ordinary citizens read the newspaper to learn the regime's official line.[12] If an ideological subgroup could place an article in the *VB*, it would lend official approval to their position and influence regime bureaucrats to act on their behalf.

Soon after the Heidelberg meeting, Lenard complained to Rosenberg of the lack of due attention accorded Aryan physics in the *VB*. Rosenberg, willing to support the Nazi physicists only under pressure, responded by inviting Lenard to name a science editor to the paper, which Lenard apparently never did. Instead, at the end of January 1936 an article entitled "Deutsche Physik und jüdische Physik" ("German Physics and Jewish Physics") appeared in the *VB*. In February, a transcript of Stark's Heidelberg speech calling Heisenberg the spirit of Einstein's spirit appeared in Rosenberg's magazine for the party faithful, the *National Socialist Monthlies*.[13]

Since the aim of Stark's and Lenard's political onslaught was control over university physics teaching, hence over the future of physics in Germany, an otherwise unknown Nazi physics student at the Berlin Technical College, Willi Menzel (not to be confused with Dr. Rudolf Mentzel), wrote the *VB* plea for the teaching of Aryan physics at German universities. The young man was obviously well coached. Quoting at length from Stark's unpublished Heidelberg address and from the unpublished foreword to Lenard's forthcoming *Deutsche Physik*, Menzel tried, as did his mentors, to distinguish a supposedly overly speculative and mathematically abstruse (formalistic) "Jewish" physics from a solidly researched and experimentally grounded "German" or "Aryan" physics. "Jewish" physics—the relativity and quantum theories—was not just wrong, it was bad—and it was bad because it was supposedly Jewish. For Heisenberg, Planck, Laue, and others who had worked at preserving decent physics at German universities—even if they did not preserve the teachers to teach it—Menzel's closing cry in an official organ of the Nazi party must have been unsettling: "We young people today want to continue the struggle for an Aryan physics, and we will succeed in making its name just as esteemed as German technology and scholarship have been for years." "Jewish" physics was to be suppressed at German universities.

While Stark's Heidelberg diatribe sank to personal attacks on individuals, Menzel's assault focused on the teaching curriculum. He mentioned Heisenberg only once, in passing—as the founder of the formalistic matrix theory—and did not refer at all to Heisenberg's Munich appointment. Menzel's mentors probably reasoned that while a student could complain about curriculum, it would not do for him to complain about a professor, no matter what that professor's position might be. Although the article was a clear setup, a stunned Heisenberg was unsure of its import and unsure whether or how he should respond. At a meeting in February with Munich's Dean Kölbl, which lasted until 1:30 AM, Kölbl confirmed Heisenberg's suspicion that the article "was expressly meant as an attack against me." Because of bureaucratic opposition raised by the article, his appointment at Munich would have to be postponed for the time being, Heisenberg told his mother, "but I can wait; this complete idiocy cannot last forever."[14]

While waiting for the irritating idiocy to pass, Heisenberg was not idle. At about the same time as the Kölbl meeting, a befriended party district leader in Leipzig advised Heisenberg to protest the Menzel and Stark articles by visiting the Saxon Culture Ministry and by demanding an audience with the Reich Minister, which he did.[15] Rosenberg—or more likely an editorial staff member—agreed to print a response from Heisenberg in the party newspaper. The article appeared, together with a comment by Stark, in the February 28, 1936, issue of the *Völkischer Beobachter.* It was so widely read that even the *New York Times* took note.[16] On the same day, Dean Rudorf of the Leipzig science faculty directed a comprehensive letter to Studentkowski, urging him to active support of the physicists' cause.

A now more worried Heisenberg wrote his mother of his apprehension about his future. "I have to have a lot of luck if anything more is to be made of my life."[17] He had begun to personalize the attack even more than was warranted. Of course, the language and the intent of the articles implied a great personal danger for Heisenberg, but they were ultimately aimed not at him but at the physics profession as a whole, of which Heisenberg was being held up as a representative. Nor did his future success in life really depend on whether or not he succeeded Sommerfeld. He began to associate his own fortunes with the fortunes of his profession.

Heisenberg took the lead in responding to Menzel in the pages of the *Völkischer Beobachter*. As he had in his Hannover address to the GDNA in 1934, Heisenberg offered party newspaper readers a carefully worded explanation of the nature and value of contemporary theoretical physics and of its importance to the education of young German physicists. This time, however, he did not dare mention any Jewish names. The goal of physics, he wrote, is not only to observe nature but to understand it. Mathematics is often the most suitable way to formulate natural laws, but theoretical systems of concepts are more essential, and these often must be adapted as science progresses.

Relativity and quantum theory served Heisenberg as examples. Further research on these theories in particular, "from which perhaps the strongest influences on the structure of our entire spiritual life will arise, is one of the most important tasks for German youth," Heisenberg proclaimed in the Nazi newspaper. "Evidently courage is not quite dead in the universities," the *New York Times* editorialized.[18] But Stark's response, preceded by an introduction by Rosenberg himself, left no doubt that party orthodoxy stood squarely behind Stark's reiterated demand: "The type of physics that Heisenberg defends ought no longer, as it has until now, exert a decisive influence on filling physics teaching chairs."[19]

Two days after the *VB* exchange, Studentkowski met with Dr. Mentzel, the REM official, to discuss Leipzig physics. A memo that Studentkowski wrote to himself after the meeting was to the point: "Professor Heisenberg remains in Leipzig. Professor Hund likewise remains in Leipzig."[20] Despite his penchant for Nazi demagoguery, Studentkowski apparently had no use for "Aryan" physics and had cultivated friendly relations with leading Leipzig scientists who, because of their reputations, could perhaps work to his advantage in Saxon affairs. Regaardless of the reason the Nazi bureaucrat supported the scientists, previous studies of this episode have overlooked the crucial role he played in the affair at this stage.

In a long reply on March 24, 1936, to Leipzig Dean Rudorf's February 28 request for assistance, Studentkowski told of successful meetings with various bureaucrats involved in the affair. First, he wrote, he had managed to find "cover" for Heisenberg in the person of Dr. Curt Lahr, head of the Saxon State Chancellery, who had direct access to the Reich plenipotentiary for Saxony. Heisenberg would not be molested as long as he remained in Saxony. Second, Studentkowski had made inroads with Rosenberg, who, as readers of the *VB* already knew, had backed Stark before declaring the entire controversy closed.

Most importantly, Studentkowski had Mentzel's ear. Mentzel's earlier conflict with Stark and the Stark-Lenard bid for control of physics appointments would, one might think, have made Mentzel eager to support the physicists. But Mentzel appeared more cautious to Studentkowski: "He does not at all approve unreservedly of Stark's attacks, but he also has certain reservations about the present methods of theoretical physics, or better, about over-reliance on them."[21] Stark's and Lenard's connections with Hess's office in Munich through the Munich-based Nazi leagues for students and lecturers, in the person of Dr. Führer, probably gave Mentzel pause. Mentzel and the REM were then seeking improved relations with Hess, Hitler's party deputy.[22] Perhaps, too, Mentzel, acting the impartial bureaucrat, sought to play both sides against each other to his own advantage. For whatever ulterior motive, he willingly granted an audience to Heisenberg—following proper bureaucratic procedures, of course—and asked Studentkowski to relay the message. Studentkowski informed Rudorf: "Mentzel requests Heisenberg through me to visit him sometime, and to make an appointment as soon as possible." Rudorf passed the letter to Heisenberg with his best wishes for an early meeting with the SS officer.[23]

Heisenberg's meeting with Mentzel, probably in early April 1936, was a turning point. According to a form letter sent by Heisenberg, Hans Geiger, and Max Wien (Willy Wien's cousin) to the entire German physics professorate shortly after the meeting, Mentzel, ostensibly speaking for Rust, had requested during the meeting a memorandum signed by "most" German university physicists describing their understanding of the relationship between experimental and theoretical physics.[24] This memorandum—in effect, a petition for theoretical physics—would provide Rust "a suitable means for his instruction, which will then enable him to alleviate the unpleasantly tense situation that has recently arisen." In other words, Mentzel (and possibly Rust) was willing to move against Stark and Lenard, but only if he had the support, in writing, of the overwhelming majority of German physics professors. Heisenberg, Geiger, and Wien wrote a carefully worded memo and attached it to their letter, requesting its return to Heisenberg's private address—with or without signature—by May 19, 1936. Not surprisingly, Lenard and Stark were livid.[25]

Geiger and Wien made ideal co-authors. Both were wellknown, politically conservative experimentalists, at once sympathetic toward modern theory and acceptable to REM officials. Wien's cousin, Willy, had been close to Stark, and Geiger, the famous co-inventor of the Geiger-Müller counter, was being considered by Mentzel as a replacement either for Debye in Leipzig or for Nobelist Gustav Hertz in Berlin. Hertz, being part Jewish, had resigned in the face of indignities.[26] Wien had already privately expressed some concern for theoretical physics and physics in general in the wake of the first round of dismissals. Encouraged by Debye, he had responded two years earlier by helping to formulate a memo on the declining state of German physics owing to regime policies. If that memo was ever actually submitted, it was almost certainly ignored. One SS Security Service (SD) officer assigned to the REM later recalled an office cabinet stuffed with similar memos![27]

Whether sent or not, the 1934 Wien memo served as the basis for the Heisenberg-Wien-Geiger petition of 1936, which, unlike its predecessor, enjoyed a huge success. Its cardinal point was utilitarian: Germany faced a critical situation regarding physics teaching and personnel that it could ill afford, especially as the four-year plan came into full swing. "The great demand for physicists in technology and the military is met by a lack of suitable candidates. Empty teaching chairs are filled only with great difficulties, and the number of physics students in the beginning semesters is much too small." Moreover, read the memorandum, theory and experiment were both essential for scientific progress and for future technological gains. Public attacks on theory were only frightening students away and damaging Germany's reputation abroad. Attacks against theory must cease in order to stimulate science and to maintain a useful physics profession in Germany.

These were arguments that few could refuse, and few did. Seventy-five physicists—nearly all of the remaining German physics professors—signed the petition. Included were theorists and experimentalists, pure and applied physicists, party members and nonmembers.[28] It was an outpouring of support for physics and especially for Heisenberg and theoretical physics that had been building since 1933. After three years of handwringing, at last, it seemed, something could be done—even if that something were merely to point out the utility of physics to regime bureaucrats.

Rust received the momentous memorandum by October 1936 and handed it to his politically adept state secretary for evaluation. The secretary, thoroughly ignorant of physics, could only respond with a criticism of his weak boss. In the "professors' conflict," he wrote in a private memo, one side—Heisenberg's—had sought help from the REM and had even referred to Rust by name in its cover letter and by title in its petition. If this were a purely scientific controversy, Rust should stay out of it; if it were political, which it was, then he, as culture minister, had even less reason "to mix in these things."[29]

Rust decided to leave matters to Mentzel, who by that time had already bested Stark on another front. The cabinet power shifts of 1936 had greatly reduced the influence of both Frick and Rosenberg, leaving Stark and Lenard without powerful supporters. But Stark ultimately did himself in through foolish mismanagement of German Research Association funds. The rabidly empirical physicist had heavily sponsored a speculative project based on an old Teutonic myth in an effort to extract gold from south German moors. The scandal gave the REM an edge in forcing Stark's resignation in November, and his replacement as head of the Research Association was none other than the indomitable Dr. Mentzel. A visit from Munich's Dean Kölbl to the Berlin REM in the same month brought welcome news, relayed to Heisenberg by Kölbl's colleague Sommerfeld: "Kölbl reported to me from the Berlin ministry that you have won in your controversy with St[ark] and L[enard] both 'scientifically' and 'morally.'"[30] It seemed that Heisenberg would at last replace his mentor.

Closely following the twists and turns of Heisenberg's political fortunes during 1936 was a sudden shift in the prospects for his physics. At the same time as he hit back at the professional assault with a highly successful memorandum of support, Heisenberg firmly believed that he had found an explanation for the appearance of high-energy cosmic-ray showers that would once again revolutionize quantum physics. The coincidence reinforced his optimism in both areas.

Since laboratory accelerators had not yet reached high enough energies to smash particles closer together than their supposed radii, high-energy physics concentrated during the 1930s on cosmic-ray events, which could do so. By 1936, every experiment on the absorption of cosmic rays in matter had seemed to indicate—mistakenly, as it turned out—a breakdown of quantum electrodynamics (QED) near a theoretically expected upper energy limit.[31] High-energy cosmic rays penetrated much further through matter than they should have according to QED. This suggested that the particles simply stopped radiating away their energy at the energy limit where the theory collapsed, and thus continued on their way through matter without slowing down. Apparently QED would not do for shower theories, since showers involved energies far above the supposed upper limit for validity of the theory. The experimental evidence seemed so strong that physicists had greeted with frank disbelief the discovery by Weizsäcker, E. J. Williams, and Lev Landau in 1934 that no good theoretical reason existed to expect any breakdown at all.[32] Some incredulous physicists—notably Oppenheimer and Lothar Nordheim in the United States—had even begun introducing cutoffs of the theory into QED in order to force the theory to stop working as indicated by the data.[33]

Heisenberg took the opposite tack, pushing the existing theory to its breaking point in an attempt to perceive what lay beyond. He caught his first glimpse of that new frontier in May 1936. Exactly one week after the stated May 19 deadline for receipt of the Heisenberg-Wien-Geiger petition at Heisenberg's private address, Heisenberg reported to Pauli from his institute that an alternative to QED yielded a vast new insight.[34] Italian physicist Enrico Fermi, then head of the internationally acclaimed nuclear research group in Rome, had recently formulated a new field theory to handle radioactive nuclear beta decay.[35] This entailed the ejection of a high-speed electron, along with (as only hypothesized by Pauli) a new elementary particle, the neutrino, from a radioactive nucleus as a neutron turns into a proton. (The emission of a positron as a proton decays into a neutron, and the distinction between neutrinos and antineutrinos, were later discoveries.) When taken out of the nucleus and applied to the high-energy collision of a proton with a nucleus, Fermi's new field, Heisenberg discovered, produced an instantaneous "explosion" of new particles as soon as the particles approached closer than a minimum length, about the size of an electron. "It thus appears to me," he wrote Pauli, "that one can understand the existence of cosmic-ray showers immediately from the Fermi beta theory."

Fermi's theory of beta decay, first proposed in 1934, had been widely studied in Leipzig and elsewhere as an explanation both for nuclear forces and for the observed properties of beta decay.[36] Heisenberg, who read Fermi's first paper in Italian, was at first very excited by it. But his enthusiasm waned when he discovered in 1934 that the resulting nuclear force was much too weak to account for the force between protons and neutrons in nuclei or even for the observed distribution of beta-decay energies. Various ad hoc proposals were put forward to fix the Fermi force, and these were intensively studied in Leipzig and Zurich thereafter. All proved unsuccessful. The Fermi force—which later became the weak force of today—remained an unsatisfactory account of the nuclear force and, at the time, even of beta decay.

Nevertheless, as a result of his studies, Heisenberg had hit upon a new and astounding property of Fermi's beta decay theory: when applied to cosmic rays, it could account for the observed creation of particles in a way that the reigning quantum electrodynamics could not, as an explosion of electrons generated by Fermi's theory. Despite his astonishment, Pauli responded with typical skepticism. He did not dispute Heisenberg's new use of Fermi's theory, but he doubted that it offered anything new. At very small distances, Fermi's theory went to infinity even faster than did QED. This feature actually unleashed the particle explosion, but to Pauli it still indicated the need for a new revolution in physics to handle high energies, not the usefulness of alternative field theories. Heisenberg actually agreed, but he believed that Fermi's formalism gave an important clue as to where to look for this revolution. He quickly submitted a manuscript on the subject in early June 1936 and took a copy with him to the annual Copenhagen physics conference later that month. Heisenberg and Pauli seemed to be tracing the same steps they had taken just a decade earlier in their search for a new quantum mechanics.[37]

But this time the need for a revolution—and the direction in which to look for it—was not easily accepted outside the Leipzig-Zurich axis. During the Copenhagen meeting, attended by numerous refugee physicists, Walter Heitler, then in Bristol, stunned the audience with his announcement of Carl D. Anderson's latest results on cosmic-ray absorption.[38] Despite theoretical arguments to the contrary, physicists were still convinced by cosmic-ray absorption data that QED was invalid for energies at which the showers of new particles begin to appear. In May, as Heisenberg was busily submitting his memo to SS Dr. Mentzel and uncovering the revolutionary shower properties of the Fermi field, Anderson was privately informing Hans Bethe and Heitler in England and Oppenheimer at Berkeley of his new data on the energy loss of electrons. His previous experiments had been set up in error. Anderson's new cloud-chamber arrangement suddenly yielded results that completely agreed with the theoretically predicted absorption of high-speed particles using the current quantum theory, including quantum electrodynamics. "As you can see," Anderson wrote cautiously in a letter to Heitler, who passed it on to the Copenhagen meeting, "these data do not show any breakdown of the theoretical formulae for [high] energies. The experiments are not accurate and the numbers measured are small, but certainly no large disagreement appears."[39] If QED did not break down after all, then there was no need for the revolution that Heisenberg was espousing, or even for the introduction of any other field theory to account for high-energy experiments.

Anderson's new experiments also clearly distinguished for the first time the two very different types of cosmic rays at sea level: a "soft," easily absorbable component, consisting of electrons and light quanta (photons); and a "hard," long-ranged component. While the soft component now showed precise agreement with QED to energies far beyond the supposed breakdown of the theory, the hard component still offered problems and was set aside for the time being. It penetrated such large layers of lead blocks that Anderson surmised that either QED eventually did break down at some extremely high energy, or, he soon came to realize, these particles might not be electrons.

Anderson's new data on the soft, electron-photon component of cosmic rays, which he published with his assistant, Seth Neddermeyer, arrived at the *Physical Review* just one day before Heisenberg submitted his paper on explosion showers to the *Zeitschrift für Physik*.[40] These two papers initiated two conflicting approaches to high-energy physics that were strong rivals into the war years and beyond. Heisenberg's theory of "explosion showers," or "multiple processes"—an instantaneous burst of new particles—attracted those who sought a new and even more radical quantum revolution. They were mainly the leaders of the first revolution, Heisenberg, Pauli, and their remaining Central European collaborators. Opposing the Central Europeans were followers of Bethe, Heitler, Oppenheimer, and their collaborators in England and the United States, who had applied the old theory to various phenomena. They now managed to develop a theory of the soft component of cosmic rays, whereby a shower would arise from a "cascade" or buildup of creation and annihilation of electrons, positrons, and photons using QED.[41] For the supporters of this rival account of high-energy showers, no new revolution was necessary at all—QED would do just fine. Heisenberg and Oppenheimer, so similar and yet so different, would never reach agreement on this or other matters during the coming years.

Yet until the Oppenheimer and Heitler groups published their detailed calculations of the formation of cascade showers in early 1937, Heisenberg's theory of explosion showers—the formation of a shower in a single event using Fermi's alternative field—remained the most plausible account of showers, and his physics remained the most likely replacement for QED. For Heisenberg it also served a practical purpose. The new theory became an object lesson in the continued productivity of the much-maligned theoretical physics in Germany. Between mountain outings and a summer military camp in 1936, Heisenberg wrote an article informing the educated German public of the potentially significant consequences of his new theory of showers. These, he emphasized, "are capable of experimental verification." Far from generating purely abstract mathematical formalisms, Heisenberg claimed that he had discovered nothing less than a new universal constant whose "introduction requires a reformulation of the entire theory, as was the case for the constants h and c." As any reader who had already experienced an introduction to modern physics knew, these two constants were the basis of quantum theory and relativity theory, respectively. Further theoretical analyses of empirical data on cosmic rays—not mathematical manipulations—"promise the most important contributions to the fundamental physical questions."[42]

The political and cultural implications of the new physics may have encouraged Heisenberg's more grandiose claims for it. However, there is no evidence that any of his German colleagues saw it that way, even while expressing their overwhelming support for his petition in defense of their profession. Even petition co-author and cosmic-ray investigator Hans Geiger remained unconvinced of Heisenberg's revolutionary theory of explosion showers. Geiger learned of explosions and their possible rival, cascades, not from Heisenberg but at a Zurich nuclear physics conference in July 1936, attended by Sommerfeld, Pauli, and Schrödinger. He preferred cascades, both in theory and in experiment. Although the possibility of explosion showers remained open, Geiger explained in a public lecture as late as 1940, other cloud-chamber photographs taken by British physicist Patrick Blackett appeared to show a buildup of showers, much as in a cascade.[43] If the vivacity of German theoretical physics was to be demonstrated by Heisenberg's new revolution, Geiger would not support the cause. Physics and politics had to be held separate at all costs.

While German scientists were less than enthusiastic about Heisenberg's new theory, Pauli, in Switzerland, once again became Heisenberg's chief collaborator and critic. Their private collaboration was not at all empirical—as Heisenberg's public may have expected from his report—but thoroughly abstract and mathematical. As usual, the two sought to push their formalism to the breaking point and beyond. As a trick for doing so, Pauli resurrected Heisenberg's 1930 lattice world, discussed earlier, whereby Heisenberg had sliced up space into cubes formed by the size of an electron. Pauli brought it with him to the Copenhagen meeting in June 1936.[44]

Heisenberg did not join Pauli's effort until after completing his required eight weeks of military training in the summer of 1936. Heisenberg was a corporal in a mountain infantry unit in Bavaria near the Austrian border. The rigorous training not only interrupted his work, but also forced his cancellation of a planned trip to the United States to attend the annual Ann Arbor summer school in physics as well as the elaborate tercentennial celebrations at Harvard University. Heisenberg's abrupt cancellation of the American trip in order to attend the military training camp made the *New York Times*. The political situation in physics was too volatile to permit his absence from Germany, he explained to his former colleague Samuel A. Goudsmit, then in Ann Arbor.[45] Apparently, adherence to German military requirements came before foreign conferences, he could have found an excuse to dodge the military duty, had he so desired. The old New Pathfinder seemed to welcome the physical challenges of military mountaineering and, he wrote his mother, the escape from responsibility: "I am physically healthy and I very much enjoy the duty itself. It is nice not to have to think for a change, but only to obey. . . . The duty agrees with me in every respect."[46]

Returning to work in Leipzig in October 1936, Heisenberg received a long letter from Pauli proving that the Fermi formalism led nowhere—with or without a hypothetical lattice world.[47] Both physicists again reacted in familiar fashion. Pauli declared that manipulations of field theories were similar to quantizing mechanical models in the early 1920s. Since both contained unobservable quantities, he suggested that Heisenberg, "once again returning from Helgoland," declare all their equations to be fundamentally unobservable, then see what happens. Applied to atomic models, it was in just this way that Heisenberg had formulated the breakthrough to quantum mechanics in 1925.

As in the twenties, Heisenberg wanted first to exhaust all lattice-world models, and he presented Pauli with various constructions for them to study and reject together. He revealed his plan in November to his former Göttingen mentor Max Born, now a refugee in Great Britain: "The showers are still occupying Pauli and me very much. I am very anxious to know how the work there will proceed. Pauli continually tries to prove that wave quantization always goes to infinite; I privately believe that Pauli is right but temporarily maintain the opposite, and in this way we get to know the mathematical properties of a nonlinear quantum field theory, which are highly interesting."[48]

Six days later, Sommerfeld informed Heisenberg of his victory over Stark. Physics and politics, though running on separate tracks, were running for Heisenberg along parallel lines.

CHAPTER 19

THE LONELY YEARS

HEISENBERG'S TWO MAJOR SUCCESSES IN 1936—HIS RESPONSE TO THE NAZI DOGMA of Aryan physics and his construction of a potentially revolutionary physics of high-energy cosmic would, in the end, be short-lived. By early 1937, the alternative cascade theory of cosmic-ray showers, based solely on the accepted physics of positrons and quantum electrodynamics, would account for nearly all of the data on cosmic-ray showers, leaving little need or expectation for any new physics. In addition, by that summer a new and even more vicious attack by the proponents of Aryan physics would make it impossible for Heisenberg to succeed Sommerfeld and nearly impossible for him to remain in Germany at all. Even Heisenberg's native optimism seemed at times outmatched by the course of events. His optimism would receive a "quantum" boost, however, during the first half of 1937, when, in rapid succession, he would meet and marry his wife, the future mother of his seven children. But even with marriage and a growing family, Heisenberg would recall the period before the outbreak of World War II as one of "unending loneliness."

Heisenberg's frequent travels abroad in this period—to Denmark, England, the United States—revealed to him Germany's marked isolation. As nearly every domestic challenge to Nazi policies failed, most Germans who found themselves in similar situations felt discouraged politically and without, wrote Heisenberg, "the slightest hope of a change from within." While he and others insisted on riding out the storm to its end, the regime continued to engender distrust of all but one's closest friends, vastly increasing the "isolation of the individual," as Heisenberg called it. But it was an isolation to which Heisenberg's personality made him particularly susceptible.[1]

Throughout his life, Heisenberg's confidence in scientific matters contrasted sharply with his lack of assurance in personal affairs. This was manifest whenever circumstances separated him from his younger male companions, even as he approached middle age. His diffidence grew especially evident during the middle 1930s, as his fortieth birthday approached, and after his only known premarital romance—a relationship with Carl Friedrich von Weizsäcker's charming younger sister, Adelheid—was brought to an abrupt end by her parents. The failed romance is revealed in his letters to his mother, recently published by one of his daughters.[2] Although Heisenberg is remembered in Leipzig as having dated a variety of eligible local socialites, little is known of Adelheid or of the reasons for their breakup. No doubt the family objected to the age difference between the professor, then 34, and the young woman, who was still in her teens and only recently graduated (1934) from a literary second-ary school in Bern, Switzerland. Her father, Ernst von Weizsäcker, probably also pre-ferred an aristocrat over an academic—even as impressive an academic as Heisenberg—as a suitor for his daughter.

During his semester break in March 1936, Heisenberg had joined Carl Friedrich and the Weizsäcker family for a mountain-climbing excursion near their home in Bern; the elder Weizsäcker was at that time Germany's ambassador to Switzerland. Heisenberg later wrote his mother that he was so distraught over his renewed encounter with Adelheid that he could not even stop in Munich on the return trip to Leipzig. Her father's resistance was the main obstacle, he wrote, but the teenaged Adelheid herself also seemed uncertain, not knowing what she wanted.[3] Two years later, she married an ennobled Army captain, Botho-Ernst zu Eulenberg, who was then on active duty with the Prussian infantry. He was later killed in action on the Russian front, leaving Adelheid alone in war-torn East Prussia with two small children.[4]

The spurned Heisenberg returned from Bern to a "horribly desolate" Leipzig in a depressed state. There, an uncertain future and a physics still swamped by cosmic-ray showers awaited him. His Munich appointment was, as of March, postponed by Stark's newspaper onslaught, and the revolutionary prospects afforded by Fermi's theory were not yet in sight. Even worse, his closest companion, Adelheid's brother Carl Friedrich, was determined to end his awkward position as Heisenberg's assistant by habilitating as soon as possible and moving to Berlin. Heisenberg's mother again worried that his life was limited to too few people, to which he responded as before: "But it just doesn't work out any other way. And in physics I have also not done it any other way."[5]

Scientific and political success later that year raised Heisenberg's spirits briefly, but by November 1936, as the days shortened into winter and a new semester began, depression cast a pall over his victory in the public battle with Aryan physics. Weizsäcker's absence and the probable severing of relations with the entire Weizsäcker family left Heisenberg lonely indeed. In his annual reflection on his life on the anniversary of his father's death, Heisenberg rededicated himself to his work, "for the sake of which I appear to have come into this world." But, hard as he tried to make work substitute for personal relationships, as he had so often in the past, he rec-ognized that now that could be only a temporary solution. As he wrote his mother, "The single life is bearable to me only through my work in science, but for the long term it would be very bad if I had to make do without a very young person next to me." Three months later, Heisenberg was engaged to a woman 13 years younger than he. The pattern was already set by his close relationships with his younger compan-ions and most recently by his attraction to Adelheid: he needed a much younger per-son to keep him in touch with the outside world and with his own youth; for marriage, only a much younger woman would suit the professor's tastes, "even though," he admitted, "I am not very young any more and no longer as lively as I want to be."[6]

While immersing himself in the antidote of physics, Heisenberg remained con-nected to the cultured social circles in Leipzig through music. One of those circles had revolved for several years around the home of Otto Mittelstädt, a Leipzig publisher. Part of Mittelstädt's house was occupied by his in-laws, the Bückings, also a publishing family. Toward the end of January 1937, Heisenberg participated in one of his frequent evenings of chamber music, this one held in the Bücking home. The piece to be played was Beethoven's Trio in G Major, which Heisenberg had known from his youth. He was accompanied on the cello by Hellmuth Bücking and on the violin by his former professors' club colleague, Erwin Jacobi, who, though dismissed from his teaching post because he was Jewish, held on in Leipzig in private legal practice.[7]

Among the small number of invited guests for the concert that evening was a young woman, a tall and slender book dealer with a warm smile and a pretty figure who had recently arrived in Leipzig. She had become acquainted with the publishers through her occupation. This was the first time she had attended one of the Bücking affairs, and, as the trio began the performance, her bright eyes met those of the pianist as his practiced fingers caressed the keys. Heisenberg and the young lady, Elisabeth Schumacher, soon struck up a conversation that did not go unnoticed by their hostess, Mrs. Bücking. At the end of the evening, as the guests prepared to leave, she politely asked, "Mr. Heisenberg, would you please accompany Miss Schumacher home?"[8] Within a week the professor and the book dealer were planning a trip together to his youth group's Bavarian ski hut (with Hans Euler along as chaperon). Two weeks later, on February 11, they were engaged, and less than three months after that, on April 29, 1937, they were married in Berlin.[9] Heisenberg was then 35 years old, his bride 22.

Elisabeth was the youngest of five children born to Hermann and Edith Schumacher. Elisabeth's father was a Bonn professor of political economy, best known for his proposals for the German annexation of occupied territories during World War I.[10] Elisabeth's brother, Fritz, was also an economist. He emigrated to England in 1939 where, in contrast to his father, he was well known as author of the book *Small Is Beautiful*, on the advantages of small-scale enterprise. A sister, Edith, married the widely read German journalist and war correspondent Erich Kuby.[11] In 1917, as the First World War ended, Professor Schumacher had assumed the chair for global political economy at the University of Berlin and moved his growing family from Bonn to the upper-class suburb of Berlin-Steglitz. The family occupied a spacious turn-of-the-century mansion in Arno-Holtz-Strasse near the botanical garden on the Fichteberg, where the professor enjoyed his daily walk.

Elisabeth's father, like Heisenberg, had married a much younger woman relatively late in life. His granddaughter's description suggests that he was typical of his generation: "Single-mindedly wrapped up in his own life: dogmatic, authoritarian and dedicated to the pursuit of his career." Life in the Schumacher home bore a striking similarity to the environment in which Heisenberg grew up: "Disciplined and regular, dictated by the professor's needs."[12] No wonder that, upon graduating from a Berlin gymnasium in 1933, Elisabeth sought her freedom at the other end of the Reich, in Freiburg, Baden, in southwest Germany. She studied German literature at the University of Freiburg, graduating in 1936. While in Freiburg she became romantically involved with the physicist Wolfgang Finkelnburg, who would later play a role in the response to Aryan physics. When Finkelnburg asked Elisabeth for her hand in marriage, she respectfully declined and moved to Leipzig, then the center of German publishing, where she entered the book trade. Although women were hardly accepted in business at that time and especially under that regime, exceptions were made if one knew the right people, and she apparently did.

Meeting another physicist so soon after moving to Leipzig naturally awakened mixed feelings for Elisabeth. Finkelnburg had spoken highly of a Heisenberg whom he knew, but she did not appreciate until later that he was referring to *this* Heisenberg.[13] As children of oppressive upper-middle-class professional homes and rebounding at the time from failed romances, they both felt from their first meeting at the s' home that they were right for each other in many ways. "That evening decisively changed our lives," she later wrote. "We both felt that we had encountered 'our fate.'" As Heisenberg portrayed it to his mother, a conversation that had started out superficially soon revealed "a close agreement of opinions between her and me on the essential things. . . . This mutual understanding . . . soon went so far that it appeared to me natural to ask Elisabeth if she wanted to remain with me."[14] This time, Elisabeth accepted without hesitation.

Of course, more than just fate or a meeting of the minds must have been at work to precipitate such a sudden and lasting attraction for each other. A rebound romance is usually a prescription for disaster. Aside from the flowering of any strong feelings of love, one of the strongest attractions seems to have been mere circumstance—each, and Heisenberg in particular, felt terribly alone and isolated in the Nazi Germany of 1937. Heisenberg had just celebrated his thirty-fifth birthday in December, and without a younger companion to serve as a social stimulus and point of reference, the world around him seemed even more foreign and unbearable in those early months of 1937 than it had at any time previously during the Third Reich. In addition, a premarital blood test later revealed that Heisenberg had been suffering from an acute case of anemia, which may have contributed greatly to his mental condition.

The lonely, lost, and emotionally floundering state in which Heisenberg existed when he first met Elisabeth was made even worse by another depressing event that occurred just before their meeting: his compulsory collection for Winter Aid on the cold, grey streets of Leipzig. The sense of stability and acceptance that Elisabeth suddenly brought to his life must have meant everything to the emotionally desperate young man.

The Winter Aid Society, a wellregarded charitable organization that provided the urban poor with food and blankets, had been subordinated to the Nazi Party in 1933. Government officials required that socially elite professors, who were state employees, demonstrate their concern for their "Volk comrades" by periodically joining student and other party organizations in soliciting donations from passersby on city street corners. This enforced public begging seemed just one more sign of the contempt in which Nazi bureaucrats held the previously esteemed professors. Heisenberg had participated at least once before without overt complaint, but this time, already shaken by Adelheid's rejection and without the support of Carl Friedrich or his youth comrades, the Nobel laureate could hardly bear the humiliation of rattling a tin can on the dismal streets of wintry Leipzig.[15]

Nazi propagandist Joseph Goebbels had ordered a special effort that year to collect additional funds in celebration of the fourth anniversary of the Reich on January 30.[16] As he shivered on his street corner, Heisenberg was overwhelmed by a feeling of "the utter senselessness and futility of what I was doing and of what was happening all around me." He later recalled sinking into and out of a disturbing mental state that, from his description, bordered on psychosis. "The houses in these narrow streets," he wrote long after the bombing raids of World War II, "seemed very far away and almost unreal, as if they had already been destroyed and only their pictures remained behind; people seemed transparent, their bodies having, so to speak, abandoned the material world so that only their spirits remained behind." Two months later, when Elisabeth went to Berlin for a few days to make arrangements for the wedding, Heisenberg barely survived until her return. "When I am by myself," he wrote her, "I now easily fall into a very strange state, which belongs neither to the past nor to the future and neither to you nor to physics, and with which nothing can be done."[17]

Just before leaving for the ski hut in early March, Heisenberg accompanied Elisabeth to Berlin to meet her family. Although Professor Schumacher was acquainted with his university colleagues Max Planck and Max von Laue, he had never heard of Heisenberg or of his political troubles. When informed of his daughter's engagement, the economist's response was to ask her intended, "How do you expect to feed my daughter?"[18] Heisenberg's income was deemed sufficient, and after her domineering father conferred his approval on the plan, the happy couple headed by car for a visit with the bride's future mother-in-law in Munich and a visit to the ski hut—whereupon their chaperon discreetly absented himself.[19]

As soon as Elisabeth and Heisenberg returned to Leipzig in mid-March 1937, plans for the wedding went forward quickly. Heisenberg enlisted his old youth comrade Wolfgang Rüdel, now a Lutheran pastor, to perform the ceremony in Berlin, but youth-movement violinist Rolf von Leyden, whom Heisenberg wanted to play, had to be written off in favor of one of Elisabeth's musical friends. As the details were being settled, Heisenberg received a telephone call from the Reich Education Ministry (REM) in Berlin with a surprise "Easter egg": the REM was ready to appoint Heisenberg to Sommerfeld's Munich position, if he still wanted it.[20]

Stark's setback and the start of the four-year plan had resulted in a reorganization at the REM that brought Rudolf Mentzel and the SS into stronger positions vis-à-vis the party organizations on whose support Stark and Lenard had relied. In January 1937, Reich Minister Rust united the two offices for university affairs under SS officer Otto Wacker, the former education minister of Baden, who turned over matters of Aryan physics to his deputy—Dr. Mentzel. SS officer Mentzel had already replaced Stark as head of the German Research Association, and was preparing to set up the

Reich Research Council to garner funds from the budget for the four-year plan when he decided to appoint Heisenberg to Munich. With the 1937 summer semester approaching (it started in May) and the dogmatists of Aryan physics in retreat, Mentzel's deputy, Dr. Wilhelm Dames, telephoned in March to offer Heisenberg the Munich post immediately.[21] By the time opponents of the plan could voice their displeasure, Dames argued, they would be confronted with a *fait accompli*.

Heisenberg refused. With wedding plans in full swing and little time to prepare for Munich lecturing, Heisenberg insisted on an August 1 appointment and the REM grudgingly acceded. Although the delay would afford Heisenberg's enemies an entire semester to mount an opposition, the optimistic physicist saw little cause to worry. Nearly the entire physics profession had rallied to Heisenberg's side against his opponents. Future plans quickly came into focus. With lecture and travel engagements set for October, the move to Munich would have to occur before then. Heisenberg and his bride decided that, after a brief honeymoon, they would stay in Leipzig until almost the end of the summer semester in July, then move into a beautiful little house that Heisenberg had purchased in the Isar River valley just outside Munich. As soon as their furniture arrived, Elisabeth would arrange their new home alone, while Heisenberg made another demonstration of his loyalty in his annual eight weeks of military training. In the meantime, the tenants living in Heisenberg's Leipzig house were given notice, and, as 38 wedding invitations went into the mail at the end of March, Euler helped his boss prepare the Leipzig living quarters and his bedroom for the new bride. The two-story house, located at Bozenerweg 14 in the suburban section of Leipzig, was damaged by Allied bombs during World War II and was eventually destroyed altogether. Elisabeth's favorite part of the house was its well-kept backyard garden, which came into full bloom just as the couple exchanged their wedding vows in April.

With his marriage approaching and an appointment at last to Sommerfeld's chair in his beloved Munich, Heisenberg's hopes soared. Although he worried openly to Bohr about the mixing of physics and marriage, he had Bohr's own successful balance of the two as a model. Bohr's example, combined with his own new power as administrator of a major institute, gave him more courage, he wrote, than Bohr could ever know.[22] Heisenberg was certain that, once he settled at last in Munich, he would be able to achieve much of the same success as physicist, administrator, and husband that Bohr, his idol and father figure, had achieved in his career—no matter that the Nazi Reich hardly compared with Bohr's Denmark. The new Munich post, he wrote in his typically flat prose, "is nice because I can now have the feeling of building up something permanent that will last as long as I am able to work at all." And to Pauli he wrote three days before the wedding: "I also have the feeling that a security will come into my life that can only be encouraging for all types of work. It seems to me as if I could now in a certain sense begin life all over again from the very beginning."[23]

Heisenberg and Elisabeth exchanged vows at 3:00 PM on Thursday, April 29, in the small St. Annenkirche in the suburb of Berlin-Dahlem. Their wedding preceded

Heisenberg with the twins, Wolfgang and Maria, 1938.

an elegant reception and dinner in the nearby Schöneberger Ratskeller. After dining on turbot filet, roast lamb, and Rhoner Hofberg Auslese 1934, the newlyweds headed by auto for southern Germany and Austria for an Alpine honeymoon.[24] Exactly nine months later, the new Mrs. Heisenberg gave birth to fraternal twins, Wolfgang and Maria, the former named for Pauli, who duly congratulated the new father on his "pair creation."[25]

Motherly Elisabeth—she would bear five more children over the next twelve years—was often insecure and much dependent on Heisenberg's drive and sense of professional purpose to buoy her during the difficult times ahead. She, in turn, apparently provided the stability and sense of belonging that Heisenberg so desperately needed then and would need even more during the coming months and years. Their marriage had been as much one of personal necessity brought on by circumstances as it was one of love or passion. Heisenberg did not easily express his feelings or readily share his personal and professional problems with Elisabeth, especially if he felt they would only upset her and the family. Their relationship never really filled the void of loneliness and alienation that often opened deep within him, nor did Heisenberg ever really allow his wife and family to displace the central position in his life that his career and duties had long occupied. His work and duties as a leading German scientist frequently took him away from home for long periods—first during the coming war, then during his intensive efforts to rebuild German science after the war. One of his daughters wrote in a memoir of her father, "He was not a father according to contemporary expectations. He was not present at a single birth and never in his life did he diaper or feed a child, even though he would have had ample opportunity to do so in our family." It was not until the 1960s, as the children left home and Heisenberg slowed his pace of work, that he and Elisabeth could at last freely travel, vacation, and simply be together undisturbed. But in many ways they could never make up for lost time. As age and illness took their toll on Heisenberg, Elisabeth quietly regretted to herself that, after all those years together, they had never really gotten to know each other "in a fundamental way."[26]

Having stabilized his personal life, Heisenberg immediately turned to building "something permanent" in his physics. But during those depressing winter days in early 1937, his work was already under attack abroad and in retreat at home. Carl Anderson's new data on the absorption of high-energy cosmic-ray electrons in matter confirmed the application of the current Bethe-Heitler theory to large energies far beyond the energy at which the theory was previously believed to break down. It was a theory that attributed the stopping of charged particles in matter solely to quantum electromagnetic processes. News of the discovery that these processes were valid after all to very high energies reached Leipzig and Zurich in January 1937, just ahead of papers by Bhabha and Heitler and by Carlson and Oppenheimer presenting a complete theory of cosmic-ray showers based upon quantum electrodynamics. In this theory a shower is the result of a cascade, a gradual build-up of particles and photons by a series of elementary electromagnetic interactions, during which the incoming particle dissipates nearly all of its initial energy.[27]

The cascade showers accounted quite well for the appearance of showers of particles in cloud chambers and for the rapid absorption of the "soft component" of cosmic rays impinging on the earth's atmosphere. But the "hard," highly penetrating component of incoming cosmic rays still defied any explanation or identification. Anderson and his Caltech assistant, Seth Neddermeyer, reasoned "that either the theory of absorption breaks down for energies greater than about 1000 Mev, or else these high energy particles are not electrons."[28] Since the newly rehabilitated radiation theory accounted quite easily for other absorption properties, Anderson and Neddermeyer settled on the second alternative and began an immediate search for a new, hitherto unknown elementary particle.

While developing their cascade theory of showers and a complete account of the soft component, Walter Heitler and his collaborator, the Indian physicist Homi Bhabha, made room for Heisenberg's alternative proposal—explosion showers derived from an application of Fermi's nuclear theory of beta decay to the hard component of cosmic radiation. But Oppenheimer was apparently not willing to concede anything to Heisenberg. Oppenheimer and his student J. F. Carlson declared that Heisenberg's theory "is without cogent experimental foundation; and we believe that in fact it is an abusive extension of the formalism of the electron neutrino field." The Fermi nuclear field was still problematic, and the suggestion that, when taken out of the nucleus and applied to cosmic rays, its infinities could somehow render characteristics of a future theory seemed to Carlson and Oppenheimer implausible and methodologically unsound. They preferred instead to study cosmic-ray events through "the more usual procedure of avoiding divergences [infinities] by the formal device of reducing the coupling between heavy and light particles for high relative energies."[29]

To prove his point and to meet the challenge of Heisenberg's almost simultaneous alternative, Oppenheimer suggested to Lothar Nordheim and his wife, Gertrude, both recently arrived German émigré physicists from Göttingen, a test of Fermi-field explosion showers for the hard component, construed to be protons. After tinkering with Fermi's force to make it yield the correct range and magnitude of the nuclear force, the outcome was as expected: Heisenberg's theory "no longer affords any explanation of showers," they declared.[30]

Heisenberg and Pauli took immediate note of the new developments but remained undeterred from "our main problem, the quantization of waves"—as Heisenberg put it.[31] Nevertheless, Heisenberg and Pauli did worry as much as Oppenheimer about accounting for cosmic-ray data. After all, their theory and its development ultimately rested on the complicated data of cosmic radiation. But their letters to each other in this period suggest that "the practical questions of cosmic radiation"—the actual unraveling of shower phenomena—were of a lower priority than their, especially Heisenberg's, primary goal, the search for "a future theory of elementary particles."[32] For them, cosmic-ray data were a means to an end, not an end itself. But for Oppenheimer and his assistants in the more practically minded American context, it was just the opposite: the theory was a means to the end of reproducing the data, that is, forcing the theory to fit the data. The irony of these preferences at that time is the reversal of the expected influence of the social criticism of science. While "Aryan" physics demanded a retreat from abstract theories to more reliance on data and experiments, the experimentally inclined quantum theorists were not to be found primarily in Germany but in the other camp, in the United States and Britain. Even more ironic was the circumstance that many of the theorists working on cosmic rays were in fact German émigrés. Although the theorists remaining in Germany (such as Heisenberg) did offer a more public appreciation of empirical and practical physics, they did not swerve noticeably in their private work from the internal trajectory of their theorizing.

As soon as Heisenberg saw Anderson's new data and the Bhabha-Heitler cascade-shower manuscript, he readily conceded nearly all cosmic-ray showers to cascades, extended the Bethe-Heitler theory to all energies, and admitted to Bhabha, then working with Pauli in Zurich, that perhaps applying the Fermi theory of beta decay to the subject of cosmic-ray showers was a "wide extrapolation from beta decay."[33] But he still held out for at least one "genuine shower formation" per 1,000 electron cascades and for the detection of neutrinos as the shower-producing hard component. Both propositions were extraordinarily difficult to test.

By early 1937, Heisenberg's explosion-shower revolution was in fragments. Pauli claimed to have proven that every shower theory encompassed infinities, no matter how one tried to avoid them. At about the time Heisenberg was collecting coins for Winter

Aid, Pauli's co-workers, Markus Fierz and Nicholas Kemmer, claimed they had proved that neither Fermi's theory, nor any relativistic variation of it, could reproduce the properties of nuclear forces, the behavior of beta decay, or yield non-infinite energies.

The newly engaged Heisenberg was undeterred. On April 26, he wrote to Pauli not only of feeling new security in his personal life, but of an indomitable optimism in his work. Although all of their attempts so far to obtain a satisfactory quantum field theory had failed, at least their efforts yielded "a feeling for the formal possibilities which lurk in the wave quantization, and I am still rather optimistic." And, he continued, "I am now also much more optimistic concerning the discovery of Fermi processes in cosmic radiation."[34] Three days later the optimist married his fiancée in Berlin.

Pauli resigned from the elated groom's physics by return mail. Heisenberg was still suggesting the existence of a fundamental minimum length in nature (the old "lattice world" of cubic cells), which, to him, seemed supported by the appearance of explosion showers in cosmic rays based upon Fermi's beta-decay field. To Pauli, Heisenberg's approach simply did not yield any progress in "*the* fundamental problem of present-day physics," the quantization of wave fields. Homi Bhabha's return visit to Zurich that spring only reinforced Pauli's "theoretical doubts" about Heisenberg's fundamental length. "I think 'we must be prepared' [an echo of Bohr] to find that the universal length of the beta decay theory will prove to be incorrect, and therefore the corresponding multiple processes [explosions] will not play any significant role at all in cosmic rays."[35] Heisenberg's entire proposal, based on the still disputed existence of explosion showers, might have to be abandoned.

Learning of Pauli's retreat upon his return from his honeymoon, Heisenberg rejected Pauli's doubts that explosions would ever be found. He called it "a terribly defeatist attitude." "On the contrary," he also wrote, "I am more convinced than ever that there the key to the wave quantization is really to be sought."[36] In a long letter to Bohr in early July 1937, ten days before moving to Munich to succeed Sommerfeld, Heisenberg reviewed the experimental evidence for explosion showers, including a new discovery by Anderson and Neddermeyer: the existence of a so-called heavy electron in the hard, penetrating component of cosmic rays.[37]

The new elementary particle was the first to be discovered since the positron and neutron earlier in the decade. It possessed the negative charge of an electron but carried much greater mass. Since, according to quantum field theory, every elementary particle is related to a specific type of field, in the years ahead the heavy electron became the subject of much research and controversy regarding its proper place in quantum field theory. To Heisenberg, it provided a welcome support of his "program for the future"—even if the new particle did not seem to fit in anywhere in his nuclear physics of cosmic rays. For him, the new particles were supposed to appear as stationary states in a unified theory of matter and fields. But the further pursuit of this program, he admitted in one of his occasional letters to Kramers, would have to await more supporting data.[38]

Heisenberg wrote his letter to Bohr while he and Elisabeth prepared for their move to Munich. Immediately after the move, Heisenberg had intended to leave for military training, but circumstances forced a change of plans. Premarital blood tests had turned up Heisenberg's anemia, and the now-pregnant Elisabeth was suffering from both an injured knee and morning sickness. The doctor forbade military duty and ordered a mountain cure for both patients.[39] The couple decided to stay in Munich for two weeks in July, then to spend most of August in beautiful Engadin in southern Switzerland before returning to Munich and their new home.

The plan came to naught. In early June, as the peach trees blossomed in Munich and heavy electrons filtered through Anderson's lead blocks at Caltech, Leopold Kölbl, now rector of the University of Munich, informed Heisenberg of an ominous development. Stark had learned of Heisenberg's appointment and had told Kölbl during a meeting in his office that he planned to take the matter to "higher authorities." The threat did not puncture Heisenberg's optimism. Stark could no longer be taken seriously, he wrote. "And in the long run, I will surely win the contest with Herr Stark, if I don't do something wrong."[40]

Five weeks later, on Thursday, July 15, the Heisenbergs arrived in Munich by automobile, as planned. They unloaded the car at the old Heisenberg home on Hohenzollernstrasse, where his boyhood battleship still stood docked on a dresser in his bedroom. From his mother's phone, Heisenberg called Kölbl to announce his arrival. The rector's only response was to ask if Heisenberg had seen the latest issue of the SS newspaper *Das Schwarze Korps* (*The Black Corps*). "There's a long article on you in it. Buy it and read it," he said. "Then we can talk."[41] Heisenberg found a copy of the July 15 issue at a local newsstand and opened it to find a full-page broadside attack on science and himself entitled "Weisse Juden' in der Wissenschaft" ("'White Jews' in Science") signed at the bottom by Stark. It was the opening of another long and lonely battle for Heisenberg.

CHAPTER 20

A FAUSTIAN BARGAIN

"THE VICTORY OF RACIAL ANTI-SEMITISM IS TO BE CONSIDERED ONLY A PARTIAL WAR. . . . For it is not the racial Jew in himself who is a threat to us, but rather the spirit that he spreads. And if the carrier of this spirit is not a Jew but a German, then he should be considered doubly worthy of being combated as the racial Jew, who cannot hide the origin of his spirit. Common slang has coined a phrase for such bacteria carriers, the 'white Jew.'"[1]

So began one of the most vicious and repulsive attacks on science, and on Heisenberg in particular, to appear during the Third Reich. It was published in the July 15, 1937 issue of the SS weekly *Das Schwarze Korps* replete with grammatical errors and non sequiturs. It precipitated an even more violent struggle between academics and Nazi demagogues for control of physics ideology and appointments, which came to an end only in the depths of World War II. Heisenberg, and others attacked in this way, ultimately prevailed in formal terms, but the episode left him even more politically debilitated and personally oriented toward the position he would take in the war. As with most political intrigues, the battle over physics had both public and private consequences.

Heisenberg's impending appointment to the University of Munich as Sommerfeld's successor had rekindled Stark's efforts to prevent the appointment and to gain control over physics teaching chairs. Stark received new encouragement during the Heidelberg celebration of Philipp Lenard's seventy-fifth birthday on June 7. With Alfred Rosenberg out of the picture and the SS ascendant, Heisenberg's opponents turned to Himmler and his organization for support in imposing their will on the Reich Education Ministry. Younger SS officers in Lenard's and Stark's circles facilitated the effort. Two in particular played key roles. Dr. Hermann Beuthe, physicist and government councilor, was Stark's right-hand man in the Physical-Technical Institute in Berlin. Dr. Ludwig Wesch, an SS *Obersturmführer* (Senior Storm Leader) who had received his physics doctorate and habilitation under Lenard, was now the old man's assistant.[2] (It is remarkable, and disheartening, how many Nazi functionaries carried the title of doctor, apparently legitimately.) Beuthe and Wesch were also members of the SD (Sicherheitsdienst), or security service, a branch of the SS under the direction of Reinhard Heydrich, later notorious as the "hangman of Lidice."

Hitler had originally established the SS (Schutzstaffel) as his elite personal bodyguard, selected from the ranks of his brown-shirted storm troopers, the SA. But after Hitler's purge of the SA in 1934—apparently in order to gain the backing of the army, which the SA rivaled in military matters—the black-shirted SS began to assume a wider range of police and storm-trooper functions. Within the SS, the SD served as a secret police. In June 1936, Hitler reorganized police activity under Bavarian police chief Heinrich Himmler, whom he elevated to the cabinet-level position of Reichsführer-SS in Berlin. Under Himmler, the Gestapo (Geheime Staatspolizei) took over secret-police work, while the SD turned to intelligence gathering and reporting on social and cultural affairs, including, in particular, science.[3]

Journalist Gunter d'Alquen, an SS officer and SD man, had worked on the editorial board of the *Völkischer Beobachter* under Rosenberg until the reorganization of the SS brought him a new opportunity: editorship of *Das Schwarze Korps*.[4] His black-shirted comrades Beuthe and Wesch probably convinced d'Alquen to publish an article submitted by the proponents of Aryan physics, titled "'Weisse Juden' in der Wissenschaft"—"'White Jews' in science."

The page-long diatribe is divided into three sections: the Nazi definition of a "white Jew" excerpted at the beginning of this chapter; a personal attack on Heisenberg, called "The Dictatorship of the Gray Theory"; and a political attack on the university professorate, titled "Science Has Failed Politically." The last section was signed "Stark" and was phrased to give the impression that Stark had been invited by the newspaper to express his views on the matter. The second part, the one on Heisenberg, although unsigned, was written nevertheless in Stark's unmistakable venomous style. He was probably assisted by his subordinate, Beuthe. The latter told Wesch that he had helped with the article and that he had gathered most of the information for the Heisenberg section.[5]

After vilifying "the Jews Einstein, Haber, and their like-minded comrades, Sommerfeld and Planck," whom the article accused again of manipulating physics appointments to exclude "Germans," the article proceeded to mount a full-scale assault on "that white Jew" and "representative of the Einsteinian 'spirit' in the new Germany"—Werner Heisenberg. Stark's perceived archenemy, Sommerfeld, had chosen Heisenberg as his successor, which made Heisenberg an heir apparent to the "white Jewish" establishment and thus a lightning rod for Stark's ruthless efforts to gain influence, at last, over appointments to German physics professorships. By exploiting the dominant hate ideology of the day, character assassination of a key individual could be used to further the political ambitions of those who had otherwise failed to achieve the influence they craved. The denunciations of the McCarthy era in the United States offer perhaps a distant parallel.

To demonstrate "how secure the 'white Jews' feel in their positions," Stark delivered a litany of Heisenberg's offensive actions, as researched by Beuthe (whose name in German means "booty"). The list was presented to make Heisenberg appear a covert enemy of the state. He had "smuggled" his article defending the teaching of relativity theory into a party newspaper; he had circulated a petition among physicists in order to influence wrongly a state agency (the REM) and to silence his legitimate critics; he had refused to join his fellow Nobel Prize winners in the 1934 declaration of support for Hitler's presidency; and his appointment to the Leipzig chair in 1927 was clearly unearned, since he was obviously too young to have accomplished anything of value—a circumstance that "proved" that he had gotten the chair only because he was backed by the "white Jewish" establishment. In addition, Heisenberg had allegedly dismissed a "German" assistant in his institute in favor of the Jewish physicists Bloch and Beck; his institute continued to harbor an inordinate number of Jews and foreigners to the exclusion of "Germans"; and so on.

For those who failed to catch the inference of this catalogue of sins, a large-print subhead calling Heisenberg "the 'Ossietzky' of physics" made the point. Carl von Ossietzky, a courageous pacifist opponent of the Nazis and winner of the 1936 Nobel Peace Prize, was at that moment imprisoned as a traitor in the Dachau concentration camp. He would die of torture and malnutrition within a year.[6] Heisenberg, "white Jew" and "Ossietzky" of physics, was only one example among "many others," the article declared. "They are all representatives of Judaism in German spiritual life who must all be eliminated just as the Jews themselves" must be dispatched.

This was a despicable and dangerous threat of violence against Heisenberg and all German physicists, especially those likely to be counted among the "many others." Its purpose went far beyond merely preventing Heisenberg from assuming Sommerfeld's position. By 1937 most of the Jewish holders of university positions had been driven from their jobs, and many were leaving the country. The battle over the dismissal policy had long been lost by the physicists. The Nazis' target now was no longer Jews, but those non-Jews who still opposed the regime and supported the work and influence of those who had left. Now that non-Aryans had been driven from influence within the Reich, any wayward thinking non-Jews had to be brought into line behind the "Volk and its Führer" or else removed from the Reich, as well, by any means necessary. This assault was, in effect, part of the last phase of the Nazi effort to gain control of the German populace, the suppression of individual opposition to the totalitarian dictatorship of the Third Reich.

Some of the "many others" mentioned in the article immediately rallied to their colleague's defense and to the defense of their profession. This time, there were no public faculty protest meetings or even a circulated petition—such acts were now too dangerous and too likely to lead to a fate like Ossietzky's. Instead, Heisenberg's colleagues submitted official letters of complaint to befriended local bureaucrats for transmission up the chain of command to ministerial authorities. They also supported Heisenberg with their official recognition. In the months immediately following the article's publication, the Göttingen Academy of Sciences elected Heisenberg a corresponding member. The Saxon Academy, populated by most of the non-Jewish members of the former professors' club, elected him deputy secretary of the mathematics and physics section at a time when the REM was attempting to coordinate academies behind the National Socialist state.

Local deans and their state overseers in nearly every instance forwarded the letters of protest to state ministries and beyond, which indicates the broad sympathy

Heisenberg and his colleagues enjoyed in this matter. Many of these local administrators carried the abbreviated title Pg., party comrade (Parteigenosse). Stark may have gained the backing of elements within the powerful SS, but he obviously did not enjoy the backing of the REM or the party faithful among academics, over whom he was ultimately attempting to gain control. They too realized that more was at stake than the fate of one individual.

Sommerfeld wrote a long letter to Munich rector Kölbl in which he complained that he and Heisenberg had been slandered. He demanded an end to Stark's ravings "in the interest of the reputation of German science"—a plea that would appeal to German nationalists. Pg. Professor Kölbl forwarded Sommerfeld's letter to the Bavarian Culture Ministry, adding a remark of his own: "It is outrageous that an active professor, a civil servant of the National Socialist state, should be attacked in this way in a newspaper."[7]

Heisenberg's Leipzig colleague Friedrich Hund, who also considered himself one of the "many others," fired off letters both to Leipzig rector Koebe and to REM chief Rust. He requested that Rust take steps to ensure that Stark "can no longer ruin the honor of our science in this way." Hund also informed Debye of this protest. Debye, now in Berlin, had already taken a copy of the SS article to a meeting of the senate of the Kaiser Wilhelm Society (the network of state-sponsored research institutes), where, he wrote Hund, "it was condemned by everyone with whom I spoke." Debye subsequently added his own letter of complaint to the growing pile.[8]

The most significant of the letters to arrive at the REM came from the victim himself. Heisenberg demanded that Minister Rust—still vacillating between the SS and the professors—take a strong and clear stand against the SS regarding Heisenberg's personal honor, a request that, ironically, Nazis would understand.

It is difficult to comprehend the painful and debilitating situation in which Heisenberg now found himself. The great, apolitical Professor Heisenberg was accused of being like the hated Jews, an enemy of the state and a subversive bent on undermining regime policies. How should we take Heisenberg's official protestations against such accusations and his demand for restored honor—that is, renewed recognition by the regime and a dropping of the charge that he was a traitorous "white Jew"? Surely it would have been a true badge of honor to be guilty of such "crimes." His official complaints notwithstanding, from what we know of Heisenberg and his views, he did not fully support the regime, but by remaining in Germany he had accepted the compromises this required—acceptance of the regime's authority under the circumstances and acceptance of the failure of his efforts to do anything about it. He was not attempting to dissociate himself from Jews for reasons of anti-Semitism; nor did he really believe that German Jews were non-Germans or that they had been enemies of the state before the state had forced them into that role. But other than expressing displeasure through bureaucratic channels and protecting those he could in his institute, these were matters of so-called private, active opposition—the rationalization outlined earlier that allowed him to accept the situation and to maintain his position as a prominent professor within Hitler's Third Reich. By its very nature, the compromise required the avoidance of any public expression of private moral and political scruples regarding the regime.

Regardless of Heisenberg's private views, by 1937 the time was long past when one could openly express any opposition in letters to Nazi bureaucrats (or even in private letters that could possibly fall into the wrong hands)—that is, if one desired to remain in Germany and continue to work as a civil servant of the state. Moreover, in this situation, as Planck had counseled long ago, Heisenberg had decided to focus only on the future—batten down the hatches and wait for what they regarded as the inevitable catastrophe to blow over.

Now, in 1937, the storm was howling right over Heisenberg's head. He was being accused of opposing the regime and allying himself with pro-Jewish subversives of like mind. If his enemies could ever make these accusations stick, they would not only render his continued work and teaching impossible, but even worse, he would be placed in grave personal danger, politically isolated, and branded a traitor.

On the other hand, if he could successfully refute these accusations and force his enemies into retreat, he apparently reasoned, he could use his victory as protection against further insult and as a basis for improving the lot of German theoretical physics, if, indeed, anything could really counter the enormous damage—scientific, political, moral—already done to it by the regime. If he did succeed, whatever private objections he harbored would be buried even deeper in his private self, where they would be even safer from discovery.

The political assault had focused on an individual; the individual now demanded a new compromise—explicit exoneration and rehabilitation in return for not vacating his public office and voiding all the compromises it required. Privately, however, that raised the painful specter of emigration. In a revealing personal letter to Sommerfeld, Heisenberg wrote of the dilemma in which Stark, as well as his own attachment to Germany, had placed him, a position that seems so reminiscent of the dilemma of which Max Born had attempted to enlighten Heisenberg nearly four years earlier. Wrote Heisenberg, "Now I actually see no other possibility than to ask for my dismissal if the defense of my honor is refused here. However, I would like to ask you for your advice in advance. You know that it would be very painful for me to leave Germany; I do not want to do it unless it must be absolutely so. However, I also have no desire to live here as a second-class person."[9]

On the day the SS article appeared, Heisenberg had arrived in Munich with his pregnant wife, Elisabeth, on their way to the Alps for a rest. Within two days of their arrival, he composed and sent an official letter, along with a copy of the SS article, to Dean Helmut Berve, his Leipzig friend and colleague. In the letter, to be forwarded through channels to the education ministry, Heisenberg demanded that the REM make a fundamental decision: "Either the ministry regards the standpoint of *Das Schwarze Korps* as correct, in which case I request my dismissal. Or, on the other hand, the ministry disapproves of such attacks, in which case I believe that I have the same right to protection that, say, the Army itself would render its youngest lieutenant in such a case." Pg. Dean Berve immediately forwarded the request to Leipzig rector Koebe, along with his own request for a stand from REM minister Rust.[10]

The Saxon Education Ministry duly forwarded the letter and similar ones to Berlin, but Saxon bureaucrat Studentkowski—previously so active in Heisenberg's defense—decided to stay neutral. For one thing, the SS was now involved, and he was not an SS member. More important, Munich's difficulty in replacing Sommerfeld and Leipzig's difficulty in replacing Debye with a suitable experimentalist meant that there was probably little chance of finding a comparable replacement for Heisenberg in Leipzig. The search for Debye's successor had lasted for more than a year before Gerhard Hoffmann, a cosmic-ray experimentalist who was last on everyone's list, was finally asked to take the post as of April 1, 1937.[11] Studentkowski did not want a repeat of such difficulties in Heisenberg's case.

With or without the numerous letters of complaint—which, of course, any Nazi agency could easily ignore, since it did not depend on professors for support—the REM remained favorable toward Heisenberg, despite the presence of many SS officers in its ranks. Stark could not be allowed to dominate REM policy, even if he could claim SS support. Accordingly, SS *Sturmführer* Dr. Otto Wacker, now head of REM office W (scholarship), immediately responded to Heisenberg's letter with a request for supporting evidence against the charges. Heisenberg forwarded a detailed rebuttal of each accusation. Wacker duly opened an investigation, but it soon became clear that neither Wacker nor his REM boss, Rust, would act until a decision had been reached in another investigation already underway.[12]

This time, Heisenberg, too, had taken the matter to a "higher authority." Exactly one week before his first letter to Wacker in July, Heisenberg had written directly to the *Reichsführer-SS* himself, Heinrich Himmler, requesting in nearly the same words as his letter to the REM a similar fundamental decision: either approval of Stark's attack, in which case Heisenberg would resign, or disapproval, in which case he demanded the restitution of his honor and protection against further attacks.[13] Not wishing to alarm his pregnant wife with the enormous gamble he was taking—of which he himself may not have been fully aware—Heisenberg did not inform her of it until much later, at which point she was overwhelmed with shock and anger.

If Heisenberg had sent his letter to Himmler through normal channels, as he did the REM letter, it probably would never have arrived. At his mother's suggestion, Heisenberg chose a safer route. As indicated earlier, Heisenberg's grandfather, Nickolaus Wecklein, former rector of the Maximilians-Gymnasium in Munich, had belonged to a hiking club of like-minded Bavarian gymnasium rectors. One of the members of this group was Himmler's father, Joseph Gebhard Himmler, assistant rector in Landshut, who died in late 1936. Heisenberg's mother had become acquainted through her father with Himmler's mother, who now lived in Munich. Mrs. Heisenberg offered to take Heisenberg's letter to Mrs. Himmler to deliver to her son, Heinrich.

According to Heisenberg's much later account of the meeting of the two mothers, which must have occurred in late July or early August 1937, Mrs. Himmler politely received the visitor in the living room of her small, respectably furnished Munich apartment. A crucifix was nestled prominently in one corner of the room with freshly cut flowers reverently arranged in front of it. Mrs. Himmler was at first rather skeptical of Mrs. Heisenberg's request, not wanting to interfere in her son's affairs. As Heisenberg later recounted it, Mrs. Heisenberg ultimately gained her confidence by saying, "'Oh, you know, Mrs. Himmler, we mothers know nothing about politics— neither your son's nor mine. But we know that we have to care for our boys. That is why I have come to you.' And she understood that."[14]

Mrs. Himmler probably gave Heisenberg's letter to her son when, records show, he visited Munich in August. But Himmler did not respond until November 4, 1937, after receiving the result of a preliminary internal investigation. No discussion of the matter appears in Himmler's surviving correspondence with his mother.[15] In his letter of November 4, Himmler simply asked for a defense against the charges. Heisenberg immediately provided a point-by-point rebuttal, almost identical to the one he had provided to Wacker.[16] First, Heisenberg explained to Himmler, the "German" in his institute had been replaced by Bloch and Beck because of his complete lack of interest or ability in modern physics. Second, Heisenberg had refused to sign Stark's declaration in support of Hitler because he had doubts about Stark himself—which were shared, incidentally, by SS functionaries—and because of his long-held belief that scientists should, as they had done in the past, remain non-political. Third, he had participated in the preparation and circulation of the Wien-Geiger-Heisenberg petition at the explicit request of REM official Mentzel, an SS officer.

The best way to settle the matter, Heisenberg suggested to Himmler in closing, was to arrange a face-to-face confrontation between himself and Stark, whom he had never met. Heisenberg was almost always successful in personal encounters, and, he naively believed, personal diplomacy still offered a good chance of inducing Stark to withdraw his charges and retreat from further interference in professional matters. The meeting never took place, but because of the severe nature of the charges, his mother's urging, and the favorable result of the preliminary investigation, Himmler decided in November 1937 to consider the case in depth. That decision and Heisenberg's written defense set in motion an intensive SS investigation that lasted more than eight months and profoundly affected Heisenberg's personal and political life.

Heisenberg also had the welfare of his new family to consider, but for him the paramount issue was whether or not Himmler would bestow his stamp of approval, thus facilitating his continued work and unhindered life in the otherwise stifling Nazi environment. Heisenberg, the eternal optimist, now turned suddenly cautious. In the summer of 1937 he entered into secret negotiations with a visitor from Columbia University in New York City. "When I thought about New York," he later wrote his mother, "it was less the thought of the intrigues of Herr St[ark] that was decisive, but more the prospect of living for many more years in an environment that makes work —for the sake of which I now exist—almost impossible."[17]

Not wanting to prejudice the two ongoing investigations into his activities, Heisenberg did not disclose the negotiations to his faculty until required by law to do so. His plan was to visit Columbia in the second half of 1938—either temporarily or permanently, depending on the situation at that time.[18] Yet despite this flirtation, the eternal optimist never truly believed that he would have to resign—nor would he do so unless Stark made matters worse by making work and teaching absolutely impossible. Heisenberg was still the ministry's official choice for Munich, and by the end of October he learned that the REM investigation had come to a conclusion in his favor. Elisabeth wrote Heisenberg's mother of the good news; she believed they would be in Munich by early 1938.[19] But, unknown to her, the SS investigation had only just begun, and Rust did not dare act until Himmler had approved of Heisenberg. Heisenberg would have to wait a while longer.

Himmler handed the Heisenberg case over to the SD "cultural division" of his personal staff. At the same time, Nazi physicists maneuvered to prevent Heisenberg's appointment to Munich even if the *Reichsführer* did stamp his approval. Munich, the "capital of the movement," was headquarters to academic organizations established by Hitler's party deputy, Rudolf Hess, in order to exert his own competing influence on academic affairs. Among these Hess organizations were the Nazi Students League; its parent, the Reich Student Leadership; and the Nazi University Teachers League.

SS physicist Wesch may have played a role in stirring up opposition to Heisenberg among Hess's Munich groups. As a Lenard pupil and cofounder and occasional head of the Nazi Students League in Heidelberg, Wesch maintained close ties to the Munich Student Leadership and to its subcommittee for natural science. Nazi science students required little encouragement to join the fray. Their journals, *Deutsche Mathematik* (*German Mathematics*), founded in 1936, and *Zeitschrift für die gesamte Naturwissenschaft* (*Journal for the Entirety of Science*), founded in 1933, produced a rising torrent of anti-Semitic Nazi science propaganda. An attempt by party member Pascual Jordan, one of the founders of quantum mechanics, to temper the radicals' fury against modern physics with his monograph *Physics of the Twentieth Century* precipitated even longer assaults on theoretical physics in the pages of both journals.[20]

Heisenberg and his publications were eventually targeted specifically, especially after Hugo Dingler, an aged Austrian Nazi physicist, decided to join the student cause. The Reich Student Leadership maintained an SD office for the evaluation of candidates for university appointments. In the matter of Sommerfeld's successor, the students were in agreement with their comrades in the Teachers League, who had already gained influence by opposing Sommerfeld's choice—Heisenberg. Interconnections among all these groups facilitated concerted action.

One historian has concluded that Wilhelm Führer, then head of the Munich branch of the University Teachers League, was the driving force behind the opposition to Heisenberg's appointment, which in turn drove the rising fortunes of the Teachers League.[21] Führer was joined by Bruno Thüring, Teachers League representative to the Munich philosophical faculty, and Fritz Kubach, historian of science and

Reich student leader for natural science, all of whom, like Hess, were headquartered in Munich. But Führer's actions were themselves instigated from even higher up, by Hess himself, and in concert with the publication of Stark's article in the journal of Hess's competition, Himmler's SS. Just one day before the article on "white Jews" appeared in *Das Schwarze Korps*, Führer received an order to reject Heisenberg as a candidate for Sommerfeld's chair.[22]

By the fall of 1937, the Munich Teachers League began proposing its own candidates to succeed Sommerfeld, all of whom Sommerfeld and colleagues rejected as unqualified. Most had little training in theoretical physics, and none could be considered suitable as a successor to the great Sommerfeld. The danger that one of these individuals, rather than Heisenberg, would nevertheless occupy Sommerfeld's chair vastly increased after a conference of university rectors in December 1937. At that meeting Wacker, apparently seeking closer ties with Hess's organizations, agreed to consider "political reliability" a specific criterion for faculty appointments. He further agreed that Hess, with the willing assistance of his Teachers League, should politically evaluate candidates for professorships. The implication was clear. Heisenberg would never succeed Sommerfeld unless the Nazi students and teachers could be convinced of his "reliability," and that was impossible without SS "exoneration."

Heisenberg himself chose not to discuss the investigation in any of his memoirs. He did not even discuss it with his wife at the time. Moreover, the SS records of the investigation were apparently lost in the war, SS functionaries burned as many documents as they could get their hands on in the last days of the Third Reich. Other sources indicate that Himmler's investigators apparently focused on two areas: Heisenberg's ideological standpoint in scientific matters, and his personal and political orientations. The SS—hardly an objective agency bent on exonerating those falsely accused of traitorous actions—employed its already infamous methods to discover the "truth." Heisenberg had to endure long and exhausting interrogations; spies were planted in his classroom and throughout the institute; the Gestapo bugged his home. The SS also used another tactic it had perfected: bringing an even more serious charge against the victim, who would then be all too eager to "confess" to the lesser, original charge in order to escape the greater danger.

The more serious charge brought in this case indicates that Heisenberg was indeed in grave personal danger. Hints regarding the charge are found only in letters surrounding the investigation. The accusation is spelled out in one such letter in November 1937 in reference to the article in *Das Schwarze Korps*: "Not everything, however, is in the article; for example Heisenberg is not clean with respect to §175; he indeed married quickly but only to cover this up."[23] The reference is to section 175 of the old Weimar criminal code and in effect until the 1970s, making male homosexuality a crime. If convicted of this crime in 1937, the offender landed immediately in a concentration camp.

The imputation requires careful handling. First, it is an accusation made by a dedicated SS functionary engaged in a campaign of character assassination. Such a source is hardly reliable or objective. Second, the SS often used the charge of homosexuality to extract confessions to lesser crimes. Third, it is true that Heisenberg did prefer the company of younger men, and one or two in particular. The investigation, however, which probably involved interrogations of some of these younger companions, apparently yielded no evidence of homosexuality; if it had, that evidence would certainly have been used against him. Moreover, if the SS functionary did think he had such evidence, it may have concerned the Wyneken affair in the Bavarian New Pathfinders, which had caused a considerable scandal in the early 1920s. Apparently it had not involved Heisenberg's group. Heisenberg did marry rather precipitously for various reasons, but there is no indication that concealing homosexuality was one of them.

The agony that such accusations must have caused Heisenberg is evident in his annual assessment of his life. In a long letter to his mother in November 1937, he expressed his feelings more openly than usual: "I wish for the coming year a clearing away finally of these horrible things, for, as unwillingly as I admit it, such a struggle poisons one's entire thoughts, and the hate for these fundamentally sick individuals who torment one eats into one's soul." A week later, as he looked forward to his first Christmas with his new wife, he had entirely withdrawn into quiet family pleasures: "In this we realize once again how important living together with decent people is."[24] Nevertheless, the accusations and investigations of 1937–1938 had a lasting effect on Heisenberg. Even toward the end of his life, long after the regime had passed, the imagined sound of Nazi jackboots and visions of black-shirted Gestapo functionaries marching up the stairs to his bedroom still occasionally awakened him, perspiring, from a fitful sleep.[25]

Heisenberg's correspondence in that period indicates several difficult trips to Berlin to further his case. At least one of these was for an official interrogation in the notorious basement chambers of the SS headquarters at Prinz-Albert-Strasse 8. A cynical sign reading "Breathe deeply and calmly" hung on the bare cement wall as a constant reminder to the victim of his or her predicament. Of the three known SS investigators assigned to Heisenberg, one worked with the Sipo (Sittenpolizei), or morals police, and all three had some training in physics. Heisenberg had even participated in the final examinations of one of them for his Leipzig doctorate in physics. Convinced by Heisenberg himself, his diplomatic Berlin supporters, and their own conscientious investigation, all three turned into strong and valuable supporters thereafter.

Perhaps the most influential of the three was Johannes Juilfs, born in 1911 and just completing his doctorate in mathematics and theoretical physics under Max von Laue at the University of Berlin.[26] After receiving his doctorate in June 1938, he was appointed as an assistant in Laue's institute. He later held the same post under Heisenberg, who succeeded Laue as head of the institute in 1943. As a member and leader of the Natural Science Group within the Nazi Students League of the university, Juilfs organized the first indoctrination camp for mathematics students in 1939. In the same year he was appointed an honorary member of SD Lead Unit Berlin in the new Reich Main Security Office (RSHA) under *Reichsführer-SS* Heinrich Himmler.[27]

In early 1938, a member of Himmler's personal staff asked Juilfs to prepare a comprehensive report on Heisenberg and theoretical physics.

Juilfs performed his duty with enthusiastic thoroughness. "Above all," he later wrote, "on the one hand I had to assimilate as wide a technical foundation as possible, and on the other hand I had to know well the technical-personal connections."[28] Conversations with Heisenberg and Heisenberg's closest colleagues pointed him toward a favorable conclusion. In gratitude, Weizsäcker helped Juilfs to reestablish himself after the war and the denazification trials by writing a physics textbook with him. Juilfs later took a teaching post in Hannover.

Thanks in part to Juilfs, the SS investigation ended positively for Heisenberg. Although the final report of the investigation has not been found, most of the bureaucratic memos on Heisenberg's views thereafter contain the same evaluations in many of the same words, which suggests they had the same source, the SS report. The earliest of these memos, a long affidavit on Heisenberg's views sent in 1939 from Himmler's office to the REM, gave Heisenberg a clean bill of political and scientific health. He was found to be neither the raving follower of "gray theories" nor the traitorous, Jewish-inspired enemy of the Reich that Stark had painted, but rather a harmless apolitical academic who reportedly even expressed favorable views regarding the Nazi regime.

However, it is impossible to determine to what extent the views attributed to Heisenberg in this report were really his own, to what extent they were formulated by Juilfs or someone else, and to what extent they were extracted under Gestapo duress. According to the 1939 memo: "For Heisenberg, theoretical physics is merely the working hypothesis with which the experimenter inquires of nature in suitable experiments. The evaluation occurs first and foremost through the experiment. The theory that is confirmed by experiment is thus the clear description of the observations made in nature using the exact means of mathematics." And, "Heisenberg's personal character is decent. Heisenberg is typical of the apolitical academic. . . . Over the course of several years, Heisenberg has allowed himself to be convinced more and more of National Socialism through its successes and is today positive toward it. He is however of the view that active political activity is not suitable for a university teacher, save for the occasional participation in indoctrination camps and the like."[29] Whether completely accurate or not, this remained the Nazi regime's official assessment of Werner Heisenberg until the end.

By the spring of 1938, as the SS report neared completion, two events suggested Heisenberg's imminent exoneration: state approval of his earlier election as section secretary of the Saxon Academy of Sciences, a state institution, and permission granted by the REM for a two-week lecture tour to England in March 1938—just a week after Germany's annexation of Austria. Neither activity would have been approved if the SS investigation were going against him.

Yet upon returning to Leipzig in early April, Heisenberg had still received no official word from the SS. As he would often do in the future, Heisenberg turned to Juilfs, who was the probable source of some disconcerting news, as quoted by Heisenberg: "The decision lies with the *Reichsführer-SS*, who in your case, in my opinion, does not want to do anything more."[30] Himmler, preoccupied with the Austrian annexation, was letting the "professors' controversy" wither away for lack of attention. Heisenberg again spoke of resignation. Nevertheless, two months later and still without an answer from Berlin, he prepared again to undergo active military training.[31]

Himmler finally acted in July 1938, just over a year after the onslaught in the pages of *Das Schwarze Korps*. But before he did, further influence was required from another quarter. Earlier that year, when Heisenberg and his colleagues had learned that the SS investigation had concluded favorably but that Himmler was ignoring the case, they took recourse in a tactic they would often use in the future: working diplomatically through respected applied physicists. Personal diplomacy was, as always, the German professor's strongest suit.

As practical men contributing to Germany's economic and military buildup, which was now in full swing, applied physicists could hardly be considered overly abstract or formalistic. A key player in this strategy, who had known and respected Heisenberg since his early days in Göttingen, was Ludwig Prandtl, the Göttingen professor of applied mechanics. As a specialist in hydrodynamics and aircraft aerodynamics, Prandtl was familiar with both modern theoretical physics and applied practical research.[32] He was also well practiced in negotiation—thus the perfect man to approach Himmler regarding Heisenberg. Prandtl did so at a banquet on March 1 celebrating Hermann Göring's new German Academy for Aeronautical Research, located in a new and spacious airplane hangar on the outskirts of Göttingen.

Conveniently seated next to the Reichsführer-SS at the banquet, the aerodynamics professor discreetly raised the topic of unjustified attacks against theoretical physics and "especially the personal distress of Herr Heisenberg." In a carefully worded five-page memorandum to Himmler on July 12 (long enough after the Austrian Anexation), Prandtl again defended Heisenberg and theoretical physics and suggested several elements of a resolution to the case.[33] First, Prandtl recommended adopting a compromise that Himmler himself had casually mentioned over dinner: that in teaching Einstein's physics Heisenberg should be asked to take great care to separate the man from his work. Second, Himmler should write a letter personally disavowing the smear of Heisenberg's character so that Heisenberg's effectiveness as a teacher could be restored. Third, for the benefit of Nazi physics students, Heisenberg should be allowed to publish an article on theoretical physics in the Nazi student publication, *Zeitschrift für die gesamte Naturwissenschaft*.

Prandtl's diplomacy finally prodded the <u>Reichsführer</u> into action. On July 21, 1938, Himmler sent an official letter to Heisenberg's private address in Leipzig informing the professor of his personal decision: "I do not approve of the attack of *Das Schwarze Korps* in its article, and I have proscribed any further attack against you." He then invited Heisenberg to join him for a "man-to-man" discussion of the matter in Berlin in November or December. He ended with a postscript admonishing the physicist in the future to separate the personal and political characteristics of the researcher from his research. He did not want or need to mention any names.[34]

On the same day, Himmler also sent a memo to Heydrich, his SD chief, enclosing a copy of Prandtl's letter. In his memo, Himmler ordered Heydrich to encourage the student science leader, Kubach, to allow Heisenberg to publish in the student journal and to refrain from further assaults on Heisenberg. Nazi physicists would have to find a way to advance their cause without the character assassination of Sommerfeld's heir. For, Himmler declared, "I believe that Heisenberg is decent, and we could not afford to lose or to silence this man, who is relatively young and can educate a new generation."[35]

Himmler's letter reached an elated Heisenberg two days later, on July 23, in the Bavarian mountain village of Fischen. Heisenberg had settled his family there while he prepared for military duty in nearby Sonthofen on the Austrian border. Heisenberg immediately thanked Himmler for his letter, "which freed me from my great concern," and readily agreed to the new compromise that Prandtl had struck with Himmler, a compromise that Prandtl had probably prearranged with Heisenberg and one that would endure throughout the period ahead.[36] Heisenberg would take pains in the future to separate the scientist from his science, but at the same time the useful results of work by Jewish researchers would be allowed in German classrooms—and could be readily exploited to German technological advantage—while any public mention of Jewish scientists themselves would not.

Most compromises are entered into as a means to an end, the acceptance of a less desirable situation for the sake of a greater benefit. In this case, the perceived benefits were both personal and professional. To Heisenberg, the compromise meant the restoration of his so-called honor, enabling him to remain in Germany and to continue his work. For his profession, it meant that the modern theories of relativity and quantum mechanics could be taught in German classrooms and exploited in German technology, even if the Jewish contributors to these sciences were ignored. Thereafter, relativity theory was called, in Heisenberg's classroom and elsewhere, by the title of Einstein's paper, "The Electrodynamics of Moving Bodies."

To Heisenberg, remaining in Germany was apparently worth almost any price, as long as he could continue to work and teach. Like many of those who were forced into emigration, his entire life and upbringing had instilled in him an unbreakable attachment to Germany that he could not easily deny, even temporarily. Only if the Nazis made life and work unbearable for him, only if they humiliated him into less than second-class status or worse—imprisonment—would he contemplate a move across the German border. (Family and children were, of course, also a consideration.) Heisenberg obviously felt an enormous attachment to Germany and an enormous drive to continue his work in Germany and for the future of German physics.

In the end, these two perspectives came together. Heisenberg came to regard his personal survival in Nazi Germany as tantamount to the survival of decent physics, and the continued survival of some elements of decent physics provided the grounds for Heisenberg's personal continuation of the struggle. Heisenberg's exoneration meant to him exoneration of theoretical physics itself. After all, Planck had given him the task of preserving an "island of existence" in the sea of anti-scientific chaos about him. And as long as he survived, that island would survive. But by seeing himself in such a grandiose rationalization for remaining in Germany, he more easily succumbed to further compromises and ingratiation with the regime.

This became clearer in the years ahead, during the struggle for two pieces of tangible evidence of exoneration that Heisenberg demanded as early as his first letter of response to Himmler in July 1938. First, he wrote, he looked forward to the proposed meeting with Himmler, so that "any sort of form could be found in which it could also be made publicly clear that the attacks on my honor were unjustified." In particular, Heisenberg insisted on publication of an article in the Nazi science student journal. Second, he wanted SS headquarters to send copies of Himmler's letter to the REM and to the rectors of the universities of Leipzig and Munich.[37] Accepting the call to Munich would serve as his ultimate exoneration and evidence of the victory of decent physics over Stark and his Nazi minions. Neither demand was easily fulfilled, and the longer they were delayed, the more importance they seemed to take on. Not until the depths of World War II did Heisenberg receive the evidence that he regarded as proof of his exoneration.

Like Faust, the personal price Heisenberg was willing to pay for his bargain with the devil could also have meant his life. In July 1938, nine days after Himmler cleared his name, Heisenberg was marching in German uniform. Six weeks later, he nearly marched off to war.[38] After successfully annexing Austria, Hitler had set his sights on the Sudetenland, a Germanic territory in Czechoslovakia that was now nearly surrounded by the new German Reich. As tensions mounted in September, active military units, including Heisenberg's, were placed on full alert. Heisenberg's superiors extended his tour of duty into October and distributed weapons to the troops. Heisenberg was sure of an impending invasion and certain that, because of the alliance arrayed against German expansionism, this attack would touch off a second world war. If war had indeed broken out then, it would have found him on the front lines, risking his life for a nation that had just subjected him to a year of inquisition and humiliation and that really cared little for him or his science.

Fortunately for Heisenberg, war was narrowly averted at the end of September by the infamous Munich treaty entered into by Hitler and Czechoslovakia's allies, Italy, France, and England. British Prime Minister Chamberlain returned to London proclaiming "peace in our time." It was the last appeasement Hitler would receive. A much relieved and now unburdened Heisenberg was back at his desk in Leipzig by mid-October.

But Czechoslovakia was now dismembered, and Germany had emerged as the most powerful state in Europe. The European Allies had reached a tentative and uneasy compromise with the Nazi regime, a compromise that Hitler would exploit and betray within a year. Maintaining his own uncomfortable truce with the regime, a meditative Heisenberg wrote his mother from his army barracks as the guns were being carefully returned to their racks for use another day: "It is strange to think how the fate of every individual and the deaths of many hundred thousand can hang on the decision of one man."[39] Within a year, this experience would become very real indeed.

CHAPTER 21

ONE WHO COULD NOT LEAVE

HIMMLER'S LETTER ALLAYED HEISENBERG'S CONCERN AND REMOVED THE SS THREAT, but it did not stop Aryan physics. Heisenberg's personal victory could not translate into a professional victory for science as a whole. In the end it could not even win him the prize he originally sought, the incident that had precipitated the whole affair—appointment as Sommerfeld's successor. Himmler's SS was only one of three competing state bureaucracies involved in academic appointments, and in this sphere it became the least influential. Of the other two agencies, Hess's party academic organizations—the students and teachers leagues—insisted on making independent recommendations to the Reich Education Ministry (REM), which made the final decisions.

Matters came to a head after the resolution of the Sudeten crisis in the fall of 1938. Sommerfeld, nearly 70, announced his intention to cease teaching after the winter semester. He hoped by this action to confront the REM with the necessity of appointing Heisenberg without delay. The REM asked the Munich faculty once again for its recommendations. By October, the faculty had two lists from which to choose. One, submitted by Sommerfeld and Gerlach, named Heisenberg, Weizsäcker, and Richard Becker, in that order. The other, submitted by Bruno Thüring, the Teachers League's university representative, named three applied physicists to the theoretical physics chair—and mediocre ones, at that.[1]

The dean, Friedrich von Faber, a backer of the Teachers League, chose the second list and forwarded it to Wilhelm Dames, the REM officer in charge of mathematics and physics appointments. The university rector, a state appointee, was noncommittal. Faber attempted in December 1938 to prod him into support with a long official letter supporting the Teachers League's choices. In order to counter the "ossification" of theoretical physics and to enable a return to the natural and the pictorial, he wrote, "under no circumstances shall men be named whose works lie in the train of thought of Einstein's relativity theory or that have been written in the spirit of pure formalism."[2]

Heisenberg's was a lost cause even before the dean's letter. SS officer Dames had been appointed to an honorary position in Stark's Berlin institute, and, like his REM boss Otto Wacker, he had joined what Heisenberg called "the opposing party." By early November 1938, Dames had already made his decision. After personally inquiring in the REM's Berlin offices about the Munich chair, Heisenberg learned that

Dames would choose Sommerfeld's successor from Thüring's list. However, to avoid alienating Heisenberg and his supporters entirely, the diplomatic Dames promised Heisenberg an appointment to a newly vacated chair in Vienna, in recently annexed Austria. Heisenberg's first choice still remained Munich.[3] And so the skirmish continued, with the Sommerfeld faction, SS, REM, and Nazi Party all jockeying for position and all determined not to lose political ground.

Hess wrote twice to Rust to inform him that, because of "the previous political behavior of Professor Dr. Heisenberg," the candidate was unsuitable for appointment to any chair, least of all Vienna's.[4] When on the basis of these letters the REM wanted to deny Heisenberg any new appointment, Heisenberg found sudden support from the man who had recently bargained to help him achieve a chair, *Reichsführer-SS* Heinrich Himmler. The *Reichsführer* promptly wrote to Rust, arguing that Heisenberg was especially suited for the Vienna chair, because he could be influenced in the proper direction by what Himmler saw as the strong Nazi character of the university, now that Austria had been annexed. He even enclosed the SS affidavit on Heisenberg's views to support his point. Himmler continually avoided the promised "man-to-man" meeting with Heisenberg—probably in order not to appear unduly influenced by the physicist. But he did meet separately in Munich with Hess and with Walter Schultze, the head of Hess's Teachers League, to try to sway them in Heisenberg's favor—either for Vienna or for Munich. Meanwhile, Sommerfeld and Gerlach were tirelessly arguing Heisenberg's case in Munich faculty meetings.[5] All was to no avail. In the end, as such things go, Dames chose the least qualified of Thüring's three candidates to serve as Sommerfeld's successor. Wilhelm Müller, the victor, was best known for his textbook on engineering mechanics and recent essays on Aryan physics.[6]

Heisenberg's contacts in Berlin informed him privately of the tangled web of backroom political intrigue—so typical of the Third Reich—that still militated against his appointment to Sommerfeld's chair. As Heisenberg reported it to Sommerfeld, party leaders Hess and Schultze could not back down under pressure from Himmler, because the party had already committed itself publicly against Heisenberg.[7] Himmler himself did not want to force the party to accept Heisenberg, because Heisenberg was not a party member, and he was at best ambivalent toward the regime. If he were placed in the "capital city of the movement," he might embarrass Himmler before the party faithful. Dames, though supportive of the Aryan physics faction when it worked to his advantage, found it prudent to counter its rising influence on REM appointments by choosing the weakest candidate from among the party's choices. Finally, the appointment of Nazi scientist Rudolf Tomaschek to the faculty of the Munich Technical College in early 1939 strengthened the district Teachers League and made Munich university officials unwilling to risk further battles with it. When provoked, the Teachers League could call down the wrath of party, state, and students on its opponents.

Müller succeeded the great Arnold Sommerfeld in the fall of 1939, just as war broke out. An enraged Walther Gerlach declared that theoretical physics was now dead in Munich. The Vienna chair was now also out of the question for Heisenberg—Austrian physicist Hugo Dingler, converted in his old age to Aryan physics, had joined the student cause to block Heisenberg's appointment.[8] As consolation, Himmler, still eager to keep the bargain, promised Heisenberg a recommendation to another highly placed chair. Such an appointment, together with the promised publication of his views in the Nazi student science journal, would constitute for Heisenberg tangible proof that the bargain still held, and that he was officially rehabilitated in the new Reich. "But it appears," Heisenberg wrote Sommerfeld, "that the last word on this has not at all been spoken."[9]

Having failed so far to achieve the Munich chair or any other, Heisenberg's victory in the SS affair also had little ultimate impact on the continued decline of his institute or on the deterioration of theoretical physics under the Reich. Wilhelm Führer, who had been recently elevated to a REM position, joined Dames in promulgating new university physics curricula and examination criteria that radically reduced concern with theoretical physics in favor of "practical" training.[10] The number of theoretical physics students and doctorates granted continued to decline sharply in Leipzig and throughout the Reich. For instance, of the 18 doctorates in theoretical physics that Heisenberg produced during his tenure in Leipzig, from 1927 to 1942, 16 graduated during the Third Reich. But only one of them, Erich Bagge, received a doctorate between the SS attack and the start of World War II. Only three achieved doctorates during the war, and of these two were foreigners.[11] REM chief Rust himself noted a similar nationwide decline and ordered Wacker, who in turn ordered Dames, to undertake a study of the possible elimination of some theoretical physics chairs. Dames, though delighted at the prospect of abridged theoretical physics, refused nonetheless to recommend reduction in his area of administration.

Throughout the losing battle for Sommerfeld's chair and the painful and humiliating investigation to which Heisenberg was subjected following the ugly SS assault, Germany was steadily sinking ever deeper into the noisome pit of Nazi barbarism. The SS and the Gestapo increased their police-state activities and expanded their use of the concentration camp to include not just detention but forced labor for all "undesirables." Deadly slave labor and genocide were not far behind. Himmler's exoneration of Heisenberg in the summer of 1938 occurred simultaneously with, in the words of one historian, a massive "tidal wave of terror" against Jews who remained in the Reich: intensified expropriation of property, arrests, expulsions, and street violence.[12] Just one month before Himmler wrote his letter to Heisenberg, he had ordered an expansion of the forced emigration of Jews, and his henchman Heydrich had issued quotas on the arrests of both Jews and "antisocials."

The descent continued into the autumn in step with Hitler's plans to provoke a war over the Sudetenland. Frustrated on that front by Allied appeasement, the regime received an unexpected opportunity to vent some of Germany's escalating war fever. On November 7, 1938, just days before the fifteenth anniversary of Hitler's Beer Hall Putsch, a distraught Jewish teenager whose Polish parents had been recently expelled from the Reich shot a third-rank German diplomat in Paris. The diplomat's death two days later served as a pretext for unleashing nationwide violence against German Jews.

Goebbels suggested that "spontaneous" protest demonstrations against Jews be held throughout the Reich. Hitler apparently concurred. Unbeknownst to Heisenberg, Himmler, while pressing for Heisenberg's call to Munich, was simultaneously arranging for the SS support of Goebbels's plan. With war frustratingly averted and average citizens lusting for action, local party and political officials, spurred on by the network of state and party organizations, had little trouble inciting mobs in every city and town across the Reich into a bestial frenzy of violence against Jews and Jewish property during the night of November 9–10, 1938. David H. Buffum, the American consul in Leipzig, described the violence as "a barrage of Nazi ferocity as has had no equal hitherto in Germany, or very likely anywhere else in the world since savagery, if ever."[13]

That terrible night came to be known as *Kristallnacht*, or the night of broken glass, for the tons of shop-window glass that littered every German street the following morning. Finally, the Nazi regime had revealed in unmistakable terms to the world and to its own citizenry its true nature. It left Heisenberg, like other nonparticipants Buffum had observed, "benumbed over what had happened and aghast over the unprecedented fury of Nazi acts." In his report to Washington, Buffum described how in Leipzig Jewish homes and apartments were invaded and demolished by ravening mobs who threw everything and everybody out onto the streets—often from upperstory windows. Men, women, and children were paraded through the streets and parks in humiliation. Hundreds of Jewish shop windows were smashed; the three Leipzig synagogues were burned to the ground; and, following Heydrich's order, police rounded up the first of several thousand Jewish men who were transported over the next two weeks to concentration camps. There they would "prove" their usefulness to the Reich by "contributing" their forced labor to the four-year plan.[14]

A visibly shaken Heisenberg wrote to his mother on November 12 that he and Elisabeth were "still completely in shock from the last nights."[15] The beautiful treelined streets of Leipzig had been turned into heaps of trash; the large Bamberger and Hertz department stores in the center of town were smoldering ruins. Worst of all, a friend had told them of a horrible scene on the morning of November 10 as entire Jewish families were dragged, screaming, to the train station, shoved onto passenger trains, and expelled from Germany. The same day Heisenberg wrote of that scene, Hermann Göring, architect of the four-year plan, signed an order systematically excluding Jews from the German economy and from schools and universities. Jews could not enter retail stores, sell goods or services anywhere in Germany, or hold any managerial or executive position.

The grim weeks and months following the frenzy of *Kristallnacht* took their toll on Elisabeth, now pregnant with the Heisenbergs' third child. To recuperate from the turmoil, the family retreated during the spring semester break in April 1939 to the peaceful, idyllic village of Badenweiler in the beautiful Black Forest region south of

286 | D A V I D   C .   C A S S I D Y

Freiburg. As if the recent events apparently still did not register within him the futility of his actions, Heisenberg continued to campaign by mail from his retreat for the Munich appointment—Müller's official call to the chair was still a year away. Tucked away in their Black Forest village, Heisenberg and Elisabeth did, however realize that the turmoil would only get worse. They decided then to search for a permanent country retreat for the difficult times ahead. Crossing over the nearby Rhine into southern France one day, they sat quietly on a hilltop bench looking back across the Rhine toward the verdant hills of their beloved German homeland. After beholding the sight for some time, Heisenberg quietly whispered to his wife, "How can I ever leave?"[16]

A friend of the family had already told them via Heisenberg's mother of a cottage for sale in the village of Urfeld, south of Munich in the Bavarian alpine foothills. The wood-framed cottage with its three bedrooms, kitchen, and large veranda had belonged to the well-known late impressionist painter Lovis Corinth. It stood at the foot of the Herzogstand mountain range. The veranda overlooked beautiful Lake Walchen and, beyond, the towering, snow-capped Isarwinkel Mountains. The Heisenbergs had not yet seen the cottage, but Heisenberg wrote his mother and the family friend to say that they were definitely interested.[17] The cottage and its price—26,000 marks, furnishings included—sounded ideal, and Heisenberg's mother began negotiations in Munich. Their Isar Valley house, which they had rented out, was now up for sale. Both sales took longer than expected, extending into the summer of 1939, when Heisenberg was scheduled to visit Columbia and several other American universities. The deal on the cottage was finally closed just before he set sail for the United States that summer. The growing family, he wrote his mother from aboard ship, would move into the cottage for the remainder of the summer immediately after Heisenberg returned to Germany.[18]

The transatlantic trip and its impact were also outgrowths of the SS newspaper assault on Heisenberg and theoretical physics two summers earlier. The onslaught had drawn the attention of the entire physics community to Heisenberg's science and to his personal views of theoretical physics and its practitioners. While Heisenberg submitted himself to SS and REM scrutiny, he did not remain passive. Although many of his close colleagues had rallied around both Heisenberg and their profession, many others had not. To encourage his supporters and to convince the wavering, Heisenberg hit the lecture circuit, speaking mainly to other physicists and those in closely allied fields. "It is probably good if I don't refuse such invitations," he wrote to his mother in November 1937.[19]

The dogmatists of Aryan physics had repeatedly called for the return of theoretical physics from its alleged overreliance on abstract formalism to a close reliance on tangible experimental data. In both nuclear and cosmic-ray physics, there existed a close relationship between theory and experiment, and the two fields contained problems and results of both intellectual and practical interest. Although Heisenberg himself concentrated on the more theoretical problems in both of these fields, starting in the summer of 1937 he tirelessly repeated two lectures emphasizing the theoretical-experimental and the intellectual and practical aspects of nuclei and cosmic rays. One lecture, entitled "The Present Tasks of Theoretical Physics," he presented to audiences in Münster, Freiburg, Stuttgart, and Ludwigshafen. The other, a more technical talk entitled "The Transit of Very Energetic Particles through the Atomic Nucleus," he gave with variations to physicists in Frankfurt, Dresden, and Bologna; to readers of *Die Naturwissenschaften* (*The Sciences*), and to the Saxon Academy of Sciences in Leipzig upon his election as deputy section secretary.[20]

In January 1938, Heisenberg reviewed the theoretical and empirical evidence in support of his still-controversial contention that a universal length must appear in nuclear and cosmic-ray physics and must be connected with the possible existence of cosmic-ray explosion showers. His only moderately technical article appeared in the pages of the *Annalen der Physik* (*Annals of Physics*), a research journal usually reserved for original technical reports of new results. In his opening paragraph, Heisenberg apologized for not meeting those standards with the excuse that the issue was dedicated to Max Planck, who, readers well knew, had only six months earlier been vilified along with Heisenberg as a "white Jew." He closed his essay to Planck by assuring his readers that "in theoretical physics it can always only be a question of a mathematical connection between experimentally observable quantities. . . . But probably a considerable expansion of the available experimental material would be the necessary precondition for the carrying out of such a program [of the universal length]."[21]

Heisenberg lectured at a nuclear physics conference in Bologna in October 1937 just days after he spoke at a colloquium on probability theory in Geneva.[22] Hess's organizations and the SS diligently opposed such traveling abroad, including travel to the United States, but the REM Congress Office, in consultation with Saxon officials, made the final decisions on foreign travel. The REM was especially favorable toward Heisenberg's foreign tours after receiving reports on his good behavior abroad—the SD and German citizen spies had kept a close watch on the traveling physicist.[23] But, of course, with a pregnant wife back in Germany, there was little chance he would speak publicly against Germany abroad or fail to return to the homeland.

Following the birth of the twins, in the spring of 1938, the REM again granted Heisenberg permission to travel abroad, this time to England, and this time with no objections raised. However, the REM resolutely refused to allow him to attend a theoretical physics conference in Warsaw. Hitler had already set his sights on Poland for his next campaign, and he did not want any German academics there to stir up trouble. At that time, the SS investigation was close to its positive conclusion. After having spent a year in painful SS and REM investigation and humiliation, Heisenberg would not jeopardize the impending exoneration over a conference. The REM intrusion would be another compromise that he would have to accept. Heisenberg regretfully complied with REM demands and canceled his travel plans to Warsaw. In his absence, the Marxist physicist Léon Rosenfeld read a French translation of Heisenberg's prepared paper to the conference, but the regime did not even permit the paper to be published in the conference proceedings.[24] According to one participant, the Warsaw conferees, who knew nothing of Heisenberg's predicament, formed a low opinion of their German colleague who, it seemed, had all too easily capitulated to the unreasonable demand of his government to dissociate himself from their conference.

Fortunately the regime had not prevented Heisenberg's important trip to England several months earlier. There he achieved a revitalization of his physics of the universal length shortly before Himmler provided the exoneration that enabled him to continue in Germany with less hindrance. The revitalization of Heisenberg's physics, even amid the worst Nazi assaults on his character, occurred just as proof emerged that rendered untenable Heisenberg's application of Fermi's beta decay theory to proton-neutron forces within the nucleus. Moreover, the elusive experimental evidence for Heisenberg's explosion showers now seemed to be nonexistent. Heisenberg's Planck paper in this same period was in fact a defense of the fundamental length, despite the slim evidence.

Hanging on to his physics, perhaps as his only hope for the future, the ever creative Heisenberg discovered a way out of the mounting evidence against his work. He suddenly shifted his focus to an entirely new type of nuclear force as an alternative to the now-failed application of Fermi's force to nuclei and cosmic-ray collisions. The alternative force was one proposed in 1935 by Japanese theorist Hideki Yukawa.[25] One problem with Fermi's force was that it was too weak to account for the strength of the force holding protons and neutrons together in the nucleus. The positive protons electrically repel each other so strongly that an enormously strong attractive force is required to make them stick together within the tiny confines of a nucleus. It also had to attract electrically neutral neutrons to the protons and to each other.

In order to meet these requirements, Yukawa invented an entirely new field, which in field theory provided a new force. He predicted that the quanta of the new field should appear as a new and much heavier charged particle. In Yukawa's theory, a neutron and a proton would attract each other by rapidly emitting and absorbing the new "Yukawa particles," which were also called mesotrons or simply mesons. Not wishing to neglect the nuclear physics of cosmic rays, Yukawa suggested further that "the massive quanta [mesons] may also have some bearing on the shower produced by cosmic rays."[26]

The discovery in 1937 of a new and previously unknown "heavy electron" in cosmic rays suddenly turned attention in the West to Yukawa's theory. The heavy electrons possessed the same charge (one positive electron charge) and about the same mass (200 times the mass of an electron) as the "massive quanta" predicted in Yukawa's theory. Intriguing as the parallels were between Yukawa's particles and cosmic-ray heavy electrons, physicists found that Yukawa's simple application of his field theory to nuclear properties simply would not work.[27] By 1937 much more had been discovered about the nuclear force than Yukawa's theory could account for.

Yukawa's ideas nevertheless inspired a flurry of theoretical activity on nuclear forces in hope of expanding and adapting his theory to account for these forces. Most of this work appeared in England, most of it was done by German émigrés, and most of it occurred during the months just preceding Heisenberg's visit there in March 1938. Pauli's former assistant, Nicholas Kemmer, who had recently left the Continent, served as a prime catalyst. Shortly after seeing Yukawa's paper in 1937, Kemmer derived all of the mathematical possibilities for Yukawa's type of nuclear field.[28] He compared their predicted properties with data on a single neutron bound to a proton, which was called a "deuteron." He convinced his colleagues Walter Heitler, Homi Bhabha, and Fröhlich that only one type of field and its associated particles, the mesons, agreed with deuteron data. Kemmer and his colleagues then managed to derive practically every available piece of nuclear data before turning to a phenomenon involving nuclear forces occurring *outside* the nucleus—cosmic rays. "We think, therefore," they wrote, "that it might be a reasonable policy to try to link up nuclear properties (forces and magnetic moments) with the cosmic-ray phenomenon of the hard component rather than with the beta decay."[29]

Of the two constituents, or components, of cosmic radiation, the easily absorbable soft component had already been identified as electrons and photons obeying quantum electrodynamics—the quantum mechanics of electric charges and the electromagnetic field. The greatly penetrating hard component had been identified as a new particle, the heavy electron. Because of its greater mass, it could penetrate greater distances of matter than could electrons and photons. Because the heavy electron and the Yukawa meson seemed so similar (and because there were no other alternatives), most physicists, like Kemmer and colleagues, proceeded on the assumption that the Yukawa meson and the cosmic-ray heavy electron were, in fact, identical. Unfortunately, they were wrong. Physicists did not realize until after the coming war that there are in fact two different types of mesons in cosmic rays. The penetrating heavy electron is actually one type of meson associated after all with Fermi's field, called the mu-meson, or muon. After penetrating long distances through matter, it decays into an electron and neutrinos. As experiments later revealed, the mu-meson is the decay product of a second type of meson, the pi-meson, or pion. It is associated with Yukawa's field. In many ways similar to the earlier explanation of the appearance of half-integer numbers through the idea of electron spin, the later discovery of the existence of two mesons in nuclear and cosmic-ray events resolved a host of puzzles and contradictions that kept theorists busy into the war years.

The theoreticians immediately encountered the same old obstacles to achieving a nuclear physics of cosmic rays: the reappearance of infinities in the mathematics and at high energies or small distances of approach. And as in Fermi theory, for distances or wavelengths smaller than a certain critical minimum length—or collision energies larger than a certain value, the energy of the meson—the number of newly created particles could conceivably multiply without bound. All this had a familiar ring. "According to the views developed here," Kemmer told his readers, "Heisenberg showers should be expected to occur."[30]

A month later, the inventor of the theory of Heisenberg showers arrived in Kemmer's office. He immediately launched into nonstop discussions with the local theorists; with his host, cosmic-ray experimentalist Patrick Blackett; and with the hordes of physicists from all over England who came to Cambridge and Manchester to hear him speak. "I hardly ever get to bed before 12:30," he wrote his wife, "get up around 8, and talk the entire day through with physicists almost without a break. It is important to me now to lose myself entirely in physics."[31]

Returning to Leipzig, Heisenberg immediately revamped his old physics with new arguments, replacing the critical length of Fermi's theory with the critical length of meson theory as a new fundamental constant of nature. The new length acted even more explicitly than in 1936 as a fundamental constant of nature, and therefore as a defining feature of both the current quantum theory and the future revolutionary theory. For Heisenberg the fundamental length marked the lower boundary, the "limits of applicability of the present quantum theory."[32] All the present quantum-mechanical methods and field theories, Heisenberg argued, ceased to apply as soon as collision energies and momenta crossed the boundary established by the new length; that is, as soon as colliding particles touched closer than the size of a Yukawa meson. Above that energy boundary, or below that distance of approach, quantum mechanics ceased to be valid, meson explosion showers burst forth, and a new and revolutionary quantum mechanics would come into play. Theoretical physics was still very much alive in Germany, if Heisenberg had anything to say about it.

Heisenberg had often discussed his revolutionary views with the sympathetic Niels Bohr. In May of 1938, he sent a copy of his manuscript, "The Limits of Applicability of the Present Quantum Theory," to Copenhagen, intending it to be something of a new uncertainty principle. While Heisenberg prepared his manuscript for publication, Hans Euler, his assistant, attempted to distill meson explosion showers from available data on large bursts of particles observed behind thick and thin lead plates in cloud chambers. The work earned Euler his habilitation later that year.[33]

That summer, as Himmler prepared his letter of exoneration and as Heisenberg contemplated the possibility of marching off to war during the Sudeten crisis, Euler and Heisenberg began a thorough analysis of all data available on cosmic rays, including cascade and explosion showers of all types, to support their physics.[34] In a scene reminiscent of his reading of Plato during the suppression of the Munich soviet republic, Heisenberg, sequestered in the Sonthofen army barracks that fall, with his machine gun at the ready, submitted a highly technical analysis of the penetrating cosmic-ray component for a special issue of the *Annalen der Physik* devoted to the seventieth birthday of another "white Jew"—Arnold Sommerfeld. Heisenberg apologized to Pauli for the circumstance that, for political reasons, only Aryans were allowed to celebrate Sommerfeld. Pauli helped organize an alternative celebration in the premier American journal *Physical Review,* along with an international boycott of the German *Annalen.*[35]

The linchpin of the Euler-Heisenberg argument for their physics was an apparent agreement between their theoretical value for the half-life of the meson, about one millionth of a second, and the result of their analysis of the decay rates of heavy electrons in air and matter. With such a short half-life, mesons could not have come from outer space—they had to be produced in the upper atmosphere, surely by the multiple, or explosion, process. By the same token, they could not survive the long journey from the upper atmosphere down to sea level within a millionth of a second unless their half-lives were somehow extended. The only way that could happen was through a process that entailed the celebrated "time dilation" of Einstein's relativity theory. According to Einstein, time slows down for extremely fast-moving particles. As a result, the mesons' clock was slowed by their motion, enabling them to reach the ground in far greater numbers than they should, if they had only one millionth of a second to survive. This argument, and its later experimental confirmation, has served ever since as a textbook illustration of the validity of Einstein's theory.

During the winter of 1938–1939, while Himmler pressed to have Heisenberg appointed to the Munich chair, Heisenberg took his latest results on another tour of the lecture circuit. He spoke at physics conferences in Hamburg and Leipzig and to the Bavarian section of the German Physical Society in the "capital city of the movement." But, of course, he carefully avoided any mention of Einstein's name when he told each audience of his new success as a theoretician in accounting for the observed mesons that survived the journey to sea level, where they appeared in experimenters' cloud chambers as penetrating cosmic rays. "This is an immediate consequence of the relativity principle—and, at the same time, a striking confirmation of it,"[36] was as far as he dared go.

Unfortunately, this confirmation and Heisenberg's plans for a revolutionary future theory did not go unchallenged. Other analyses of available data, undertaken mainly abroad, seemed to confirm instead a discrepancy between the theoretical and experimental half-lives for the cosmic-ray meson. Multiple processes and explosion showers still eluded direct detection at sea level, the penetrating component was observed once again to penetrate much further than it should, even with Yukawa forces, and the discovery of extended bursts of cascades led to the suspicion that Euler's explosion bursts were really a multitude of simultaneous cascades, rather than a single explosion event. British, French, and American teams confirmed the suspicion.[37] As a result, one American team told the June 1939 Chicago cosmic-ray symposium (which Heisenberg attended), "We shall adopt the more interesting and extreme position, and deny the existence of explosions until we are forced to recognize them."[38]

Most physicists who denied the existence of explosion showers believed at the same time that quantum electrodynamics, which rested on the reigning quantum mechanics, might encounter a boundary of validity at some very high energy, but this boundary would have nothing to do with Heisenberg's critical length associated with the rest mass of particles or with the corresponding relatively low critical energy.[39] Quantum mechanics should be fully applicable to the empirical mysteries of both nuclear and cosmic-ray physics. Practically alone in his views by 1939, Heisenberg believed in the existence of just such a length, establishing a boundary, and the need

LEFT TO RIGHT: *Oppenheimer, Fermi, E. O. Lawrence at the construction site for the Berkeley accelerator, 1939.*

(COURTESY ESVA, AIP NIELS BOHR LIBRARY)

for a new revolution in physics, similar to the one that had led to quantum mechanics during his earlier glory days, to handle nuclei and cosmic rays. Although some empirical data were open to interpretation either way, most of the physicists opposing Heisenberg were associated with the Oppenheimer group in the United States and the Heitler-Bhabha-Kemmer group in Britain. The controversy continued into the war years, until the leaders of both sides were drawn into wartime nuclear research on opposite sides of the war. It resumed briefly after the war, when new experimental evidence and theoretical innovations soon showed that, in fact, no new revolution was required after all. The old physics could be fixed to work just fine.

In pursuit of his purposes, Heisenberg brought his revolutionary new shower theory to the United States in the summer of 1939. REM officials now offered little objection to travel by the recently exonerated physicist, even as his hosts strongly objected to both his scientific and his nonscientific views.[40] During his month-long travels to New York, Chicago, Ann Arbor, and Indiana, American physicists and their émigré colleagues had a last look at the great physicist who had chosen to stay in Germany and who would in coming years assume a prominent position in the rival German nuclear research effort. Their impressions of him at that time helped to shape their attitudes toward him both during and after the war. There was the question of his re-argued science, which differed so greatly from the practical, empirical, nonrevolutionary thrust of their own work. But the main controversy surrounding Heisenberg was why he wanted to remain in Germany. The *Kristallnacht* of the previous November, his ill treatment at the hands of the vicious Nazi regime, and the impending world war seemed more than sufficient reason for emigration. Mystification at his continuing to live in Germany may even have affected his colleagues' perception of his science. In their view, if Heisenberg had been in the United States during the previous year, he would surely have seen that American experimentalists and theorists had no need for his physics. American science was, after all, on an ascendant trajectory, while German research seemed in obvious decline.

These unarticulated views may have informed the response to Heisenberg's newly developed arguments for the universal length at a cosmic-ray symposium held in Chicago at the very university where, three years later, Enrico Fermi would produce the world's first sustained nuclear chain reaction. Most of the papers presented during Heisenberg's session of the symposium opposed the identification of the meson with the heavy electron and contained arguments against his proposal of a universal length. According to Heisenberg's recollection of the meeting, the animated discussion following his session soon degenerated into a shouting match between himself and J. Robert Oppenheimer, the doyen of West Coast physics and the future head of the Manhattan Project.[41]

The reasons for Heisenberg's refusal to leave the Reich were much more complex than his American hosts or the Warsaw conferees before them probably realized. In both instances sentiment toward Heisenberg tended to be negative, and from the vantage of retrospect it is a sentiment with which one may well agree. Of course, Heisenberg did have ample opportunity to emigrate: after the SS newspaper assault, two major American universities had offered him tailor-made positions. Heisenberg had refused both; as incredible as it seemed, he did not want to leave Germany. Heisenberg's secret negotiations with physics professor George Pegram of Columbia University had led to a lucrative offer by the summer of 1938. But Heisenberg had by then received Himmler's exoneration and had accepted the bargain it entailed. He informed Sommerfeld from a village near his military barracks: "I have written to Columbia University that I want to remain in Germany, and that I would like to come over there sometime, but only for a brief period."[42] Many wondered if the real reason for staying in Germany was his sympathy for the Nazi regime.

Arthur H. Compton, head of the American Physical Society, had learned of Columbia's inquiries, probably from Pegram, who was also an official of the society. During Heisenberg's American travels a decade earlier, Compton's brother, a Princeton University professor, had offered him a position there, but "for patriotic reasons" Heisenberg preferred to return to Germany at that time. "Because of the recent developments we think that he may now be interested in an offer," Compton now informed his son, who was in Germany shortly after the 1937 SS assault. The son was instructed to repeat the invitation, this time for the University of Chicago, and

Compton prepared to meet any demands Heisenberg might make.[43] Again Heisenberg refused—he still would not leave Germany. When Heisenberg finally arrived for his month-long visit to the United States in 1939, his colleagues were more than a little curious about what possible reason, other than devotion to Nazism, he might have for wanting to stay in Germany, especially when faced as he was with the terrible social and professional conditions of 1939 and with an impending war.

Those reasons were still patriotic and, it seems, still predicated on the hope that Hitler and his henchmen would somehow be replaced or would become more reasonable in due course. And he still felt that it was his duty to help preserve what little remained of decent culture and science. Recollections abound of Heisenberg's statements during his American tour about the approaching war and his decision to stay. Heisenberg himself recalled expressing the conviction that Germany would lose the war and that he would be needed there to pick up the pieces.[44] However, British physicists Nevill Mott and Rudolf Peierls (an émigré) write, "in the recollection of his colleagues, he appeared to foresee a German victory. Was this a failure in communication, or did the views appear to him, or to the others, in a different colour in retrospect? One knows the fallibility of human memory."[45]

Beyond patriotic attachments to national survival and postwar revival, adversity had already bred tenacity. Heisenberg's colleagues could not know that absolutely no rational argument, however cogent, could outweigh the psychological grounds for his decision to return to Germany. It is difficult for anyone to leave their native country and move themselves and their young family to another country where the customs, the language, the politics are all quite different, even foreign. Few left Germany voluntarily. Heisenberg had already survived six and a half years of the Nazi regime. He had decided long before, even as the situation worsened, that he would not leave his homeland unless the authorities made work and teaching absolutely impossible for him personally. Moreover, his perceived duty to his students and to the future of his profession in Germany coincided with and reinforced this decision. Nor, apparently, did he seriously contemplate moving his wife and family to safety abroad. Just before he boarded the ship for the United States, he had closed the deal on the mountain cottage. While he debated with his colleagues in the United States, his family anxiously awaited his return to Germany so that they could move to their mountain retreat ahead of the gathering political storm.

By first threatening his work and his teaching and then officially protecting both, the SS affair had actually reinforced Heisenberg's decision to stay in Germany and to accept his lot under the regime. With the SS affair settled, promises of protection and support from Himmler, a newly acquired mountain refuge, and his potentially revolutionary science on an upswing, Heisenberg boarded the ship for New York thoroughly content with himself and with his decision to return to his troubled country. "I have the feeling," he wrote his mother, "that now everything is in place [in my life], as far as this depends on me. Of course many difficulties could still come from the outside. But I will deal with them much more easily than with the inner difficulties."[46]

After lecturing in Chicago to overflow audiences and spending a tiring two weeks of endless seminars and discussions with numerous faculty and students at Purdue University in Indiana, Heisenberg accompanied former Göttingen students and meson theorists Lothar and Gertude Nordheim to Ann Arbor for a few sticky days during a sweltering mid-July. There he stayed in the home of Samuel Goudsmit, an organizer of the annual Ann Arbor summer school in physics. Talk among the many students and faculty attending the summer school soon turned, inevitably, to emigration. Heisenberg recalled one such conversation in his memoir, *Physics and Beyond*. To Fermi's arguments for emigration, Heisenberg responded that he had long since gathered about him a circle of young people whom he saw as the hope of the future, and "if I abandoned them now, I would feel like a traitor. . . . [Besides] I don't think I have much choice in the matter. I firmly believe that one must be consistent. . . . People must learn to prevent catastrophes, not to run away from them. Perhaps we ought even to insist that everyone brave what storms there are in his own country."[47]

Other participants remember the same conversation somewhat differently. Goudsmit, then on the Michigan faculty, recalled, "Enrico Fermi and I asked him the question many others had asked: 'Why don't you come here?' He answered: 'No, I cannot, because Germany needs me.' He believed that the Hitler excesses, of which he strongly disapproved, would soon blow over. He felt that he would be needed to repair the damage made by the regime."[48] Max Dresden, another participant (then a Michigan student), recalled that Heisenberg seemed at first to waver in answering the question put by Fermi and Goudsmit. At that point, Mrs. Fermi, who had been driven from Italy by German-inspired anti-Semitic laws, said that anyone must be crazy to stay in Germany, whereupon Heisenberg launched into his vehement objection.[49] The patriot obviously would not budge.

Heisenberg returned to a hot New York for the last week in July, lecturing at Columbia and visiting his Uncle Karl and Aunt Helen in the suburbs. During his stopover, Pegram, the fatherly experimentalist who had done everything he could to get Heisenberg to join the Columbia faculty, tried one last time to convince him to stay, and Heisenberg tried once again "to get him to see my point of view."[50] For one final time, the answer was no. Heisenberg set sail in early August, leaving a thoroughly puzzled Pegram on the dock as the nearly empty luxury liner *Europa* steamed across the Atlantic toward the German Reich. One month later, the German Reich was at war.

CHAPTER 22

WARFARE AND ITS USES

"ON SEPTEMBER 1, 1939, WAR BROKE OUT; ON THE DAY AFTER, TOWARDS EVENING, OUR son Heinrich was killed." On this sad note Ernst von Weizsäcker, then state secretary in the German Foreign Office in Berlin, began his memoirs of World War II.[1] His son, Lieutenant Heinrich von Weizsäcker, platoon leader in the Ninth Infantry Regiment, fell the evening of September 2 on the Tucheler Heath near Danzig. His brother Richard, in the same regiment, watched over him until morning, when he was brought back to Stuttgart for a funeral presided over by his other brother, Heisenberg's colleague and confidant, Carl Friedrich. "And there he now lies," wrote his father, "under the wooden cross from the Tucheler Heath"—one dead among the millions to follow during the next six years, the carnage sparing neither innocent nor educated nor witting belligerent.

Two days after Hitler unleashed his army into neutral Poland, England and France declared war on Germany. For the second time in less than a generation, Europe—and soon the world—was at war. Its savagery and brutality could barely have been imagined in those early days when Heinrich fell so far from home.

The Weizsäckers were not the only family among Heisenberg's cultured acquaintances with fathers and sons at, or on their way to, the front. The musical Mr. was on the front lines that day as the German army smashed its way through Poland toward the Vistula. By December he was back in Leipzig with an Iron Cross and orders to join the forces at the western frontier, whence Hitler would soon unleash his second Blitzkrieg.[2]

Many of Heisenberg's physics colleagues also reported for duty that September. On September 16, seven aging experimentalists turned up as ordered at the Army Ordnance Office in Berlin, toting their military knapsacks packed with underwear and toiletries for the front.[3] They were more than a little relieved to learn that they were not headed for the front. Kurt Diebner, army research expert for nuclear physics and explosives, and his assistant, physicist Erich Bagge, had ordered them instead to a meeting on the potential applications of a recent German discovery—nuclear fission. During the meeting, the scientists explored the technical means of exploiting fission, both controlled and uncontrolled. It was soon clear that much more research was required. Dr. Bagge suggested to the assembled experimentalists that his Leipzig mentor, Professor Heisenberg, be included in their newly formed "uranium club" in order to provide a theoretical foundation for their work.

Bagge's mentor, meanwhile, had been waiting impatiently since early September for his marching orders.[4] The slightly built physicist looked much younger than his 37 years, and when he put on the uniform of his reserve infantry unit, the innocence betokened by his friendly face, blond hair, and disarming smile contrasted with everything that grey uniform and its Nazi insignia represented. Ever since he had nearly marched into battle during the Sudeten crisis a year earlier, Heisenberg had been convinced that war was inevitable and that he would be in it; like his acquaintances, he was ready for the fight. He was sure that orders to the front must be on their way. The orders finally did reach him later that month—but they did not direct him to the front. Bagge arrived in Leipzig on September 25 to inform the physicist that he was to attend the second meeting of the Uranium Club in Berlin the next day and that Bagge had arranged for Heisenberg's mobilization, not for the infantry but for research under the auspices of Army Ordnance.[5] Heisenberg traveled to Berlin that night and joined the uranium club the next morning. Unlike their counterparts in the previous world war, Heisenberg and other German scientists would fight their war not in the trenches, but on the research front.

On September 26, Heisenberg, Otto Hahn, Carl Friedrich von Weizsäcker, and several other nuclear scientists reported as ordered for the second meeting held in the research office of Army Ordnance (*Heereswaffenamt*) on Hardenburgstrasse, directly across from the Berlin Technical College. Once again the scientists reviewed the practical means of exploiting fission, and the theorists provided what they knew about the fission process. But they all agreed that continued research, both theoretical and experimental, was required before the possibilities could be definitely determined. After discussing the types of questions to be answered, the scientists dispersed across Germany to their various institutes. With funding provided by Army Ordnance, they began to implement a research program developed and coordinated from Berlin by Bagge and Diebner. The German nuclear fission project had begun, and it was already fully under the control of the German Army.

Both German and Allied interest in nuclear energy, controlled and otherwise, had quickly mushroomed after the 1938 discovery of fission—the splitting apart of a heavy nucleus with the release of enormous amounts of energy- by Otto Hahn and Fritz Strassmann in Berlin. After the annexation of Austria, a longtime colleague of Hahn's, Austrian Lise Meitner, had fled to Sweden just before the big discovery. Meitner and her nephew, Otto Frisch, who had also fled the Reich, soon showed from the outpost in Stockholm how fission could occur on the basis of Niels Bohr's recent model of heavy nuclei. Upon absorbing a stray neutron, the nucleus, viewed as a heavy liquid drop, would begin to vibrate so violently that it became unstable and eventually split into two smaller nuclei with the release of several neutrons and a lot of energy. Frisch told Bohr in Copenhagen of the discovery, and Bohr brought the news to America in January 1939. While Bohr and John Wheeler worked out a complete theory of nuclear fission in Princeton, a nuclear research team in Paris under Frédéric Joliot, the son-in-law of Marie Curie, confirmed in April that on average more neutrons were released per fission than were absorbed. A chain reaction could occur, releasing an enormous amount of energy in a very short time—in other words, an explosion.

Physicists on both sides of the coming war alerted their governments to the prospect of a new weapon made possible by the discovery of fission. In March 1939, several months before Heisenberg arrived for his visit in the United States, George Pegram and Enrico Fermi had contacted the U.S. Navy about the remote possibility "that uranium might be used as an explosive." But like most scientists, Fermi was skeptical that an uncontrolled chain reaction could actually be achieved. The navy shelved the idea until, later that fall, after the outbreak of war, Einstein's famous letter to President Franklin D. Roosevelt, written in August at the urging of Leo Szilard and a now-enlightened Fermi and Pegram, reached Roosevelt's hands. Several months later, Frisch and Rudolf Peierls, both German refugees in England, alerted the British government to the possibilities regarding nuclear fission. By mid-1940, as the German army blitzed across Europe, two Allied nuclear fission projects were already under way—one in the United States, the other in Britain. They and the French team later merged to form the Manhattan Project—the Allied effort to construct an atomic bomb.[6]

Meanwhile, several German scientists likewise had alerted their superiors to nuclear developments. Two Göttingen professors informed the Reich Education Ministry, which turned their letter over to Abraham Esau, the head of the physics section in the Reich Research Council. Esau organized a study committee. Two other professors in Hamburg, Paul Harteck and Wilhelm Groth, had informed Erich Schumann, head of the weapons research office in Army Ordnance (and a descendant of the composer), of the possibility of an enormous new explosive. A skeptical Schumann had handed the matter over to his explosives expert, Dr. Diebner, who had earlier worked on nuclear physics. Diebner had enlisted Bagge to the cause.[7]

Since Diebner and Bagge were the least skeptical about the practical potential of nuclear fission, by the outbreak of war in September 1939 Germany was the only nation in the world with a military project to exploit the new discovery. The Reich also controlled the world's largest supply of uranium ore, having seized the rich Joachimsthal mines in occupied Czechoslovakia. Aided by its own work and by the open publication of results in Allied nations until as late as June 1940, Germany was also privy to all the necessary basic research. Moreover, Siegfried Flügge, one of Heisenberg's former pupils, broadcast Germany's interest in nuclear energy with a widely read article, "Can the Energy Content of Nuclei Be Made Technically Useful?"[8] No wonder that in years to come Allied scientists would be convinced of Germany's head start in the race for an atomic bomb.

Judging from his rate of production during the first few months of the war, Heisenberg was indeed working with enormous energy on the theoretical possibilities for exploiting fission. The prospect of performing such research surely came as no surprise to him. During his visit to the United States in the summer of 1939,

Heisenberg had had long discussions with Fermi and Pegram, both of whom were by then well aware of the theoretical possibility of an explosive and, perhaps further induced by their encounter with Heisenberg, would soon approach Einstein concerning the letter to Roosevelt. Heisenberg recalled in his 1969 memoir, *Der Teil und das Ganze* (*Physics and Beyond*), that during their conversation in Ann Arbor Fermi raised the prospect that after the outbreak of war scientists in all nations would "be expected by their respective governments to devote all their energies to building the new weapons."[9]

Heisenberg recalled conceding the truth of Fermi's point: "You are only too right in what you say about our participation and responsibility." But he offered that no matter how feverishly governments and scientists strove to achieve an atomic bomb, "for the present I believe that the war will be over long before the first atom bomb is built." In other words, as Heisenberg recalled his position in the summer of 1939, he and other German scientists would readily work on a nuclear project established by their government, but they would probably not achieve a bomb—not because they would refuse at this point to produce one on moral or political grounds, but because the technical difficulties were so great that the war would end before they could overcome them. These circumstances and his remembered reactions to them would change over the coming years and decades. As with his research and activities before the war, each step of Heisenberg's participation in the German nuclear project must be seen in its own context and in the light of the inherent tensions within an individual attempting (however questionable this might be) to make the best of life under a vicious, antiscientific dictatorship now at war.

Heisenberg's approach to the first phase of the nuclear project is obvious—he immediately immersed himself in it. Within three months of receiving orders to report for the Berlin meeting, he produced the first of two parts of a secret, comprehensive theoretical report to Army Ordnance entitled "The Possibility of the Technical Acquisition of Energy from Uranium Fission."[10] Using the scant available data on nuclear properties and the basic Bohr-Wheeler fission theory, available openly in the journal *Physical Review*, Heisenberg surveyed every aspect of the practical exploitation of fission in a uranium "machine." The conclusion of his detailed report confirmed the possibility that a controlled fission reactor was technically feasible and that one of the uranium isotopes (forms of uranium), when obtained in sufficiently enriched form, would constitute a tremendous nuclear explosive, with energy much greater than that released by TNT. Heisenberg's report immediately made him the leading German expert on nuclear fission, and it served as a basic guide for the German project throughout the war.

For his fellow researchers, Heisenberg's report also confirmed and reiterated many of the essentials. As Flügge had already noted in his article, natural uranium consists of two main isotopes, uranium 238 (U-238) and the very much rarer uranium 235 (U-235), less than 1 percent of natural uranium. As work during the 1930s had shown, isotopes of an element possess the same number of protons in their nuclei but different amounts of neutrons. This is reflected in the numbers 235 and 238, which each give the total number of protons and neutrons in the nucleus.

In a flash of genius, Bohr had realized that each isotope responds quite differently to bombardment by neutrons.[11] The rarer U-235 is easily fissionable by very low speed "thermal neutrons"; the more abundant U-238 fissions only with difficulty and only for very high-speed neutrons. It also absorbs neutrons at certain specific (resonant) energies, producing the unstable isotope U-239. Since a fissioning nucleus emits neutrons of all energies or speeds, a controlled chain reaction may be best achieved in a lump of natural uranium by slowing down the faster neutrons to thermal speeds with a so-called moderator—a substance that slows the fast neutrons without absorbing too many of them. At the lower energies, the neutrons will not be absorbed by U-238 but will simply bounce off, continuing to bounce off other U-238s until the neutrons find one of the rare U-235 nuclei. When they do find a U-235, they will probably fission it, producing in the process two or three more neutrons, each of which can then go on to produce more fissions and more neutrons. The chance of fission improves if the U-235 content of natural uranium can be enriched by separating out some of the more plentiful U-238. The entire process increases in speed and energy release as more U-235 is available. If the entire lump of uranium is composed of U-235, the chain reaction will become uncontrolled and an explosion occurs within a fraction of a second. But the first step down the road of nuclear energy was to obtain a controlled reaction in what we now call a nuclear reactor.

Using very preliminary nuclear data, Heisenberg theoretically examined the use of several types of moderators with different amounts of natural uranium in two basic arrangements: spherical and cylindrical configurations of alternating layers of uranium oxide and moderator—a so-called nuclear pile. Size and shape were crucial—the optimal arrangement would prevent the circumstance that too many neutrons could escape the pile before they found and fissioned another U-235 nucleus. Pure carbon and heavy water (water in which each hydrogen nucleus has an extra neutron) seemed to Heisenberg the best moderators to slow the neutrons in order to escape capture by the plentiful U-238, and so cause the U-235 nuclei to fission. The use of both moderators in a cylinder (or cube) filled with alternating layers of uranium oxide, heavy water, and carbon seemed at first the best configuration. But such a device required enormous amounts of each substance. Assuming a pile of about one cubic meter in volume, Heisenberg predicted that a chain-reacting layer configuration could be achieved with 600 liters of heavy water, 1,000 kilograms of pure carbon, and 2,000 to 3,000 kilograms of pure uranium oxide.[12]

Heisenberg also predicted that, because of the absorption of neutrons by U-238, the reaction would reach equilibrium by itself at a temperature high enough to generate large amounts of electricity, if the reactor were used to heat steam to drive an electric dynamo. He did not realize that if more material were used than the minimum amount necessary, equilibrium would require a much higher absorption of neutrons through the presence of a control substance—otherwise, the chain reaction would increase without stopping, creating a very messy and deadly meltdown. The earliest German piles never contained such controls; neither did they go critical.

By enriching the U-235 content of natural uranium, a smaller, mobile reactor could be built at a higher temperature, which, Heisenberg later suggested, could be used to drive German tanks and submarines. If enough U-235 were separated entirely from a block of natural uranium and compressed into a ball, fission would be nearly instantaneous: an explosion would occur. But he did not calculate—at least not in this report—exactly how much was needed or how big the ball would have to be for the so-called critical mass. An assessment of the critical mass was a crucial step toward building a bomb. Isotope enrichment was the only way to obtain a mobile machine, Heisenberg told the German army in his report, and isotope separation was "the only method for producing explosives, the explosive power of which exceeds that of the strongest available explosives by several powers of ten."[13]

In the second part of his secret report, submitted at the end of February 1940, Heisenberg seemed less optimistic about the practical realization of the possibilities opened by nuclear fission.[14] He did not again mention an explosive and cautioned about the engineering. First, the enrichment and separation of rare isotopes such as U-235 were beyond Germany's (or any nation's) technical capabilities at that time. Since an element's isotopes are chemically identical to each other and differ only very slightly in mass, highly sophisticated techniques are required to separate and identify each isotope. In the coming years, the regime's racist policies would blind the scientists to a crucial alternative. Since the late twenties, Nobel laureate Gustav Hertz had been perfecting the gaseous diffusion method of isotope separation. But in 1935, because his famous uncle Heinrich Hertz, the discoverer of electromagnetic waves, was of Jewish descent, he was forced out of his position as head of the physics department at the Berlin Technical College. Hertz managed to remain in private industry in Berlin until the end of the war, but the Germans never developed his isotope separation method. It was one of the successful methods used by the Allies in the Manhattan Project and later, thanks to Hertz, by the Russians.[15]

Furthermore, although Germany possessed large quantities of uranium ore, it still lacked techniques to process it on an industrial scale into usable uranium oxide and eventually into metal plates, cubes, and powder. Nor did Germany possess the heavy water required for a self-sustaining critical reactor. The moderator problem was made even worse by Heisenberg's new conclusion, in which he was encouraged by imprecise data and the preference of others for heavy water. He had determined that the more plentiful element carbon, even in the form of pure graphite, probably would not do: the estimated cross section—the effective size of a carbon atom as seen by a fast neutron, which would slow the neutron by collision—was much too small. "It has therefore become doubtful," Heisenberg declared in his second report, "whether the uranium machine could be built with pure carbon." Calculations by Weizsäcker's Berlin assistants promptly supported this conclusion. In a series of experiments using purified industrial graphite, Hans Bothe and his assistant in Heidelberg mistakenly confirmed the inappropriateness of graphite a year later, and the Germans did not reconsider carbon as a moderator until late in 1944.[16] By then they were far behind the Allies. What they had failed to realize, however, is that the graphite must be in an ultrapure form, without any other elements present, far purer than even pure industrial graphite. Fermi's Chicago pile first went critical in December 1942 with an ultrapure graphite moderator to slow the neutrons and cadmium control rods to slow, or stop, the reaction after it went critical.

One of Weizsäcker's assistants predicted that the two possible configurations for the reactor—horizontal layers or concentric spherical shells—would both go critical with only uranium and heavy water. Since it was smaller, a "spherical machine" seemed preferable, he wrote, requiring only about 720 kilograms of uranium oxide and 400 liters of heavy water, packed in seven alternating layers to a radius of about one meter. Heisenberg's coworkers at the university's physics institute in Leipzig immediately began constructing shells for a spherical machine.

Another piece of the puzzle fell into place with the discovery of alternatives to U-235 as the fissionable isotope. Otto Hahn's Berlin team had already discovered that U-239, derived from U-238 by the absorption of a neutron, decays in 23 minutes to the new element 93 (uranium being element 92), which the Hahn team named "Eka Re" (now called neptunium). In a secret report, Weizsäcker suggested to Army Ordnance that Eka Re should be fissionable by thermal neutrons; since it was easily separated by chemical means from uranium, it would enable the construction of a very small machine or a very explosive bomb.[17]

Weizsäcker was on the right track. Before the Allies finally banned the publication of fission research results, an American research team in the June 15, 1940, issue of *Physical Review* reported that Eka Re is itself unstable, decaying in 2.3 days into the long-lived element 94, now called plutonium.[18] Everyone soon realized that this new element would be equally (or even more) suitable as an explosive and was easily obtainable from the transformation of U-238 in a working natural uranium machine. In theory, at least, a working reactor would produce not just energy but a second type of material, plutonium, for an atomic bomb.

It is difficult today to comprehend the motives and rationale that allowed Heisenberg and his colleagues to place their great abilities so easily at the service of the German army at war. What would compel Heisenberg to report immediately to the research section of Army Ordnance on the workings of a new energy-producing device and to confirm the possibility of a powerful new explosive in technical reports marked "*Geheim*" (secret) in his own hand?

Critics and supporters of Heisenberg have been sharply divided. Two former members of British wartime nuclear research, Rudolf Peierls and Nevill Mott, offer the following explanation: "It is reasonable to assume that [Heisenberg] wanted Germany to win the war. He disapproved of many facets of the Nazi regime, but he was a patriot. . . . Most citizens of most countries at war participate in the war effort when called upon, and the few who do not require exceptional courage and exceptional strength of conviction."[19]

Others have offered the opposite assessment of Heisenberg. In their view Heisenberg did display exceptional courage by gaining a leading scientific position on the project, thereby taking a large share of responsibility for the direction of research. From this position he was able to suppress information that might have led to a bomb, and he further sabotaged the project by slowing it down and keeping other, less scrupulous scientists from constructing a weapon that would indeed have enabled Hitler to win the war.[20]

The latter account of Heisenberg assumes a level of control that he never really possessed. Although he did develop many of the theoretical aspects of the project and did direct the Leipzig and Berlin branches of the undertaking, the overall project was in the hands of Army Ordnance until 1942 and under the Reich Research Council and other agencies thereafter. In addition, the project itself eventually split into two independent branches, one under Heisenberg, the other under Diebner. Recently a number of captured German documents pertaining to the German fission project have surfaced in former Soviet archives. They were returned in 2004 to the archive of the Max Planck Society in Berlin-Dahlem. In his recent book *Hitlers Bombe*, economic historian Rainer Karlsch made a number of stunning assertions after having examined these and other documents. He argued that the SS later gained control of Diebner's work at the German army's Gottow research station near Berlin and that, with the support of Walther Gerlach, then the administrative head of German fission research, Diebner and his team managed to achieve a chain reaction at Gottow before the end of the war. Karlsch also claimed that Diebner produced two small nuclear explosions at two other locations in Germany, likely involving a small fusion reaction, and that it appears from the reports that several hundred prisoners or slave laborers died as a result. A recent analysis of the soil at Gottow showed a higher than normal level of radioactivity, but this could also have arisen from normal reactor research or from Soviet activities there immediately following the war. The evidence for these assertions is inconclusive, and the story is technically not credible. Nevertheless, these reports and statements by Diebner and Gerlach after the war suggest that they were working feverishly toward such results in the closing months of the war. This may have resulted in the reported explosions, which for technical reasons could not have involved either fission or fusion reactions, but might have involved (if they happened at all) perhaps a type of "dirty bomb."[21]

One of the advantages of biography is that an individual's actions may be seen within the context of his entire life, not just the period surrounding those actions. In Heisenberg's case, the outbreak of world war and his immediate entry into work on nuclear fission occurred after seven years of living under the Third Reich, and after several years of personal attacks by Nazi physicists that had resulted by 1939 in an explicit compromise with the regime that allowed him to remain in Germany. The issues of patriotism, participation in the war on Germany's side, and the desire to defend the German nation—if not the Hitler regime—against defeat had been settled for Heisenberg long before the outbreak of war, as they were for most of his cultured colleagues and acquaintances. The interpretation offered by Mott and Peierls, coincides perhaps most closely with this view, while more extreme positions seem out of step with what we know of Heisenberg and his activities, strategies, and compromises during the prewar years.

But Heisenberg's prewar years also suggest that for him there were other inducements to undertaking work on nuclear fission, aside from patriotic motives. As with his Allied counterparts, scientific curiosity and a more utilitarian goal were also evident. The outbreak of war and the interest of German Army Ordnance in nuclear fission suddenly offered Heisenberg and the German atomic scientists a unique opportunity to prove their worth at last to their rulers. At the same time, he and they would have the protection from ideological meddling and economic cutbacks offered by a significant government project run by the victorious German army. After seven years of a depressing, losing battle against regime intrusion, and now faced with a weakened profession, the continued ravings of "Aryan" physicists, and the further blockage of Heisenberg's professional ambitions, these were valuable gains.

Heisenberg did take the lead, not in preventing the project from achieving a nuclear weapon, but in seizing the practical opportunity with his comprehensive two-part report to Army Ordnance on applied nuclear fission theory, produced in amazingly short order. In the process, he apparently convinced himself again that, amidst the technological and material conditions prevailing in early 1940, he could advance toward a useful energy-producing machine during the course of the war—useful to himself, to his profession, and to Germany—while disregarding the possibility of an explosive, which, he then believed, lay in the far distant future. He could ensure continued recognition of himself and of nuclear research by tantalizing regime officials with the prospect of a bomb, without concerning himself at this point with the prospect of actually building one. Even isotope separation should be supported, he argued, although, should it succeed, it would enable the extraction of a readily explosive isotope.

These are the reasons, it seems, for the incongruous circumstance that Heisenberg referred explicitly to isotope separation and to an enormous nuclear explosive in the conclusion to his first report to German Army Ordnance, while the paper itself actually concerned only the application of nuclear fission theory to the construction of a reactor. Patriotism, professional utility, scientific curiosity, and support of the German war effort united to produce the extraordinary effort that Heisenberg invested in nuclear fission research during the following early months and years of the war.

Physicist Peter Debye painted a similar picture of the motives and attitudes of the German atomic scientists in a long conversation in Berlin with Warren Weaver, an official of the Rockefeller Foundation. Within days of the outbreak of war, Army Ordnance had invoked military prerogatives to wrest control of fission research from Abraham Esau and the Reich Education Ministry. Seeking to centralize research in Berlin, Schumann took over Debye's Kaiser Wilhelm Institute for Physics in Berlin.

Debye, a Dutch national, refused either to resign or to accept German citizenship and informed the Rockefeller Foundation (which had built his institute in 1936) of the turn of affairs.[22] Weaver headed for Berlin to see for himself.

According to Weaver's log of his meeting with Debye in February 1940, Debye and the uranium club were well aware that the army hoped to achieve an "irresistible offensive weapon" from nuclear research. The researchers themselves conveyed their own, very different aim to Debye: "With D[ebye] they consider it altogether improbable that they will be able to accomplish any of the purposes the Army has in mind; but, in the meantime, they will have a splendid opportunity to carry on some fundamental research in nuclear physics. On the whole D[ebye] is inclined to consider the situation a good joke on the German Army."[23]

Heisenberg recalled, after the war, "I, like several of my colleagues, was told to work on the technical exploitation of atomic energy"—a somewhat embellished story. Elsewhere he wrote: "The official slogan of the government was 'We must make use of physics for warfare.' We turned it around for our slogan: 'We must make use of warfare for physics!'"[24] At first Heisenberg did doubt the practical realization of the theories that he readily provided the authorities, as would be indicated by his reaction when he realized in 1941 that his theoretical predictions would come true. And for the next three years the only fundamental research recorded by the curious theorist did consist of applied nuclear reactor theory. Heisenberg did invoke the rationale of using warfare for physics, but he did so while pursuing just the opposite: making physics useful for warfare in order to render it and himself acceptable to the rulers of the Reich. Whether or not they thought they could or would build the weapon the army wanted, Heisenberg and his compatriots saw themselves as walking a very difficult fine line. We, in retrospect, see them diligently performing basic research that they would have performed at this stage whether or not any such line existed, in their minds or anywhere else. Who, we may ask, was fooling whom?

Until more pile components became available, most of the Uranium Club work focused on confirming the details of Heisenberg's theoretical predictions and obtaining precise measurements of various properties of the materials. This information would be used to improve theoretical estimates and to guide reactor design. Three technical problems also required solving: the scientists had to develop suitable methods of isotope enrichment and separation; they had to obtain large quantities of heavy water and uranium oxide; and they had to discover the right geometry and size for a self-sustaining critical pile.[25]

Of the nine task-oriented research groups scattered among German laboratories and coordinated by Diebner, Heisenberg worked closely with two, one at his own physics institute at the University of Leipzig and the other at the Kaiser Wilhelm Institute for Physics, headed by Debye, who was soon removed from the post. By October 1940, Heisenberg was dividing his weeks equally between Leipzig and Berlin, then about two and a half hours apart by train.[26] After removing Debye, Schumann had appointed Diebner, the project administrator, to be provisional head of the institute, located in the quiet, tree-lined western suburb of Berlin-Dahlem. But unfortunately for Diebner, Debye's institute staff, most of whom were close to Heisenberg, were still in place. They included Weizsäcker and his assistants, as well as experimentalists Karl Wirtz, Erich Fischer, Fritz Bopp, and Debye's faithful technician, Herr Gretschmer. The staff considered the energetic Diebner, a man with strong party connections but a weak grasp of nuclear theory, unworthy to administer their work. Weizsäcker and Wirtz increasingly involved Heisenberg as an outside advisor in institute affairs, hoping that one day the famed physicist would supplant Diebner as permanent head of the institute.

Heisenberg, for his part, used his Berlin connection to gain an influential role in all of the theory and most of the main reactor experiments in Germany, which were performed by the Berlin and Leipzig groups. In order to house the Berlin (B) series of radioactive cylindrical reactor models, in October 1940 the Berlin team constructed an outbuilding—named the Virus House to keep away the curious—on the grounds of the neighboring Kaiser Wilhelm Institute for Biology. At the same time, Heisenberg and his Leipzig team prepared the L series of spherical reactor models, which lasted through 1942.

Under Heisenberg's supervision, Robert Döpel, Fritz Kirchner's successor as Leipzig professor of radiation physics, directed the experimental work at the Leipzig institute. He was assisted by his talented wife, Klara, a lawyer, and by the institute's able technician, Wilhelm Paschen, who actually built the various contraptions. Karl Friedrich Bonhoeffer collaborated in the resolution of heavy-water problems. But Gerhard Hoffmann, Debye's successor as Leipzig professor of experimental physics, had little to do with Leipzig pile research, even though he was a charter member of the club. Heisenberg's disdain for Hoffmann's former pupil, Diebner, who administered the entire German research effort, did not endear the theoretician to him, nor would the often difficult Döpel countenance interference from the crusty old experimentalist.

The Uranium Club soon suffered shortages. In early 1940, Heisenberg put in a request to Diebner for up to a metric ton (1,000 kilograms) of pure uranium oxide, and experimentalist Paul Harteck asked for up to 300 kilograms for his pile research in Hamburg. But because it had not been deemed important until then, only 150 kilograms of industrially pure uranium oxide existed in all of Germany. Diebner assured the impatient Heisenberg that by the end of June 1940 he would have his metric ton. The Berlin Auer Company was working at top speed on the uranium ore from the seized mines at Joachimsthal.[27] Later that year it would tap the stores of captured ore mined from the Belgian Congo.

Heavy water also posed a problem. In a letter to Heisenberg in January 1940, Harteck asked if anything was being done to procure heavy water. Again Heisenberg had the jump on everyone else. Meeting with Diebner just after the New Year, Heisenberg recommended the construction of an industrial heavy-water plant as soon as he had experimentally confirmed his theoretical prediction that heavy water would make a suitable moderator. The Döpel's and Heisenberg supplied the confirmation in August, but by then German army conquests had again provided an outside source.[28]

In April 1940, Germany, attempting to outflank Britain, marched into Denmark and Norway. On May 3, German troops captured the Norwegian town of Vemork, near the world's only heavy-water production plant, which was operated by Norsk Hydro-Elektrisk. Rather than build their own plant, the Germans would simply use Norway's. With improvements in electrolytic production demanded by the Germans, the plant was producing 300 liters of heavy water a month by the time British and Norwegian commandos put it temporarily out of service in 1943. By then, Germany had extracted probably just enough moderator to make a reactor.[29]

Hitler's conquests also provided German scientists with a cyclotron. Thanks to the Rockefeller Foundation, a cyclotron had just been completed in Niels Bohr's Copenhagen institute, and a second was nearing completion in Frédéric Joliot's Paris laboratory. A cyclotron would enable the researchers to measure the necessary nuclear constants and to create fissionable artificial elements such as plutonium, otherwise obtainable only from a reactor.

When Paris fell to the German army, Schumann and Diebner immediately headed for Joliot's lab. By April 1941, Bagge and Wolfgang Gentner were working alongside Joliot—who remained in Paris where he was active in the Resistance—to complete the cyclotron. Heisenberg urged Bagge to make the most of the opportunity by performing some fundamental nonfission research with the esteemed Joliot.[30] With the successful exploits of the German army, everything seemed to be in place. German scientists now possessed all the uranium, heavy water, and cyclotrons they would need—a circumstance that seemed to bother no one except Allied scientists, even though the invasion and exploitation of neighboring countries had made it possible.

The astonishing successes of Hitler's Blitzkrieg also left Leipzig relatively undisturbed until the war came home in 1943. Of course, reminders of the war were everywhere: frequent nighttime air-raid alarms (but few actual raids at first) and annoying shortages of food and fuel. Elisabeth Heisenberg, now coping with five children, had to find necessities by herself, while her husband split his time equally between Leipzig and Berlin. Yet the Heisenbergs managed to live a more or less unperturbed life. There were the births, baptisms, and usual illnesses of their children, musical evenings with their friends the Jacobis (who somehow survived in Leipzig to the end), Christmas celebrations, and the annual blooming of Elisabeth's flower garden.[31]

During the first two summers, from school's end until late fall, Elisabeth, the children, and a nursemaid moved to their Urfeld retreat in the Bavarian Alps. Heisenberg lived alone in the Leipzig house or with his in-laws, the Schumacher family, in Berlin-Steglitz, not far from the Dahlem institute. While working intensely on fission research in two cities, Heisenberg seemed satisfied with his personal life. "If one can keep life in one's small circle in order," he told his mother, "one must be content."[32]

The university also remained in relative order after Helmut Berve, Heisenberg's friend from the professors' club who had been demoted as dean, became university rector in 1940. In his inaugural address, Berve quoted Goethe to the student führer: "Bilde Künstler, rede nicht" ("Educate artists, hold your tongue"). Academic matters, not politics, predominated under Berve. Heisenberg pursued his nuclear research, his lecture duties, and his supervision of two doctoral candidates and was generally left alone.[33] Party member Berve even blocked the promotion of one associate professor who attempted to advance on the strength of his party credentials rather than the quality of his academic research. The injured man complained bitterly to the office of party ideologue Alfred Rosenberg in 1942: "The rector, dean, and a large portion of the faculty are humanistically oriented and try to do everything to label revolutionary efforts as unobjective."[34]

But there could be no doubt that the regime was long since very much in control of German life, academics, and nuclear research, and there were constant reminders of its power. On entering the university, students were divided into close-knit indoctrination squads under a Nazi student leader, who oversaw nearly every detail of their personal lives and studies. Party members among advanced students and faculty were required to wear party badges everywhere, even in the laboratory. On two bitterly cold days in January 1942, the Gestapo publicly rounded up some of the remaining Jews in Leipzig—men, women, and children—stripped them of their coats, and drove them 18 kilometers in an open truck to a small town for brief internment. We now know this was the first leg of what would be their final journey to the eastern death camps. Heisenberg's acquaintance Carl Goerdeler, the former mayor of Leipzig who would take part in the failed attempt in 1944 to assassinate Hitler, recorded the events after watching helplessly from his home.[35]

The regime's influence on institute affairs, and Heisenberg's attempts to counter it, are illustrated by the case of Edwin Gora. A Polish student of German descent, Gora was studying theoretical physics in Warsaw when Germany invaded Poland. Warned of the imminent arrest of intellectuals, he returned to his hometown in southern Poland and wrote to Heisenberg, the leading German theorist, of his predicament. Heisenberg invited the young man to Leipzig and helped him to enroll at the university and to obtain a job as a tram conductor. But in 1941 the Gestapo ordered Heisenberg to bar Gora from the institute, they had received a less than favorable report from Gora's hometown regarding his attitude toward the Reich. Heisenberg complied without resistance, but with encouragement from his wife, who felt a motherly compassion for the young man, Heisenberg quietly took Gora under his wing, gave him private physics instruction in their home, and eventually enabled him to pass his doctoral examination under Friedrich Hund in 1942. Gora worked until the end of the war with Walther Gerlach in Munich, before moving on to the United States.[36]

Despite his assistance to Gora, Heisenberg, who worked so hard to preserve the insularity of his institute, was in the end often powerless to prevent the tragic impact of events on his young colleagues. Especially calamitous were the fates of his brilliant assistant, Hans Euler, and Euler's close friend, Bernt Olof Grönblom, a promising Finnish physics student. As noted earlier, Euler, a Soviet sympathizer, had survived in Germany under Heisenberg's protection. Like many "fellow travelers" of that era, Euler was badly shaken by Stalin's 1939 pact with Hitler and by the German-Soviet partitioning of Poland later that year. When Stalin invaded Finland a year later, Grönblom left Leipzig to defend his homeland, leaving Euler a changed man. Stalin's actions left the sensitive Euler completely unsettled about himself and his political orientation in a world now dominated by two ruthless dictators. Heisenberg recalled inviting Euler to join the Uranium Club, fearful that Euler would be drafted despite his fragile health. To Heisenberg's surprise, the unbalanced Euler had already signed up for the Luftwaffe.[37] After flight training, he served as a meteorologist on a reconnaissance plane, flying missions over Crete, Egypt, and England in 1940 and 1941. Heisenberg's efforts to persuade his assistant to return to work, with or without joining the club, were of no avail.

Euler's letters to Heisenberg in this period are heartbreaking for their carefree tone. In his last note, on June 16, 1941, Euler wrote, "We often recall to each other the ocean and the mountains beneath the sun in the south and the heat over Africa, and we will probably still do that for a long time afterward when we sit together in our new surroundings [at our next assignment]."[38] Seven days later, while helping to carry out Operation Barbarossa—Hitler's surprise attack on the Soviet Union—Euler's squadron lost contact with his plane over the Azov Sea, near the Crimea.

Aided by Euler's mother and sister, Heisenberg desperately tried to locate his lost assistant through military channels. He even inquired of British physicist Patrick Blackett, whose sympathies with the Soviet Union were no secret and who may have had contacts there. The search continued after the war, but no trace was ever found of Euler or his plane. Two months after Euler's disappearance, Grönblom fell defending his homeland. Deeply disturbed by both losses to the war, Heisenberg later wrote a moving memorial for his student in the proceedings of the Finnish Academy, saying, in part, "The more outstanding his first achievements were in the field of science, the more reason we have to mourn the loss of a young man who was suddenly torn from us and his work because of a higher duty."[39]

CHAPTER 23

A COPENHAGEN VISIT

WHILE AWAITING THE ARRIVAL OF MORE URANIUM AND HEAVY WATER IN 1940, Heisenberg's Leipzig and Berlin research teams tested paraffin and regular water as possible moderators, substances that would slow the fission neutrons enough to escape capture by U-238 and thus fission U-235. Each fission, set off by the absorption of a single neutron, produces on average two or more neutrons. Each of these released neutrons, if sufficiently slowed, could go on to produce more fissions of the rare U-235 isotope in a piece of natural uranium, each producing in turn more neutrons, and so on. An energy-producing chain reaction would occur in a self-sustaining, energy-producing nuclear reactor.

Heisenberg's research teams tested the two moderator candidates in nuclear piles of alternating layers of moderator and small amounts of natural uranium in the form of uranium oxide, which they called for security reasons "preparation 38," U_3O_8.[1] With only a general notion of the best layer configuration, the scientists had to find through trial and error the best geometry and the optimal amounts of material needed. The Berlin team tried alternating horizontal layers of powdered preparation 38 and moderator in a cylindrical aluminum tank, 1.4 meters in height and diameter. It was immersed in water in the *Brunnengrube* (well hole), a water pit dug inside the Virus House, an insulated wood-frame house standing under cherry trees, near the Kaiser Wilhelm Institute for Biology. The water in the 2-meter-deep hole absorbed and reflected neutrons escaping from containers lowered into it on chains by a crane spanning the 3-meter-wide pit.[2]

Under the able direction of Wirtz and Fischer, the Berlin group performed five experiments through mid-1942 using paraffin and uranium oxide as a solid or ground into a metal powder. Heisenberg and his Berlin coworkers reported their findings in detailed, secret technical reports to army research. The result: none of the models worked. Most of the neutrons emitted by a small source in the center of the pile were absorbed by the contraptions rather than multiplying in fission.[3] In the spring of 1942, the four Leipzig experiments had yielded quite a different result: the Germans' first positive neutron multiplication.[4] But after this promising start, they never managed to achieve a chain reaction. Fermi's Chicago group surpassed the Germans within months in the achievement of the world's first self-sustaining chain reaction.

Back in early 1941, after first trying paraffin and water as moderators (model L-1), Heisenberg's collaborator Robert Döpel, who was closer to the power source than was the Berlin team, switched to precious heavy water as the moderator for model L-2. It was arranged in concentric, spherical, aluminum-lined shells, alternating with uranium-powder shells. After Döpel and institute technician Paschen placed the concentric shells within two aluminum hemispheres, they bolted the ball (radius about 40 centimeters) shut, then winched it into a water tank in the basement of the Leipzig Physics Institute. On inserting a weak neutron source through a connecting tube into the center of the ball, they measured the neutron flux as a function of radius. Model L-2 proved a dud. Heisenberg calculated that the source neutrons were all absorbed within the sphere, mainly by the considerable aluminum.[5]

To confirm this result and the findings by others that powdered uranium metal oxide, "38-metal," was superior to preparation 38, Heisenberg and his Leipzig team turned to metal powder and heavy water late in 1941. But only enough metal powder was available in Leipzig to make one spherical shell surrounded on both sides by heavy water. Careful measurements of this model, L-3, reported in early 1942, indicated a much smaller loss of source neutrons. Heisenberg and Döpel were convinced that one more layer of uranium metal, a total of 755 kilograms, with 164 kilograms of heavy water, would yield a genuine multiplication, and proof at last that Heisenberg's "machine" really would work.[6]

But the dangers of combining 38-metal and water had been overlooked. When water (heavy or not) and uranium meet, they produce flammable hydrogen gas. According to a formal report filed by Döpel, one day in December 1941 Paschen and his apprentice, F. Zumkeller, were pouring the powder into one of the hemispherical shells when heavy water somehow leaked in. An enormous flame suddenly shot out of the sphere, singeing the ceiling and burning Paschen's hand so severely that he could not work for nearly a month.[7] Döpel, his wife standing by with a fire extinguisher, gingerly poured the remaining powder into the sphere for the L-3 measurements.

More 38-metal finally reached Leipzig in early 1942, and sometime in late spring, pile model L-4 began multiplying neutrons at the rate of 13 percent. "A simple expansion of the layer arrangement described here would thus lead to a uranium burner," the ecstatic Heisenberg team coolly reported to Army Ordnance. "With that," wrote Heisenberg and Wirtz after the war, "[we] proved the possibility of an independently working, energy-producing uranium burner."[8] Nuclear fission research was no longer just a politically useful or theoretically interesting exercise; the likelihood of controlled—and even uncontrolled—fission suddenly became very real indeed.

Years later Heisenberg recalled, "It was from September 1941 that we saw an open road ahead of us, leading to the atomic bomb."[9] The project was at that time reporting on model L-2 and probably only just beginning L-3. Moreover, every German effort to extract from natural uranium the rare fissionable isotope U-235, the explosive material needed for a bomb, had so far proved a failure. A reactor and a bomb were still far beyond reach.

But in a report dated "Berlin, August 1941," Fritz Houtermans of Manfred von Ardenne's Berlin research institute (funded by the German Post Office!) obtained an extremely important result: a theoretical confirmation of the plutonium alternative. In a secret report he confirmed Weizsäcker's result that when U-238 absorbs a neutron, it produces U-239, which decays into element 93 (uranium is element 92 on the periodic table). This element should decay into element 94 (plutonium), which, wrote Houtermans, should be as fissionable by neutrons as U-235. The implication for what Houtermans called "the theme of our work" was clear from Weizsäcker's and Heisenberg's earlier reports. Once a natural uranium reactor was finally up and running, it could act as what we now call a "breeder reactor," producing through absorption of neutrons by U-238 the stable yet highly fissionable element now called plutonium. Since plutonium differs chemically from uranium, Houtermans wrote, "it can therefore be separated [from uranium] by normal chemical methods."[10] In other words, a working reactor would produce fissionable material that could be easily extracted from the reactor and used either for a mobile energy-producing machine or for a new "irresistible" offensive weapon. The first of the two Allied bombs dropped on Japan in 1945 was powered by the separated uranium isotope U-235, the second by reactor-bred plutonium.

Heisenberg apparently learned of Houtermans's results shortly after Houtermans submitted his report. A letter from Heisenberg to his professors' club colleague Hermann Heimpel, dated October 1, 1941, strongly suggests that Heisenberg did perceive an open road to a deadly explosive and that he was already mindful of the possible consequences. Thanking historian Heimpel for a copy of his book *Deutsches Mittelalter* (*German Middle Ages*), Heisenberg wrote, "I really liked the passage in your book about the mind-set of the Middle Ages in contrast to our epoch. In this connection it suddenly came to me that such a transformation could occur once again in the near future. For perhaps we humans will recognize one day that we actually possess the power to destroy the earth completely, that we could very well bring upon ourselves a 'last day' or something closely related to it."[11]

Heisenberg discussed the newly opened road with his trusted Berlin institute staff. Sometime in August or early September, they determined that Heisenberg should discuss the turn of events with Niels Bohr in German-occupied Denmark. A September lecture series on astrophysics at a German propaganda institute in Copenhagen, in which Heisenberg had agreed to participate, provided a splendid opportunity. On September 15, 1941, Heisenberg traveled to occupied Copenhagen for the official purpose of participating in a lecture series at that institute, along with Carl Friedrich von Weizsäcker and other German scientists, and with the unofficial intention of meeting with Bohr.

Heisenberg probably met with Bohr on the evening of September 16.[12] Their meeting is still shrouded in controversy and questions. Michael Frayn's popular award-winning play *Copenhagen*, which centers on this meeting, has inspired much additional thought and debate.[13] Although the official circumstances of the trip and some of its immediate consequences are well documented, the only indication of the content of the unofficial meeting comes from postwar accounts by the participants, their colleagues, and the colleagues of their colleagues. Given the intense feelings and tensions of the early postwar period, the veracity of all these reports is open to some question. Most of the German accounts were offered in defense of Carl Friedrich's father, a high official in the German Foreign Office, whose subdivision administered the German cultural propaganda institute in Copenhagen and who was tried and convicted at Nuremberg in the Foreign Ministries Case. These accounts are obviously products of the aims for which they were written. Rightly suspecting Gestapo surveillance of their meeting, Bohr and Heisenberg themselves did not dare commit any of their discussions to paper at the time.

Given the setting of the meeting (German-occupied Denmark), the occasion (Heisenberg's lecture in a propaganda institute), and the topic (nuclear fission, controlled and otherwise), it may be little wonder that Heisenberg's visit greatly disturbed his former mentor. Heisenberg felt he had failed to communicate with Bohr. Bohr and Heisenberg had been close friends and colleagues for nearly 20 years. If Bohr came away from their meeting in great distress, it may well have been because Heisenberg said something distressful.

In a carefully worded statement about the visit written in 1948, apparently an early draft prepared for the Weizsäcker trial (it differed from the less informative official defense exhibit that he submitted), Heisenberg again recalled his realization in 1941 that the production of an explosive, an atomic bomb, was now a real possibility.[14] Heisenberg remembered that his most important talk with Bohr occurred one evening as they strolled along a tree-lined path in the large and secluded Fælledpark, just behind Bohr's institute. The Danish professor and his former assistant had often talked together while walking along these quiet paths. Aside from its other attractions, the venue offered the advantage of escaping whatever bugs the Gestapo had installed in the institute—discussing secret nuclear research was treasonous for the German and life threatening for the Dane. The boyish-looking Heisenberg recalled opening the discussion with the taller and more distinguished-looking Bohr by asking whether Bohr believed that "as a physicist one has the moral right to work on the practical exploitation of atomic energy."

An obviously startled Bohr responded by asking whether Heisenberg believed that atomic energy could be practically exploited in this war. "Yes, I know that," Heisenberg answered. However, he claimed that he was referring only to a machine. Because of the technical difficulties involved, he told Bohr, a bomb could not be produced before the war was over.

Unraveling Heisenberg's postwar account of this meeting, his intentions, and Bohr's reactions requires a fuller appreciation of the broader context of the meeting.[15] But again our sources allow only speculation on crucial points. Heisenberg's remembered question on morality is a case in point, for available sources give no indication that he had ever raised it before. But it was also not quite so certain until then that a reactor could actually be constructed, that it would soon be within reach, and that, once working, it could easily provide an alternative route to the atomic bomb through the production of plutonium. Since Bohr had served for years as a father figure to his youthful charges, especially to Heisenberg, and had often discussed philosophical and ethical concerns with his younger colleague, it seems reasonable that Heisenberg might have turned to Bohr when faced with an ethical dilemma in his research. But during the past few years, particularly during the SS affair, Heisenberg had sought ethical advice on science in the political arena, not from Bohr, but from his German academic elders, Max Planck and (probably) Max von Laue, both of whom were in Berlin and more accessible than Bohr. Though in semiretirement, Laue was still vice director of the Kaiser Wilhelm Institute, as he had been since the days of Einstein. Yet among the scant surviving records there is no indication that Heisenberg or his colleagues approached Planck or Laue about the morals of nuclear research, or that they even fretted over them.

Aside from any advice that he may or may not have been seeking, Heisenberg probably had other aims in seeing Bohr. One accusation, which later arose in Copenhagen, is that Heisenberg was in fact acting as a German spy in an attempt to determine from Bohr if the Allies were working on a nuclear weapon and, if so, how far along they were. No direct evidence for this is available, although a newly discovered report by Weizsäcker following the visit does indicate an interest in Bohr's knowledge about applications of nuclear fission.[16]

Heisenberg himself seems to suggest another purpose for his visit in a letter to B. L. van der Waerden written after the war. Heisenberg implied that he was attempting to stave off an Allied crash program on nuclear fission research.[17] However untenable and naive such an aim may have been, it may find support in a consideration of Heisenberg's trip to the United States two years earlier and especially of the circumstances surrounding his 1941 visit to Denmark.

Heisenberg's 1939 trip to the United States occurred within several months of the publication of the basic Bohr-Wheeler theory and the French confirmation of neutron multiplication in atomic fission. As indicated earlier, on at least two later occasions Heisenberg recalled discussing the possibility of nuclear explosives with Fermi in the United States, a discussion that may have occurred with others as well.[18] As reported years later by Heisenberg, both participants had expressed a ready willingness at the time to engage in fission research for their respective governments. In September 1941, Heisenberg was working as hard as possible on nuclear energy, and he could expect that the Allies were doing just the same. Elisabeth Heisenberg wrote that throughout the war her husband "constantly tortured himself" with the thought that the better supplied Allies might develop the bomb and use it against Germany.[19]

Moreover, at the time Heisenberg visited Bohr, the German Reich had reached its greatest extent. Most of continental Europe was under Nazi occupation, the German army was plunging into Soviet Russia, and an end of the war may have seemed in sight. It was easy to suppose that if the war ended soon with the German army in place, or if it bogged down at that point—as had World War I in the trenches of France—the United States, which had not yet entered the war, would have enough time and resources to build a nuclear weapon, which the Allies would surely use on

Germany. At least one secret German report in early 1942 indicated that the Germans somehow knew about secret American pile research—and they now knew where that research could lead.[20]

After the war, Heisenberg wrote that he learned after the September visit that Bohr was in contact with Allied scientists. The Gestapo had intercepted a secret message from Bohr to British scientists and had delivered it to Heisenberg, probably because it reported on Heisenberg's visit. Heisenberg may have suspected Bohr's contacts with the Allies even before his visit. Perhaps, as his postwar letter to van der Waerden suggests, Heisenberg was trying to avert an Allied crash program and an ultimate nuclear attack on Germany by letting the Allies know through Bohr that the Germans—who believed throughout the war that they were ahead of the Allies—were still a long way from constructing an explosive.

Whatever Heisenberg's aims and intentions, his understanding of Bohr's frame of mind in German-occupied Copenhagen and of how he himself would be perceived in Denmark was woefully incorrect and misguided. He had last seen Bohr in 1938. Two years later, the German army overran the Danish kingdom practically without firing a shot. With the German Reich well entrenched in Denmark and across most of Europe by September 1941, it must have been cold comfort for Bohr to hear Heisenberg's amoral qualifier about the prospect for nuclear weapons: "At this point, it is certainly only a question of the exploitation of energy in machines; the production of bombs would probably require such an enormous effort that the war would be at an end before they could be made."[21] Even as Heisenberg himself remembered it, only time and effort stood between him and the bomb, and Bohr could regard neither to be insurmountable in September 1941. Nor could he have been very pleased at the much sooner prospect of a new energy source to power the German economy and to drive German ships and submarines around the world.

Until the autumn of 1943, the German occupation forces maintained the fiction that they had no intention of nazifying Denmark, they did not want a rebellious populace that would drain their forces from other invasions. For the most part German commanders left Danish Jews alone, and the German army and occupation authorities were under strict orders to avoid offending the Danes as much as possible. Bohr, his institute, and his cyclotron were also undisturbed, apparently, according to Bohr, on encouragement from Carl Friedrich's father in the Foreign Office.[22] The Germans allowed Bohr's institute to function as normally as possible—with continued American dollars from the Rockefeller Foundation.[23] But such niceties could not disguise the fact that proud Denmark had been reduced to a colony of the Nazi Reich, and that it was being subjected to incessant propaganda by its occupiers. Even a German propaganda expert had few illusions about Danish resentment: "A feeling of quiet rage prevails here, which only comes to the fore when the Danes believe themselves alone and unobserved."[24]

Several months before Heisenberg arrived in Denmark, and just days before Carl Friedrich and a party of German scientists were arriving for a visit, Danish communists and other anti-German Danes had been summarily arrested and deported to Germany, an action that incensed the Danes further toward any Germans. Carl Friedrich himself had already made the German scientists thoroughly unwelcome. During a visit to Bohr's institute the previous March, Weizsäcker had reportedly insulted Bohr by bringing the head of the local German Culture Institute to meet him.[25]

The newly opened Culture Institute was a propaganda arm of the Culture Division in the German Foreign Office. In March 1941, Carl Friedrich had spoken there and elsewhere in Copenhagen to Germans and local sympathizers. According to surviving Reich Education Ministry records, Weizsäcker's trip was so successful that the German occupation office in Denmark requested he return in the fall, this time accompanied by Professor Dr. Heisenberg. Weizsäcker was in contact with Heisenberg right after the March visit and probably informed Heisenberg of the plan—if he was not already privy to it.[26] Certainly by mid-July Heisenberg knew that Carl Friedrich and the Culture Institute were planning to demonstrate the support of German sympathizers among Danish scientists by organizing a conference on astrophysics to be held in late September 1941.

On July 22 Weizsäcker wrote to the German Academic Exchange Service to confirm himself, Heisenberg, and several other German scientists as invited speakers, and—aware of the lingering opposition to Heisenberg within the government—to argue for Heisenberg's participation. After conferring with Heisenberg and Weizsäcker in early August, the Reich Education Ministry scheduled the conference for September 18–24. Heisenberg, claiming personal commitments, would be in Copenhagen September 15–21. But whether naively or malevolently, Carl Friedrich again insulted the Danes by cordially inviting them to attend his and Heisenberg's lectures on solar physics and cosmic rays, and in the odious German Culture Institute. Already imbued with "quiet rage," Bohr and his colleagues did not appreciate the scientists' participation in a crass propaganda campaign. Heisenberg later surmised that his meeting with Bohr "did not have the intended effect [because] Bohr evidently disapproved of my taking part in an astrophysics conference at the 'German Culture Institute.'"[27]

Although it is not clear from the available documents which came first—the plan to speak with Bohr about fission or the plan to speak at the propaganda institute—the timing of the latter seems to precede Houtermans's results and the perceived open road to nuclear weapons. If so, this raises a question: What was Heisenberg doing in occupied Copenhagen in the first place? If he did decide to go there even before he contemplated approaching Bohr about nuclear research, it might be argued that he wanted to assure himself that Bohr and his institute were unmolested by the German occupation. But surely Weizsäcker had already determined that, and others could have kept Heisenberg apprised of the situation. Instead, the most likely answer seems to be that Heisenberg was indeed joining his friend and colleague Carl Friedrich in a conscious or unconscious propaganda effort instigated by the Foreign Office subdivision under Carl Friedrich's father. Still eager to prove his reliability to regime officials, and to obtain indications of trust through permission to travel abroad after the SS affair, Heisenberg readily joined Carl Friedrich in carrying out the propaganda effort, or at least allowed himself to be drawn into it. There does not seem to be any other compelling reason for him to have visited occupied Copenhagen before or after the dilemmas of nuclear research became acute.

This interpretation is supported by further Reich Education Ministry documents. In his postwar affidavit, Heisenberg argued (as he would in other such cases) that his lecture at the Culture Institute was the smallest compromise possible to gain permission to visit Copenhagen. Although some German opposition to his touring abroad did still exist, it was not insurmountable. In particular, the REM office of Wilhelm Führer, earlier prominent in blocking Heisenberg's Munich appointment, had to approve all foreign travel, and he had no intention of approving Heisenberg's travel anywhere, unless pressured. During a hastily arranged meeting with the REM in early September, Führer demanded that party headquarters pass final judgment on whether or not to allow the Leipzig professor out of the country. The party quickly consented after the senior Weizsäcker's Foreign Office suggested that the trip could be used as a test case of Heisenberg's suitability for future propaganda lectures.[28] For Heisenberg, his lecture trip to Copenhagen could be seen as a minor personal victory; for the regime, it would be a test of the professor's reliability as a precondition for future exploitation; for the Danes, it was nothing more than crass propaganda. Bohr's wife, Margrethe, never wavered in her opinion of the episode: "No matter what anyone says, that was a hostile visit!"[29] Bohr and Heisenberg were never as close thereafter as they had been before the war.

What of the "intended effect" that Heisenberg had hoped to achieve by his meeting with Bohr? If the effect was an Allied moratorium, or even a joint boycott of applied nuclear research, the visit could not have achieved any such thing, even if Heisenberg had been well received in Copenhagen. Isolated in occupied Denmark, Bohr had only little inkling of the progress in nuclear research on either side of the war, and because of this, Allied scientists probably would not have accepted his assessment of a boycott offer from German scientists. Weizsäcker, for instance, reported in March 1941 that "concerning the more technical questions [Bohr] knew a great deal less than we." Perhaps this and the German scientists' actions on their September trip are the origin of the view expressed in postwar Copenhagen that the Germans were on a spying mission.[30] No wonder Bohr was so thoroughly disturbed to learn from Heisenberg six months after Weizsäcker's first visit that German scientists already saw an open road to an atomic bomb. Nevertheless, he still seemed skeptical two years later. Responding in 1943 to hints from British physicist James Chadwick (discoverer of the neutron and Bohr's secret liaison with the British nuclear research team), Bohr wrote (in deliberately cryptic English): "Above all I have to the best of my judgment convinced myself that in spite of all future prospects any immediate use of the latest marvelous discoveries of atomic physics is impracticable."[31]

LEFT TO RIGHT: *Bohr, Elisabeth Heisenberg, and Heisenberg at the Acropolis, Athens, c. 1956.*

Several months after Bohr's letter to Chadwick, the German occupation turned uglier. In one of the most spectacular rescue operations of the war, the Danish underground conveyed the part-Jewish Bohr, his family, and virtually the entire Jewish population of Denmark to neutral Sweden, just ahead of a planned roundup and deportation of Jews to the German death camps. A British plane flew Bohr to England, where he met immediately with Chadwick. The little-known summary report of a British nuclear committee meeting before Bohr's arrival describes Chadwick's impression: "Chadwick . . . says that Heisenberg has visited Bohr in Copenhagen. He also says that he himself has been in communication with Bohr within a month or so, and that Bohr believes that there are no military possibilities. He thinks that perhaps Bohr has been sold this idea by Heisenberg."[32]

If Bohr felt he had been led astray by Heisenberg, he would have thereafter resented Heisenberg's wartime visit all the more. Perhaps this is one source of the anger that emanates from the recently released drafts of letters to Heisenberg written by Bohr over a period of years beginning in 1957. But there was an even stronger reason.

In response to recent debates over Heisenberg's 1941 visit invoked by the play *Copenhagen*, the Niels Bohr Archive in Copenhagen decided in 2002 to release previously withheld unsent drafts of letters written by Bohr to Heisenberg. Perhaps because of their strong wording, uncharacteristic for Bohr, he had chosen not to send them.[33]

Bohr wrote these letters in response to the 1956 best seller *Heller als tausend Sonnen (Brighter than a Thousand Suns)*, a history of the atomic bomb written by the Swiss journalist Robert Jungk. In contact with Weizsäcker, Jungk claimed to portray "the actual personal attitudes of the German experts in atomic research." He presented the German experts as having prevented the development of a German bomb. He claimed further that Heisenberg had met with Bohr in Copenhagen in 1941 in order to propose to the Allies a boycott of nuclear weapons research: "By the expedient of a silent agreement between German and Allied atomic experts, the production of a morally objectionable weapon was to be prevented."[34]

Jungk sent a copy of his book to Heisenberg in December 1956, asking Heisenberg for more information about the 1941 visit to Bohr in Copenhagen. Heisenberg responded on January 18, 1957 with a four-page letter in which he expanded upon his 1948 affidavit for the Nuremberg trial. Heisenberg prefaced his memory of the meeting by repeating his assertion of 1948, "We knew that one could produce atom bombs but overestimated the necessary technical expenditure at the time. This situation seemed to us to be a favorable one, as it enabled the physicists to influence further developments." Then, he wrote, concerning the meeting with Bohr,

> This talk probably started with my question as to whether or not it was right for physicists to devote themselves in wartime to the uranium problem—as there was the possibility that progress in this sphere could lead to grave consequences in the technique of war. Bohr understood the meaning of this question immediately, as I realized from his slightly frightened reaction. He replied as far as I can remember with the counter-question, "Do you really think that uranium fission could be utilized for the construction of weapons?" I may have replied: "I know that this is in principle possible, but that it would require a terrific technical effort, which, one can only hope, cannot be realized in this war." Bohr was shocked by my reply, obviously assuming that I had intended to convey to him that Germany had made great progress in the direction of manufacturing atomic weapons.[35]

Jungk published these passages as part of a longer excerpt from Heisenberg's letter in the Danish and English translations of his book, appearing in 1957 and 1958, respectively, prompting Bohr's angry draft letters to Heisenberg in response. As Gerald Holton has pointed out, in his letters Bohr contradicted and corrected Heisenberg on every point. In his first and most detailed draft Bohr wrote, in the English translation provided by the Bohr Archive, "I think that I owe it to tell you that I am greatly amazed to see how much your memory has deceived you in your letter to the author of the book. . . . Personally, I remember every word of our conversations, which took place on a background of extreme sorrow and tension for us here in Denmark." Several sentences later he wrote:

> I also remember quite clearly our conversation in my room at the Institute, where in vague terms you spoke in a manner that could only give me the firm impression that, under your leadership, everything was being done in Germany to develop nuclear weapons and that you said that you . . . had spent the past two years

working more or less exclusively on such preparations. . . . If anything in my behavior could be interpreted as shock, it did not derive from such reports [of the prospect of a bomb] but rather from the news, as I had to understand it, that Germany was participating vigorously in a race to be the first with atomic weapons.[36]

If scientists could control the development of the bomb by arguing that it would not be ready in time for this war, in Bohr's view the Germans were not doing so.

In subsequent drafts Bohr, noting again that "I carefully fixed in my mind every word that was uttered," reiterated his clear impression: "I did not sense even the slightest hint that you and your friends were making efforts in another direction." The whole affair, he wrote, "is a most awkward matter for us all."[37]

So, how is it that after Heisenberg's visit Bohr saw no immediate practical use for nuclear fission, and that, according to Chadwick, "Bohr believes that there are no military possibilities," probably as a result of Heisenberg's influence? The most likely way to reconcile these two assessments seems to be that Bohr did believe that Heisenberg was working on an atomic bomb, but that Bohr, perhaps influenced by Heisenberg, did not believe that the work would succeed in providing "any immediate use" owing to practical difficulties. This seems compatible with Heisenberg's intention of bringing up the prospect of such a weapon, the many difficulties in achieving it, and his claimed desire to stave off an Allied crash program that could well succeed if the war lasted long enough. But one cannot ignore the fact that the visit occurred during one period of very trying circumstances, while Jungk's book and Bohr's response occurred in quite a different, though perhaps equally trying period— a period of post-Hiroshima cold war in which Denmark and all of Europe would be the likely nuclear battlefield in the event of a hot war between East and West.

Moreover, it was the fear that Heisenberg and the Germans were building the atomic bomb that drove the intensity of the Manhattan Project, eventually bringing about the nuclear age. The old wounds opened by Jungk's book in the new era of nuclear distress surely brought for Bohr a flood of painful memories and unresolved anger that flowed onto the pages of his unsent letters to the man who, as Bohr saw it, had dared to exploit the German occupation of Denmark in order to raise the prospect of an atomic bomb in Hitler's military arsenal. For his part, that man had returned home pleased at least to receive the imprimatur of the German Foreign Office as a traveling spokesman for the Reich.

During the months following his meeting with Bohr, Heisenberg revealed his own reactions to the lately proven potentialities of fission research. This occurred in the context of a shift in the institutional and political framework of the uranium project in the wake of Germany's changing fortunes at the front. By the end of 1941, the German Blitz had run its course. The invasion of Russia, begun in June, was bogged down outside Leningrad and Moscow by December. Previously confident of total victory by Christmas, Germany had squandered most of her raw materials. As the predicament became a crisis during that bitterly cold winter, Hitler took over as operational commander. For the first time, he ordered the full mobilization of the German economy in support of the war effort, along with the full exploitation of occupied territories. Germany would now use every means at its disposal to wage total war. As a result of this new state of affairs, Erich Schumann, head of army research, informed the uranium research directors in December 1941 that henceforth the Army Ordnance Office could support their efforts only "if a certainty exists of attaining an application in the foreseeable future."[38]

Ironically, just as the United States was entering the war following the attack on Pearl Harbor and launching a crash program to build the bomb in a supposed race with Germany, Schumann was calling the German scientists to Berlin for a meeting to decide whether their efforts were worth continuing at all. During that meeting, on December 16, 1941, which Heisenberg probably attended, the uranium researchers agreed to prepare a comprehensive report on their progress and on the prospects of their research for General Emil Leeb, the head of Army Ordnance. Schumann agreed to call a conference of all uranium researchers for the end of February 1942 to evaluate the status of the project.

The only available copy of the 144-page memorandum to General Leeb on progress in nuclear fission, dated simply "February 1942," lacks a title page and authors' names. Heisenberg probably did not help to write it, but some of the wording and ideas appear in his reports and lectures of that time. Apparently well aware of Allied research, the scientists' recommendations are clear: "In the present situation preparations should be made for the technical development and utilization of atomic energy. The enormous significance that it has for the energy economy in general and for the Wehrmacht in particular justifies such preliminary research, all the more in that this problem is also being worked on intensively in the enemy nations, especially in America." The authors considered time no longer a problem, for recent reactor experiments in Leipzig and Berlin implied that "success can be expected shortly." But the building of a nuclear weapon for the Wehrmacht depended on the development of new isotope-separation techniques or the generation of the new element plutonium in the first working reactor. The report concluded that progress toward achieving a working reactor was being hindered less by scientific problems than "by problems pertaining to the acquisition of materials"—problems that, the researchers felt sure, the military sponsors would not find difficult to solve.[39]

The army chose to ignore the scientists' qualified optimism. Leeb and Schumann slashed uranium research funding, reduced activities to Diebner's army laboratory in the Gottow suburb of Berlin, and abandoned research altogether at the Kaiser Wilhelm Institute for Physics, returning the institute to its sponsoring society. The scientists clung nevertheless to their three-day conference on nuclear technology at the institute, scheduled to start on February 26, 1942.

Recognizing a sudden opportunity to regain control of uranium research after two years, Abraham Esau of the rival Reich Research Council scheduled a separate, conflicting series of non-technical lectures for the opening day of the Army Ordnance conference at the council's mansion, the House of German Research, in the neighboring suburb of Berlin-Steglitz. With the army withdrawing from research and leading nuclear researchers eager for a new sponsor, Esau lined up an impressive panel of scientists—Heisenberg, Otto Hahn, Hans Bothe, Hans Geiger, Paul Harteck, and Klaus Clusius—to deliver short lay lectures on nuclear energy development. Their audience was to be the top echelon of army, government, and SS officers—Heinrich Himmler, Hermann Göring, Martin Bormann, Albert Speer, Wilhelm Keitel, Erich Raeder, and others. But fate was unkind to Esau. Apparently through an error, Esau's secretary enclosed the wrong list of lectures with the invitations to the dignitaries. Instead of receiving Esau's list of eight non-technical lectures, they received the list of reports for the rival Army Ordnance conference and found themselves invited to hear 25 technical talks on such arcana as neutron diffusion lengths, enriched isotopes, and neutron multiplication and absorption factors in the latest Leipzig model.[40] Most of the dignitaries declined to attend.

With REM chief Bernhard Rust as chair, military researcher Schumann opened the research council's session in the cozy lecture hall of the former private mansion with a talk titled "Nuclear Physics as a Weapon." Hahn followed with "The Fission of the Uranium Nucleus," after which Heisenberg presented his favorite subject, "The Theoretical Foundations for Energy Acquisition from Uranium Fission."[41] While emphasizing reactor construction, Heisenberg mentioned the possibility of weapons development, but as in the Leeb memorandum, he neither discouraged government support for fission research nor encouraged the government to expect a weapon in the foreseeable future. Heisenberg was still walking a fine line.

Heisenberg did attest, however, that a uranium "machine" could soon be built to generate enough power to drive battleships and submarines. Moreover, enough pure U-235 would constitute an "explosive of totally unimaginable power." But he hastened to add that separating U-235 from a block of raw uranium was difficult and required sophisticated techniques that were still unavailable. An alternate route to weapons lay through the uranium machine: "As soon as such a machine is in operation, the question of how to obtain explosive material, according to an idea of von Weizsäcker, takes a new turn. In the transmutation of the uranium in the machine, a new substance comes into existence, element 94, which very probably—just like U-235—is an explosive of equally unimaginable force. This substance is much easier to obtain from uranium than U-235, however, since it can be separated from uranium by chemical means."[42]

The implication of Heisenberg's argument seemed to be that if the regime would leave scientists alone, ideologically and professionally, while at the same time supporting and protecting them through reactor research, it would help itself progress toward a more distant but no less cherished goal—the development of a powerful new explosive. Whether or not Heisenberg believed that this goal could actually be reached before the war ended, at the very least the regime could be assured in a few years of a vast new energy source to power the German economy. As historian Mark Walker writes, "Tailored both to his audience and to the times, Heisenberg's talk illustrated clearly and vividly the war-like aspects of nuclear power."[43] From another perspective, it also illustrated the extent to which Heisenberg, the young man who had once thrilled at the precipices of mountain peaks, was willing to flirt with the catastrophic consequences of atomic research for the sake of what he believed would be beneficial to himself and to German science.

Heisenberg's renewed commitment to this dangerous strategy of enticing Nazi bureaucrats with the potentialities of nuclear energy in order to gain personal and professional advantages was not the effort of a loner. It coincided closely with a major campaign launched by the German Physical Society, aimed at Reich officials, to encourage more material support for physics education and research.[44] With Germany now headed for total war, the society's main argument was the familiar one: physics and physicists were making tangible contributions to the war effort. Heisenberg fully intended his February 1942 Berlin lecture as a contribution to this campaign to make warfare serve physics by demonstrating how physics could serve warfare.

Two months after he delivered his Berlin talk, Heisenberg's colleague, Wolfgang Finkelnburg, now vice president of the German Physical Society (and his wife's former suitor), congratulated Heisenberg for his efforts. His lecture to the Reich Research Council and the press reports about it seemed to be having a "satisfactory effect," he wrote. "I have received various inquiries from party officials with questions about the war relevance of theoretical physics and especially about the relevance of your work." Heisenberg concurred: "In general, the interest of the highest officials in modern physics now seems to have become quite great."[45] A month earlier, the authorities had suddenly confirmed their interest in Heisenberg and his physics. At the end of April 1942 one of Himmler's two promises to Heisenberg was at last fulfilled: Heisenberg received a call to Berlin to succeed Debye in the directorship of the Kaiser Wilhelm Institute for Physics and to assume a concurrent professorship in theoretical physics at the University of Berlin.[46]

As the leaders of the Reich, unbeknownst to Heisenberg or to anyone else, prepared their final assault on humanity—the Final Solution agreed on at the Berlin-Wannsee conference in January 1942—Heisenberg and theoretical physics basked in the prospect of imminent rehabilitation and full recognition by some of those same leaders. The fine line had now become a tightrope.

CHAPTER 24

ORDERING REALITY

As THE URANIUM PROJECT HEADED DOWN THE ROAD THAT COULD LEAD TO A BOMB, AND as the German Reich tightened its genocidal grip over most of Europe, Heisenberg's intellectual activities moved away from abstract physics and back toward nontechnical matters of interest to himself and to his public following. He was developing in particular a concern for grand philosophical issues that went far beyond the philosophy of physics. As usual, he presented his nontechnical thoughts primarily through the medium of public lectures, both at home and abroad. Between May 1941 and the end of 1942, Heisenberg produced five lectures and one book-length manuscript on philosophical issues. The manuscript was published nearly 50 years later as *Ordnung der Wirklichkeit (The Order of Reality)*.[1] He delivered the lectures to educated pro-German Hungarians in Budapest, to Zurich university students in neutral Switzerland, to Leipzig faculty and students, and to radio listeners and newspaper readers throughout the Reich. The lectures seem either preparatory to or derivative of the large manuscript.[2] During this period, Heisenberg did not produce a single nonpractical scientific paper and wrote only a few surviving letters touching on scientific matters.

Heisenberg was working in that period to enhance both nuclear fission research and the recognition of his profession. Elisabeth Heisenberg reports that his intensive work on the 161-page philosophical typescript occurred only during his brief vacations with the family in Urfeld in 1941 and 1942, when he had privacy and time away from his other work and duties. Elisabeth typed at least one of the two existing versions of the untitled work, which they simply called "Philosophie," and they gave it to close, trusted friends as a present for Christmas 1942. Despite the obvious peril of committing private opinions to paper at the time, Heisenberg included several passages that were somewhat critical of the regime. He declined to publish the work after the war, declaring it to be too personal. Moreover, the conditions under which it was written and of which it was a part had rendered postwar publication anachronistic.

As noted earlier, Heisenberg usually did not engage in serious philosophical inquiry or exposition without a reason. Often the reason derived from professional matters that were then reflected in his addresses to non-specialist audiences. Most of his philosophical publications derived from these public lectures. However, the motivations for Heisenberg's 1942 philosophy manuscript do seem much more personal in nature, relating directly to his age and to his situation. This was a man who had devoted his life to rapid scientific achievement, who was engaged in a constant struggle to preserve what he valued, and who had spent most of his private life in youthful set-tings and among younger men who glorified the virtues of youth. For such a man, his fortieth birthday, arriving in the middle of an intensifying war, was probably some-thing of a trauma. Heisenberg's manuscript, written during the months before and after he turned 40 in December 1941, opens with the stated goal of discovering how his life's work "harmonizes with the whole," meaning the whole of what he would call reality: "He who has dedicated his life to the task of going after the individual connections of nature will be confronted over and over again with the question of how those individual connections fit harmoniously into the whole, other than the whole presented to us by [everyday] life or the world."[3]

In an earlier time of chaos and confusion, Heisenberg had felt a desperate need for an order and harmony that could provide the basis for stability in his life. He believed that he had experienced some sort of stabilizing central order during one of his early youth-movement meetings. Now, as he headed into middle age in the midst of war and dictatorship, Heisenberg seemed to require a new type of order, one that would reassure him of the significance of his life's efforts by giving him and his sci-ence meaning in the context of a larger, transcendent scheme. The mature man estab-lished this new order himself under the influence of his readings. Writing in 1941 and 1942, he argued that a whole does exist, that the reality that it represents is organized in a hierarchical order, and that it is the task of the scientist, especially himself, to comprehend his science in the context of this hierarchy: "In every period the attempt has been undertaken to submit our knowledge of reality to a general order."[4]

Heisenberg adopted a hierarchical, almost mystical order from his favorite author, Johann Wolfgang von Goethe, Romantic writer, poet, and sometime natural philoso-pher. In his Budapest lecture, delivered in April 1941, Heisenberg described Newton's and Goethe's incompatible theories of colors emitted by a prism irradiated with white light as being not right or wrong in themselves but as referring to two complementary types or layers of reality: the physical and the spiritual. At the end of the lecture and at the outset of his manuscript, he expanded the layers of reality to include the nine vertically ordered layers presented, from bottom to top, by Goethe himself: acciden-tal, mechanical, physical, chemical, organic, psychic, ethical, religious, genial.[5] The order of these layers moves from what Heisenberg calls the objective to the subjec-tive, from areas of reality that lie outside of us to those that lie wholly within human experience, the genial or creative powers capping the hierarchy. In his elaboration of each level in his manuscript, Heisenberg placed his own life's work, quantum physics, just below the organic level in the realm of the chemical, because it dealt with atoms. The man of knowledge who wanted to comprehend more fully what Heisenberg called the "grand connections" must climb this ladder of realities—much as a moun-tain climber struggles alone to reach the top of an enormous mountain, much as a yogi struggles to attain a knowledge of nirvana.[6]

Sitting alone in his Urfeld nest—alone even when surrounded by his large family —the physicist passed his fortieth birthday with a new understanding of the place of his life's work in the broader scheme of things. Faced again, as he had been several times in the past, with the stark contrast between the beautiful, genial world of his understanding of nature and the horrible death, destruction, and ugliness of the Nazi dictatorship at war, he found an order that far transcended that ugly world. Science and all that he treasured in it and in life were layers within a grander, transcendent hierarchy that made them all worthwhile.

Despite the escapist quality of his exercise, it may seem admirable to find Werner Heisenberg, the great twentieth-century scientist, contemplating the limits of his science, acknowledging that alternative ways of looking at nature and our place in it may be as valid as the cool, often exploitive, rationality of those who, since Bacon, have ascribed power to knowledge. Yet there is also a disturbing quality to an order of reality that is conceived as a vertical hierarchy with science, rationality, and the individual buried somewhere far below the top. In a way, Heisenberg's hierarchy of reality layers seems overwhelming, even depressing, to contemplate. Given the years in which the work was written, one cannot help wondering to what extent the hierarchical situation in which he lived had a direct impact on such views.

Perhaps unconsciously acknowledging the connection, in the closing pages of his long essay Heisenberg considered the immediate reality of the war and the role of the individual—himself—in it. As described in these pages, probably written in 1942, the war for him was not merely a struggle for power and territory initiated and pursued by a ruthless dictator, but an expression of more fundamental "movements in the foundations of human thought": a shifting of the layers of reality over the heads of individuals in such a way that the evil side of the irrational, the "dark demons" now loose upon the world, took on a greater role.[7] It is not clear whom he considered the dark demons to be, nor whether he expected their role to be permanent or merely temporary.

Heisenberg also made several statements that would have gotten him into deep trouble had they become known at the time. For instance, he lumped National Socialism together with Bolshevism as a "strange sort of this-worldly religion." He admired the "Anglo-Saxon" enemy for producing "the first great figures of the early modern period" that led to modern science and to the knowledge of objective reality (at the lower end of the hierarchy, however). Contemplating the contemporary situation, he even provided a moral imperative for his readers: "We must make it clear to ourselves over and over again that it is more important to treat others humanely than to fulfill any sort of professional, national, or political duties."[8]

Nevertheless, to fulfill his humane duties to others, Heisenberg the individual at the same time had to sustain the role of his public persona—the "active opposition" that fostered the illusion of living at once in two separate worlds. In his essay, he reaffirms this by separating the larger public world from the smaller world of the private being. The grand "movements of thought" occurring in the war far transcended the actions and influence of mere individuals such as himself. The individual "can contribute nothing to this, other than to prepare himself internally for the changes that will occur without his action."

The helplessness of the individual before the forces of national and international struggle thus established, Heisenberg closed his essay with his own personal recommendation: while striving to help others, he wrote, one can do little more than to accept one's fate within the broader circumstances of one's life. The individual is conveniently relieved of any responsibility for what is going on outside himself: "For us there remains nothing but to turn to the simple things: we should conscientiously fulfill the duties and tasks that life presents to us without asking much about the why or the wherefore. We should transfer to the next generation that which still seems beautiful to us, build up that which is destroyed, and have faith in other people above the noise and passions. And then we should wait for what happens. . . . Reality is transformed by itself without our influence."[9]

And that is what Heisenberg did upon moving to Berlin in 1942. Even when playing a leading role in a research project integrated into the hierarchy of the German war effort, even when serving as a cultural representative to occupied and oppressed territories, even when acceding to the demands of Nazi functionaries, he assured himself that his actions really made no difference at all on the grand scale of reality. On the other hand, in the smaller world of himself, his colleagues, and his family and friends, his actions might help to preserve something of "that which still seems beautiful to us." In other words, he had convinced himself of what he wanted to believe all along: that he could live and work as a subject of this system, but not be a part of it and thus have no responsibility for it. He would need such an outlook as the scientific head of German fission research in Berlin in 1942.

Heisenberg's transfer to Berlin signaled shifts in the setting and status of German uranium research as the authorities subordinated Germany's economy and science even more directly to the war effort. Germany's waning fortunes on the Russian front and her rapidly diminishing supply of finished goods demanded a radical change of priorities, with production taking precedence. During the terrible winter of 1941–1942, Hitler appointed Albert Speer, his astute architect, to succeed the late Fritz Todt as head of arms production in the four-year plan, which was administered by Reich Marshal Hermann Göring, the commander of the Luftwaffe. In March 1942, Speer had Hitler decree that the needs of the German economy were subordinate to those of arms production—thus to Speer himself.[10] Among the potential elements of arms production at that time was a recently orphaned research project on the possible exploitation of nuclear energy.

After the Army Ordnance Office decided to relinquish most of its control over nuclear research in early 1942, a hodgepodge of government bureaucracies grabbed for the plum. Reich Education Minister Bernhard Rust sought to enhance his standing by immediately assigning the project to his own Reich Research Council (RFR), now headed by SS Colonel Rudolf Mentzel. Dr. Mentzel turned it over to the physics section of the RFR, still headed by Professor Esau. But the Education Ministry and its Research Council collided with the Kaiser Wilhelm Society (KWG), which administered a network of government research institutes. Until its confiscation by Army

Ordnance, one of these institutes had been Debye's Kaiser Wilhelm Institute for Physics in Berlin-Dahlem.[11] The KWG vigorously sought to regain control of Debye's institute from Army Ordnance. When KWG president Albert Vögler informed Speer of the potentialities of nuclear research, military and otherwise, Speer, already appreciating the importance of science for the war effort, began to take a personal interest in nuclear matters. In 1942 he had Hitler name Göring head of the new Reich Research Council to run under the four-year plan and to foster war-related research. By the end of the year, Göring had appointed Mentzel head of the managing committee of the RFR; at Mentzel's suggestion, he gave Esau the dual titles of head of the physics section of the RFR and Reich Plenipotentiary for Nuclear Research.[12] Although this left the German nuclear project at the end of 1942 temporarily suspended between competing bureaucracies—Speer and the KWG on the one hand and Göring, Esau, and the RFR on the other—the nuclear project was nevertheless now fully integrated into the newly established hierarchy. Research was to serve the war effort and it would be dominated by two men—Speer and Göring.

While Vögler and his staff schemed at the beginning of 1942 to diminish the influence of the Education Ministry in KWG affairs, scientists in Debye's old institute were maneuvering to keep Heidelberg physicist Hans Bothe from replacing Kurt Diebner as head of the Kaiser Wilhelm Institute. The institute's staff still wanted Heisenberg. After discussing the conditions of work and residence with Vögler and Ernst Telschow, the KWG manager, Heisenberg finally signed a contract to head the institute in June 1942.[13]

Because the status of the institute was still uncertain in 1942, as was Debye's association with it, Heisenberg took a leave from his Leipzig chair rather than resigning outright.[14] Not until March 1943 did Army Ordnance, the RFR, and the KWG finally reach a formal agreement. While Speer's office controlled priority ratings and resources, henceforth the KWG would administer the Kaiser Wilhelm institutes, but the RFR would fund nuclear research. Army Ordnance would transfer most of its equipment to the KWG but would continue to fund nuclear research in the Gottow research station, just south of Berlin, to which Diebner now retreated, and in Harteck's and Clusius's Hamburg and Munich institutes.[15] Bureaucratic niceties prevented the KWG from naming Heisenberg as director of the physics institute—officially, Debye was still on leave. The KWG fell back on semantics and named Heisenberg not director *of* the institute but director *at* the institute. Until the summer of 1943, Heisenberg, who commuted weekly from his Leipzig home, registered his official residence as a rented room in the Schumacher home and listed his official job description as director at, not of, the institute. Dualities could exist on more than one level of reality.

Since Einstein's day, an appointment to the directorship at, or of, the Kaiser Wilhelm Institute for Physics automatically meant a simultaneous appointment to the faculty of the University of Berlin (but with no teaching obligations). This brought the REM into the game. Both of Heisenberg's nominations—to the Kaiser Wilhelm Institute and to the Berlin faculty—again raised the specter of political opposition.

The SS offered no objection; it regarded the appointments as fulfillment of Himmler's promise to have Heisenberg suitably placed somewhere other than in Munich. But Rudolf Hess's former party bureau and Alfred Rosenberg's ideology office remained strongholds of anti-Heisenberg Aryan physics. The party's University Teachers League and Rosenberg's Main Office for Scholarship still claimed the right to evaluate and approve every academic and institute appointment.[16]

In the spring of 1942, the KWG and REM offered Heisenberg the dual Berlin appointment that Einstein had once held. Ignoring the implications of his acceptance, Heisenberg and his supporters regarded approval of his dual appointment in the heart of the German Reich as a major victory for theoretical physics over its lingering ideological opponents. Since Heisenberg had earlier regarded politically motivated professional attacks to be personal in nature, he could now regard his personal and professional success as a triumph for his entire profession. The identity of himself with his science under the wartime dictatorship was complete. As Elisabeth Heisenberg put it years later, "The only thing that gave Heisenberg any satisfaction in this matter was the fact that his summons to Berlin had to be viewed as a clear victory of modern physics over 'German physics.' . . . I know how important this was to him."[17]

Heisenberg's colleagues did all they could to ensure his appointment to Berlin as part of their broader, ongoing campaign to achieve greater recognition of the value of theoretical physics. To demonstrate the Physical Society's support, Carl Ramsauer and Finkelnburg, supported by Ludwig Prandtl, submitted a memo to the REM describing the decline of German physics compared with Anglo-Saxon physics and arguing the importance of increased support for research, especially nuclear research, for the war effort.[18] Coming on top of the February 1942 rounds of lectures on nuclear energy, just as the tide of war was beginning to turn against the Reich, the arguments of the physicists began to have an effect. Nearly every top-ranking official expressed a new appreciation of the contribution of scientific research to the war effort. Dr. Erxleben, who was the head of Scholarship Observation and Evaluation in Rosenberg's Scholarship Office, declared in September 1942, "It appears to us urgently necessary to do everything to effect an upswing of atomic physics research"—a 180-degree turn from the sentiment of even a year earlier.[19]

The more diplomatic of Heisenberg's colleagues managed to neutralize the remaining pockets of ideological opposition through direct negotiations, or what they called "religion debates." One such debate had already occurred in 1940; Finkelnburg, a member of the Teachers League, organized a second in Munich on June 25, 1942, in support of Heisenberg's call to Berlin. In light of the REM's positive reaction to the Physical Society's memo, Finkelnburg easily routed REM functionary Wilhelm Führer and convinced Gustav Borger, the moderator of the meeting and head of the Teachers League Office for Scholarship, of the value of Heisenberg and his science to the German cause.[20]

Correspondence that summer among Borger's office, Erxleben's office, and party headquarters indicates complete support for Heisenberg. Reporting to the party on the

Munich meeting and citing the evaluations of some of Heisenberg's former SS investigators, now on his own staff, Borger echoed the earlier SS conclusions that Heisenberg's personal behavior was exemplary and that "his political position is in no way to be designated as argumentative. He is doubtless the unpolitical academic type." Erxleben delivered in turn his recommendation that the party withdraw from scientific conflicts. With a war raging, the competitive status of German physics was suddenly far more important: "Under no circumstances can we allow atomic physics research in Germany to remain inferior to work being done abroad. Professor Heisenberg's accomplishments in this field doubtless justify his call to the Kaiser Wilhelm Institute."[21] Heisenberg would have his institute, but his appointment to the teaching chair, from which he began lecturing in the fall of 1942, still required official approval.

To put the matter to rest, Heisenberg's supporters held a final "religion debate" with representatives of Aryan physics in November 1942 in the Tyrolean mountain village of Seefeld. Thanks to the raging war and the persistent efforts of Finkelnburg and others since 1940, the representatives of Aryan physics were already in retreat; they could not carry their argument either on scientific grounds or on the practical grounds of the war effort. During the meeting, which Heisenberg attended as an observer, the parties adopted five points of agreement drafted at the previous meeting, effectively defusing the political influence of Aryan physics, but they also elicited several further compromises from the theoreticians, especially regarding the avoidance of Einstein's name in public.[22] Three months later, the REM officially appointed Heisenberg to a chair for theoretical physics at the University of Berlin and to the directorship at the Kaiser Wilhelm Institute for Physics. Friedrich Hund replaced Heisenberg as Leipzig professor of theoretical physics.[23]

Heisenberg had at last attained one of the most important and visible positions in all of German physics, and after nearly a decade of struggle, German physicists had at last achieved a victory over their ideological opponents. They regarded these events as outstanding achievements, which at first glance they were. They had lifted recognition for theoretical physics from the depths of the SS affair of 1937, in which a theorist could land in a concentration camp merely for teaching modern physics, to nearly universal appreciation in 1942 by leaders of the most influential power centers in the Reich. Deft diplomacy, combined with the practical needs of the war and a decline in the need for ideology as a weapon of control and political advance, had made this possible.

Nevertheless, in many ways the victory was in the end too little, too late. Irreparable damage had been done to German physics during the previous decade; ideology had been silenced but ethnic hatred had not. Moreover, as historian Alan Beyerchen has pointed out, the victory simply meant that the regime no longer saw the struggle over modern physics as a conflict between camps loyal and disloyal to the regime, as it had before the war. Instead, it saw the conflict as one between two loyal factions within the system, each representing a different approach to scientific research. If physicists regarded Heisenberg's appointment as a monumental victory for the recognition of modern or "Jewish" physics, the regime saw it merely as the acceptance of a controversial approach that promised greater practical benefits for the war effort.

Heisenberg and his colleagues painstakingly encouraged regime officials to regard nuclear research as an important potential source of practical applications. This was evident in Heisenberg's February 1942 lecture to regime officials, discussed earlier. In it, he emphasized the potential benefits of nuclear fission, both controlled and uncontrolled, while echoing the caution of the army's Leeb report concerning technical hurdles. It was evident again when, on June 4, 1942, Heisenberg and other nuclear scientists met in the Kaiser Wilhelm Society's Harnack House to brief the newly appointed Speer and three military heads of weapons production on their work. The meeting occurred just before Hitler transferred the Reich Research Council and its nuclear subdivision to Göring's four-year plan, over which Speer exercised considerable influence.[24] Apparently repeating much of his February lecture, Heisenberg emphasized the need for a cyclotron and requested more funding for isotope separation.

Although the Leipzig L-4 experiment was at that moment multiplying neutrons at the rate of 13 percent, in contrast to the February lecture the plutonium alternative to isotope separation did not receive much notice. Heisenberg apparently did not want to awaken undue optimism, nor did he want to be ordered to build a bomb, since it could not be achieved without enormous effort. There is no indication that he withheld information on moral principle. Even though the German army was scoring successes at that very moment in North Africa and on the Russian front, the Battle of Britain was, for the Germans, all but lost, and Allied bombs were already striking northern German industrial regions. A new type of bomb would certainly have interested the weapons procurers among Heisenberg's listeners. The available version of Speer's office journal for 1942 contains no mention of a bomb discussed at that meeting. It states simply, "That evening there was a lecture in Harnack House on atom smashing and the development of the uranium machine and the cyclotron."[25] Speer recalled that Heisenberg seemed most disturbed by the lack of technical and financial support given to nuclear research and by the more massive backing to this field probably then available in the United States.[26]

During the discussion after the lecture, Speer surely did ask Heisenberg about a bomb. "His answer was by no means encouraging," Speer stated long after the war, perhaps in an attempt to defend himself against accusations from some quarters that he had not pushed hard enough for a bomb to save the Reich. The scientific solutions had been found, Heisenberg seemed to indicate, but technical difficulties prevented their realization in the foreseeable future. Field Marshal Erhard Milch, in attendance, recalled asking during the meeting how large a bomb would have to be to destroy a large city, such as London. To the astonishment of his audience, Heisenberg reportedly replied, "About the size of a pineapple"—perhaps referring only to the U-235 content.[27] The incredulous officials must have thought the scientist insane.

But Speer was still intrigued enough after the meeting to ask Heisenberg for a tour of his new institute, only blocks away. During their walk to the institute on the cool early summer evening, Heisenberg reportedly requested only moderate increases in support of the nuclear project: the construction of a radiation- and bomb-proof bunker, a cyclotron, and high-priority ratings for the acquisition of materials. Speer became convinced that the project would make only a modest contribution to the war effort. Accordingly, Speer gave other projects—Wernher von Braun's rocket research in particular—top priority. His report to Hitler on the atomic energy conference was less than enthusiastic. Point 15 on Speer's list of topics discussed with Hitler on June 23, 1942, states only, "Reported briefly to the Führer on the conference on splitting the atom and on the backing we have given the project."[28] Speer continued to support the project on a modest scale for the rest of the war, hoping to gain at least a new energy-producing machine to power ships and the economy. Ironically, at almost the same moment, American science administrators Vannevar Bush and James B. Conant were informing President Franklin D. Roosevelt of their conviction that an Allied bomb would be ready just in time to be used in the war.

On the day Speer reported to Hitler on the nuclear conference, just one week before Heisenberg took up his duties in Berlin, the last Leipzig model, still submerged in its water tank, began to emit a tiny stream of bubbles. Robert Döpel, suspecting a leak somewhere in the sphere, had it hoisted out of the water. When the hapless technician, Wilhelm Paschen, gingerly opened one of the inlet valves, air rushed in, then suddenly reversed, showering him with bits of burning radioactive uranium powder. This spurt was followed by an intense flame that melted the aluminum and set more uranium on fire. Döpel, Paschen, and apprentice F. Zumkeller managed to get the fire temporarily under control and lowered the sphere back into the water to cool.

Heisenberg, summoned to the scene by his staff, appeared briefly, saw that all was under control, and left to direct a seminar. He was summoned again soon afterward when the sphere began to heat up in the tank. Watching the sphere in the water, Heisenberg and Döpel saw it suddenly begin to swell. Both men leaped for the door, escaping from the room just as the ball exploded in a burst of flame and smoke, destroying the laboratory and raining burning uranium powder over the entire area. The two hemispheres of the reactor, held together with hundreds of bolts, had been literally ripped apart. Summoned by the dazed staff, the fire department soon arrived to put out the numerous fires ignited by the powder. After managing this task with great difficulty, the fire chief thanked the shaken professors for their fine display of explosive "atomic fission."[29] The incident marked the obvious end of the successful Leipzig series of reactor experiments—and just in time for Heisenberg's move from Leipzig to Berlin.

CHAPTER 25

PROFESSOR IN BERLIN

HEISENBERG ASSUMED THE DIRECTORSHIP AT THE KAISER WILHELM INSTITUTE IN BERLIN, along with his professorship at the university, on July 1, 1942. While his wife and growing family remained behind in Leipzig and Urfeld, Heisenberg immediately turned to the Berlin reactor experiments. At the end of the month he outlined a large-scale semi-technical experiment to achieve a controlled chain reaction in which the dangerous uranium metal powder was to be replaced by much safer and easier-to-handle uranium metal plates.[1] Cadmium control rods would also be used to slow and stop the reaction by absorbing the neutrons produced by the fission process and thereby hindering the chain reaction. A jacket of carbon was wrapped around the cylindrical tank in order to slow down and reflect any neutrons escaping from the contraption.

Henceforth, under Heisenberg's direction, the Berlin project would combine the more difficult theories of cylindrical reactor design with the results of different amounts and geometries of horizontal metal plates separated by the institute's 1.5 metric tons of heavy water in the hope of at last obtaining a self-sustaining chain reaction. It was an experiment based essentially on a process of theoretically guided trial and error—hardly at the level of sophistication already exhibited at that time by the Allied Manhattan Project.

Heisenberg's ambitious plans for these experiments depended, however, on the industrial production of the metal uranium plates from metallic uranium oxide and the completion of the underground bunker laboratory to house the experiments. Neither of these was available when Heisenberg took over in July 1942. An upgraded priority classification would have speeded plate production, but despite Heisenberg's request to Speer during their meeting the previous month, Speer had given the project the lowest classification, *kriegswichtig* (important for the war), that still allowed it to function. In addition, the cost of producing finished uranium plates from raw uranium ore was so great that it consumed most of the RFR uranium research budget for fiscal years 1943 and 1944.[2] Purifying and casting the metal into plates apparently presented little difficulty, but uranium is one of the densest metals known and cutting the plates with specially designed hardened machine tools was difficult and costly, a problem exacerbated by the incessant bombing of German industries and the shortages of supplies and resources. Because of the cost and the delay in plate production, full-scale reactor experiments did not begin again under Heisenberg until late in 1943—nearly one and a half years later—and just when conditions quickly worsened with the Allied bombing of Berlin. It wasn't until nearly a year afterward, well into 1944, that Heisenberg and his coworkers, having realized that they were on the wrong track, switched from metal plates to the more efficient metal cubes. Diebner and his independent reactor group, sponsored by the army and later the SS, had been using uranium cubes since 1943.

During the summer of 1942, as they submitted plans for the large-scale reactors, the Berlin team under Heisenberg's leadership completed the last of the B-5 series of preliminary experiments, using less suitable paraffin as a moderator to slow the neutrons. While awaiting the plates for the B-6 experiments to arrive, Heisenberg gradually turned to other matters.[3] The uranium project had now served its initial purpose for Heisenberg. His call to Berlin and the personal support accorded the project by Speer and Göring, the controllers of the economy, constituted concrete acknowledgment of the practical importance of theoretical physics and of Heisenberg's leadership role. He had also established that nuclear fission could be exploited for practical purposes, in both a reactor and a bomb. Heisenberg could now safely turn to less dire scientific matters and to his role as an important German professor living in war-torn Berlin.

Commuting to Leipzig for weekends with his family, Heisenberg found his new life as a prominent Berlin professor a busy one. He lectured on theoretical topics twice a week at the university, directed graduate research, and oversaw a lecture series at the institute on his favorite subject, the high-energy physics of cosmic radiation. He edited and published the lectures in a monograph completed in June 1943.[4] In addition, as he had in Leipzig, Heisenberg moved within the established circles of the cultural and intellectual elite of Berlin, the capital of the German Reich. On September 1, 1942, Luftwaffe commander Göring made him a corresponding member of the German Academy for Aeronautical Research. REM officers approved his election to the science section of the Prussian Academy of Sciences in April 1943—exactly ten years after the same section had expelled Albert Einstein. And in November 1942, at the invitation of Prussian finance minister Johannes Popitz, Heisenberg attended a meeting of the famous Wednesday Society, a group that had functioned since the Wilhelmine era as one of Berlin's socially elite men's clubs.[5]

The Wednesday Society selected its members from among the leaders of Berlin's cultural, academic, administrative, and military life. The group met every several weeks (on a Wednesday, naturally) in the home of one of its members. In addition to refreshments, the host provided a general lecture on his work. In 1942, the society's 28 members included such noted figures as the surgeon Ferdinand von Sauerbruch; General Ludwig Beck, chief of the army general staff; Ambassador Ulrich von Hassel; and professors Eduard Spranger, Wolfgang Schadewaldt, and Jens Jessen. As these names suggest, the Wednesday Society also served during the Third Reich as a meeting place for many members of the conservative Prussian military and professional opposition to Hitler and as a breeding ground for the unfortunate conspirators of the failed coup d'état of July 20, 1944. Members were chosen for their sympathy with the views of the non-Nazi German and Prussian cultural elite, which though patriotic and nationalistic, insisted on moral rectitude. Heisenberg, whose sentiments harmonized to an extent with the members of the society, was already well known to Spranger, a former member of the Leipzig professors' club, and to the future conspirators (but not club members) Carl Goerdeler, former Leipzig mayor now in Berlin, and Adolf Reichwein, a former youth-movement member. Heisenberg attended his first meeting as a member in December 1942.[6]

Mrs. Schadewaldt, whose husband managed to escape death during the reign of terror that followed the failed assassination of Hitler, later recalled a meeting of the society in their home on March 17, 1943. After the close of the meeting, Heisenberg, Hassel, and Sauerbruch lingered in conversation. "And then Heisenberg, in somewhat subdued terms, and Sauerbruch, in his spirited manner, grumbled about 'Schimpanski,' that was the code name for Hitler."[7] (Schimpanski is close in sound to chimpanzee.)

According to most studies of the assassination attempt, most of the conspirators had opposed National Socialist policies, in word if not in deed, since the early years of the regime, but they did not object to German national and military revival. By the outbreak of war, their Prussian scruples regarding Nazism had led them to conspire against the Reich. But events did not favor a coup d'état until the demoralizing defeats of the winter of 1942–1943. As the German army went into retreat and captivity on the Russian front and in North Africa and as the Allies intensified their bombing campaign, the general populace began to realize that the promised military triumph was turning into another long and humiliating defeat. Members of the professional classes were emboldened to allow their anti-Nazi scruples free rein at last and to consider alternatives to Nazism. Many wanted a restoration of the monarchy, and some, especially the courageous Count Klaus von Stauffenberg, saw assassination of Hitler as the only way to gain military backing for a coup d'état. Even if the coup attempt failed, they reasoned, it would demonstrate to the world that despite Hitler and his evil Reich, a decent Germany still survived. Everyone in the Wednesday Society knew of its anti-Hitler orientation and most knew of the festering conspiracy. It is uncertain, however, how many members actually took part in it as words turned to deeds in 1944. Elisabeth Heisenberg claims that Reichwein, who was not a member of the society, came to Heisenberg's institute in 1944 to ask him to join the conspiracy, but Heisenberg, unwilling to conspire, respectfully declined.[8]

According to the published records of the Wednesday Society, Heisenberg attended most of its meetings and hosted two of them, including its last, on July 12, 1944, in the KWG's Harnack House. The institute's gardens, near the Virus House, provided berries for the occasion, and the KWG supplied wine. Heisenberg's lecture that evening on the constitution of stars included an explanation of the process of nuclear fusion, developed earlier by Heisenberg's émigré colleague Hans Bethe. Ten members attended, among them the conspirators Beck, Jessen, Sauerbruch, and Ludwig Diels. Diels wrote in his diary, "The mood is subdued. Jessen, who is a defeatist, contributes especially to this."[9]

Heisenberg submitted the minutes of the meeting and a copy of his lecture to Popitz on July 19, then left for southern Germany and a visit with his family in Urfeld, perhaps in order to have an alibi. He arrived in Urfeld the next day to learn that the attempt had failed. Stauffenberg, a wounded war veteran who had occasional access to Hitler's war room, had entered the room with a suitcase containing a bomb. He placed it under a table being used by Hitler and a group of army officers, then left the room. One of the officers, seeing the suitcase, moved it out of the way. It exploded moments later, killing three officers and wounding and enraging Hitler.

During the reign of terror that ensued, many of the members of the Wednesday Society were rounded up, summarily tried, and executed by guillotine (or allowed to commit suicide because of their status). Even those, in or out of the society, who knew of the plot but did not participate were tried and executed. Max Planck lost his eldest son, Erwin. With less than a year of his reign remaining, Hitler had destroyed many of the last leaders of the Wilhelmine upper cultural, social, and aristocratic classes. Heisenberg again miraculously escaped death. There is no indication that he was even interrogated. The high-ranking regime officials—Speer, Göring, Himmler—who had supported him previously may have protected him from suspicion, perhaps for fear that their judgment, and loyalty, might be called into question.[10]

Even before they had any hint of opposition or of a conspiracy among the elite, the authorities in Berlin held Heisenberg and his coworkers, along with everyone else in such positions, under constant surveillance lest they stray from the fold. Immediately upon moving to Berlin in July 1942, Heisenberg had named two party members at the institute to serve as party representatives. A short while later, the SS security service representative in charge of "political questions" at independent research institutes insisted on inspecting Heisenberg's institute. Heisenberg used the inspection to his advantage. When two SS physicists who were keeping regular watch over the institute showed up one day, Heisenberg treated them cordially, showed them some interesting experiments, and impressed them with his usual charm. Already positively disposed toward their surveillance target, one of them later claimed that they offered to inform Heisenberg of any governmental actions that might interest him.[11]

Heisenberg's university institute also harbored Johannes Juilfs, an SS officer and an assistant in theoretical physics. Juilfs had earlier played a crucial role in Himmler's investigation and exoneration of Heisenberg. Heisenberg's penciled notes to his institute secretary on certain government letters he received indicate that Heisenberg made use of the man whenever he was engaged in delicate dealings with SS and party officials. But Juilfs demanded certain concessions, including in particular the avoidance of Einstein's name. In the fall of 1942 Heisenberg, upon receiving a complaint from Juilfs, who had apparently received a complaint from Sommerfeld's publisher, wrote to Sommerfeld requesting that he delete as many references to Einstein as possible from the latest edition of his forthcoming textbook. Sommerfeld, aware of Heisenberg's situation, politely complied—approval of Heisenberg's Berlin university appointment had not yet been received. In contrast, when Max von Laue was accused by Mentzel of using Einstein's name to excess in a lecture delivered in neutral Stockholm, the physicist steadfastly refused to recant.[12]

Until the Allied bombing of German cities began in earnest in 1943, Heisenberg's wife and children stayed in their Leipzig home while Heisenberg spent most of his time in Berlin. This arranagement ended after one of the Allied thousand-bomber raids practically leveled Berlin in early March 1943. The Heisenberg twins, Wolfgang and Maria, were with their father in Berlin to celebrate Grandfather Schumacher's seventy-fifth birthday. Heisenberg was at one of his social gatherings one evening when the bombs began falling. As he drove frantically from the city center to the Steglitz suburb, hellish walls of flame on both sides of the road etched the scene forever in his mind. He arrived to find the roof of the Schumacher house, hit by two incendiary bombs, entirely engulfed in flames. Having assured himself that his children and in-laws were safe, he ran next door, where a young woman was frantically crying for help. Her elderly father was fighting a losing battle against the flames of their burning home. The athleticism of the middle-aged professor served him well in this crisis; he was, fortuitously, wearing a leotard track suit that night. Heisenberg managed to scale the outside wall to the roof of the burning house, leapt across a pool of flames, and led the old man to safety.

Everyone in the Schumacher household survived, but the children, temporarily sent to stay with the former head of the Emergency Association, Friedrich Schmidt-Ott, were so terrified by the experience that Heisenberg decided to move his family permanently to their Urfeld summer cottage. After renting the Leipzig house in April 1943 and shipping the furniture to Berlin for storage, Elisabeth Heisenberg, together with her five young children, a distant cousin whose non-Aryan husband had "disappeared," and her child, moved into the tiny Urfeld cottage for the duration. After the Schumachers moved south that summer, Heisenberg, completely alone in Berlin, moved into the bachelor quarters of Harnack House, which he shared with an influential member of the Foreign Office. It was here that he twice entertained his Wednesday Society colleagues.[13]

Eight months after the move to Urfeld, the Heisenbergs' Leipzig house on the quiet tree-lined street, with its beautiful flower garden and shady backyard, fell victim to another heavy bombing raid, which also damaged the physics institute. The same series of raids bombed Heisenberg's sickly mother out of her Munich apartment on Ainmillerstrasse. With Heisenberg's help, she found new rooms in the Bavarian border town of Mittenwald. Her son also arranged, at the height of the war, for a truck to move her furniture. She might have joined the family in the Urfeld cottage had they not just then come down with scarlet fever.[14]

Not until long after the end of the war would Elisabeth Heisenberg and the children see their husband and father again except for brief vacations and holidays. They would have to face the last, most difficult years of the war alone in their mountain hideaway. Heisenberg's determination to stay in Berlin was taking a toll on his private as well as his professional life. Although Heisenberg regarded his appointment to his prominent positions in Berlin as an enormous victory for German physics, Himmler's second promise, the publication of Heisenberg's views in a Nazi science journal, was not yet fulfilled. To Heisenberg, every promise of Himmler's 1938 letter exonerating him of the accusations of treason required fulfillment as proof to him of the defeat of Aryan physics and the rehabilitation of himself and of theoretical physics.

Despite the obvious victory his appointment signified, Heisenberg used every conceivable strategy to make Himmler's second promise a reality. The task was not easy, for the journal that was to publish the article, the *Zeitschrift für die gesamte Naturwissenschaft (Journal for the Entirety of Science)*, served as the official organ of the Reich Students League, which, like the University Teachers League, remained a stronghold of anti-Heisenberg sentiment in 1942. Working through Juilfs, who was a former member of the Students League, and probably through his SS security service (SD) connections among the surveillors of his institute, Heisenberg managed to have Himmler order SD chief Turowski to accept an article from Heisenberg at about the same time that Heisenberg was summoned to Berlin.[15] Heisenberg had drafted the article two years earlier and submitted it to Himmler's infamous Reich Main Security Office for transmittal to the journal's editor, Bruno Thüring. Himmler's office also ordered SS officer Dr. Fritz Kubach, head of the science section in the Students League, to publish the article. But Thüring, who had earlier opposed Heisenberg's call to Vienna, was not in the SS and would not budge without further pressure and extracted compromise.

Heisenberg confronted Thüring during the Seefeld "religion debate" in November 1942. After the encounter, Thüring wrote to Heisenberg that he had the distinct impression that publication of the article "is supposed to be a type of rehabilitation for the . . . attacks directed against you"—a rehabilitation that Thüring strongly opposed.[16] Another letter from Heisenberg to Himmler's office (Himmler himself was "away on business"), in which he complained of the journal's intransigence, brought results. Heisenberg's article, "The Evaluation of 'Modern Theoretical Physics,'" finally appeared in the October 1943 issue of the Nazi journal and was eagerly read by physicists throughout the Reich.[17] The article upheld the compromises achieved earlier, especially those regarding Einstein's name and the separation of the physicist from his physics. Heisenberg reassured a concerned Theodor Vahlen, now president of the Prussian Academy of Sciences, which would appoint Heisenberg to membership several months later, that in this article he presented the viewpoint that the special theory of relativity "would have arisen even without Einstein."[18]

The article and its publication showed how far Heisenberg was willing to go for the sake of attaining a new token of rehabilitation of himself and of theoretical physics, which for him were closely identified. It was as if recognition and respect, which he always required, had been withheld for so painfully long and, when finally given, were so tenuous that no amount of recognition accorded him or his physics could ever make up for this or reassure him that their status had not somehow declined once again. There were many opportunities for reassurance in 1943 and 1944.

When Heisenberg received the prestigious Copernicus prize of the Reich University of Königsberg, Heisenberg wrote to Dr. Borger, the head of the Teachers League's Office for Scholarship, who had moderated the Seefeld "religion" conference, that he was especially pleased with this award, because it could be regarded as even further official rehabilitation of theoretical physics. "Hopefully this development . . . will continue in the future."[19] The development had already begun. In 1942, Heisenberg's protégé Carl Friedrich von Weizsäcker was asked to accept the theoretical physics chair in German-occupied Strasbourg. In March 1943, the Nazi party newspaper, *Völkischer Beobachter*, asked Heisenberg for an article celebrating Planck's eighty-fifth birthday. In October, Göring recommended Heisenberg to Hitler for the War Service Cross, First Class. And in 1943 and 1944 Planck and Heisenberg were featured in front-page stories in *Das Reich*, Goebbels's cultural propaganda newspaper.[20]

Reassurance of his standing also seems to have been one of the personal and professional motives behind Heisenberg's many cultural tours abroad throughout the war. Travel by professors beyond the Reich's original boundaries, especially on professional business, required explicit approval of the Education Ministry's foreign office and of the German Congress Center and Foreign Currency Bureau, an arm of Goebbels's propaganda ministry. In granting approval, the Education Ministry relied on the evaluation of its own member in charge of foreign exchange and on the party's Teachers League representatives at the professor's university. Every permission to travel on official business could thus be seen by the recipient as concrete evidence that the authorities held him in high regard and trust.

But of course the authorities saw such permission in a quite different light. Germany had long promoted foreign propaganda using noted scholars who participated either wittingly or unwittingly. While the Teachers League strongly preferred those of the proper ideological persuasion, the Congress Center was primarily concerned with professors who could be exploited for the German cause; it did not demand ideological purity. Philosopher Hans-Georg Gadamer, one of Heisenberg's Leipzig colleagues from the professors' club, later commented on his own wartime trips abroad, one of which was to Portugal in the company of Carl Friedrich von Weizsäcker: "I did not fully recognize that thereby one was being used for purposes of foreign propaganda, for which a political innocent was sometimes suitable. Such instances were an escape with mixed feelings."[21]

Weizsäcker also accompanied Heisenberg, by now no innocent, on many of his foreign travels. Their 1941 trip together to German-occupied Copenhagen had been a test case of Heisenberg's suitability for future exploitation. It was regarded as such a success that the elder Weizsäcker's Foreign Office advised the REM that from the "political culture point of view" a further presentation by Heisenberg was extremely desirable.[22] Heisenberg probably knew of this recommendation. If he did, the co-optation seems complete: just as the German nuclear project made a case for its members and their science by pointing out their utility to the war effort, the Foreign Office argued for Heisenberg's unhindered travel throughout the Reich and even outside the Reich by pointing out his usefulness to the German propaganda cause. At the same time, Heisenberg could see such permission as further approval of himself and as an opportunity to pursue his own purposes during these largely voluntary visits. He could thus regard the compromises required by the authorities as of little significance. Naturally, those whom he visited during these trips often took quite a different view of the visit and the visitor.

After 1941, most of Heisenberg's travels resulted from solicited or unsolicited invitations, usually asking him to lecture, as he did in Copenhagen, at one of the infamous German cultural propaganda institutes. As noted earlier, the culture institutes, set up on occupied soil by Ernst von Weizsäcker's Foreign Office, were created for the express purpose of extolling the virtue of German rule through the example of German culture. Fritz von Twardowski, head of the Political Culture Division in the Foreign Office through 1943, later admitted this function.[23] But in most instances, nationals in occupied countries identified the cultural visits with imperialism. Most felt insulted by an invitation to attend a Heisenberg lecture in a propaganda institute; acceptance would have signified collaboration. At least one foreigner later accused Heisenberg, among the several German visitors to his homeland, of possessing "the least understanding of the situation." Yet Heisenberg was also a man who could write in his private papers: "It is more important to treat others humanely than to fulfill any sort of professional, national, or political duties."[24] At least he seemed to know what was right, even if he could not always carry this out.

Between the SS attack on him in 1937 and the Teachers League's "religion" retreat in the summer of 1942, Heisenberg undertook five recorded trips abroad: north to England and south to Geneva and Bologna in 1938, to the United States in the summer of 1939, to Budapest in German-allied Hungary in 1941, and to Copenhagen in the same year. Heisenberg was able to make these trips largely because of the pressure exerted on party bureaucrats by Weizsäcker's Foreign Office. After Heisenberg moved to Berlin in 1942, he had little trouble traveling as he wished, but he still perceived every approved trip as a personal victory, in addition to being a means of proving himself to authorities and maintaining contacts with foreign physicists.[25]

None of his trips after July 1942 seems to have been self-motivated. Every trip originated with an "invitation" communicated directly to Heisenberg from the REM or through his new Berlin dean, Ludwig Bieberbach, a leading proponent of the Nazi "Aryan mathematics" movement. The surviving examples of invitations are worded in a way that strongly urges acceptance. Heisenberg could have refused, with some difficulty, but duty and the opportunities it presented required no urging, even when he would be traveling into an unsavory situation, even when his visit was being used to bolster faltering support for the Reich. Heisenberg readily accepted an invitation conveyed on May 25, 1943, from Hans Frank, General Governor of occupied Polish territories, to lecture in a Cracow cultural propaganda institute and to stay as a guest in Frank's Schloss Wartenberg.[26] Just two months before Frank offered his invitation, he and his henchmen had liquidated the Cracow Jewish ghetto; only days earlier, in mid-May 1943, they had annihilated the heroic Warsaw ghetto. The Lodz ghetto was still in existence when Heisenberg arrived that December. If Heisenberg did not know of these events before his arrival, he could not have escaped rumors about them during his stay.

Between July 1942 and the end of the war, Heisenberg made another seven recorded foreign trips, in addition to the December 1943 visit to Cracow. Two were to neutral Switzerland, in November 1942 and December 1944. He made a return trip to Budapest in the company of Planck and Weizsäcker at a time (November–December 1942) when Hungary's allegiance to the Reich had begun to waver. He was in Pressburg (Bratislava) in Reich-aligned Slovakia also at a time (March–April 1943) when pro-German sentiment had begun to falter. He traveled to the occupied Netherlands in October 1943. And he made two trips to Copenhagen, in January and April 1944, after Bohr's departure.[27]

Reich Education Ministry guidelines strictly controlled the activities of its emissaries. A professor had to submit lists of everything he carried into and out of a country; he was required to report to local occupation authorities (so he could be watched); and, upon his return, he had to prepare a written report of his activities, including his observations of the political situation. Compliance ensured control. Three visits for which considerable documentation, as well as controversy, are extant are Heisenberg's trips to Denmark in 1944 and the trip to the Netherlands in 1943. Documentation is no guarantor of truth. No matter how thorough the written record, how vivid the memories of witnesses, they are overshadowed by the terrible circumstances of the times. Matters of life and death, murder and torture, resistance and collaboration, manipulation and self-delusion are all at play against an extraordinary background of war, plunder, oppression, and genocide. It is difficult enough to reconstruct past events occurring in normal, peaceful times with some degree of accuracy. How much more do the horrors, feelings, and fears of those terrible times hinder our confidence in any source describing events of that period?

Nevertheless, the available sources do provide a partial account of Heisenberg's travels and the circumstances in which they occurred. His January trip to Denmark came at a particularly dark moment in Danish history. Having subjected the Danes to occupation since 1940, the German authorities suddenly dropped their facade of benign intentions in the fall of 1942 after the monarch fled and Hitler decided to incorporate Denmark fully into the Reich. The German command replaced the civilian governor with SS officer Dr. Werner Best. As resistance and sabotage increased in reaction to nazification, Best declared a state of emergency, imposed martial law, and in cooperation with Berlin prepared to round up Danish Jews on October 1, 1943, for transport to the German death camps. Thanks to a German industrialist, the resistance learned of the plan, and with the help of the entire population and the cooperation of the Swedish government, within several days Niels Bohr, his family, and nearly all of the estimated 8,000 Danish Jews were ferried in private boats across the narrow Øresund Straits to neutral Sweden.[28]

Soon after this heroic and spectacular escape, the resistance inquired of Bohr whether his institute should be blown to bits to prevent its use by the Germans. Bohr declined the offer, and, as expected, German troops occupied the building on December 6, 1943. They used the pretext that Bohr and his institute were conducting anti-German propaganda and contacting the Allies.[29] The troops immediately imprisoned the institute's only resident, Professor J. K. Bøggild, a cosmic-ray researcher. They subjected his colleague, a Dr. Olsen, to five days of Gestapo interrogation, and barred the remaining members from the premises. The Danes informed Hans Suess, a German heavy-water expert who was passing through Copenhagen on his way home from the captured Norwegian plant, of the takeover. Suess informed Heisenberg of the Copenhagen events after returning from Christmas vacation. At the same time, Heisenberg learned of Reich Research Council plans to strip Bohr's institute of its cyclotron and other equipment and to ship them to Germany.[30]

Working through Walther Gerlach, Abraham Esau's successor as head of Research Council physics, and Dr. E. Six, Twardowski's successor as head of cultural propaganda, Heisenberg obtained permission to fly to Copenhagen on January 24, 1944, accompanied by Kurt Diebner, to help arrange for the disposition of Bohr's institute. Diebner could throw the weight of the Army Ordnance Office behind any plan worked out with the institute staff. Officially, the two were traveling as members of a committee to investigate the charges against Bohr's institute. Heisenberg's uneasiness is evident in a letter to his mother written the night before he left. "I am not at all happy about the trip . . . [but] it is probably necessary that I try to determine what is right and, if possible, to rectify the situation."[31]

Several sources, all open to question, exist that describe what transpired on this journey. One is a lengthy report written in Danish by the institute members soon after the institute was released from Nazi control in 1944. Its purpose is unclear, but it casts Heisenberg in an ambiguous light at best. Other accounts include anecdotal postwar recollections casting Heisenberg in a much harsher light and a January 8, 1944, appeal to Heisenberg concerning Bohr's institute from Swedish physicist Hans von Euler with emendations at the end in Heisenberg's hand.[32]

According to the Danish report prepared by the institute members, Heisenberg wrote to them on January 10 informing them of his arrival in 14 days. Shortly after their arrival, Heisenberg and Diebner met at the institute with Bohr's colleagues, physics professors Christian Møller and Hans Jacobsen, whom Heisenberg knew through their earlier work on nuclear forces and cosmic rays, and with the Gestapo officer in charge of the institute. Committee member Heisenberg demonstrated to the officer that Bohr was not engaged in any underground activities by walking over to his file cabinet and seemingly randomly pulling out one of his own letters from among Bohr's voluminous correspondence.[33]

As intermediary, Heisenberg then discussed three proposals for the institute, which are outlined in the report. They are compatible with four proposals that Heisenberg noted, with pros and cons, at the foot of the letter from Euler. He wrote them probably in preparation for the meeting. The three proposed scenarios were: (1) German scientists would take over and run the institute for war research; (2) the German forces would return the institute to the university under the conditions that no war research would be carried out in the institute and all research would be published; and (3) the institute would be returned to the university, but the Germans would strip the institute of the apparatus they needed. According to the report, the Danes naturally rejected all three proposals, refused any conditions at all, and demanded Bøggild's immediate release from prison before discussing anything further. Heisenberg warned them that under the current circumstances, unless conditions were imposed, the authorities might not release the institute at all. An impasse was avoided the next day when the senior Weizsäcker's Foreign Office, probably notified of the situation by Carl Friedrich, informed the university rector that the institute would be released without condition on February 3. Bøggild was released from jail that afternoon, and, after filing a report on the harmless nature of the institute, Heisenberg flew back to his duties in Berlin.[34]

But the Copenhagen authorities would not let the matter rest so easily. Two months later Heisenberg received a strong request from the Reich Education Ministry to accept an invitation from the German occupation administrators in Denmark to lecture at the newly reopened German Culture Institute in Copenhagen.[35] He could surely have refused without much difficulty. Instead, he readily accepted. Heisenberg returned to occupied Copenhagen, delivered his lecture as requested, and, to the dismay of the Danes, even dined publicly with the brutal Reich commissar, SS *Obersturmbannführer* Dr. Werner Best. This return trip, his warm thanks to Best, and a positive report to the REM on the culture institute were, he apparently somehow convinced himself, payments demanded by the Reich in exchange for its reduced interference in the affairs of Bohr's institute. One 1949 report, based on Danish assessments at the time, declared, "It is thought that Heisenberg is not a Nazi but is an intense nationalist with the characteristic deference to the authorities in control of the nation."[36]

Heisenberg's trip to the occupied Netherlands is even more questionable and certainly more difficult to interpret. Nazi racists considered the non-Jewish Dutch, like the non-Jewish Danes, to be an Aryan people and therefore suitable for full incorporation into Hitler's Reich. But while the Danish government remained in place until 1943, the Dutch monarch and government had fled to London immediately upon the country's occupation in 1940. This provoked instant, large-scale resistance to the German occupation.

The Reich commissar for the Netherlands was the infamous Dr. Artur Seyss-Inquart—an Austrian Nazi who had earlier presided over Hitler's annexation of Austria and then served as second-in-command to Hans Frank in the Polish territories. He was later tried and executed at Nuremberg. Like the first governor of occupied Denmark, Seyss-Inquart at first attempted to co-opt what he called "obliging circles in the field of economy, especially agriculture, culture, art, and science."[37] The Dutch responded with grudging acquiescence and apparent collaboration in some quarters. But attempts to nazify Dutch society and to introduce anti-Semitic measures as early as 1941 provoked strikes and riots, which were met, on orders from Himmler, with brutal suppression.

Mass executions, the internment of Dutch army officers, and the start of Jewish deportations to the death camps in the summer of 1942 provoked open resistance, sabotage of Nazi targets, and the organized concealing of Jews and other endangered persons. According to historian Werner Warmbrunn, students and faculty members at Dutch universities "kindled and spearheaded ideological resistance to National Socialism." He estimates that students, subject to being sent into forced labor in Germany, made up as much as one-third of the persons executed during the Nazi occupation.[38]

German policies met with particularly strong resistance at the University of Leiden, where Heisenberg would have his most remembered encounters. Unlike the essentially passive reaction in Germany and elsewhere under German rule, Leiden students organized a mass strike of classes upon the dismissal of Jewish professors in November 1940. A year later the majority of the faculty responded to a German attempt to replace an outspoken faculty member by handing in their resignations and refusing to cooperate further with the authorities. Scores were subsequently taken hostage, and many were executed in reprisals for underground actions. Physicist Hendrik B. G. Casimir, then a researcher at the Philips Company in Leiden, reported that at the university's famous Kamerlingh Onnes low-temperature laboratory, where at least one Jew was concealed, "things went on much as usual." As for the rest, writes Warmbrunn, "Leiden ceased to exist as a university."[39]

Both university protests and repressive measures to quash them spread across the Netherlands in 1943, as the deportations to the death camps continued. By September they were complete, while the remainder of Dutch Jews, among them Anne Frank in Amsterdam, held out in hiding. A month later Werner Heisenberg, successor in Berlin to Dutch national Peter Debye, arrived in the Netherlands for what must have been an extremely tense encounter with his Dutch colleagues.

Considerable official correspondence and documentation pertaining to Heisenberg's visit are available, as is some ostensibly private correspondence between Heisenberg and his colleagues in the Netherlands. Letters between Heisenberg and correspondents in occupied territories were, however, never written as private communications, for the authors knew that the Gestapo could and very likely would read the contents before they were delivered. The disturbed conditions in the Netherlands at the time make it difficult today to interpret these sources. With the end of the deportations apparently in sight, on June 15, 1943, the REM forwarded an invitation to Heisenberg, ostensibly from the Dutch Education Ministry and unnamed Dutch scientists, to visit the Netherlands in the fall. The REM urged acceptance, and Heisenberg did so but inquired which scientists were involved. Two months later, he received a letter from his old Copenhagen colleague, H. A. Kramers, now in Leiden, with an explanation of the origin of the invitation—or as much of an explanation as he cared to put into writing.[40]

According to Kramers's account, the chief of education in the Dutch ministry wanted to improve the conditions of Dutch professors by arranging for personal contacts with foreign colleagues. Kramers, apparently in close touch with the education chief, discussed this with his physics colleagues, Wander J. de Haas, Casimir, Ralph Kronig, and others. They agreed to invite Swiss experimentalist Paul Scherrer and German theoreticians Heisenberg and Richard Becker to visit in the fall.[41] Taking this account at face value, aside from the intellectual stimulus of foreign colleagues, the Dutch physicists apparently felt that if the newly rehabilitated German theorists took a personal interest in Dutch science, they might use their influence with the German authorities to protect Dutch laboratory equipment and to improve the Dutch scientists' working conditions. At any rate, an official in the Main Department for Science, Education, and Promotion of Culture in the office of the Reich commissar made little effort to conceal *his* reasons for approving the invitation. In a letter to Heisenberg in September, he wrote, "We desire that in this way the rather unstable relationships we have in the scientific field may be strengthened once again. You will probably have heard that the majority of Dutch university professors reject or mistrust our political views and ideas, but on the other hand they have no desire to sever connections in the professional sphere."[42]

Heisenberg's Dutch travels from October 18–26, 1943 took him to the Education Ministry in Apeldoorn; to the Kamerlingh Onnes Laboratory in Leiden, where he met with Kramers, Casimir, and Willem Hendrik Keesom and delivered a colloquium lecture on elementary particles; to Utrecht, where he stayed with Léon Rosenfeld; to the universities of Delft and Amsterdam, where he lectured; and to a meeting with Seyss-Inquart at occupation headquarters, where he argued for the preservation of the Kamerlingh Onnes Laboratory.[43]

Back in Berlin, Heisenberg continued to work with his contacts to try to prevent German interference in Dutch research, and he reported to the REM his official observations. According to his report, the Dutch expressed "a blunt rejection of the German viewpoint. . . . However, on a purely scientific basis, a collaboration with Dutch colleagues is perfectly possible."[44] Although this statement may express Heisenberg's actual views, it seems equally an echo of the views expressed by the occupation official in the Main Department for Science. Dutch interest in scientific collaboration had been strengthened by the visit—precisely the outcome the official had desired. At the same time, the stated possibility of scientific collaboration was a good argument for the preservation of Dutch laboratories and research.

During the war, two physicists Heisenberg had visited in the Netherlands warmly thanked him for his interventions on behalf of themselves and their science.[45] Nevertheless, immediately after the war and in subsequent writings, several scientists recalled distressing private encounters with Heisenberg during his visit. According to a June 1945 report filed by G. P. Kuiper, a Dutch-American member of the Alsos mission, an Allied science intelligence unit, Casimir had told Kuiper that during a private walk with Heisenberg in Leiden, Casimir learned that Heisenberg knew about the

German concentration camps and about Germany's plunder of occupied territories, yet "he wanted Germany to rule." As Kuiper reported it, Heisenberg said: "Democracy cannot develop sufficient energy to rule Europe. There are, therefore, only two possibilities: Germany and Russia. And then a Europe under German leadership would perhaps be the lesser evil."[46]

These statements circulated widely among Allied scientists after the war and contributed to their negative assessment of Heisenberg in those years. Like other evidence, they require closer scrutiny rather than face-value acceptance. Human frailty being what it is, recollections can be distorted by painful memories of wartime experiences; occasionally they are distorted by efforts to protect those who acted less than nobly in a difficult situation by accusing others of even worse behavior. In any case, if Heisenberg did make this statement, then, as Casimir has pointed out, he showed an appalling lack of sensitivity, to say the least. In view of the bitter hostility harbored by the Dutch toward Germany at that time (especially in Leiden), the expression of such a sentiment would, as one writer puts it, have forever poisoned Heisenberg's relations with his Dutch colleagues, however friendly and grateful for favors they may appear in their wartime letters.[47] And whether or not Heisenberg actually said those things, the very fact that Casimir and Kuiper claimed he did could hardly have helped Heisenberg's case during the postwar evaluation of his actions during the Third Reich.

But what were Heisenberg's true feelings about the war? We have already seen that he wanted to protect Germany from defeat and that he regarded the war as a shift of power far beyond the control of mere mortals like himself. In October 1943, when he visited the Netherlands, the Western Allies were still only a weak presence on the European continent, while the Russian army, after the crushing defeat of the German Sixth Army at Stalingrad, was already advancing toward Poland. Heisenberg at that moment may well have seen the war as a struggle over which dictatorial system would ultimately rule Europe, and it looked very much to him, as it did to many Germans, as if Germany would lose. Ever since the trauma of the Bavarian soviet republic, Heisenberg could imagine nothing worse than the conquest of Germany by Soviet troops—no matter that others were already suffering the terrible consequences of conquest by Hitler's troops. Heisenberg's choice of Hitler's regime as the lesser evil became even more obvious after the war: having refused to leave Germany during the 12 years of Nazi rule, he reacted to early cold-war fears that Soviet forces might overrun Western Europe by making contingency plans for his immediate emigration to the United States.[48] Heisenberg's lately elevated status in Nazi Germany apparently evoked highly questionable compromises in his sensitivities and perceptions, in addition to those in his private and professional life and work.

CHAPTER 26

RETURN TO THE MATRIX

HEISENBERG'S WORK IN APPLIED NUCLEAR FISSION HAD AT LAST SERVED ITS PURPOSE upon his appointment to the top of the German physics profession in Berlin. Shortly after arriving in Berlin in July 1942, the brilliant theorist immediately turned to his "real work," theoretical high-energy particle physics. Nuclear engineering had been for him only a means to an end. As soon as he had completed the plans for a large-scale reactor, had submitted a long-term order for uranium metal plates, and had obtained sufficient funding and proper war classification, the project could coast along while its scientific director delved again into the abstract intricacies of elementary particles.

Just two months later, Heisenberg submitted the first part of what, by 1945, became a four-part work on a new and widely studied fundamental theory of elementary particles. He wrote to the ailing Hans Geiger, to whom he dedicated the first paper, that he had been working on these new ideas off and on for quite some time.[1] Heisenberg's bachelor life in Berlin and his return to academic activities enabled him to formulate the first two parts of the work so rapidly that he completed them by the end of October 1942. The third and fourth parts appeared over the next two years in collaboration with his colleagues in occupied Denmark and the Netherlands.[2]

Heisenberg's prewar researches in quantum field theory, undertaken in part with Pauli, had led him into the study of cosmic rays, the highest energy particles then available for research. When an extremely high-energy cosmic ray strikes the earth's atmosphere, or a metal plate inside a cloud-chamber detector, it induces a shower of newly created particles and photons. This effect was to be explained on the basis of quantum field theory. Heisenberg's researches had previously convinced him and others of the inadequacy of field theories for this task. Infinities and divergences plagued all three of the available theories—quantum electrodynamics, Fermi's theory of beta decay (relating to what is now the weak force), and Yukawa's meson theory (relating to what is now the strong, or nuclear, force).

The small size of elementary particles and the close approach of the particles to each other in a cosmic-ray collision—which triggered the particle shower—indicated to Heisenberg during the late 1930s that the difficulties in quantum field theory could be resolved only if a universal minium length, a new fundamental constant, were introduced into the theory. But every attempt to build such a constant into available field theories had failed. In a 1938 paper and in his section of a later-published report with Pauli intended for the ill-fated 1939 Solvay Congress, Heisenberg argued that the new fundamental length not only should occupy a central place in the future field theory—accounting for the existence of elementary particles—but also should mark the "limits of applicability of present quantum theory."[3] For Heisenberg and Bohr over a decade earlier, classical mechanics had lost validity at the atomic scale of events, below which a new quantum mechanics was required. Now, according to Heisenberg, quantum mechanics itself broke down when applied to events occurring within regions smaller than the size of an elementary particle.

Despite these arguments, a great many calculations and equally numerous letters exchanged with Heisenberg throughout the 1930s had never really convinced Pauli of Heisenberg's program for the fundamental length and field theories. As in the 1920s, he believed that something entirely new was needed. Pauli had already suggested that Heisenberg, as he did when formulating the 1925 breakthrough to quantum mechanics, should focus only on observable quantities and attempt to exclude all unobservable variables from the theory. Heisenberg now attempted to do so, at the height of the World War. His effort led to what became after the war his widely studied new theory of elementary particles, the so-called S-matrix theory.

As the war raged in 1942, Italian physicist Gian Carlo Wick, who had previously worked with Heisenberg in Leipzig, returned to Germany for a visit. The visit stimulated Heisenberg's researches at a time when the uranium project was beginning to coast into the latter war years. After stopping in Munich to see Sommerfeld in June, Wick appeared before Heisenberg's last Leipzig seminar, consisting of just four students. He then accompanied Heisenberg to his new job in Berlin before returning to Rome in early July.[4] According to a footnote to Heisenberg's first publication, Wick brought with him reports on his own work as well as on work being conducted by Gregory Breit, Neils Bohr, Georg Placzek, and Rudolf Peierls along similar lines. Perhaps fearing competition, Heisenberg rushed his first S-matrix paper, titled "The 'Observable Quantities' in the Theory of Elementary Particles," to the *Zeitschrift für Physik* by early September.[5]

In his new approach, Heisenberg used his hypothesized fundamental length to define the allowed changes in the momentum and energy of two colliding high-speed elementary particles. This limitation would help identify the properties of the collision that were observable in present theories. Because the identified properties are measurable in laboratory experiments, they must also appear in any future theory. Heisenberg considered events to be observable if they occurred at distances larger than that set by the fundamental length. Those at smaller distances were unobservable. For two colliding particles, this yielded four sets of observable quantities with which to work: two of these were the properties of the two particles as seen in the laboratory long before they collide with each other; and two were their properties long after the collision. During the collision, they approach within a distance less than the fundamental length and are thus unobservable. These four sets of observable properties could be arranged in a table, or in this type of work, a matrix, which Heisenberg called the scattering- or S-matrix.

Although Heisenberg could not actually specify the four elements of the S-matrix, he demonstrated that it must contain in principle all of the information about the collision.

In his second paper, completed in October 1942, Heisenberg further showed that the S-matrix for several simple examples of scattering of particles yielded the observed probabilities for scattering. It also gave the possibility for his favorite phenomenon—the appearance of cosmic-ray explosion showers, an instantaneous burst of shower particles in a single collision.

As it had in 1925, Heisenberg's retreat to observed quantities had made for an entirely new approach to problems at the subatomic level. But unlike 1925, in the end no new revolutionary physics emerged. The reason was the inability to relate laboratory events with the events occurring within the small space of the collision where the particles come closer than the Heisenberg's critical length. This length marked the boundary below which all available theories failed. Heisenberg's S-matrix was, in effect, a theoretical orphan; it could not be derived from any present theory nor, thought Hciscnbcrg, even from a future theory. Thus, as Pauli, now in Princeton, put it in 1943, the S-matrix was "an empty concept."[6] But for Heisenberg, who could leap to solutions without bothering with the intermediate steps and without even articulating an underlying theory, it was, he believed, a concept that would reappear of necessity in the future theory, of whatever type. Despite the war and its devastating effects, the tantalizing prospects of a future theory inspired Heisenberg and others into the early postwar years to give these ideas serious consideration.

The most ardent postwar supporters of Heisenberg's new approach had learned of it directly from Heisenberg during his wartime travels. Gregor Wentzel and Ernst Stückelberg, who made extensive use of the theory, heard about it during Heisenberg's two wartime visits to Switzerland. Christian Møller, who brought the theory to the attention of Allied scientists after the war, corresponded with Heisenberg and discussed the theory extensively with him during Heisenberg's visit to Copenhagen in April 1944. And one evening in October 1943 Heisenberg presented his new theory to an informal colloquium in Kramers's home near Leiden in the German-occupied Netherlands. The university was closed at the time, and official colloquia were banned. Unlike the more unpleasant aspects of this visit, records and memories of the talk and dinner abound.[7] During the discussion of Heisenberg's talk, Kramers made the insightful remark that if the actual elements of the matrix could ever be determined without a complete theory, they would yield a so-called "analytic function"—that is, a function containing real and imaginary parts. (Imaginary numbers involve the square root of a negative number, a quantity that exists only in the imagination.)

Following the discussion, dinner, and the inevitable musical interlude that followed, Kramers worked through most of the night on Heisenberg's views. Back in Berlin, Heisenberg wrote immediately that he had grown "more and more enthusiastic" about Kramers's remark "because I believe that with it one can really arrive at a complete model of a theory of elementary particles."[8] Heisenberg suggested to Kramers that they collaborate on a paper on the subject, a suggestion he made at the same time as he was reporting to the Reich Education Ministry that scientific collaboration with the Dutch would be quite possible.

But Kramers declined to write the suggested paper with Heisenberg. He argued that the post was too slow between Leiden and Berlin for such work—a letter either way took about a month to arrive—but he did ask that his suggestions be acknowledged, which they were.[9] Apparently shaken out of an intellectual depression by Heisenberg's visit, Kramers now desired the intensity of direct collaboration. But Kramers probably also felt that it would be inappropriate to collaborate publicly with the German physicist. Heisenberg obviously did not agree. As he progressed with Kramers's idea during the following months, he repeated his inquiry about collaboration at least three times. Finally, as Heisenberg prepared to publish in March 1944, Kramers told him, "I feel that the moment is not right for a joint publication—the moment is indeed right for joint work, but I would have to give up my scientific program entirely for the next months."[10] Heisenberg submitted his third S-matrix paper, with an acknowledgment to Kramers. But just two days after declining to publish with Heisenberg, Kramers presented his thoughts on the theory to a Utrecht symposium.[11] He would publish, but not with Heisenberg.

Late in 1944 Heisenberg prepared a fourth installment of his theory—a paper dealing not just with two colliding particles but with many interacting particles, all giving rise to an even more complicated S-matrix. He presented the essentials of the paper in Zurich at the end of 1944, but the paper could not be published before the presses stopped at war's end. There was a sudden spurt of interest in the analytic S-matrix during the early postwar years, thanks to the work of Møller, Kronig, Wentzel, Harald Wergeland, and Heisenberg's own postwar summaries of his wartime S-matrix research.[12]

The enthusiasm for the S-matrix soon died, however, after Pauli's Princeton student, S. T. Ma, discovered S-matrix solutions (zeros) that did not satisfy Heisenberg's prediction that they would correspond with the observed elementary particles. Pauli renounced the S-matrix during the 1946 Cambridge meeting of the British Physical Society. Interest soon returned to the usual field theories when the process of so-called renormalization suddenly rendered quantum electrodynamics useful after all, even to very high energies and small distances of approach much shorter than the supposed fundamental length. The hitherto mysterious elements of the S-matrix could be calculated harmlessly from the old quantum field theory. Although there were many difficulties in applying the renormalization technique to other types of field theories, Heisenberg's original S-matrix program languished in the corners of interest until it experienced a brief, though intense, revival during the early 1960s.[13]

Back in the throes of war, as Heisenberg busily prepared his S-matrix papers and ingratiated himself and his physics with the authorities, Carl Ramsauer and the German Physical Society launched a new campaign in 1943 for even greater acknowledgment and appreciation of physics by the authorities. Ramsauer's strategy, involving "the self-mobilization of science," entailed all of the usual tactics employed by the scientists: memos to bureaucrats, personal diplomacy, and lectures to high-ranking officials.[14] Göring's German Academy for Aeronautical Research provided a convenient forum for the last of these devices, and academy member Heisenberg was a useful and willing participant in this forum.

Ramsauer's new plan, as he explained it to Heisenberg, consisted of a three-part "logic": the premise that German physics would be a decisive factor in Germany's future; the assertion that German physics "was quantitatively and qualitatively far surpassed by American physics"; and the conclusion that new measures had to be undertaken that would require far more than the current financial and organizational support available for research. Ramsauer presented his syllogism in a widely disseminated address to the aeronautical academy in early April 1943. One of his concluding recommendations called for "the most rational use of the existing physicists, particularly the setting up of new guidelines for the military use of physicists."[15] One month later Heisenberg, together with Otto Hahn, Hans Bothe, and Klaus Clusius, made the case before the academy for nuclear physicists in particular.

Organizing the lectures by the scientists required some diplomacy. By agreement among Albert Vögler, Rudolf Mentzel, and Abraham Esau, nuclear fission research had been split between the Kaiser Wilhelm Society (KWG), the Reich Research Council (RFR), and the Army Ordnance Office. Albert Speer's Ministry for Armaments and War Production and the KWG largely supported the work of the KWG institutes. But late in 1942, Göring had placed Esau in charge of all nuclear research outside Army Ordnance and had transferred Speer's KWG nuclear research to Esau's subsection of the RFR. Esau certainly did not appreciate KWG scientists, Heisenberg and his colleagues, acting as independent spokesmen for nuclear research.

For his part, by March 1943 Heisenberg equally resented Esau's attempts to direct his research. That month, Army Ordnance finally withdrew entirely from nuclear research and transferred control of Diebner's Gottow research team to the Reich Physical-Technical Institute (the German bureau of standards), also headed by Professor Esau.[16] Within days of Diebner's transfer, Esau confiscated for Diebner's use about 600 liters of heavy water that Heisenberg had stored in his institute's air-raid shelter for the B-6 series of large-scale experiments. Esau still had to be placated, however, and to neutralize his objections to the scientists' lectures before Göring's aeronautical academy, Heisenberg proposed that Esau chair the meeting, which he did. Then, to prevent his further meddling, Heisenberg pressured Speer to have Esau removed. Speer in turn pressured Göring and Mentzel, who replaced Esau at the end of the year with a man more to Heisenberg's liking, the Munich experimentalist Walther Gerlach.[17] Heisenberg and the physicists had clearly regained their influence.

Heisenberg's lecture to Göring's aeronautical academy several months earlier, on May 6, 1943, bore the relatively neutral title "The Acquisition of Energy from Nuclear Fission."[18] As in his 1942 talk to Speer, Heisenberg performed a balancing act before the scientific academy and the larger audience of regime officials who would hear or read his talk. On the one hand, Heisenberg revealed privately, he wanted to emphasize the practical significance of nuclear research "in order to help the work achieve a favorable status as well as practical support."[19] On the other hand, in view of the shocking debacle at Stalingrad and the Allied bombing raids that were systematically reducing German cities to smoldering rubble, he did not want the unlikely prospect of a new bomb in the foreseeable future to result in either the project being taken away from the scientists or the scientists being ordered to produce what they could not soon deliver.

Heisenberg deftly steered between the two extremes. Without mentioning the plutonium alternative, he argued that fission energy could be obtained only by using natural uranium with a moderator or by enriching the U-235 content of uranium by complicated—and expensive—processes. He admitted that if enough U-235 could be separated from natural uranium and concentrated into a small enough ball, fission would take place almost instantaneously, whereby "a correspondingly large amount of energy is released explosively." This comment was the closest he came to reminding his audience of the bomb. The remainder of his talk focused on the development of a uranium "burner," but his conclusion covered both possibilities—a burner and a bomb —while again dampening any undue optimism about the latter by noting the enormous technical and practical difficulties yet to be overcome. Atomic energy in large amounts was technically feasible, he declared, but "on the other hand the practical realization of this goal would naturally face great difficulties in this tense wartime economy."[20]

Heisenberg's lecture and the physicists' campaign had the desired effects. Steady funding and a sufficient "urgency classification" to keep his nuclear project going under the severest wartime conditions were ensured for the duration. Although defeat loomed larger and Hitler began touting the imminent use of *Wunderwaffen*, secret weapons that would turn the tide of war, pressure for immediate production of a bomb and efforts to wrest control of the project from the scientists surrounding Heisenberg had been discouraged. The scientists even achieved further favors from the regime: better coordination of all war-related research through a new planning office in the Reich Research Council and temporary exemption from military service for as many as 5,000 scientists, engineers, and students to allow them to pursue war-related research and studies.[21]

The preservation of young German scientists for the future was one reason offered after the war by Max von Laue to explain the willingness of German scientists to continue research on the project under Hitler. The opportunity to use his influence to protect young scientists was also one reason Elisabeth Heisenberg gave for her husband's decision to accept the prominent Berlin posts.[22] After the war, however, criticism surfaced regarding Heisenberg's apparent failure to use his influence to the fullest to rescue or to protect as many people as he might have. One does indeed wonder why Heisenberg and his colleagues did not turn their influential associations with such powerful Nazi figures as Himmler, Speer, and Göring to more decisive efforts at rescuing endangered persons. One postwar critic said of Heisenberg years later: "He saved physics; he did not save physicists." Kramers's biographer writes that Kramers asked Heisenberg several times to intercede for friends interned in concentration camps but that "there is no indication that this had any effect."[23]

Throughout his life, Heisenberg saw himself as primarily responsible only for his own circle of friends, colleagues, and students. Although he had taken on the task of preserving decent German physics, he did so primarily within his own professional circle and through his own personal survival and advance. This trait was evident already in his perception of himself as the leader of his youth group, in his reaction to the dismissal policy of the early Nazi regime, in his desire to preserve an island of students and assistants in Leipzig, and in his efforts to extract from the authorities military exemptions for physicists (but not nonphysicists) during the war. It reached its extreme in his wartime manuscript "The Order of Reality," wherein he viewed the individual as a helpless pawn of historical forces. The ideological and political blinders of the Third Reich had continually encouraged such a distorted perception, to which Heisenberg was already prone. At the same time, with his rehabilitation achieved at last, they encouraged Heisenberg and his colleagues indeed to emphasize professional matters above all else, as if these were somehow independent of the fates of individual practitioners.

Although the available written record is again of varying insight and reliability, when Heisenberg did act, he did so as he had in past situations—usually offering too little, too late. With concerted, large-scale efforts out of the question, records do indicate at least five attempts by Heisenberg to save individuals from threatening situations. Two postwar affidavits by Heisenberg refer to four other cases, and Elisabeth Heisenberg recalls two more instances—all in addition to the Edwin Gora episode discussed earlier. All of the cases not involving institute members appear to have occurred during or after 1943, when Heisenberg felt he had gained the full confidence of the regime. Whether for lack of sensitivity to the dire predicaments of those outside his circle or for lack of comprehension of his potential options for action, or both, his efforts seem from this distance pitifully weak as responses to the life-and-death situations of those he sought to help.

The records show Heisenberg making a variety of brief approaches to different officials on behalf of endangered persons. In March 1943, he forwarded a letter to Himmler from a Münster mathematician on behalf of the French mathematician Elie Cartan.[24] He also encouraged an SS supporter in his institute to act on behalf of a Dr. Wetzel imprisoned in Stuttgart for incautious remarks; he appealed to Army Command Headquarters via the Kaiser Wilhelm Society to have returned from the front a promising physics student who was also an infantry officer; he worked through his Foreign Office roommate in the KWG's Harnack House to try to obtain the release of a foreign scientist imprisoned by the Gestapo. In the last of his S-matrix letters to Heisenberg in 1944, Kramers requested Heisenberg's help for a young jurist friend of the family who had been imprisoned in Buchenwald. Heisenberg responded, "Whether I can do something for Herr D. is unfortunately very doubtful, but I will try."[25] There is no indication of what he attempted on the young prisoner's behalf, nor of what became of him.

The most tragic and far-reaching of the situations in which Heisenberg attempted to help involved Samuel A. Goudsmit's aged parents. Goudsmit, one year younger than Heisenberg and the Dutch co-inventor of electron spin, was still an awestruck physics student when he first met the great Heisenberg during the period in 1925 when Heisenberg was formulating the foundations of matrix quantum mechanics. Since the two physicists worked along similar lines, they occasionally corresponded during the intervening years. After receiving his doctorate under Paul Ehrenfest in 1927, Goudsmit left

Samuel A. Goudsmit.
(COURTESY ESVA, AIP NIELS BOHR LIBRARY)

the Netherlands for the University of Michigan. He and Heisenberg often met there when Heisenberg went to Michigan to visit or to lecture at the Ann Arbor summer school in physics. They had met there last in the summer of 1939.

Although Goudsmit had moved permanently to the United States, his Jewish parents had remained behind in The Hague. Goudsmit obtained American visas for them to enter the United States, and they had just received their travel papers when tragedy struck. During the Nazi deportation of the Dutch Jews in 1943, the Goudsmits were taken out of their home, loaded onto a cattle car, and transported to the Auschwitz death camp. Goudsmit's friend and colleague Dirk Coster, who had been instrumental in rescuing Lise Meitner after the annexation of Austria in 1938, appealed for help to Heisenberg by letter. Heisenberg responded by writing a letter to Coster on February 16, 1943, to be shown to the authorities. In it, he described Goudsmit's friendly hospitality to visiting Germans and his own anxiety regarding the safety of Goudsmit's parents.[26] It is uncertain what effect Heisenberg expected this letter to Coster in the Netherlands to have in rescuing the Goudsmits from Auschwitz. Perhaps he did not know that they had already been deported. There is no indication that Heisenberg tried any other avenue of rescue. Possibly due to the notoriously slow post between the Netherlands and Germany, Heisenberg wrote his letter too late. Five days before the date of Heisenberg's letter, Goudsmit's father and his blind mother had died in the Auschwitz gas chamber, on his father's seventieth birthday.

Goudsmit returned to the Netherlands less than a year later. He was now the scientific head of the Alsos Mission, the secret Allied science intelligence unit that, by then, was after the German nuclear energy project and its leading scientist—Werner Heisenberg. Standing in tears in the wartime ruins of what had been his boyhood home,

Goudsmit wrote several years later, "I was gripped by that shattering emotion all of us have felt who have lost family and relatives and friends at the hands of the murderous Nazis—a terrible feeling of guilt."[27] It was also a feeling of rage, rage at the Germans and surely rage at Heisenberg for having failed to help. He did not mention Heisenberg's letter in this passage. It is uncertain when he learned of it, but even if he did know at the time, it would not have altered his influential assessment of Heisenberg after the war as a great physicist with a deep and tragic character flaw. Even years later, Goudsmit's opinion of the episode had not changed much. In an obituary for Heisenberg, written in 1976 for the American Philosophical Society, Goudsmit was much kinder but still angry with the late physicist: "Heisenberg was asked in 1943 to intercede in the case of acquaintances who were being sent to a concentration camp. He responded merely with a vague letter. I doubt that he could have done anything else. I doubt that I or most of the physicists I know would have done better under the same circumstances."[28]

While the restored recognition accorded Heisenberg and other nuclear scientists did not greatly affect their rescue of threatened persons, it did noticeably affect their outlook regarding the nuclear research effort. Heisenberg and his colleagues were now confident of their nuclear science, of their control of their profession, and of their lead in utilizing nuclear energy. While Allied scientists, with the much greater material and organizational support of the United States government and military, were now settled in the New Mexico desert constructing the uranium and plutonium bombs, Heisenberg and the German scientists were convinced that the Allies could not have progressed any further in applications of nuclear fission than the Germans had. In a report to Mentzel and Göring shortly after the aeronautical academy lectures in May 1943, Abraham Esau, in his last days as head of nuclear research in the Reich Research Council, drew that conclusion from the scientists' papers and lectures. Transferring the report to Göring, Mentzel echoed Esau in declaring that, while nuclear research for technical reasons would not lead soon to a machine or an explosive, as Heisenberg had argued, "it is certain that the enemy powers could not present us with surprises in this field."[29] The continual loss of physics personnel following Hitler's rise to power and Ramsauer's case for the overall superiority of Anglo-American physics as a motive to build up German physics notwithstanding, Esau's report and subsequent events make it clear that German physicists could not believe themselves inferior in the field where they had predominated for so long—especially now that they had won new respect from their superiors. And, by the same token, the "enemy powers" believed until late 1944 that German science had indeed advanced at least as far as Allied research. For the Germans, the long struggle to preserve German physics could not have been for nothing.

CHAPTER 27

ONE LAST ATTEMPT

HEISENBERG DID NOT DEVOTE AS MUCH ATTENTION TO NUCLEAR TECHNOLOGY AFTER moving to Berlin as he had before. Nevertheless, as head of the Berlin pile research team, he did set the research program, and by the end of July 1942 he had settled on a new large-scale reactor experiment that he hoped would approach criticality. It would consist of 1.5 metric tons of heavy water and 3 metric tons of uranium metal plates arranged in horizontal layers within a cylindrical metal tank. Heisenberg had based the experiment's design on the results of his preliminary Leipzig and Berlin attempts. He also wanted to compare the results of this arrangement with its theoretically predicted properties.[1] Heisenberg's research program remained on course for nearly two and a half years. But during that time the course appeared increasingly at variance with the promising reactor designs of his chief rival in reactor construction, Kurt Diebner.

Working at the army's Gottow weapons research station, the inventive Diebner had hit upon the alternative idea of arranging the uranium metal, not in plates, but in cubes suspended in a cylindrical tank of heavy water. This allowed more contact between the uranium and the heavy water; presumably, more of the neutrons released in uranium fission would thus be slowed by the heavy water to energies enabling further fission of the rare U-235 rather than captured by the more plentiful U-238.

Diebner's first attempt, using frozen heavy water to support the metal cubes, yielded about 36 percent neutron multiplication—three times the best Leipzig result. Any multiplication indicates that more neutrons are produced than absorbed; the greater the multiplication, the closer to a chain reaction. In his second attempt, undertaken in 1943 under the auspices of Esau's Physical-Technical Institute, Diebner suspended his uranium cubes on thin wires in liquid heavy water—some of which had been commandeered from Heisenberg's institute. This contraption yielded nearly 110 percent multiplication.[2] More cubes and water might go critical. But Diebner's experimental destiny collided again with Heisenberg's when Allied bombers leveled the Degussa Company, the manufacturer of Diebner's uranium cubes. Degussa's subsidiary, the Auer Company, was the sole remaining producer of uranium metal, and Auer already had a long-standing contract to deliver Heisenberg's metal plates.

Heisenberg had made a brief evaluation of Diebner's Gottow experiments in early 1943. He grudgingly admitted during his aeronautical academy lecture in May that, compared with the Leipzig layer arrangement, the Gottow group offered "a

somewhat improved apparatus . . . in which the neutron multiplication was somewhat higher."[3] But he remained confident of his planned plate experiments.
Apparently he even promised Vögler, the head of the Kaiser Wilhelm Society, that the
"uranium machine" would be up and running by the end of the year—if the plates
were delivered soon. They were not. Mentzel of the Reich Research Council, an Auer
representative, and a now lame-duck Esau were still arguing at the end of the year
over production priorities. They finally agreed that "production of cubes for Gottow
should not interfere with plate production."[4] Heisenberg would have his plates, but
because of his low war-priority rating, they did not arrive until January 1944.
Heisenberg did not give up the less-efficient plates for Diebner's more fissionable
cube arrangement until nearly a year later.

By the time the uranium plates finally arrived in Berlin, Vögler was beginning to
wonder what had become of the promised machine. Karl Wirtz, the head of the actual
pile construction in Berlin, had only just begun assembling the first of the B-6 experiments in the Berlin institute's newly completed underground bunker. He and his team
were assisted by several members of Bothe's Heidelberg group. They moved to Berlin
after Bothe decided to concentrate instead on cyclotron construction. The elaborately
equipped bomb- and radiation-proof bunker in which they now worked had 2-meter-
thick walls of iron-reinforced concrete. Inside were a main laboratory with a water-
filled pit, rapid air and water pumps in case of accident, a workshop, remote-control
apparatus for handling radioactive materials, and heavy-water tanks with a purifica-
tion system—but no deliberate protection from radiation, or even a means to shut
down the reactor should it go critical.[5]

Wirtz and his team arranged the one-centimeter-thick uranium plates in the mag-
nesium-alloy cylinder (1.24 meters wide by 1.64 meters high) and lowered the vessel
into the pool of water, which acted as an absorber and reflector of neutrons. Then they
filled the vessel with heavy water and measured the neutron flux as a function of
radius when a small constant neutron source was lowered into the center of the
device. After four trials, they found an optimal separation of 26 centimeters between
the five uranium plates, which resulted in an encouraging 206 percent multiplication.
Wirtz and Heisenberg repeated the configuration as experiment B-7 in December
1944. This time, however, instead of using water as the main reflector and absorber
of neutrons, they wrapped graphite around the container; it was the first time graphite
was used in a German reactor experiment. The neutron multiplication factor rose even
higher, but not enough to hope that similar pile configurations would ever lead to a
critical state. The plate geometry had failed to achieve a chain reaction, but
Heisenberg and Wirtz, who would have suffered life-threatening radiation if it had
gone critical, could console themselves that "[w]ith the layer experiments a satisfactory agreement was achieved with the theory, which was ever more refined in the
course of time (Heisenberg, Weizsäcker, Höcker)."[6]

Meanwhile, the intensified bombing of Berlin was having an increasing impact
on Heisenberg and his project. By the summer of 1943 the Allies were conducting

round-the-clock bombing raids. Max von Laue wrote to Heisenberg from Pomerania describing the "uncanny experience" of hearing bombers flying overhead toward Berlin for three solid hours—"uncanny mainly because nothing was done against them."[7] In July, Speer ordered all war research institutes to identify places to which they could move should the air raids make work impossible. Although the nuclear project was safely nestled in its new concrete bunker, there was no guarantee of a steady flow of electricity, water, and supplies to keep the project and its members going, nor were there living quarters within the bunker for all the project members. Heisenberg looked for a safe haven to the south and west in order to be closer to his family in Urfeld and closer to the Western Allies than to the Soviets in the east when the war finally ended. Physicist Walther Gerlach, who would replace Esau as administrator of nuclear research, had earlier taught in the university town of Tübingen to the southwest. Apparently on a tip from him, Heisenberg learned of a group of peaceful little villages nestled among the hills of the Swabian Alps region of the Black Forest, just south of Stuttgart and Tübingen.

As conditions worsened daily in the summer and fall of 1943, Heisenberg decided to dispatch to the south all personnel who would not be needed for the forthcoming B-6 series of experiments. By the end of the year, about a third of his 55-member Kaiser Wilhelm Institute staff, including assistant director Max von Laue, were settled in the Black Forest. They moved into a large and nearly vacant textile factory at Weiherstrasse 1 in the picturesque town of Hechingen, where they set up offices and rooms for the measurement of materials and the construction of apparatus.[8]

The institute members were not alone for long, however. Otto Hahn, the codiscoverer of fission, and his staff moved to nearby Tailfingen after Hahn's Berlin chemical institute suffered a direct hit and burned to the ground. Weizsäcker and his family escaped to Hechingen just ahead of the Allied bombing and capture of Strasbourg. But not until January 1945 would Heisenberg, Wirtz, and the remaining staff move permanently to their Black Forest retreat.

The fissioning of his institute and the uprooting of his family put a new strain on Heisenberg. When not traveling to German-occupied countries or writing his S-matrix papers, Heisenberg shuttled for the next year and a half by train and auto between his three widely dispersed homes in Berlin, Hechingen, and Urfeld. The strain increased with the difficulties his family encountered in their new quarters. No sooner had Heisenberg's wife, their now six children, and their live-in relatives in Urfeld recovered from the December bout with scarlet fever than they had to contend with a partially collapsed roof caused by a heavy snowfall. The mayor of Kochel, the nearest large town, was unable to release materials for the repairs, so Heisenberg had to obtain permission from the mayor of Leipzig to transport roofing tiles from the ruin of their Leipzig house to Urfeld.[9] In addition to the distrust of the Berlin professor's family by the local Bavarian peasants, Elisabeth Heisenberg recalled the constant hunger and illness of their Urfeld existence, the latter caused by the cold (the house was intended only as a summer residence) and the former by the impossibility of vegetable farming (the house was situated on a rocky slope near Lake Walchen, at the foot of a mountain).

At first Heisenberg had intended to move his family to Hechingen, as other institute staff members had done, but the Berlin researchers who had descended on the small town ahead of him had snapped up all the available housing, leaving Heisenberg himself with nowhere to live but a rented room in the home of a befriended family. The Heisenberg family furniture, which had been trucked to storage in Berlin, now landed in the basement of the Hechingen family's house. Isolated from the institute families and denied Hechingen's agricultural advantages, Elisabeth came to resent her husband's "Swabian idyll," which, she wrote, "was always a slight cause of dispute between us." Heisenberg's musical concerts for the Hechingen residents surely did not help matters.[10] As life became daily more dangerous and as Germany careened toward its inevitable defeat, the couple's brief moments alone together in their Urfeld "eagle's nest" grew ever more precious. Two weeks before the not-unexpected D-Day invasion (of occupied France, on June 6, 1944), Heisenberg wrote to his mother from Berlin: "We experience each beautiful day that is given to us as a gift from the Good Lord, for which we are thankful," and when he and Elisabeth were able to be together for a few days in Urfeld and listen to the play of the children, "we don't want to think of anything but this happiness, for these could indeed be our last moments together."[11]

Back in Berlin, Heisenberg welcomed the news of the D-Day invasion not for its likely ending of the Reich but for its hastening the end of the war "one way or the other." Later, in Hechingen, he waxed romantic in his thoughts about the future. After the war ended, he opined, "the sun will continue to shine as it has before, we will be able to make music and to do science, and whether or not we live richly or modestly, it will make no great difference." But by the end of 1944, as Germany's prospects grew grimmer, Heisenberg poured out to his mother in a revealing letter his gratitude for his past life and his feeling of helplessness before a future that might well bring their deaths:

Even when I take into account all of the misfortune that surrounds us today and that has existed in my life as in everyone's life, I have been on the whole unbelievably lucky and I am thankful that I could be so long on this remarkable and often so wonderfully beautiful earth. I would be happy if I could still see my children grow up, if I could once again experience a harmonious life and be able to work in it. But even if that should not be determined, I want to be thankful for that which fate has given me. . . . I have the feeling that I still have many tasks to perform here, but none of us knows how he will get through the last and strongest blast of the hurricane that still stands before us. In any event, even here I gladly give my life trustfully into the hands of that higher power that has led it until now.[12]

As British, French, and American troops smashed their way inland from the Normandy beachhead, and as British and American scientists worked feverishly at Los Alamos to assemble the first atomic bombs, Allied intelligence agents scrambled for reliable information about the progress of German nuclear research. Through reports received from Debye and refugee scientists, as well as from inside sources, Allied scientists were only too well aware of the secret research under way at the Kaiser Wilhelm Institute in Berlin and of Heisenberg's role as head of the main reactor project. As early as 1942, the danger that Germany might actually succeed in building a bomb had seemed so great that two of Heisenberg's former colleagues had suggested that the Allies kidnap him when he went to Zurich to lecture at the end of the year.[13] At the very least, they had suggested, Wentzel or Wick should "interview" Heisenberg in Zurich to extract whatever information they could about the German nuclear effort. Nothing came of either suggestion; at the end of 1942 the Allies had only just begun to establish an international intelligence network. But the suggestions did apparently inspire special interest in Axis science, especially regarding nuclear fission.

In 1942, soon after establishing the Office of Strategic Services (OSS), the precursor of the Central Intelligence Agency (CIA), the United State's General William (Wild Bill) Donovan began recruiting agents for atomic intelligence. His top atomic spy was Morris (Moe) Berg, a man of many talents. An erudite, multilingual Princeton graduate who was familiar with physics, Berg had played until 1942 as a catcher for the Boston Red Sox baseball team. He came to Donovan's attention through his propaganda broadcasts to Japan—in Japanese—and through his diplomatic efforts in Latin America to counter anti-American Nazi influence. Donovan recruited the catcher to catch European nuclear intelligence.[14] Berg's first assignment was in Yugoslavia; he was next assigned to Italy in 1944, where he was to operate behind enemy lines to capture nuclear physicists Wick and Edoardo Amaldi in order to determine the progress of Axis science. Capturing the two proved unnecessary after the Allies broke through the Gustav Line and liberated Rome just before D-Day. Berg was in Rome within days interrogating his subjects about German research.

Berg's experience and abilities made him a natural to focus on German affairs. But OSS activities conflicted with those of other agencies of the armed forces, and Berg's mission to Italy conflicted with that of the Alsos mission. General Leslie Groves, the military chief of the Manhattan Project, had already dispatched the small Alsos mission (*alsos* is Greek for "grove") to Italy to capture Italian atomic scientists and materials as the Allied front lines advanced and to obtain from the Italians as much information as possible about a German bomb.

In order to minimize conflicts, Groves's superiors placed him in command of all American nuclear intelligence. Late in 1944, Groves assigned Berg to Allen Dulles's OSS office in neutral Switzerland, and he dispatched a reconstituted Alsos mission to London for the Normandy invasion. As it had been in the past, Groves's mission was under the field command of Colonel Boris T. Pash, a Russian-American veteran of anti-Soviet battles known for his often ill-considered bravado. But the new Alsos unit included a scientific section for the first time—headed by Samuel A. Goudsmit. Goudsmit was chosen for his familiarity with European physicists, physics, and languages and for his lack of familiarity with the Manhattan Project, should he be captured. Accountable only to Groves, the small entourage of the Alsos mission rolled across northern Europe with the Allied armies, confiscating and examining every scrap of material even vaguely related to German science and even vaguely hinting at the whereabouts of the premier German physicist who had failed to help when needed.[15]

Berg, working in Zurich independently of Goudsmit and Pash, had already obtained a lead on Heisenberg's location. Because Switzerland, while neutral, was surrounded by Axis countries, spies of every stripe flooded the country during the war. According to one count, Germany alone had 23 organizations operating in northern Switzerland.[16] Berg established a liaison with Paul Scherrer, a professor of experimental physics at the Zurich Polytechnic, and a fervent anti-Nazi, who was eager to help Berg in every way he could. From the postmark on a letter from Heisenberg to Wentzel in 1944, Berg learned of the Hechingen outpost and relayed the information to Goudsmit. But Berg's greatest service to his country was his role in an operation even more radical than the kidnapping proposed earlier. In a later account to a friend, no doubt embellished, Berg claimed that he was ordered to have Scherrer invite Heisenberg to lecture in Zurich in December 1944. At the slightest hint that Heisenberg was constructing an atomic bomb, Berg, standing ready with a loaded pistol, was to assassinate the scientist.[17]

Although Scherrer may not have known of Berg's intentions, the first part of the scheme went off as planned. Heisenberg knew Scherrer well from their common interests in cosmic rays and from an earlier wartime visit to Zurich at Scherrer's invitation. Heisenberg readily accepted the second invitation but insisted that he would lecture only on a nonpolitical subject. He knew he would be carefully watched by spies on both sides of the war and in general avoided public lectures on political issues. He chose to lecture on the fourth installment of his S-matrix theory.

Carl Friedrich von Weizsäcker, whose wife was Swiss, accompanied Heisenberg to Zurich, where about 20 people, including Berg and several pro-German Swiss scientists, attended the lecture in the Polytechnic's physics institute, an institute where Einstein had once studied and taught and from which Heisenberg's colleague Pauli had fled to Princeton. Although Heisenberg was now practiced in treading warily before a lecture audience, he apparently managed to get himself into trouble at a private dinner party in the Scherrer home. Berg said he sat next to Heisenberg with open ears and a loaded pistol. But he was disappointed: Heisenberg's main indiscretion, later reported all the way up to President Roosevelt, was a defeatist remark about Germany's failing war fortunes. This time there was no talk of a future Europe "under German leadership," but according to Goudsmit, who apparently heard it from one of the Swiss scientists, Heisenberg supposedly did make an equally incriminating remark: "How fine it would have been if we had won this war."[18]

Pro-German spies reported Heisenberg's defeatist remark to the Gestapo, which brought it to the attention of the Berlin SS. Under the conditions of German total war, defeatism of any sort was construed as treasonous and could lead to the same fate as that meted out to the White Rose, the group of Munich student antiwar protesters—

execution. Unbeknownst to Heisenberg's family in Urfeld, the SS planned a full investigation of both Heisenberg and Carl Friedrich von Weizsäcker. SS officer Mentzel informed Gerlach, and Gerlach warned Heisenberg. Fortunately, Gerlach was able to defuse the issue. When an SS general appeared in Gerlach's office to lodge the complaint against Heisenberg and Weizsäcker, Gerlach feigned horror at the charge and promised that Heisenberg would receive a severe reprimand for his behavior. That apparently satisfied the general.[19] Heisenberg had once again eluded danger and death. But the catcher in Zurich had caught enough to steer Goudsmit and Pash to the area of the Black Forest just south of Tübingen.

During the weeks following Heisenberg's Zurich trip in December 1944, the massive air raids on Berlin continued without cease. Near panic gripped the city as the Soviet army rolled relentlessly westward. Amidst constant bombing, mounting rubble, and frequent power failures, Gerlach, who had moved his offices from Munich to Heisenberg's Berlin institute, finally ordered all nuclear pile research transferred out of Berlin. During his year as Göring's so-called "plenipotentiary for nuclear research," Willy Wien's successor as Munich's professor of experimental physics had supported both of the competing pile designs, Diebner's cubes and Heisenberg's plates, until Heisenberg finally acceded to the "more favorable arrangement." He claimed he wanted to see which of the two designs would prove more likely to produce a chain reaction before the war was over.[20]

With the European war just three months from its conclusion, at the end of January 1945 Wirtz and the remainder of Heisenberg's Berlin team had assembled their largest pile experiment to date: hundreds of cubes cut from the B-7 plates and suspended on aluminum wires from the lid of the reactor cylinder, which was then filled with the institute's 1.5 metric tons of heavy water. The vessel lay, wrapped in a mantle of pure graphite, in the institute's bomb-shelter water pit, ready for neutron multiplication measurements that would constitute experiment B-8.[21]

No sooner had they assembled this appliance, however, than Gerlach ordered it and Diebner's contraption dismantled and shipped south—better to have a slight delay in research than to have men and materials fall into Russian hands. Gerlach, Wirtz, and Gerlach's newly appointed assistant, Diebner, in German army uniform with a revolver strapped to his hip, left Berlin the next day accompanied by several trucks. They headed for Hechingen, where Heisenberg awaited them, but they got only as far as Diebner's new outpost in Stadtilm in the province of Thüringen, about halfway across Germany toward Hechingen in the southwest. Gerlach abruptly decided to stop there and have the apparatus reassembled under Diebner's direction in the desperate hope of achieving a chain reaction as soon as possible. A worried telephone call from Wirtz to Heisenberg in Hechingen brought Heisenberg and Weizsäcker to Stadtilm as quickly as the Allied bombing and strafing of Germany's transportation system would allow. Erich Bagge's arrival from Hechingen with a convoy of trucks to transport Heisenberg's uranium and heavy water the rest of the way finally convinced the plenipotentiary of the preferability of a Heisenberg reactor to a Diebner reactor. Diebner had won Gerlach's favor for his innate experimental abilities, but Heisenberg was the more powerful. And with trucks at Heisenberg's disposal, Gerlach apparently did not want to risk further delay due to an unseemly fight over materials. Heisenberg's equipment and materials finally arrived in Hechingen at the end of February 1945, four weeks after they had left Berlin—and just two months before the end of the war. Within a month, according to Karlsch, Diebner's team managed to detonate a crude nuclear device on a military training field near Ohrdruf in Thüringen. Such an achievement seems impossible, though perhaps not for lack of trying.[22]

During the final months of the war in Europe, Heisenberg's team, together with Weizsäcker and members of Bothe's group, worked feverishly on what would be their last attempt to achieve a critical reactor. The work took place in the nearby picturesque village of Haigerloch, which, as Gerlach knew from his earlier days in Tübingen, offered ideal protection for the experiment. A huge rock formation, topped by a Renaissance church and monastery, dominated the center of the town. At the base of the rock a small cave had been dug horizontally into the side of the rock to serve as a wine cellar for the local innkeeper. It now served the scientists as an air-raid-proof "atom cellar." Advance teams from Berlin had already disposed of the innkeeper's wine (one way or another), enlarged the cave, dug the water pit, attached water and power cables, and assembled the winches and heavy equipment needed to handle Germany's last pile attempt.

Even as Germany disintegrated in devastation and chaos around them, the scientists worked calmly and steadily at their task. The physicists' successful campaign in previous years to enhance the status of their science now worked to their advantage. Himmler had already ordered the release of another 14,600 scientists from active military duty; and Bormann excused nuclear researchers, including Heisenberg, from all but minimal participation in the Volkssturm, the people's army, Hitler's last line of defense and a vehicle of his last desperate attempt to maintain control.[23]

Ironically, amid total war and facing imminent defeat, the scientists and the Nazi leaders suddenly committed themselves to a small-scale technical research effort that at that stage seemed to offer no additional practical benefit to the war effort. At the beginning of the war, they had believed that nuclear engineering research could give them a new war machine and possibly even a new and immensely powerful explosive. Now, at the end of the war, they hoped that the simple experimental model in Haigerloch would go critical before the collapse. The motives for this hope among both the scientists and the Nazi officials are difficult to pin down. In addition to scientific curiosity and the exhilaration that success would bring in the face of defeat, several indications suggest that both groups were operating on the mistaken belief that German research had advanced much further than had Allied research. Probably the Nazi leaders hoped to use the secret of nuclear fission as a bargaining chip in negotiating a conditional surrender with the Allies. They were in for a nasty shock when the Allies expressed absolutely no interest in such an offer.

Goudsmit (CENTER, REAR) *and the Alsos mission dismantle
the last reactor attempt in Haigerloch, 1945.*

(COURTESY ESVA, AIP NIELS BOHR LIBRARY)

The scientists, especially Heisenberg, were also looking to the future in the last months of the war. They believed that Germany, as it had in the years following the defeat of World War I, would look again after this war to its leading science—and in particular to its leading scientists—as the remaining pillar of German competitive greatness, whatever defeats and humiliations might be suffered in other areas. Heisenberg wrote to a former student in April 1944, "It is indeed very important that after the war we take part once again in the competition of research."[24] Heisenberg also wanted badly to reassure himself that, as a successful scientist producing useful results—as would be proved by a working reactor—he and his profession would not be disregarded and abused, as they had been under Hitler, by whatever regime succeeded Hitler. These goals would be realized if they could only achieve a chain reaction before the war reached its inevitable end.

By early March 1945 Heisenberg, Wirtz, and their team of technicians began the final assembly of experiment B-8 in the innkeeper's "atom cellar." Two weeks later they winched the graphite-covered lid supporting the chains of uranium blocks into place over the cylindrical reactor vessel, then slowly filled the vessel with heavy water. As the pumping progressed, the neutron multiplication rate increased. It seemed to Heisenberg and Wirtz that at long last the pile might go critical. Then, in the middle of their excitement came the sudden realization of their extreme peril—they had ignored all but the most rudimentary safety precautions. A block of neutron-absorbent cadmium was at hand, ready to be tossed into the tank should the reaction get out of control. Only now did the scientists begin to wonder seriously if that would be enough to halt the reaction in time. It is testimony to their determination—and desperation—that no one tried to stop the experiment. Everyone was determined that Germany should be the first to achieve a sustained chain reaction, regardless of the danger to themselves.

They watched nervously as the remaining heavy water flowed unchecked into the tank—but, alas, it was not enough. The experiment had yielded the highest multiplication rate yet, 670 percent, but Heisenberg quickly calculated that they still needed nearly 50 percent more uranium and heavy water for the reaction to become self-sustaining.[25] Perhaps more of both could be found at Diebner's Stadtilm outpost—but it was too late. American troops were already advancing through Thüringen in east-central Germany; by early April they were within miles of Stadtilm. Diebner finally abandoned his outpost on April 8 and headed south to join his mentor, Walther Gerlach, who had retreated to Munich.

On that same day, U. S. Secretary of War Henry Stimson met with General Groves in Washington to decide what to do about the German scientists in the south.[26] Since entering northern Europe in 1944, the Alsos mission had absorbed every available bit of information about the German project from Joliot's Paris laboratory and from the papers left behind in Weizsäcker's hastily evacuated Strasbourg institute. Goudsmit, Pash, and the Alsos team then crossed the Rhine with the Allied armies in February 1945. At the end of March, as experiment B-8 lay in its cave-protected water pit, the Alsos mission entered the old university town of Heidelberg, situated northwest of Stuttgart. As the U. S. Army set up a forward command post, which remained in place for decades, the Alsos team established its Advance Base, South.[27]

After seizing and interrogating Bothe and Wolfgang Gentner in Heidelberg, Goudsmit and his staff contemplated their next move. By then they knew the locations of all target scientists and laboratories and had relayed to Washington their conclusion that Hitler's promised secret weapons did not include an atomic bomb.[28] Groves demanded that they be absolutely certain and that they capture all the remaining project members. Unfortunately, most of their targets were within the areas earmarked at the Yalta meeting of Allied leaders for invasion and occupation by French and Russian forces. Mission priorities suddenly shifted, as a result: instead of gathering intelligence in an attempt to thwart the German bomb effort, the Alsos team was now bent on snatching up German scientists, papers, and equipment before the Russians and the French could take them into custody.

The three Western Allies had agreed that the entire region south of Stuttgart should be occupied by the French, but Colonel John Lansdale, attached to the Alsos mission, had different ideas, as he reported several weeks later: "Our feeling was that the individuals and materials down there should be seized by the Americans in advance of the French, or if that were impossible, destroyed to the fullest extent."[29] Atomic scientists and equipment were simply too valuable to allow them to fall into the hands of any other nation, even the French. Groves and Stimson, meeting in Washington, considered a full-scale American invasion of the south. An army operations commander, approached in Heidelberg by Colonel Pash, recommended an airborne assault or, at the very least, the carpet bombing of the entire region. But the French were advancing too rapidly for either plan to be enacted. In the end a local Heidelberg commander assigned a combat engineer battalion to the ever-zealous Colonel Pash, who took off immediately with a convoy of jeeps and armored cars and arrived in Haigerloch and Hechingen less than an hour after the French forward line swept through the area on April 23.

Pash and his men promptly set to work. They arrested Wirtz, Bagge, Weizsäcker, and Laue, found and confiscated their papers and equipment, began dismantling the Haigerloch pile, and blew up the alloy containment vessel of the last German reactor experiment. They then moved to Tailfingen to arrest Otto Hahn, the codiscoverer of fission. By the time the French commanders realized what was happening, the Germans' heavy water and uranium were on their way to Alsos mission headquarters in Paris, and the prisoners and their papers were on their way to Heidelberg for interrogation and study by Goudsmit and company.

But three important targets still remained at large: Gerlach and Diebner in Munich, and what they called "target number one"—Heisenberg—who, interrogations revealed, had left Hechingen for Urfeld shortly before Pash's arrival. In the last weeks of the war, with a nearly critical reactor, no way to obtain more materials, and defeat and occupation at hand, Heisenberg's first priority remained what it had been for the past year, even above family and personal safety: to ensure the survival of his scientific staff and equipment for the future. After burying the uranium cubes, to be retrieved later, he waged what he called a never-ending battle for the lives of his institute members. They were apparently endangered both by a lack of food and by the zeal of the local populace to fulfill Hitler's order to fight to the very last German.[30] There were many instances across Germany of lynch mobs going after anyone who counseled surrender to invading Allied troops.

As the French lines advanced toward Hechingen on April 19, Heisenberg installed his staff, along with whatever food supplies remained, in the textile factory basement for protection against bombing and artillery. Then he set out on the only transportation available, a bicycle, to attend to his second priority, his family. The Nobel laureate bicycled first to nearby Kleintissen, where his brother and family, whom he had seen only rarely since the outbreak of war, had settled for the duration. After staying with his brother for a few days, he then embarked on an incredible marathon bicycle trip all the way across war-torn southern Germany to Urfeld, a distance of about 250 kilometers (150 miles). Pedaling only at night to avoid marauding German army units and low-flying Allied aircraft, both of which shot at anything that moved, he made it to Urfeld in an amazing three days.

There the situation was chaotic. Eisenhower had ordered U.S. Army units to turn from their advance toward Berlin and head south in a vain search for the purported

"Bavarian redoubt"—a rumored stronghold where Hitler's most fanatical followers would make a last stand. With the complete breakdown of civilian and military order, Waffen SS units retreating from the American advance were rampaging through the Lake Walchen area in a last frenzy of pillage and murder. One night they hanged 17 soldiers from a German recuperation company in the woods near the Heisenberg home for "desertion."

The Heisenberg family situation was in dire straits. Eliabeth Heisenberg was still fighting a losing battle with family illnesses, lack of food, and house repairs when her husband suddenly reappeared. Soon after he arrived, one of his sons became so ill with what appeared to be appendicitis that Heisenberg had to drive him over snow-covered, bombed-out roads to the nearby military hospital that had just lost 17 of its patients. The doctors determined that the ailment was not appendicitis. Heisenberg also managed to move his aged mother to Urfeld from her apartment in Mittenwald. Scouring the nearby village, Heisenberg gathered a stockpile of groceries and fuel for the family before settling in at the cabin to await the end of the war and the arrival of the American Seventh Army.

The end for Heisenberg differed from what he expected. On April 30, the same day Hitler and Eva Braun, his new bride and former mistress, killed themselves in their Berlin bunker as Russian troops closed in, U.S. Army headquarters in Heidelberg dispatched two teams to Bavaria in search of the remaining scientific targets. One team, led by the Heidelberg Alsos commander, was to locate and capture Gerlach and Diebner in Munich; the other, in an "alpine operation" under the command of the indomitable Colonel Pash, was to capture Heisenberg in Urfeld, which was still under enemy control.

In a report to Washington and later in an action-packed monograph, Pash described the execution of what he regarded as "the most important single intelligence mission of the war."[31] Pash's task force of ten men and four vehicles arrived in Bavaria on May 1 and advanced to the town of Kochel, which lay on the opposite side of a mountain from Urfeld. A reconnaissance the next day revealed that a bridge on the road around the mountain to Urfeld had been destroyed, cutting off the road to vehicles. Determined that "nothing was going to stop me from getting to Urfeld that day," Pash led a foot patrol over the snow-covered mountain. Exhausted from the climb, they arrived in town by late afternoon and promptly engaged in a shooting match with a small German force, killing two Germans and scaring off the rest.

While Pash and his patrol were holding their positions in Urfeld, two high-ranking German officers rode into town on motor scooters and attempted to surrender their battalion-sized unit to them. Obviously outnumbered, Pash bluffed his way out by demanding that the officers bring their entire force into town the next day; then he beat a hasty retreat to Kochel. That night the combat engineer unit repaired the bridge to Urfeld, and at 6:00 AM on May 3 Pash's team rolled into Urfeld, followed later that day by an infantry battalion from the Kochel area to take the German forces prisoner. Arriving in town and deploying his team, Pash and two of his men climbed the hill to Heisenberg's cabin and found their target sitting calmly on his veranda overlooking the lake. Heisenberg politely invited them in, introduced them to his stunned wife and curious children, who had obviously not expected their husband's and father's arrest, or at least not before the Americans had captured the Urfeld area.

As Heisenberg quietly gathered his belongings and papers, the sound of small-arms fire sent Pash rushing out the door and down the hill, waving a pistol in the air as he ran. A small German unit had attacked and quickly retreated. Fearing imminent attack by a much larger force, possibly the surrendering battalion, Pash loaded his prize, along with his papers and belongings, into an armored car and hastened with his men back to Kochel. The next morning Heisenberg commenced a bone crushing trip to Heidelberg in the back of Pash's jeep. Three days later, on May 7, General Alfred Jodl, chief of operations of the German High Command, and Admiral Hans-Georg von Friedeburg, German U-boat commander, signed the instruments of unconditional surrender at Reims, ending the war, the Third Reich, and German wartime uranium research. Heisenberg would not see his family again for nearly nine months.

CHAPTER 28

EXPLAINING THE PROJECT: FARM HALL

THE ALLIED ARMIES, SWEEPING ACROSS GERMANY AND THE REST OF EUROPE IN THE first half of 1945, brought the long-awaited collapse of the Nazi dictatorship. Their arrival also brought an end to German nuclear research, and captivity to the German nuclear scientists. V-E (Victory in Europe) Day in May was followed three months later by the capitulation of Japan under the shadows of the mushroom clouds over Hiroshima and Nagasaki.

The perceptions and rationales that Heisenberg and many other Germans accepted before and during the war enabled them to continue their work and daily lives under Hitler's regime. But as the circumstances changed with the collapse of the Reich, the rationales collapsed with it. While many Germans reacted with shock and bewilderment in the wake of defeat, they also began to construct a new rationale tailored to their new world and the questions that arose about their past.

The postwar era brought with it the realization of two terrible truths. The first came with the disclosure to the world of the utter depravity of the Nazi regime, exemplified by the unspeakable horrors of the Nazi death camps. The second truth to be faced was the awesome destructive fury of nuclear weapons, a fury unleashed through the genius of scientific and technological research. Both of these truths had ramifications that went far beyond the immediate experience of the war, changing forever our perceptions of human progress and human potential. They taught us to be skeptical of so-called modern, enlightened societies, however cultured, and, for many, to be wary of modern science, however promising. While each person, and especially each nuclear scientist on both sides of the war, struggled to come to terms in his or her own way with one or both of these terrible lessons, Heisenberg, having much to explain, took a leading role in publicly articulating the reactions of leading German scientists.

As newspapers around the world blazoned reports of Nazi atrocities uncovered by Allied troops during the spring of 1945, the public remained unaware of nuclear weapons until the destruction unleashed on Japan in August. The German nuclear scientists, who thought their research at least equal to, and probably far ahead of, Allied research, supposed that the Alsos mission had captured them in early May in order to tap Germany's superior knowledge. The day after Heisenberg arrived at the Alsos outpost in Heidelberg, he was ushered to an interrogation on his work by his erstwhile colleague, Samuel A. Goudsmit, the scientific head of the Alsos mission.

Heisenberg and Goudsmit had last seen each other in Ann Arbor shortly before the war. Much had happened since. Now, as Goudsmit faced the man he had looked up to as a young physicist but who had apparently made little or no effort to rescue his parents from transport to Auschwitz, the German physicist seemed to him despicably haughty and self-involved. Heisenberg, for his part, seemed to welcome the attention the Allies accorded him for his wealth of nuclear knowledge. The extraordinary efforts that Pash had made to arrest him reinforced such pretensions. When asked about his nuclear research, Heisenberg was so confident of its significance that he offered to instruct the Americans on uranium fission.[1] Goudsmit, knowing of Allied progress, though not about an imminent bomb, politely thanked him for the offer.

On impulse, Goudsmit repeated his question of six years earlier: "Wouldn't you want to come to America now and work with us?" Heisenberg repeated the answer he had given earlier: "No, I don't want to leave. Germany needs me."[2] To Goudsmit, this seemed further evidence of Heisenberg's overweening self-importance. But with most of Germany in ruins and its economy near collapse, Goudsmit could hardly have expected any other response at that point from a man so attached to his country that he had remained at his post throughout the past 12 years.

Heisenberg was interned. Of the fourteen leading nuclear scientists rounded up in the flurry of Alsos strikes, four were sent—willingly or not—to the United States to help with American research. Goudsmit remanded the rest—including Hahn, Laue, Weizsäcker, Bothe, Harteck, Wirtz, and Horst Korsching—to American military authorities, who held them incommunicado for two months in a series of prisoner-of-war camps in France and Belgium. Hahn and Laue were included, although they had not worked on the nuclear project, in the hope that they would have a positive influence on the reconstruction of science in postwar Germany. All of their families were left to fend for themselves.

Heisenberg, Diebner, and Gerlach joined their seven fellow prisoners at Chateau du Chesnay near Versailles, now a detention center known as "Dustbin." Despite this inauspicious omen, the Allied military eventually treated their new captives surprisingly well, providing them with adequate food, English-language newspapers, weekly physics colloquia, and even a jogging track around a flower garden.[3] Nevertheless, the supposed reason for their internment—to allow the Allies "to catch up"—seemed to the scientists hardly sufficient to keep them so long. Laue in particular could not understand why he, who had not worked on fission, should be held against his will. To their inquiries, the British officer in charge, Major T. H. Rittner, replied only that they were "detained for His Majesty's pleasure."[4] Thereafter they called themselves the "detainees."

Scottish physicist R. V. Jones, professor of natural philosophy in Aberdeen and head of intelligence for the British Air Staff, had been following German science since the start of the war. He had assisted the Alsos mission on both occasions when it landed in Europe—the first in Italy, the second in Britain. But America's irksome decision in the last months of the war to exclude the other Allies from sharing the mission's nuclear booty inspired Jones and his staff to begin looking out for their own interests. When an American general reportedly expressed the opinion that the best solution to the problem of German nuclear physics was to shoot all the German nuclear physicists, Jones took action.[5] Not only were executions, or even war-crimes trials, out of the question, but the British seemed in awe of their prestigious detainees. Jones graciously offered to relieve the Americans of responsibility for the scientists. Apparently wanting not the scientists themselves but only their silence, the Americans agreed, on condition that the scientists be kept out of Russian or French control. The Russians and especially the French were already diverging from the Americans and British on postwar policy toward occupied Germany.

Fearful that the prisoners would be captured by the other Allies or sent to the United States if they remained on the continent, Jones arranged to move them to Britain. As an intelligence chief, he knew of a country safe house called Farm Hall in the tiny village of Godmanchester near Cambridge and a large Allied air base. The house had been used by MI6 agents as a staging area for parachuting into German-occupied territory. In early July, after outfitting the house and grounds with secret microphones, Jones had his ten German scientists flown under heavy military guard from their camp in Belgium to their new home in England. There they remained until Jones figured out what to do with them.

The British knew they could not hold the scientists forever, but they did not want to turn them loose in England for fear that they might learn too much about British research from less security-conscious colleagues. By the end of the year, the British had decided that only a revival of the German economy and a measure of political and cultural autonomy in the British zone of occupation were consistent with British commitments and German social and political stability. Science and technology were envisioned as crucial elements of the intended revival. With the radioactive dust of the atomic blasts now settled and the British zone firmly under British control, on January 3, 1946, six months to the day following their internment and in accord with British law, a transport plane flew the detainees to less restricted detention in a northern German town in the British occupation zone. There they could move about during the day, but they had to return to British quarters at night. Several months later, they were all released. Most, including Heisenberg, settled in the undisturbed university town of Göttingen, intended by British authorities to serve as a crystallization point for the revival of West German science.

During the entire period of the scientists' stay at Farm Hall, a team of bilingual British agents assigned to what was known as Operation Epsilon monitored the scientists' conversations via the concealed microphones. They recorded on shellacked metal disks only those conversations that they deemed of special intelligence interest. The interest extended to matters of morale, political orientation, loyalty to the Western Allies, and, after Hiroshima, their knowledge of nuclear fission. The recorded conversations were then transcribed and translated into English. Although they considered the possibility, the scientists gave no indication that they were aware of the concealed microphones. Major Rittner summarized their conversations and excerpted long passages from the English translations in weekly or biweekly reports to his superiors. Copy number 1 went directly to the military head of the Manhattan Project, General Leslie Groves. The existence of these classified reports remained a secret until Groves published his memoirs, appropriately titled *Now It Can Be Told,* in 1962.[6] Thirty years of efforts by interested scholars and others to gain the release of these reports finally succeeded in February 1992 with the declassification of the British and American copies of the reports. The reports were soon published in edited and unedited versions, as well as in a retranslation back into German.[7] Unfortunately, the complete original German transcriptions were lost, while the original recordings were reshellacked at Farm Hall and the disks reused.

Nevertheless, Major Rittner's reports with their verbatim excerpts from the conversations at Farm Hall provide a unique and valuable insight into the German scientists' state of mind before Hiroshima, their reactions to Hiroshima, and their shock at the realization that, rather than being far ahead of the Allies, they were in fact far behind. How did they explain this to themselves, to their countrymen, and to their former enemies? And, as their return to postwar Germany approached, how should they prepare for rebuilding postwar West German science?

The detainees, wrote Rittner in his first report, "are pleased with the treatment they are receiving but completely mystified about their future."[8] Their sumptuous English country manor was located on a rolling grassy estate surrounded by flowering hedges, large trees, and an unobtrusive fence. For their pleasure, several tennis courts were located in the rear; a well-tuned grand piano stood in the parlor; and they had books, newspapers, game paraphernalia, a radio, even the *Physical Review.* The detainees whiled away their hours with relaxation, lectures to each other on nonnuclear work, and walks about the grounds. Rittner and his staff of officers and house workers, carefully chosen for their ignorance of Allied bomb research, provided their prisoners with new clothes, shoes, and hearty English meals. The royal treatment prompted one ungracious officer to comment that the prisoners were living better than the average English family—to say nothing of the average German family, or the average family across most of war-torn Europe.[9]

The detainees' only real complaint was that they were not permitted to communicate with their wives and families. All contact with the outside world was prohibited until the evening of August 6, 1945, when they were abruptly made aware of the reason for "His Majesty's pleasure." On that evening, Rittner wrote, he informed Hahn, the codiscoverer of fission, that the British Broadcasting Corporation (BBC) had announced the detonation over Japan of what was being called an atomic bomb. A very upset Hahn was finally calmed down "with the help of considerable alcoholic stimulant." He then joined the other scientists as usual in the manor dining room where the evening meal was served punctually at 7:45 PM. Pandemonium reigned when Hahn informed his colleagues of the news. The shocked and disbelieving scientists huddled around the radio at 9:00 PM to hear a more detailed report from the BBC—but those crumbs of information only deepened their perplexity. If the Allied scientists had really been successful, and it seemed they had, then German nuclear superiority had been a mere fantasy. "At any rate," Hahn told Heisenberg, "you're just second-raters and you might as well pack up." "I quite agree," he replied.

The news was devastating. Each man responded to it in his own way. Walther Gerlach, the Reich's last administrator of nuclear research, behaved like a defeated general and apparently suffered a nervous breakdown of sorts. Heisenberg and Weizsäcker, who shared a bedroom next to Gerlach's, feared he might attempt suicide and looked in on him that night to assure themselves of his safety. The angered younger physicists, long chafing at the bottom of the power hierarchy, accused their elders of mismanagement; Hahn and Laue, initially shaken by the news, washed their hands of the whole affair.

Heisenberg quickly set to work calculating, raising some of the biggest points of contention for posterity: Did Heisenberg ever know before Farm Hall that only about 50 kilograms of fissionable yet extremely rare Uranium-235 were required to create a critical mass that would explode as an atomic bomb? If so, why didn't he pursue it? If not, why not? The answers are still ambiguous. Recently discovered Soviet documents suggest that such a calculation by Heisenberg exists among captured German documents, but it has not been found. Right after the news of the Hiroshima bomb, Hahn and Heisenberg discussed this very point, according to the verbatim report.

> HEISENBERG: I still don't believe a word about the bomb, but I may be wrong. I consider it perfectly possible that they have about 10 tons of enriched uranium, but not that they have 10 tons of pure U-235.
>
> HAHN: I thought that one needed only very little 235.
>
> HEISENBERG: If they only enrich it slightly, they can build an engine that will go, but with that they can't make an explosive that will—
>
> HAHN: But if they have, let us say, 30 kilograms of pure 235, couldn't they make a bomb with it?
>
> HEISENBERG: But it still wouldn't go off, as the mean free path is still too big.
>
> HAHN: But tell me why you used to tell me that one needed 50 kilograms of 235 to do anything. Now you say one needs 2 tons.
>
> HEISENBERG: I wouldn't like to commit myself for the moment . . .

After learning more details from the BBC reports that evening about the bomb and the Manhattan Project, they continued.

> HAHN: In 1939 they had made only a fraction of a milligram [of U-235]. They had identified the "235" through its radioactivity.
>
> HEISENBERG: That would give them 30 kg a year.
>
> HAHN: Do you think they would need as much as that?
>
> HEISENBERG: I think so certainly, but quite honestly I have never worked it out, as I never believed one could get pure "235."

LEFT TO RIGHT:
*Heisenberg, Max von
Laue, Otto Hahn, c. 1947.*

Shortly after this conversation Heisenberg attempted to work it out, producing a rough calculation of the critical mass of U-235 required in order to yield the reported energy of the Hiroshima bomb. The result came out to be "about a ton." This rough calculation, as Jeremy Bernstein has pointed out, was filled with errors and incorrect assumptions.[10] If Heisenberg had previously calculated that a small amount was required, as Hahn apparently remembered, it would probably have been in preparation for his two initial reports on the subject in late 1939 and early 1940, or else for the meetings of 1942. If so, Heisenberg must have forgotten anything about this; perhaps he never did make such a calculation. Several days later Heisenberg made a much better calculation with a much more accurate result, and it appears from his presentation of it to his Farm Hall colleagues that it was quite new to him. On the whole this suggests that Heisenberg probably never did make this calculation. Were three to five years of war enough to cause him to forget it entirely?

The news of a successful Allied bomb turned a glaring public spotlight on several profound questions for the Germans. As Laue expressed it in a letter to his son the next day (the letter was posted later): "The main question is naturally why we in Germany did not achieve a bomb."[11] Put another way: why was the German achievement, whether or not they were ultimately aiming for a bomb, so slight in comparison? Beyond this, another question has been asked ever since, one that may have no direct answer: in view of the incredible death and destruction wrought by both the atomic bomb and by the Hitler regime, what scruples, if any, did German nuclear scientists bring to their wartime work? It is a question that can be asked of both sides in the war.

The scientists spent the evening of August 6 and most of August 7 pondering and debating the reasons for their shortcomings. We have already observed such factors as their over-confidence, their fear of being forced to build a bomb and not succeeding, the error regarding pure graphite, the more immediate aims of their leaders in the midst of total war, the destruction of German infrastructure, and the scientists' reliance on Heisenberg, whose primary interests and expertise lay elsewhere and who saw this project mainly as a means to personal and professional ends. At Farm Hall, the scientists' immediate reaction was to remark on the enormous scale of the Manhattan Project reported by the BBC, a scale that rendered it one of the world's largest scientific projects to date. Their reaction was recorded by the hidden microphones.

HAHN: Of course we were unable to work on that scale.

HEISENBERG: One can say that the first time large funds were made available in Germany was in 1942 after that meeting with Rust, when we convinced him that we had absolutely definite proof that it could be done. . . . On the other hand, the whole heavy-water business [reactor construction], which I did everything I could to further, cannot produce an explosive.

HARTECK: Not until the engine is running [to produce plutonium].

WEIZSÄCKER: How many people were working on V-1 and V-2 [rockets]?

DIEBNER: Thousands worked on that.

HEISENBERG: We wouldn't have had the moral courage to recommend to the government in the spring of 1942 that they should employ 120,000 men just for building the thing [bomb] up. . . . I would say that I was absolutely convinced of the possibility of our making an uranium engine, but I never thought we would make a bomb, and at the bottom of my heart I was really glad that it was to be an engine and not a bomb. I must admit that.[12]

As the scientists discussed the reasons for their relatively poor progress, British news reporters were reporting that Germany had lost the race against the Allies for the atomic bomb. In view of the harm to their reputations abroad and at home if their work were viewed as a lost race against the Allies, the scientists insisted to Major Rittner that "no such work had been carried out." Perhaps as a means to alleviate their frustration, Rittner suggested that they compose and sign a memorandum for the press and public "setting out details of the work on which they were engaged." Heisenberg immediately began composing such a statement together with the diplomatic Weizsäcker, his closest and most trusted colleague among the scientists, while strolling unmonitored on the grounds after lunch. Gerlach and Wirtz later assisted in drafting the statement, an early version of which, in Heisenberg's hand, survives on the pages of an English military school exercise book. On August 7 a final draft was formulated, typed, signed by all, and handed to their captors. It was probably not released to the public at the time.[13]

Explaining why Germany never achieved even a chain reaction, let alone an atomic bomb, Heisenberg and his co-authors argued in their August 7 statement that by the end of 1941 they had come to the conclusion that they would be able to build a "machine," a reactor. "On the other hand," they continued, "it was the view of the researchers that the conditions for the production of a bomb were at that time not available within the framework of the technical possibilities then accessible in Germany." Thus, they were not engaging in any race with the Allies to build a bomb, mainly because the available resources did not permit it. Instead, they continued, "The further work therefore concentrated on the problem of the machine for which, in addition to uranium, heavy water is required." That work was slowed by the limited supply of heavy water, but they nearly achieved a chain reaction by the time the war ended. This harmless sounding statement contrasts with Bohr's remembered impression of Heisenberg's aims in 1941, while neglecting the many other substantive reasons for the project's lack of progress.

But there was something more in this statement that may now take on a new meaning. While at Farm Hall one of the younger physicists, Erich Bagge, wrote in his diary on August 10, "The story [in the memorandum] found wide-ranging but not complete acceptance." It had been signed by all only after difficulties with the younger physicists had been resolved.[14] Until now, this seemed to refer to the lingering hope that a bomb, and not just a reactor, could be built. However, this and a comment appended to the above-quoted statement in the memorandum might take on new meaning in light of the recent claim that a crash program to build the bomb did exist under Dieber, and that it actually succeeded. The appended comment states in translation, "Regarding the question of the atomic bomb, let it be stated further that no researches, for example by other groups in Germany, which had as their goal the immediate production of the bomb have become known to the undersigned. If however such attempts should be found to have been undertaken, they were in any case carried out by dilettantes and are not to be taken seriously." No wonder Bagge reported difficulties. Gerlach, who co-authored the memo, signed the statement, as did the younger Bagge and his superior, Diebner; and Gerlach and Diebner were supposedly directly involved in the crash program that led to the reported explosions (whether nuclear or not). These suppositions, if true, may help to clarify Rittner's comment in his accompanying report from Farm Hall that there was considerable discussion regarding the wording of the memorandum, "in the course of which Diebner remarked that he had destroyed all his papers, but that there was great danger in the fact that Schumann had made notes on everything. Gerlach wondered whether Voegler had also made notes."[15]

In the end all ten scientists did sign the memorandum. Max von Laue, who had remained aloof from fission research throughout the war, joined the others in endorsing the statement, though he noted on the document that he had not worked on fission. He repeated the gist of the main argument in a letter written on August 7 to his son, a historian, who was sequestered in Princeton during the war to avoid the German draft. In late September, Laue forwarded a copy of the statement to his son for distribution in the United States.[16]

With moral issues being raised by British reporters and their own competencies as scientists in doubt, by the evening of August 6 the German scientists were already exploring the moral dimension, led primarily by Weizsäcker. Weizsäcker declared at Farm Hall: "I don't think we ought to make excuses now because we did not succeed, but we must admit that we didn't want to succeed." And just prior to this: "I believe the reason we didn't do it was because all the physicists didn't want to do it, on principles. If we had wanted Germany to win the war we could have succeeded." To which Hahn replied, "I don't believe that, but I am thankful we didn't succeed." But later that night, according to Rittner's paraphrase, Heisenberg told Hahn, "he feels himself that had they been in the same moral position as the Americans and had said to themselves that nothing mattered except that Hitler should win the war, they might have succeeded, whereas in fact they did not want him to win."[17] The upshot seemed to be: in order to protect their competencies in the public arena they would emphasize the material conditions of war in their memorandum the next day, while they would invoke moral scruple as a primary reason for their poor showing, for now, only in the private sphere. Laue reported the emerging dual argument in his August 7 letter to his son, "All of our uranium research was directed toward the achievement of a uranium machine as an energy source, first because no one believed in the possibility of a bomb in the foreseeable future, and second because fundamentally no one of us wanted to put such a weapon in Hitler's hands."[18] The official Farm Hall statement did not mention Laue's second reason, and there was no mention of the plutonium alternative. These arguments have served ever since as the foundation of the German scientists' position regarding their wartime work.

In order to comprehend more fully the position Heisenberg and his compatriots presented in their Farm Hall statement and its derivatives—"so violently debated in all scientific circles ever since," as Groves put it—one needs the perspective afforded by hindsight.[19] First of all, whatever their failings as scientists and as citizens, the German scientists were not entirely responsible for the moral character of their country. It is true that, as noted earlier, the mere fact that these world-renowned scientists continued to live and work in Germany after the moral and political affronts of the early years left them already politically and morally compromised and lent the regime an unwarranted and false credibility. They compounded their failing by continuing to seek out and to accept collaborative accommodations with the regime—a regime that continually demonstrated its utter disregard for decency of any sort from the very beginning and clearly held the scientists and their science in outright contempt.

Nor were the scientists alone in their eagerness to prove their value to their government by creating the weapons of war. Ever since Archimedes built catapults for the king of Syracuse, ever since Francis Bacon declared that knowledge is power, science has been, at times, the handmaiden of every nation's economic, military, and political interests. Not until after the introduction of chemical warfare in World War I and nuclear warfare in World War II, and the prospect of hydrogen bombs during the cold war, have moral scruples really played any role in the willingness of scientists to arm their respective nations.

Secondly, however, much of the fury of the "violent debate" regarding Heisenberg and German wartime research appears to be fueled (for some, at least) by the circumstance that many of the Allied scientists found it necessary to console themselves with the argument that they had built the bomb in a race to counter the far greater evil of an atomic bomb in Hitler's arsenal. While certainly valid, this rationale could not disguise the fact that their efforts had resulted in the deaths of hundreds of thousands of civilians. The Smyth report, the official account submitted to Congress on the Allied nuclear effort, attempted to counter public criticism by stating the obvious: "This weapon has been created not by the devilish inspiration of some warped genius but by the arduous labor of thousands of normal men and women for the safety of their country."[20]

Amid the throes of their own moral anguish, many of those men and women of the Manhattan Project were simply appalled to perceive a conspicuous lack of any similar soul-searching on the opposing side, among those whose work had created the fear that drove the Allied effort at least until late 1944. Instead, the Allied scientists perceived a shocking failure on the part of the German scientists to acknowledge that they too had been working just as hard as they could on nuclear fission for their country and, like scientists everywhere, had been willingly exploited, not just by any government, but by some of the most depraved leaders of history. American physicist Philip Morrison probably spoke for many of his colleagues in 1947 when he wrote, "No different from their Allied counterparts, the German scientists worked for the military as best their circumstances allowed. But the difference, which will be never possible to forgive, is that they worked for the cause of Himmler and Auschwitz, for the burners of books and the takers of hostages."[21]

Unknown to all but a handful of people, the German scientists at Farm Hall were not only avoiding admission among themselves of their complicity in working on fission under the Hitler regime; but, beyond that, beginning on the very day following the destruction of Hiroshima, at least one of the German scientists had the audacity to congratulate the German scientists for their moral superiority for having not built the bomb! According to the Farm Hall transcript, Weizsäcker stated on August 7, "History will record that the Americans and the English made a bomb, and that at the same time the Germans, under the Hitler regime, produced a workable engine. In other words, the peaceful development of the uranium engine was made in Germany under the Hitler regime, whereas the Americans and the English developed this ghastly weapon of war."[22]

Heisenberg and his colleagues expounded the Farm Hall position, both its technical and moral aspects, in the months and years following their return to Germany. But the times had changed, and so too did the nuances of that position in response to the demands upon those who now appointed themselves to the task of rebuilding German science out of the ashes of political and military destruction.

CHAPTER 29

EXPLAINING THE PROJECT: THE WORLD

SOON AFTER RETURNING TO GERMANY AND SETTLING IN GÖTTINGEN IN THE BRITISH occupation zone, Heisenberg began to present to the public his side of the story of German nuclear research. In December 1946 British authorities permitted him to publish a brief, non-technical summary of German nuclear research for his German compatriots. An article appeared in the German journal *Die Naturwissenschaften* (*The Sciences*) with a partial translation in the British journal *Nature*. Beginning in 1947, Heisenberg sat for a series of interviews by German newspapers, and in 1948 the science editor of the *New York Times* interviewed him in response to Goudsmit's reports in the United States on German war research and the Alsos mission.[1]

In each of Heisenberg's accounts, early 1942 is depicted as the turning point. During the previous year, 1941, Heisenberg's Leipzig team had established that an atomic bomb was possible in principle, apparently upon the first inklings of a chain reaction, and his Berlin team had learned that a reactor could, in principle, breed plutonium for a bomb. But the technical hurdles to be overcome in achieving a reactor or a bomb were still enormous and costly. By 1942, as the war situation worsened, the Army Ordnance Office had decided to forgo most of its nuclear research effort, since the office could not be sure that the effort would lead soon to a useful weapon for the war at hand. Because of that decision, as well as the reduced capacities of German industry and the technical obstacles still to be surmounted, "all hope of making bombs was given up," Heisenberg told the *New York Times*.[2] Although Heisenberg had tantalized Reich officials in early 1942 by hinting that a running reactor would produce equally fissionable plutonium, after the war he insisted that he wanted only to ensure their continued support. He dampened any expectation of an imminent weapon by stressing the technical difficulties. The strategy worked, he said—even though such a strategy was unnecessary, since he seems to have believed that the difficulties really were insurmountable in the short term.

For Heisenberg, the decisive meeting occurred with Albert Speer on June 6, 1942. After that meeting, Speer had ordered that the project be continued only on a modest scale but that the researchers should work, in Heisenberg's postwar words, for "the only attainable goal": "to build an energy-producing uranium burner for powering machines." Of course, the production of a burner to power, say, submarines would have been no small contribution to the war effort. But in his 1946–1948 accounts, Heisenberg also seems to imply the Farm Hall moral argument—that the scientists had made a conscious decision to control the research in order to prevent Hitler from obtaining the bomb, and to deter the regime from ordering them or anyone else to build one. As he put it in the pages of *Nature*: "The German physicists had conscious-ly worked from the very beginning toward maintaining control over the project, and they used . . . their influence to direct the work in the sense depicted in this report." Beyond that, even if they had not decided against building a bomb, he indicated they were still immune from any moral concern. Since the project had never progressed much beyond its status in 1942, Heisenberg wrote in 1946, he and his colleagues were therefore conveniently spared "the difficult moral decision" of whether or not to build atom bombs for Hitler.[3]

Heisenberg and his colleagues had good reason after the war to portray their proj-ect as they did. In order to reestablish German science, to ensure that scientists could never again be disregarded and abused by their government, and to counter public criticism of their wartime behavior, it was essential that, once again, they acquire as much influence as possible, first in the British zone, then within the emerging West German state. As previously, emphasizing the prestige and utility of nuclear research and technology was the surest means of establishing themselves as vital to Germany's science and to its economic revival. To realize their goal, suspicions of their having worked to arm Hitler with nuclear weapons had to be addressed. The Allied occupa-tion forces had placed denazification and the control of nuclear energy at the top of their priorities list. Heisenberg, not a party member and ideologically victimized by Stark, sought and gained the confidence of the occupation authorities in the matter of denazification.

In addition, after the war, Heisenberg and his supporters took great pains to dis-tance themselves from former Army Ordnance researchers and from anyone else who openly admitted having worked toward (or having achieved) the goal of an atomic bomb under Hitler. During the late 1940s, Heisenberg's circle also began an intensive public campaign to establish a West German nuclear power program, a campaign that continued until the rescinding of Allied science control laws and the granting of sov-ereignty to the West German Federal Republic in 1955 as part of the NATO alliance. With West German self-rule imminent after 1950 as the cold war heated up, Heisenberg and the nuclear scientists pushed for the establishment of a cabinet-level ministry for nuclear energy policy. At the same time, they now acted to mobilize pub-lic opinion against the acceptance by the German government of a NATO plan to equip the West German army with tactical nuclear weapons.

The scientists succeeded on both scores. They ensured that the West German army would remain nonnuclear; while practicing the launch of nuclear weapons in the event of war, the Americans would provide the weapons only when actually needed against a Soviet invasion. In addition, the scientists successfully negotiated with Washington for permission to begin a full-scale nuclear reactor program, a program that by the late 1960s was the most successful in the world. West Germany was by then the leading exporter of nuclear technology. Heisenberg later wrote with satisfaction in his memoirs, "The fact that in wartime no attempt was made in Germany to construct atom bombs, although knowledge of the principles existed, probably had a favorable effect on these [Washington] negotiations."[4]

As far as was publicly known in the late 1940s, no actual attempt, beyond the theoretical stage, *had* been made to construct an atomic bomb in Germany. A strong difference of opinion emerged between German and American scientists as to why the attempt was not made. The loudest and most divisive debate occurred between Heisenberg and the former Alsos science head Samuel A. Goudsmit, then professor of physics at Northwestern University. Goudsmit offered his highly influential views in a series of articles and in a monograph, widely read among American scientists, titled *Alsos* and published in 1947. Their debate raged through the pages of the *New York Times* and in a fascinating exchange of correspondence.[5]

In many ways Goudsmit was bitterly disillusioned concerning Germany, German science, and one German scientist in particular, Werner Heisenberg. Moreover, the broader concerns that he and his colleagues faced regarding science in the United States were quite different from those the Germans were facing. As the cold war deepened, the paramount issues for American scientists were those of secrecy, administration, and the relationship between science and the military. Goudsmit expressly intended his account of the failed German project—"failed," ironically, because it did not produce an atomic bomb, or even a reactor—as a case study of what can go wrong, an example of "how incompetent control (which is not restricted to totalitarian countries) can kill scientific progress in a short time."[6] If Heisenberg was arguing the competence and success of the German scientists in preserving their science and their scruples under Hitler, Goudsmit was arguing just the opposite—each, in part, for his own contemporary audience. And indeed each audience has tended ever since to subscribe to the respective views offered by Goudsmit and Heisenberg.

According to Goudsmit, a variety of factors caused the death of science in Nazi Germany. Nazi doctrine removed essential personnel from the laboratory and the classroom and weakened the scientists' adherence to modern scientific theories. The organization of German science and its support systems was disastrous in its lack of coherence and cooperation. The scientists themselves, who had grown accustomed to leading the world in modern science, became convinced that their superiority was absolute and therefore grew complacent: if they could not make an atom bomb, neither could the Allies. And finally, said Goudsmit, the German scientists indulged in an excess of hero worship, such as that practiced by "the smug Heisenberg clique," which overlooked less heroic but more practical-minded technicians, such as Kurt Diebner or the self-made Manfred von Ardenne.[7]

The German researchers had concentrated on a reactor, said Goudsmit, because they believed that, uncontrolled, it would eventually explode. But even then, they believed, the Allies were far behind. In Goudsmit's opinion, the

Germans had completely missed both fast-neutron fission and the plutonium alternative. If they had seen them, they, like the American scientists, would have pressured their government for more support. Thinking themselves far ahead, wrote Goudsmit, in actuality German scientists had only the vaguest notions of how a uranium bomb or even a reactor actually works, as shown by the lack of control rods and protections against radioactivity in their experiments. They were obviously far behind the Allies in such technical efforts as isotope separation and moderator testing and production.

Heisenberg vehemently objected to Goudsmit's account on nearly every score. In long exchanges with Goudsmit, in letters to and interviews with the *New York Times*, and through his surrogates C. F. von Weizsäcker and B. L. van der Waerden, then in the United States, Heisenberg vigorously maintained the advanced state of German war research.[8] Possibly through his American Uncle Karl, who lived in New York, Heisenberg gained the backing of Waldemar Kaempffert, the German-American science editor of the *New York Times*. In an interview by Kaempffert in response to Goudsmit's *Alsos*, Heisenberg, speaking "with an objectivity that is convincing," insisted that the destruction of German industry and unresolved technical problems forced the German scientists to give up "the idea of devising an atomic bomb and to concentrate on the development of atomic power for industry." Three days after the interview appeared, Goudsmit wrote a letter to the *Times* taking issue with Heisenberg's account: "Heisenberg stresses the lack of industrial resources during the second half of the war. The book, 'Alsos,' points at the lack of vision of the German scientists." Kaempffert angrily replied that "liars do not win the Nobel prize"—a remark that prompted Goudsmit's publisher to inquire of Einstein whether in fact Nobel laureates do lie.[9]

Of course, there were glaring errors in Goudsmit's sometimes angry, sometimes oversimplified account of the German wartime research effort, but Heisenberg in particular was concerned that the research effort should be seen not only as morally untainted but also as highly competent, despite its poor showing. He had made it his personal mission to preserve the high quality of modern physics in Germany in the face of adverse conditions. He and his colleagues could not afford to appear to be incompetent fools if they were to be influential in West German science affairs; the more they were thought to have known about atomic bombs, the nobler they would seem to contemporaries for not having attempted to build them. He defended the obvious hero worship, the formation of a clique around himself, as a means of excluding "unscrupulous persons" from influence on the course of uranium research. Heisenberg's 1941 visit with Niels Bohr was now described as an effort to convey to the Allies that the Germans knew about the bomb but would not pursue it. They would work on nuclear energy only to gain funding and recognition and to save young physicists from the draft.[10]

Bohr's views on that visit now became crucial to Heisenberg's case, and Uncle Karl again assisted his nephew. He had earlier befriended Bohr during one of Bohr's many fundraising trips to the United States. With his uncle's help, Heisenberg managed to reestablish contact with Bohr and received permission—from both Bohr and the British—to travel to Copenhagen in 1947. With his wartime motives and behavior in question abroad, Heisenberg apparently wanted to discuss the situation with the influential Bohr and, more importantly, to learn what Bohr remembered of their 1941 encounter. Accompanied by an Allied control officer, Heisenberg made the trip just as Goudsmit's book appeared in the United States.[11]

After all that had occurred in Denmark during the war, and still angry with Heisenberg about his 1941 visit to Copenhagen, a cordial Bohr proved much less supportive than Heisenberg had expected. Bohr flatly refused to discuss the details of the visit, and Heisenberg did not report much of what transpired during this encounter. Bohr had taken the 1941 meeting, Heisenberg reported, merely as an indication of German progress on nuclear fission research. Brushing Heisenberg off, Bohr told him to get in touch with Goudsmit. Heisenberg would have to refute Goudsmit without Bohr's backing. A year later, the diplomatic B. L. van der Waerden, Heisenberg's self-appointed "attorney," composed an aide-mémoire in English on the German position and presented it to Bohr.[12] There is no record of a response. Bohr's relationship with Heisenberg remained civil but strained thereafter.

Heisenberg did get in touch with Goudsmit. Although it is unclear how much, if any, Heisenberg read of the book, soon after the publication of *Alsos* Heisenberg wrote to Goudsmit attempting to explain the difficult psychological situation the Germans had to face during the war.[13] It was a terrible moral dilemma. On the one hand, he claimed, the German scientists were well aware of the "horrible consequences" that a German victory would mean for Europe; on the other hand, they did not wish to see Germany defeated—not because of patriotism but because of "the hate that National Socialism had sown," a compromising statement at best. The dilemma led the scientists to pursue "a more passive and modest posture," he argued. This was a reference to the position outlined in his earlier essays "Active and Passive Opposition" and "The Order of Reality"—to help on a small scale where it is possible, to appear to work in support of the regime even if one was not, and otherwise to do research that will perhaps prove useful later.[14]

Heisenberg's letter elicited an angry five-page, single-spaced typed response from Goudsmit, repeating many of the arguments in his book. Passive opposition, Goudsmit told the German physicist without ado, was simply a self-serving rationalization, fabricated ostensibly for the pursuit of an impossible goal—the mere preservation of relativity and quantum theory under Hitler. "How could you ever hope to be successful? How could you ever think that these were important issues?"[15] The two argued back and forth in public and private exchanges over the following year. Heisenberg consistently maintained Germany's scientific success, despite Nazi policies, and the German scientists' moral dilemmas, while Goudsmit was unrelenting on Germany's scientific failure and the scientists' compromising position toward the Hitler regime.

In 1948, at the request of American occupation authorities, Heisenberg and Wirtz published a technical account of the German project in a series of U.S. Army reports on German science and technology commissioned by the U. S. Army's Field Information Agency, Technical, the so-called FIAT reports.[16] The authors argued, of course, that Germany had been well advanced in reactor engineering, but they did not even hint at any broader issues. Writing the account enabled Heisenberg to reexamine available research reports, and at Heisenberg's insistence Goudsmit, too, reexamined copies of the captured reports in Washington.

As a result of this exercise, Goudsmit corrected his most obvious errors, conceding that the Germans had, in fact, been aware that a bomb differed from a reactor and that they had also been cognizant of the plutonium alternative. But, Goudsmit wrote in the *New York Times*, Heisenberg's claims of advanced theoretical knowledge notwithstanding, the reports "show clearly that their scientists had only a very vague notion of the working of the atomic bomb, and their ideas about a uranium pile were in a very preliminary stage." The reason for their meager progress was, again, their lack of vision. And their lack of vision was the direct result of "the stifling atmosphere in which scientists work under a totalitarian regime."[17]

This, of course, had been Goudsmit's fundamental point all along. Again and again Goudsmit made the same point to Heisenberg and his emissaries: what he really wanted to see from Heisenberg, Hahn, and other leading scientists were articles about the frustration of scientific progress under a totalitarian system of government (which, by extension, could apply to an autocratic American government). He insisted that they should stop extolling the greatness of German science and acknowledge its decimation by the Nazis—a demand that they were hardly in a position to fulfill. In fact, their position was almost impossible to maintain under any circumstances: trying to distance themselves from the Nazi regime while at the same time claiming that they had done great but harmless work under, and for, it.

That Heisenberg would even attempt to defend the pursuit of decent science under the Nazi regime, or believe such were possible, seemed outrageous to many American scientists. Goudsmit had already declared of Heisenberg: "He fought the Nazis not because they were bad, but because they were bad for Germany, or at least for German science."[18] Carl Friedrich von Weizsäcker learned of the outrage among Americans (or refugees in America) firsthand when, in 1949, he met in Chicago with émigré physicists James Franck and Maria Goeppert-Mayer, both of whom—in view of the increasing secrecy in the U.S., military influence on nuclear research, and the drive to build the hydrogen bomb—were very concerned with ethical issues. Franck was especially critical of the German scientists, Weizsäcker reported to Heisenberg. In Franck's view, even the defense of decent physics and the acquisition of support and draft deferments could not justify the compromises Heisenberg had made with the Nazi regime.[19]

Goudsmit and Heisenberg never did settle their quarrel. Years later the two men, now old, met one last time in an attempt to heal old wounds. During his last trip to the United States, in the spring of 1973, Heisenberg lectured at the Smithsonian Institution in Washington, D.C., and Goudsmit, who had long since moved to the Brookhaven National Laboratory on Long Island, traveled to Washington to meet him. Goudsmit once again admitted the technical errors in his book's portrayal of a backward German project and apologized for any personal injury he had caused the German physicist. A complete reconciliation, however, was impossible. Heisenberg died three years later.

In his obituary of Heisenberg for the American Philosophical Society, Goudsmit offered his assessment of their controversy after decades of reflecting on it. In Goudsmit's view, Heisenberg had failed to realize that German physics was already in precipitous decline relative to physics in other nations even before the Nazis came to power. The United States in particular was rapidly outpacing Germany. In addition, the American research system of cooperative university departments, large-scale industrial research, and close collaboration between experimentalists and theorists was much more conducive to the progress of contemporary physics, especially nuclear physics, than was the German tradition. Thanks in part to American fascination with European science, which made possible Heisenberg's own many trips to the United States, American physics was already surpassing German physics when the Nazis began driving many of their best scientists from Germany. Heisenberg's efforts to maintain an illusory German lead in contemporary physics were thus completely misplaced. "If Heisenberg had realized this," wrote Goudsmit, "he would not have taken the German failures so personally."[20] Perhaps he would not have been so willing either to enter into the debilitating compromises he endured by convincing himself that he was indeed personally responsible for the preservation of German physics.

The concern of many scientists in the United States during the late 1940s over the issue of government control of research was eventually settled to their disappointment in the tightened secrecy criteria of the cold-war era, and in the government-mandated H-bomb program. Whatever the lessons of the past, some scientists would work nonetheless to fashion weapons of mass destruction that were becoming ever more prevalent and ever more powerful. The prospect of mutually assured destruction became more assured indeed with the invention of intercontinental ballistic missiles, which could carry the instruments of destruction around the globe within minutes. Concern for moral issues and the social responsibility of the scientist mounted everywhere, especially in the United States. Some American scientists felt satisfaction once again that at least they were building bombs, not for a Hitler or Stalin, but for the protection of American democracy.

The satisfaction was soon challenged by the publication of Robert Jungk's history of the atomic bomb, *Brighter than a Thousand Suns*. It appeared in German in 1956 and in Danish and English in 1957 and 1958, and provoked Bohr, as noted earlier, to unusually angry draft letters in response both to the book and to Heisenberg's version of their meeting in 1941.[21] Those years were ones of vehement debate in West Germany over nuclear weapons and reactor technology, and Jungk's book took up the German scientists' cause.

According to Jungk, the German scientists—meaning those around Heisenberg —so distrusted the regime and other, less-scrupulous physicists that, pursuing their opposition to the hated regime, they continued to work on the project but secretly sought to control the outcome out of moral scruple, "to divert the minds of the National Socialist service departments from the idea of so inhuman a weapon." They gladly welcomed the authorities' conclusion that an atomic bomb could not be made under wartime conditions in Germany and were content to concentrate on trying to build a reactor, while awaiting the inevitable defeat. As Heisenberg put it in his letter to Jungk, and published in part in the Danish and English translations, they were con- vinced that atomic bombs could be achieved "only with enormous technical resources."[22] Jungk pointed out that, while the Germans demurred, the Americans were making that extreme effort and did succeed in providing their government with a weapon of awesome destruction that was promptly used on Japan. The moral impli- cation was clear. Jungk, acknowledging the help of C. F. von Weizsäcker, published a near-verbatim repetition of Weizsäcker's appalling private statement at Farm Hall: "It seems paradoxical that the German nuclear scientists, living under a saber-rattling dictatorship, obeyed the voice of conscience and attempted to prevent the construc- tion of atom bombs, while their professional colleagues in the democracies, who had no coercion to fear, with very few exceptions concentrated their whole energies on production of the new weapon."[23]

Weizsäcker and Heisenberg, heavily involved at that time in the German scien- tists' dual campaign against weapons and for reactors in West Germany, had to pres- ent to the German public an image that was morally untainted by nuclear weapons— even morally superior to the builders of bombs—if they were to be influential in West German affairs. In his four-page letter to Jungk after receiving a copy of Jungk's book, Heisenberg gave no indication that he objected to Jungk's near quotation of Weizsäcker's Farm Hall moralizing.[24]

Jungk's book, widely read in the United States at the height of cold-war tensions, immediately revived the debate over German wartime research. German physicist and editor Paul Rosbaud, who by later accounts had supplied the Allies with inside infor- mation on the German uranium project during the war, wrote in his review of the book for *Discovery*: "Out of all of the theory [regarding the research] emerges a strange picture in which it sometimes appears that the German physicists alone have no actu- al or moral guilt for the A-bomb."[25]

Max von Laue, a former detainee at Farm Hall, wrote to Rosbaud denying the moral element in the story and repudiating Jungk's portrayal of the German scientists as morally driven. Referring to the Farm Hall discussions on August 6 and 7, 1945, he wrote, in a near repeat of his assessment to his son at the time: "The version (*Lesart*) was developed that the German atomic physicists really had not wanted the atomic bomb, either because it was impossible to achieve it during the expected duration of the war or because they simply did not want to have it at all. The leader in these discussions was Weizsäcker. I did not hear the mention of any ethical point of view. *Heisenberg was mostly silent* [his emphasis]."[26]

After reading Jungk's book in 1964, a group of science students at Cornell University asked their mentor, Hans Bethe, to give a talk on the social responsibilities of scientists and engineers. In his talk, which was later published in the newsletter of the Society for the Social Responsibility of Science, Bethe reviewed the public record of the Goudsmit-Heisenberg controversy and noted its seeming lack of focus on moral issues: "Neither Goudsmit nor Heisenberg indicated that conscience played any part in the German failure to develop the atomic bomb."[27] Their debate had revolved around the failure of German science.

Without access to the full record, Bethe had apparently overlooked the strong ethical, and perhaps moral, argument that Heisenberg was making, as indicated, for instance, in his letter to Goudsmit on the German scientists' dilemma. Bethe's talk brought an objection from Heisenberg, who again emphasized his scruples. In his letter to Bethe, Heisenberg acknowledged that German physicists were morally no better nor worse than their American counterparts, but he still maintained that they did not build bombs because they did not want Hitler to win the war. On the other hand, certain that Germany would eventually lose, they did not "wish a total and obliterating defeat of Germany." Apparently this sentiment, voiced earlier, was their justification for working to build a reactor to power the German economy before and after defeat. Again Heisenberg claimed that the controversial 1941 meeting with Bohr proved their moral concern. With the technological and administrative turning points in 1942, they were thus relieved that they could concentrate on developing a reactor without fear that they would be ordered to build a bomb. Jungk's overemphasis on morality apparently inspired all parties to stress moral scruples when telling their stories.

The 1967 publication of David Irving's account of the German project, published as *The Virus House* in Britain and as *The German Atomic Bomb* in the United States, again revived the debate.[28] British author Irving, recently convicted in Austria of Holocaust denial, managed to obtain from the German participants a wealth of previously unavailable primary sources. His portrayal, the first source-based, nonparticipant history of the German effort, was not surprisingly sympathetic to the Germans. Nevertheless, he too doubted that moral scruples played a role at any stage of the research. He believed that curiosity drove the German scientists in their work and would have driven them all the way to a bomb, had circumstances and the right administrators allowed it.

This prompted Heisenberg to offer another round of interviews and book reviews, in Germany and the United States, on the German project. Heisenberg expressed satisfaction that "Irving's investigation confirms the German report in all important points" in the pages of the *Bulletin of the Atomic Scientists* and in the influential *Frankfurter Allgemeine Zeitung* (*Frankfurt General Newspaper*). But he believed that

Irving's interpretation of motives was faulty. Irving "does not recognize sufficiently how deep a mistrust could exist, yes, often has to exist, between human beings in a totalitarian state, even between those who work closely together." Because of this, as well as the war situation and the engineering hurdles, "German physicists did not insist on pursuing, by means of practical measures, a path which they could hardly have trodden during the war."[29]

In a preface to Heisenberg's review of Irving's book, *Bulletin* co-founder Eugene Rabinowitz took direct issue with Heisenberg's contention that distrust, hence some measure of scruple about the use of their work, had determined the scientists' behavior. He conceded that distrust of the regime probably did exist and probably did make the German scientists reluctant to provide weapons to Hitler. But the possibility that Germany might lose the war and that defeat would mean the end of the great revival of the German nation begun by the Nazi regime at first made the scientists much less reluctant. This changed after 1942, wrote Rabinowitz: "As the war dragged on, and the likelihood of German defeat loomed more and more ominously for all who retained a modicum of rationality, the scruples of the leading German physicists became stronger, and the alibi of developing not an atom bomb, but a postwar reactor, actually became reassuring."[30]

The controversy lingers to this day, bursting into public discourse whenever new evidence or new interpretations appear, perhaps because the deeper issues and the painful history still remain unresolved. To a large extent, Goudsmit, Rabinowitz, and others are correct. Many today do feel there was a profound failure by Heisenberg and others to be completely candid about their attitudes during Hitler's rule and especially during the war—a failure to explore their errors as well as their successes; to point out the human frailty as well as the human resilience of the scientist and the citizen in this the world's first major encounter with the nightmare of genocidal dictatorship in an advanced industrial nation; to debate not merely the stifling of science by governments but the stifling of the human spirit itself. Certainly it was unacceptable for Heisenberg and his close colleagues to claim that they had consciously delayed the project because of moral scruples. It was not much better for Heisenberg to say that he might have built the bomb had that been attainable during the war, but otherwise to absolve himself of any moral or ethical failing. After all, had he and his colleagues not worked on reactors to power the German war machine? Had he and they not allowed themselves to be exploited by a monstrous regime? Had their very work on any aspect of nuclear fission not instilled the fears that drove the intensity of the Manhattan Project to complete its work, at least through the end of 1944? How did this all come about? How did these highly educated scientists, blessed with the best of moral culture and learning and the highest ideals of scientific inquiry, find themselves in this situation?

These are all very profound and important questions that the scientists themselves, through candid reflection, could perhaps have helped us to answer better than could any postwar biographer. Perhaps it was too much to expect a soul-searching confession from Heisenberg and his colleagues rather than a rationale carefully tailored to the changing postwar situation. How many scientists in other nations who have worked and who continue to work today on instruments of mass destruction have bared their souls to posterity, or even to themselves?

What is remarkable in Heisenberg's case is that, despite their frustration with him, many of Heisenberg's severest American critics remained sympathetic and more than politely cordial toward him, even while publishing the most devastating public repudiations. It was as if they recognized how much they shared his difficulties, how much scientists everywhere are caught up in the universal dilemmas created by the rise of contemporary science in concert with the contemporary global power structure: that scientists everywhere, no matter how devoted they may be to the search for truth and universal understanding, are, for many reasons, invariably drawn into work for their governments, and that many will serve their governments by fashioning the weapons of war and destruction.

The closing paragraph of Samuel Goudsmit's obituary of the man he had so admired, and yet so reviled, expresses the frustration as well as the pity that many others must have experienced. "Heisenberg was a very great physicist, a deep thinker, a fine human being, and also a courageous person," he wrote. "He was one of the greatest physicists of our time, but he suffered severely under the unwarranted attacks by fanatical colleagues. In my opinion he must be considered to have been in some respects a victim of the Nazi regime."[31]

CHAPTER 30

THE LATER YEARS

HEISENBERG RETURNED IN JANUARY 1946 TO A GERMANY ON ITS KNEES. BOMBS AND artillery had reduced nearly every city and hamlet to heaps of rubble. Roads, railway lines, river passages, and bridges had been severed or destroyed altogether; gas, water, and electric lines had been cut. Nearly a quarter of German housing was lost, and industrial and agricultural production was at a standstill. Severe shortages of food, clothing, and shelter engendered fears of starvation and epidemic. Everywhere children begged for food while their parents rummaged through garbage for whatever they could find.

As Germany's economy and infrastructure lay in ruins, her population increased dramatically. Millions had been lost to death or were in Allied captivity, but many millions more had arrived in Germany—former slave laborers and those who had fled or been expelled from Eastern countries under Soviet domination. "[The Germans] are immeasurably depressed," Max von Laue wrote to his son in Princeton after returning to Germany. "The complete suffering of war makes itself felt only now." The psychological trauma of defeat, coupled with the shock of acknowledging that the man who had vowed to lead Germany to greatness was an unspeakably depraved criminal, added moral devastation to the physical destruction suffered by the once proud German nation.[1]

Death stalked Heisenberg's family and friends as it did many others. So many of the older generation died, it was as if fate did not intend that generation to survive into the postwar era. The conspiratorial members of the Wednesday Society, which had embodied the strengths and weaknesses of the Wilhelmine and Weimar eras, were only the most well known to perish in the last months of the war. In the only letter he received from home while in British captivity, Heisenberg learned of the sudden death of his 74-year-old mother in July 1945. She had broken her hip in a fall in Urfeld and died of complications several days later in a hospital in the nearby town of Bad Tölz. Heisenberg had been close to his mother right up to the end and had brought her to Urfeld in the last days before he was captured. In his lonely captivity, the news of her death was a heart-wrenching blow. "It was difficult to get over it as I was so alone," he wrote.[2]

Fearing that the physicist might be kidnapped by one of the other Allied powers, the British did not allow Heisenberg to travel to Bavaria, then in the American zone of occupation, until several months after his return to Germany. After heading for Urfeld, he made a tearful pilgrimage to his mother's grave. Under the dire postwar conditions, the family had not been able to provide a proper funeral. "It was very sad,"

Elisabeth and Werner Heisenberg in Göttingen, 1947.

he wrote his Uncle Karl after visiting the grave, "for it was nothing more than a heap of earth beneath a small wooden cross without a name."[3]

Within a year after the death of Heisenberg's mother, her sister died; so did Heisenberg's Osnabrück aunt, Guste, and his Uncle Karl and Aunt Helen, who both died in the same month. In the science world, the aged Hans Geiger and Gerhard Hoffmann died in the days following the capitulation, and the 90-year-old Max Planck died in Göttingen after losing all his children to the war, sickness, and the executioner. Robert Döpel's wife and an institute technician were killed in the last bombing raid on Leipzig, which also destroyed Heisenberg's already damaged house and most of the physics institute.[4] Kaiser Wilhelm Society head Albert Vögler and Reich Education Ministry chief Bernhard Rust committed suicide; Rudolph Mentzel, Johannes Stark, Philipp Lenard, and the senior Weizsäcker faced criminal trials. Every former member of the Nazi party faced a denazification tribunal; Heinrich Himmler and Hermann Göring committed suicide; Alfred Rosenberg was executed along with other war criminals at Nuremberg; and Albert Speer received a 20-year prison sentence.

With the I. G. Farben chemical complex broken up as a criminal enterprise, Heisenberg's brother, who had worked for the Agfa film and chemical company, his wife, and their four children lived meagerly after the war as Black Forest pottery makers. Elisabeth's brother, Fritz Schumacher, returned to Germany with the British forces to help begin Germany's economic recovery. Her sister and elderly parents waited patiently for better times as they prepared to welcome home their captured relatives at the Bavarian country cottage to which they had fled during the war.[5]

The Heisenbergs in Osnabrück had also fled to the countryside to escape the bombing. Heisenberg returned to the demolished city in early 1946 to find the house he had visited so often as a child in ruins. Sadly, he picked through the ashes of the once warm and cozy kitchen for a memento of his childhood, collecting a piece of tile from the old wood-burning stove near which he had once contentedly played. In the ashes of this safe and happy place to which he could never return, he saw the end of an era, both in his own life and in the life of the German people.[6]

His mother and father now gone, Heisenberg had just turned 44 when he arrived from England in early 1946 in the north German town of Alswede, near Lübeck, a village left undamaged by the ravages of war. He was still a detainee of the British Crown. There he and the other detainees were housed in a confiscated confectionery store to which they had to report every evening until moved to their permanent locations. A month after Heisenberg arrived at the confectionery store, his determined wife negotiated the chaotic railway and border system to travel across two occupation zones and nearly the entire length of Germany, finally reaching the husband she had not seen for nearly nine months.[7]

The weeks and months following her husband's capture in Urfeld had been dire ones for Elisabeth. While Heisenberg feasted on British officer's rations, sported on the tennis courts of the Farm Hall mansion, and formulated the motives of his wartime work, his wife, who had already borne six of their eventually seven children and whom he had left at the height of the war for bachelor quarters in Berlin, had to cope alone for months. Heisenberg openly worried about the "Urfelder," but there was little more he could do for them. Colonel Pash, who had arrested Heisenberg in May, returned to the house several times with groceries, but these ran out by summer 1945.[8]

Within a few weeks of Heisenberg's mother's death in July, Elisabeth's distant cousin who shared their Urfeld home also died. Elisabeth was left alone in the lakeside cottage to care for her cousin's son as well as her own six children. She no longer had her husband's Berlin salary and, in any case, there was little on which to spend it in the mountainous Urfeld area. The Sommerfeld family, still living in Munich, helped her as best they could, but it would be nearly 18 months before the Heisenbergs could be reunited under better circumstances. A photograph of Heisenberg and his wife, taken shortly after their reunion, shows a gaunt and determined Elisabeth. Heisenberg is energized, ready to face the postwar world with his warm smile, radiant eyes, and still-blond hair that made him look much younger than his forty-something years.

During his captivity in Farm Hall, Otto Hahn learned that he had been awarded the Nobel Prize for the discovery of fission; his colleagues Lise Meitner and Fritz Strassmann were not included. Of the ten detainees, the three Nobel laureates—Hahn, Heisenberg, and Laue—emerged as the leading spokesmen for German science. During their captivity, they had met twice in London with British scientists and had once driven into the German hills with their British control officer to discuss the future of German science. With the permission of the United States, the British had decided to settle the scientists permanently in the British occupation zone that stretched across the northwestern section of Germany. Following the Cambridge model, the scientists emphasized the necessity of reestablishing their institutes in a university town. Göttingen was the obvious choice for its academic tradition and many research institutes, and it had come through the war nearly unscathed. Only its proximity to the Soviet occupation zone caused Heisenberg some worry.[9] Göttingen was to become for the British a scientific and technological center that would eventually bring about the revival of German science, first in its own zone, then in the emerging West German state. Not wishing to subsidize or colonize its occupation zone indefinitely, the British more than any other Allied power intended the scientific revival to occur hand in hand with an economic and political revival that would eventually lead to German autonomy.

Some of the non-laureate detainees were less enthusiastic about the British plan. During the weeks following their arrival in Germany, Harteck and Diebner returned to their old Hamburg institute, and Harteck eventually emigrated to the United States. Gerlach went to Bonn and later returned to his chair for experimental physics in Munich. At the end of February 1946, Heisenberg and Hahn were the first to arrive in Göttingen. They were joined soon after by Laue, Weizsäcker, and several other former detainees.[10]

The British arranged for Heisenberg and Hahn to reestablish their old Kaiser Wilhelm institutes for physics and chemistry, respectively, in the empty rooms of Ludwig Prandtl's former Aerodynamics Experimental Institute. The institutions were still part of the state-run Kaiser Wilhelm Society and thus independent of any university. The plan nearly foundered at the start. In Hechingen, the French, still irked at the American seizure of nuclear equipment and scientists from their occupation zone, were holding instruments and technicians from the two institutes and were unwilling to release them to the British. With only three coworkers, no laboratory materials, and no association with the University of Göttingen, Heisenberg was miserable. To add to his troubles, the flood of refugees in the small town again made it impossible for him to find decent living quarters; he was reduced to sleeping on a straw sack in a dingy tenement.[11] For the moment, the Urfelder would have to remain in Urfeld.

In the midst of Heisenberg's Göttingen misery, the aged Sommerfeld revived the old question of who should be his successor in the Munich chair for theoretical physics. He had submitted yet another list of three candidates to the surviving Munich faculty: Heisenberg, Weizsäcker, and Friedrich Hund.[12] Heisenberg once again gratefully accepted, but he again needed the consent of higher authorities—this time the Allies—and their approval was uncertain. They had refused initially to allow Gerlach to return to his Munich chair.

Within a year it was settled: Heisenberg stayed in Göttingen but by his own choice. His institute was finally thriving, but more important, in the summer of 1946 British troops had vacated a beautiful mansion that they had confiscated on the Hainberg, the hill just outside Göttingen where Heisenberg had strolled with Bohr so many years ago. British science officers arranged for the house to go to Heisenberg. Elisabeth, in another display of almost superhuman determination, managed by herself to arrange for trucks to cross through three different occupation zones to gather their furniture from Hechingen and Urfeld and to take it, along with the six children, to Göttingen (where a seventh child was born). Her relative's child was apparently sent to live with other relatives. Together at last, the Heisenbergs moved into the Hainberg mansion in September 1946. It was so large (and they had so little money) that they rented out two of the rooms to Göttingen students, one the son of Johannes Popitz, the executed Wednesday Society conspirator, the other a Weizsäcker family friend.[13]

Once the Heisenbergs had settled in Göttingen and with Heisenberg's institute flourishing, a move to Munich lost any remaining attraction. They still owned the Munich house they had bought in 1937, but since it was undamaged by the war American officers had confiscated it for their use. In addition, American authorities were not then nearly as supportive of German science as were the British. Fritz Bopp, a former Sommerfeld pupil and a collaborator on the uranium project, eventually succeeded the great Sommerfeld in Munich in 1947.[14] Over a decade later, amidst a reorganization of West German nuclear energy policy, Heisenberg finally moved, along with his institute, to his beloved Munich, where he remained for the rest of his life. Long before the move, his institute was renamed the Max Planck Institute for Physics and Astrophysics, an institute within the Max Planck Society, the network of federally funded research institutes that replaced the Kaiser Wilhelm Society in the early postwar years. Upon the move to Munich, Heisenberg was named an adjunct professor at the university, but he never again held an academic teaching chair. He remained director of the non-academic state-supported institute until he was forced by illness to retire in 1970.

Despite the pleasant radiance that emerges from postwar photographs of Heisenberg, several of Heisenberg's colleagues throughout the later years observed that he seemed to suffer from a perpetual depression.[15] Aside from the toll taken by advancing age, he seemed particularly depressed by several factors: that his actions during the Third Reich were not understood abroad; that German physics, for which he had worked and suffered so long, was now indeed eclipsed by American physics; and that his own research was not as successful or as well received as it once had been. His response to depression was as it always had been—work and ever more work.

When not coping with Goudsmit and other critics of German war research, Heisenberg focused his energies after the war on two major concerns: science and science policy. Until the Western Allies ended the formal occupation in 1955 and granted sovereignty to West Germany as part of the NATO alliance, science policy issues took precedence and so absorbed him during the late 1940s and early 1950s that it is hard to believe he had much time, or emotional energy, for anything else. His most well-known new physics, a proposed unified field theory, did not fully emerge until the late 1950s.

Heisenberg again regarded his efforts to influence science policy as a service to German science, and again he convinced himself that he was the most qualified to assume this burden. His experiences with the Third Reich had deepened rather than mitigated his perceptions of duty. He seemed determined that German science should return again to world-class standing and that German scientists, himself naturally included, should never again be disregarded or abused by government authorities. In the international arena, he avidly lobbied for the establishment of a European accelerator facility, the Centre Européen pour la Recherche Nucléaire (CERN) in Geneva, to rival the new American machines, and he served actively on its governing policy committee.[16] In 1952 he became president of the Alexander von Humboldt Foundation, a federal agency that brought foreign postdoctorates to West Germany to perform research and in the process to broaden themselves and reestablish their countries' contacts with German science. As a one-time foreign postdoctoral student in Denmark, Heisenberg knew the value of work abroad and came to cherish the Humboldt Foundation presidency above all the many offices he held. It was the last official post from which he resigned when illness set in during the final years before his death.[17]

Domestic policy issues consumed even more of Heisenberg's time. Until preparations began to end the military occupation and establish the West German state in the former French, British, and American zones, Heisenberg focused his efforts on occupation problems. The occupation itself was coordinated in Berlin through the Allied Control Authority, which was headed by the military governors of the four occupation zones. After the military occupation ended in 1949, the civilian Allied High Commission maintained control until 1955. The four Allies did not fully relinquish their control rights until German reunification over 35 years later.

At least two Allied policies directly affected German science: denazification and Allied Control Law 25, the control of scientific research.[18] Denazification, declared to be a dominant objective of the occupation, was intended to rid German public life of Nazi influence. It was enforced to widely varying degrees in the different zones, but all Germans, in whatever zone, were required to fill out a questionnaire about their political past. Former members of the Nazi party or party organizations had to appear before a military tribunal to explain their activities. In the British and American zones, the tribunals were soon turned over to local courts. For many, an appearance before a tribunal was likely to result in the loss of a job, since former party members were generally excluded from civil service posts, which included all teaching positions. Anti-party feeling ran high, and many German academics found it impossible to work. After one science teacher lost his job due to party membership, Max von Laue complained angrily to his son that denazification as practiced by the Americans in particular made "every use of reason impossible." Realizing finally that the Western zones could not be turned into a power vacuum and that their economy and society could not be revived if every single party member was excluded, the U.S. occupation command practically ended denazification as a broad-ranging policy after 1947.[19]

People called before denazification tribunals were permitted to submit affidavits from prominent persons testifying to their behavior. The testimonials were appropriately nicknamed "Persilscheine" (whitewash certificates), after the popular laundry soap Persil, whose slogan is "not just clean but pure." Heisenberg was frequently sought out as a writer of such certificates by his friends and coworkers and by those to whom he owed a debt—those who had helped him during the SS affair, as well as receptionists and lower functionaries in Rust's and Himmler's offices who had enabled access to their bosses. But he refused to supply testimonials to those whom he did not know or did not care to support. He wrote an evaluation of Stark on behalf of the prosecution at Stark's trial.[20]

Although it is uncertain how effective Heisenberg's certificates were, the occupation authorities seem to have greatly valued Heisenberg's judgment. The British and American authorities had already decided, or had become convinced, that Heisenberg and the other atomic scientists were far too valuable to the revival of German science to be held in any but the highest regard. Heisenberg himself could claim that he had been victimized by the Nazis. Having never joined the party or its affiliations, his own questionnaire was both clean and pure.[21] Moreover, most of the denazification proceedings were completed before the controversy in American scientific circles over Heisenberg's wartime activities. Most local authorities in the British and American zones probably took little note of the squabble.

Countering the Allied law controlling science was a more difficult matter.[22] At the top of the projects prohibited by the law was research in applied nuclear physics, which included every form of nuclear reactor research and isotope studies, as well as cyclotron construction and experimental high-energy physics. All research had to be cleared in advance by a science officer, and all results and publications had to be submitted to the officer for review. The prohibited nuclear research included the very subjects to which Heisenberg had intended to return. In particular, unlike the Allied emphasis on product research and development, Heisenberg had planned that nuclear reactor technology would be the foundation of the revival of German physical science, and he tried to argue that progress in this area would lead to the revival of the entire German economy. He believed too that the prestige and momentous impact of nuclear energy must not be denied to West Germany or to West German scientists. Heisenberg, Hahn, Gerlach, Weizsäcker, and others campaigned vigorously for the promotion of reactor research, taking their message to the public as well as to the occupation authorities.[23]

Control Law 25 was finally relaxed in 1949, but non-theoretical applied nuclear research was not permitted in Germany until as late as 1955—the Allies did not want the Germans playing with nuclear fission under any guise. Once it did become legal, however, both the government and Heisenberg's institute jumped at the opportunity. Within a decade West Germany was the world's leading exporter of nuclear technology. At the same time, as noted earlier, Heisenberg and his closest colleagues vigorously opposed the 1955 NATO plan to equip the West German army with battlefield nuclear weapons. The scientists mobilized the German populace in ways they had never considered previously, and the plan was so soundly defeated in 1958 that the West German army remained non-nuclear. The prospect of nuclear weapons in German hands had already caused enough problems for the scientists.[24]

Heisenberg's domestic governmental policy efforts spoke directly to a historic, century-old conflict within German culture that resurfaced at the end of the war: the sometimes virulent competition between the previously independent and now federalist-minded states, or *Länder*, and the central government, or Bund, over responsibility for supporting scientific research. Financial support was tantamount to administrative control. Heisenberg and his supporters allied themselves with the centralist faction. Chancellor Konrad Adenauer's establishment in 1955 of a cabinet-level "federal ministry for atomic questions" to promote nuclear energy development was their most striking achievement. But Heisenberg and the centralists were much less successful in forging a strong national authority over other branches of scientific research. The delicate balance today between Bund and Länder regarding research funding is the hard-won result of decades of careful negotiations and contractual agreements.

For more than a century, the regulation and support of education and culture, including science, had been the prerogative of each Land, and since Bismarck the prerogative had been law. Most research was done at universities, and universities were under the authority of the cultural ministry in each *Land*. Two interstate organizations had challenged this arrangement in the early decades of the twentieth century. In the throes of industrial revolution, Kaiser Wilhelm II had expressed his personal interest in promoting science and technology by establishing the Kaiser Wilhelm Society, the network of pure research institutes in various scholarly and scientific fields supported directly by the Reich government. Second, during the early years of the Weimar Republic German scientists had established the competing Emergency Association of German Scholarship in support of impoverished laboratories and researchers. It bypassed the federal government by funneling public and private funds directly to individuals and university laboratories through association-appointed committees of scientists and administrators. Grants were awarded on the basis of peer review of project applications, a process they called the self-administration of science.

Most of the original Emergency Association administrators enjoyed close ties to the division for higher education in the large and powerful Prussian Culture Ministry. But soon after coming to power in 1933, Hitler had reduced the semi-autonomous states to powerless administrative districts. When he elevated the Prussian Culture Ministry to the status of Reich Education Ministry, Emergency Association administrators became minor functionaries controlled by party and SS bureaucrats. Rudolf Mentzel's science section, which oversaw the Emergency Association, grew out of the Prussian division for higher education.

After the war, most members of the former Prussian higher education division migrated to the culture ministry of Lower Saxony in the British occupation zone. In concert with American Zone colleagues, the Prussian bureaucrats soon founded a new Emergency Association for the promotion of science.[25] They were determined to rebuild German university science and to avoid the errors of the past in the democracy of the future. In their view, the major error had been neither their own disdain for the

Weimar democracy nor the blatant support for Hitler exhibited by some of their leaders. Their problems had their roots in the failure of the Weimar constitution to prevent the creation of the Reich ministries, which encroached on state prerogatives and ultimately on the administrators' personal spheres of influence (as though Hitler would have paid the slightest attention to any constitutional checks on his power).

This time the bureaucrats were determined to keep the emerging postwar federal government off their cultural and scientific turf. So-called self-administration served once again as their chief ideological weapon. Their position found powerful support in the conservative Christian Democratic Union, which was emerging as the dominant political party in the British zone and in West Germany. They also found themselves supported by the state cultural ministers and university rectors and by the policies of the occupation forces. A 1948 aide-mémoire from the military governors to the German Parliamentary Council, which was deliberating the constitution of the new government, declared, "The powers of the federal government shall be limited to those expressly enumerated in the constitution and, in any case, shall not include education, cultural and religious affairs."[26]

The Allied attitude, especially that of the Americans, who were inherently more oriented toward decentralization (federalism) than the British and who oversaw the most federalist state of all, Bavaria, was hailed by the states and by Kurt Zierold, the administrator of the higher education department in the culture ministry of Lower Saxony.[27] Zierold was a jurist and former official of the Weimar-era Emergency Association. However, in 1946 the centrally oriented British, with advice and influence from Heisenberg and Hahn, reincarnated the Kaiser Wilhelm Society with Otto Hahn as president. Within two years, the society of federal research institutes was operating on an inter-zonal basis with inter-zonal funding and a new name: the Max Planck Society. This revival served to induce Zierold and the Western culture ministers to reincarnate the Emergency Association for the repair and support of university laboratories. The plan naturally met with little objection from the increasingly powerful conference of West German university rectors, of which Zierold happened to be president.

The rectors and culture ministers officially recognized the new association in 1949, just as the Parliamentary Council laid the foundations for the West German Federal Republic in its new constitution, or Basic Law. According to its bylaws, the new Emergency Association called once again for the self-administration of science. Its supporters even believed this notion to be legally sanctioned by Article 5 of the Basic Law, which declared simply, "Art and science, research and teaching are free." The Emergency Association interpreted this article to mean free from interference by federal bureaucrats. Not surprisingly, the new organization located its headquarters in Hannover—in the department for higher education in the culture ministry of Lower Saxony.

The administrative territory staked out by the Emergency Association was challenged just two months later by the founding of the very different German Research

Council (*Deutscher Forschungsrat*), or DFR, headed by Werner Heisenberg. The DFR naturally placed its headquarters near Heisenberg's Max Planck Institute in Göttingen—a town that was conveniently located also in the state of Lower Saxony.[28]

The DFR's roots, like those of the Emergency Association, lay deep in Germany's past, but it inclined in the opposite direction: toward federal authority, elite scientists, and research policy and planning, rather than toward the direct funding of research. Like the Emergency Association, the DFR enjoyed powerful patrons among the occupation authorities—specifically in the Research Branch of the British Control Commission, with which Heisenberg and Hahn were closely associated. It also had the backing of the more centralist Social Democratic Party (SPD), reemerging in West Germany, and of Konrad Adenauer, the new federal chancellor.

In December 1948, Carlo Schmid, the head of the SPD delegation to the Parliamentary Council, read into the record a letter received from Heisenberg and three other members of an advisory science council established by the British Research Branch. It urged that the organization and promotion of scientific endeavors be assigned to the federal government rather than to the individual states in the new Basic Law. The arguments of the scientists are not surprising: "The individual state cannot bear the responsibility for and the financing of German scientific research, which has long since outgrown the boundaries of individual states." The needs of large-scale projects, such as nuclear reactors and accelerators, or large networks of institutes, such as the Max Planck Society, could not be met by the meager resources of individual states. They also declared, "We must look with horror at where an attempt to limit the life-sustaining field of science has already led in the last years." In other words, local bureaucratic control, not the creation of the Reich, had enabled the nazification of German science.[29]

The scientists' letter and the parliamentary debate had two effects: first, the phrase "promotion of scientific research" was inserted into article 74 of the West German Basic Law, which enumerated the concurrent powers exercised by the federal government and the states. Second, the DFR was formally created in March 1949 by the Max Planck Society and the Western academies of science, who placed it under the direction of their favorite physicist: Heisenberg. The original 15 self-appointed members of this council were all wellknown, elite academic scientists close to Heisenberg and his Göttingen circle, and most had been associated in some way with Heisenberg and his nuclear fission project during the war.[30] But not one of these council members was associated with technology, industrial research, or education at universities and technical colleges—the very fields the DFR intended to manage. This lack of experience, the perceived elitism, and the apparent condescension toward technology were tactical blunders, naturally alienating leading figures in those fields.

In his address to the inaugural convention of the DFR, Heisenberg went so far as to proclaim the new council "the sole representative of all German science" and the professional representative of science and technology in Germany and abroad. The council's bylaws accorded it the power not only to advise federal and state govern-

ments and to represent German science at international gatherings but also to participate "in the financing of scientific research, in particular in the solicitation and distribution of public funds for research purposes." Having thus ignored its competitor,
the Emergency Association, in these sweeping tasks, the council proceeded to make
clear in its bylaws the inferior status it accorded to the association: "In the fulfillment
of these tasks the German Research Council will rely foremost upon the Emergency
Association of German Scholarship."[31]

Needless to say, the Emergency Association did not appreciate being cast as a
supporting player in the DFR's show, it was all too reminiscent of the attempted
usurpation of power by the Reich Research Council. Zierold complained privately to
British authorities that the DFR was a superfluous organization, consisting wholly of
effete snobs who had no intention of reviving German science where it belonged—at
the universities and technical colleges. Democracy itself—a concept to reckon with
during this era—became a pawn in this game. Zierold argued that a self-appointed
committee of elite academics could have little acquaintance with democratic principles, to which Heisenberg replied that the regional culture ministers and university
rectors could hardly set themselves up as the guarantors of democracy after their
"shameful" behavior during the Third Reich: "an all-too-eager submission to the state
authorities."[32] No matter that the elite scientists had hardly distinguished themselves
for their resistance to the Nazi regime or that Heisenberg's organization now made no
secret of its eagerness to submit to federal authority.

State and federal authorities immediately came to the rescue of their respective
organizations. Within a month of the founding of the DFR, the western culture ministers issued a joint declaration, the Königstein Agreement, the first sentence of which
struck to the heart of the matter: "The states of the three Western zones consider the
promotion of scientific research as fundamentally a task of the states," hence of the
culture ministers.[33] Article 74 of the Basic Law notwithstanding, the declaration
remained in force for over a decade and became the basis of all state-federal negotiations in science and research for decades.

Not to be trumped by the culture ministers, Chancellor Adenauer requested and
received a memo from the DFR on its intentions.[34] The memo, written by Heisenberg,
called for recognition of the DFR as the sole scientific advisory panel to the chancellor and for the establishment of a small service bureau in the chancellor's own
office—presumably as close to his desk as possible—to coordinate direct advice from
the scientists and to sort out the research policies of the federal government vis-à-vis
the states. Heisenberg discussed his strategy in a letter to a DFR colleague: "The
Federal Chancellor is the only strong personality" who is in a position "to carry out
our wishes for the centralized direction of research."[35] No stranger to realpolitik,
Adenauer deftly played both ends against the middle. While refusing to supply federal funds or even to grant official recognition to the DFR, he gladly welcomed
Heisenberg's service bureau as an opportunity to control these elite scientists while
outmaneuvering the culture ministers. By appearing to listen to the scientists' wishes,

he could co-opt their allegiance, while pointing to their organization as an excuse to ignore the pressure exerted by the culture ministers.

Despite their competition, both policy groups managed to register notable successes during the two years of their simultaneous existence, a crucial period for the Federal Republic. While the Emergency Association raised over a million marks for university and local research labs through its fund-raising efforts in German industry, Heisenberg and the DFR gained U.S. Marshall Plan money for German science, even though the Marshall Plan had originally excluded science from support. Heisenberg also arranged for the admission of the Federal Republic to UNESCO's International Union of Scientific Councils and applied for a further relaxation of new science control laws issued by the Allied High Commission in 1950. He presented several detailed memos to Adenauer on the organization and funding of research in the present and future West German state. Heisenberg told his man in the chancellor's office, "It is . . . a fundamental question of the international competitive ability of German research and thus, ultimately, of the German economy." As might be expected, in each memo Heisenberg's foremost concern was that Germany be prepared for immediate development of nuclear energy as soon as the Allies permitted it.[36] No wonder he could not be perceived in those years as having had anything to do with nuclear weapons during the war or of having compromised his competency in nuclear research.

Yet pressure was also mounting on the DFR to "fuse" with the Emergency Association. German industry, academies, and scientists realized that the embarrassing competition for funds and control was hindering, rather than helping, the recovery of German science and technology. The culture ministers at first categorically demanded that the DFR be joined with the Emergency Association, but the general membership of the Emergency Association rejected a temporary agreement proposed in 1949 because it did not go far enough—they really wanted the total subordination of the DFR.[37] By 1951 the pressure was unbearable: industrial donors especially refused to contribute to one organization, the Emergency Association, for purposes claimed by another organization, the DFR. Adenauer, for his part, allowed the controversy to continue as a way of diverting the scientists and administrators from the increasing power of his chancellery.

Thanks to the intervention of Heisenberg's colleague and war-time nuclear administrator Walther Gerlach, who was now rector of the University of Munich and a leading figure in the Emergency Association, both sides gradually reached an agreement on fusion. Just as fusion seemed imminent, however, Adenauer suddenly tendered his official recognition of the DFR—but only if it remained an independent organization. Heisenberg, taking the lure, responded with a new memo on the organization of German science in which the DFR's independence was assured.[38] Only Gerlach's renewed diplomacy could save the fusion, set for early August 1951. Again Adenauer blocked the effort with an eleventh-hour promise of new research funds, and again only outside pressure on the reluctant Heisenberg, exerted by scientists, industry, and the academies of science, forced his acquiescence.[39] The two bodies finally joined in August to form the German Research Association (DFG), the present-day equivalent of the U.S. National Science Foundation and National Endowment for the Humanities, combined. According to prior agreements, the DFR, upon joining with the Emergency Association, became the distinguished "senate" of the new German Research Association. Heisenberg was elected to the presidium of the DFG and to the chairmanship of its influential committee on nuclear research, with the understanding that the senate would continue to pursue the goals of the now-defunct DFR.

Until as late as 1969, the federal government supported only such interstate research organizations as the Max Planck Society and only those fields that most contributed to national prestige and influence—namely, nuclear energy, space research, and computer technology. Matters changed in the late 1960s when, in the midst of a supposed technology gap with the United States, marked by a brain drain and fears of being flooded with high-technology American products, the more centrally oriented Social Democrats came to power for the first time in the federal government. The new government began to pursue a more broad-based effort to stimulate German science and technology through federal initiatives and closer federal-state-industry cooperation.[40]

Heisenberg himself regarded the whole affair as a defeat both for him and for German democracy—and as further cause for general despair. In the 1951 closing report of the DFR, he described what he saw as the cause of the defeat, which, in fact, applied equally to both the DFR and the Emergency Association. "We do not have an old democratic tradition," he wrote, "and we Germans are in general grateful when we can turn over the responsibility for public life to our superior authorities."[41]

Heisenberg's intense concern with science policy issues throughout the postwar era did not hinder his scientific concerns, which, as before, flourished simultaneously with political events.[42] The simultaneity was evident at the start: Heisenberg's Farm Hall interlude initiated a flurry of post-detention publications that set the stage for his later research. The contemplative leisure afforded at Farm Hall and the shared detention with Max von Laue and Carl Friedrich von Weizsäcker stimulated Heisenberg's continued work in two areas: the phenomenon of superconductivity, which he discussed with Laue, and the problem of hydrodynamic turbulence, which he explored with Weizsäcker. His papers on superconductivity during the late 1940s were the less successful of the two efforts. The phenomenon, still in some respects a mystery, entails a sudden drop in the electrical resistance of a material to zero at a definite critical temperature, a very low temperature. Although Heisenberg succeeded in demonstrating that superconductivity can be seen as a type of phase transition, similar to the condensation of a liquid from vapor, he failed to account for the transition.[43]

Pursuing his own interests in astrophysics during and after the war, Weizsäcker examined the properties of turbulent rotating masses of hot gases as a model of spiral nebulae and as an account of the formation of planets in our early solar system.

Weizsäcker's work apparently encouraged Heisenberg's Farm Hall return to hydro-dynamic turbulence, the subject of his doctoral thesis, as well as a joint paper with Weizsäcker on spiral nebulae. During the late 1940s, Heisenberg published several studies and talks on a new statistical theory of turbulence.[44] Nevertheless, the main impetus for Heisenberg's return to turbulence was probably an even more fundamental concern: quantum field theory.

In one of his papers just before the war, Heisenberg had suggested the treatment of explosion showers produced by cosmic rays hitting the atmosphere or metal plates in cloud chambers as a spray of "droplets" unleashed by a turbulent field made up of matter. The explosion was produced when the collision of the particles occurred in a region that is smaller than a universal minimum length. The result would be a burst of mesons and neutrinos that could penetrate large blocks of lead. Unfortunately, detection of Heisenberg's predicted showers could not be confirmed, and most physicists, especially those in the United States, doubted Heisenberg's program for a future theory based upon the existence of a fundamental length.

Soon after the war, two events revitalized quantum research. First, a number of theoreticians in the United States invented the process of so-called renormalization. Put briefly, the infinities and divergences plaguing quantum electrodynamics could be simply defined away, rendering the theory applicable to all energies and even to the very smallest distances of approach between particles. Again, it seemed, there was no need for any new theories.[45]

Second, in 1947 Cecil Powell and coworkers in Bristol, England, finally unraveled the puzzle of cosmic-ray mesons and in the process confirmed the existence of Heisenberg's explosion showers, now called multiple processes. Heisenberg's showers were predicated on a so-called strong force in Hideki Yukawa's theory of nuclear forces. This force that binds protons and neutrons together to form an atomic nucleus could be envisioned as arising from the exchange between a proton and a neutron of a heavy elementary particle—the pi-meson, or pion. Powell discovered that the meson usually observed in cosmic-ray experiments is not Yukawa's pion at all but rather the mu-meson, or muon, belonging to Enrico Fermi's theory of beta decay, which entailed another force of nature, the weak force. Pions and muons are produced in multiple processes, and they seem to arise from field theories that contained mathematical infinities at distances of extremely close approach. But, unlike quantum electrodynamics (QED), the infinities in these theories, hence the explosion showers they release, could not be defined away.

Powell's discoveries, as well as difficulties with the wartime S-matrix, revitalized Heisenberg's quantum field physics for the study of high-energy elementary particles. This work culminated in 1958 in his proposal of a new unified theory of three of the four forces of nature (except gravitation) that, though beset by difficulties and never accepted by most physicists, implied for Heisenberg the onset of his long-sought new revolution in quantum physics. As Heisenberg struggled for his policy preferences during the late 1940s and early 1950s, in a series of papers presented to the Göttingen

Academy and to the newly established *Zeitschrift für Naturforschung* (*Journal for Natural Research*), Heisenberg evaluated the situation in quantum field theory and set his program for future, revolutionary advance.[46]

In field theory, he observed, particles and their interactions are represented by fields that satisfy one or more wave equations, which must in turn satisfy the requirements of special relativity theory. But whenever these equations are quantized, the physical properties of the fields diverge to infinity. These infinities seemed to arise from so-called local interactions, interactions occurring at arbitrarily small distances. They could be avoided either by a cutoff of short distances (small wavelengths), or by resorting to the S-matrix, which avoided events within the problematic region of approach, or by Heisenberg's preferred method—the introduction into the theory of a minimum length as a lower boundary. But, he discovered in 1951, the latter proposal allowed for violations of causality arising from special relativity theory, which, unlike 1928, was now to be maintained. Yet the causality principle for local events seemed incompatible with the quantization of relativistic equations. The S-matrix was one way to avoid this dilemma, by dancing around the local events, but the S-matrix could hardly be considered satisfactory. It connected events long before and long after a collision, without illuminating anything about the actual collision. The S-matrix, Heisenberg declared in his first postwar paper, was meant only to provide a general "mathematical framework of quantum field theory."[48]

To Heisenberg, the situation was, once again, similar to that of the early 1920s, the years just preceding quantum mechanics. This time he believed that explosion showers and turbulence provided the critical clues to the revolutionary new field theory. Previous field theories treated elementary particles and the forces between them as distinct field entities. The dissipation of the turbulent field into a shower of myriads of particles at short distances suggested to Heisenberg that the new field theory should deal, not with individual particles and fields at all, but with one general overall matter field. Distinct elementary particles would appear as stationary energy states of this general matter field. Like a liquid flowing in a channel, the field would become turbulent when confined to a "bottle" smaller than the universal minimum length.

For the mathematical representation of these properties of the matter field, Heisenberg chose a simple, relativistic quantum wave equation studied by Max Born and Leopold Infeld during the 1930s. For the wave function he chose one with mathematical properties introduced earlier by Paul Dirac. The appearance of the universal length in the equation controlled the onset of explosion showers, while the length itself determined the masses of the elementary particles condensed from the matter field.

Amazingly, the new theory worked. As so many times before, the ingenious Heisenberg had incorporated the main difficulties of current theories into a new and potentially revolutionary theory, a theory that would be truly unified because all of matter and forces could be reduced to one simple set of equations for one unified field encompassing every form of matter and force. It was indeed a revolutionary theory.

Heisenberg spent over eight years exploring the problems and possibilities of his new theory. The properties of various formulations of equations, which were, in mathematical parlance, "nonlinear," were among the most difficult equations to solve. Their relationships to various models, laws (conservation and causality) and properties (symmetry and mass) of elementary particles had to be explored. By 1957, Heisenberg had modified his new matter field to form a field containing eight components, or dimensions. But what kind of field equation would it satisfy?

As in the early 1920s, Heisenberg, once again in Göttingen, turned increasingly to his old friend and colleague Wolfgang Pauli, now back in Zurich, for advice and criticism. Their correspondence, recently published, attained even greater intensity than before. This was especially the case when in February 1957, amidst the scientists' opposition to German nuclear weapons, Heisenberg fell ill and retreated with Elisabeth to the town of Ascona on Lake Maggiore in Switzerland. A mathematical "battle of Ascona" broke out between Heisenberg and Pauli over the technical arcana of relativistic unified field equations.[49] Heisenberg insisted on invoking an earlier proposal, apparently made by Dirac, to extend the number of wave functions allowed. Pauli just as resolutely refused. After six weeks of what Heisenberg described as a painful battle, Pauli finally capitulated. On his return home, Heisenberg stopped off in Zurich for a medical checkup and a mopping-up operation on Pauli, who finally conceded "boring unanimity" of opinions. Heisenberg returned to Göttingen to continue the work and to tangle in public with Adenauer over weapons and reactors.

Nine months later, Heisenberg returned to Switzerland, this time for policy meetings at the CERN accelerator in Geneva, and stopped in Zurich for a stimulating visit with Pauli. Within a few weeks of the visit, Heisenberg recalled, he happened on a very simple field equation that seemed to satisfy every symmetry property demanded of it. Pauli was elated; a joint paper on the Heisenberg-Pauli equation, the basis of a unified field theory, seemed in order as soon as the mathematical consequences could be worked out. Fittingly, it would be their first joint publication since they had laid the foundations of field theory almost exactly 30 years earlier.

The paper was never published. As mathematical difficulties mounted, Pauli left for a prearranged two-month visit to the United States, while Heisenberg, his family, and his institute prepared to move from Göttingen to their new quarters on the northern outskirts of Munich. With the mathematics still unsolved, the two physicists decided to publish the equation only as a so-called preprint, a preliminary communication of results to be sent to selected physicists. The distribution was set for February 27, 1958. A 14-page typescript was prepared in English and duplicated on a mimeograph.[50]

Three days before the planned distribution preprint, Heisenberg announced the new formula in a lecture at the physics institute of the University of Göttingen. An eager reporter in the audience relayed word of a sensational new "world formula" around the world. One enthused press agent proclaimed, "Professor Heisenberg and his assistant, Wolfgang Pauli, have discovered the basic equation of the cosmos!"[51]

Two months later, more than 1,800 listeners turned out to hear Heisenberg reveal the secret of the cosmos in the same auditorium on the occasion of Max Planck's one-hundredth birthday. During his highly technical talk, Heisenberg carefully wrote his new equation on a transparency projected onto a screen in the darkened room. As the two-foot-high symbols slowly appeared on the huge screen, flashbulbs popped all over the hall. Just as an equation-filled page from one of Einstein's (unsuccessful) field-theory manuscripts had made it onto the front page of the *New York Times* in 1949 under the heading "New Einstein Theory Gives a Master Key to the Universe," Heisenberg's so-called world formula found its way onto front pages throughout Germany. In both instances, the more incomprehensible the purported key to the cosmos and the more public the physicist, the greater the public's fascination with both the physics and the physicist.

The public quickly regained its senses, however, after Pauli's sudden renunciation of Heisenberg's world formula. Ever the critic, Pauli had grown increasingly doubtful until, two weeks before Heisenberg offered the formula to the eager Planck celebration, he refused any further support of the theory in a strong letter to Heisenberg and in a two-paragraph statement in English that he distributed to 67 leading physicists.[52] But the letter and Pauli's renunciation did not deter Heisenberg. He presented his formula to receptive audiences all over West and East Germany.

A July conference on elementary particles at CERN brought Heisenberg face to face with his critic for the first time since the renunciation. Pauli was chairman of the session at the CERN conference in which Heisenberg was scheduled to present again his new field equation. Pauli opened the session on fundamental ideas in field theory with the remark, "What you will hear today is only a substitute for fundamental ideas." During the discussion he declared Heisenberg's work to be "mathematically objectionable," then he proceeded to tear it apart.[53]

Over a decade later, Heisenberg was still smarting. "Wolfgang's attitude to me was almost hostile," he wrote. "He criticized many details of my analysis, some, I thought, quite unreasonably."[54] Most physicists, especially those in the United States, were already doubtful of Heisenberg and his theory and did not consider it further. Nevertheless, it was pursued, improved, and modified thereafter by Heisenberg and his disciples in Munich, especially by Hans-Peter Dürr, Heisenberg's immediate successor as institute director.

The CERN meeting was the last that Heisenberg would see of the man who had worked so closely with him throughout his career and who had so greatly influenced his many contributions to physics. Pauli returned to Zurich after the CERN conference. Four months later he died suddenly of cancer at the age of 58.

Heisenberg's move at last to Munich in September 1958 as director of the Max Planck Institute, 22 years after his initial call to succeed Arnold Sommerfeld at the university, and now 56 years of age, marked the beginning of the last phase of his life. It was a phase characterized by extensive travel, continued work on field theory, gradual withdrawal from science policy affairs, greater involvement with his institute and family, increasing concern with placing his work in philosophical perspective, and the sad deaths of many of his teachers, colleagues, and competitors: Max von Laue, Erwin Schrödinger, Niels Bohr, Otto Hahn, Lise Meitner, Paul Scherrer, Max Born, Hans Kienle, and J. Robert Oppenheimer. Sommerfeld had died in 1951, Einstein four years later.

Faced with his own mortality and the likelihood that he would never regain the stature he had once enjoyed, Heisenberg now increasingly attempted to place his life's work in a permanent intellectual tradition. His fiftieth, fifty-fifth, and sixtieth birthdays brought him renewed concerns about his advancing age and his ability to continue first-rate physics. Under the influence of his longtime friend and colleague Carl Friedrich von Weizsäcker, his preferred intellectual tradition derived from ancient Greek philosophy. One former student recalled that Weizsäcker and Heisenberg began every lecture course, no matter what the subject, with a reference to Greek philosophy.[55]

By the winter of 1955–1956, when Heisenberg delivered the prestigious Gifford Lectures on physics and philosophy at the University of St. Andrews in Scotland, he had already distinguished contemporary elementary particle physics from nineteenth-century atomism. For him, the latter was a form of repugnant mechanistic materialism derived from the atomic theories of Democritus and Leucippus; the former held closest affinity to the work of the sagacious Aristotle. The underlying matter field of Heisenberg's unified field theory bore similarities to the notion of substance in Aristotelianism, an intermediate type of reality. Measuring the properties of elementary particles seemed closest to the Aristotelian notion of "potential" (potentia), since, as he had declared in his uncertainty paper, the particle comes into full being only in the act of measurement.[56]

By the 1960s, for Heisenberg particle qualities had succumbed to the symmetry properties of field equations, and Aristotle had succumbed to Plato. The Platonic atoms of his remembered youth were now fundamental. "The particles of modern physics are representations of symmetry groups and to that extent they resemble the symmetrical bodies of Plato's philosophy," he declared in one of his last publications.[57] In his 1969 memoir, written as a Platonic dialogue, he claimed that Platonism had dominated his thinking throughout his career. Toward the end of the memoir he wrote of his happy days in the old Urfeld cottage during the 1960s when—with Colonel Pash and the war far behind him—"we could once again meditate peacefully about the great questions Plato had once asked, questions that had perhaps found their answer in the contemporary physics of elementary particles," a physics that found its meaning in the ancient idealism and transcendent philosophy of Plato.[58]

The Platonic contentment was interrupted more and more by illness. A liver condition caused increasing weakness, dizziness, and depression. The slender young man who had dazzled audiences with his scientific bravado and friends with his physical courage had turned by the late 1960s into an obviously aging physicist to whom life and physics were no longer as kind as they had once been. As he closed his memoirs in 1969, he knew that the end would not be long in coming. In

Heisenberg surveying the Urfeld scenery, late 1960s.

his last sentences, he turned to what had been the one sustaining force in his life—the wondrous beauty and harmony of classical music. As he recounted listening to two of his sons and a colleague play Beethoven's youthful Serenade in D Major one sunny afternoon in an institute in the beautiful Bavarian countryside of the land he so loved, he closed by extolling how the Serenade "brims over with vital force and joy. . . . Faith in the central order keeps casting out faintheartedness and weariness. And as I listened, I grew firm in the conviction that, measured on the human scale, life, music and science would always go on, even though we ourselves are no more than transient visitors or, in Niels' words, both spectators and actors in the great drama of life."[59]

A few years later, Heisenberg fell ill again and was hospitalized. Exploratory surgery indicated advanced cancer of the kidneys and gall bladder for which little could be done. Chemotherapy helped delay the inevitable. In 1975 his condition worsened. He was hospitalized again and returned home too weakened to recuperate.

Werner Karl Heisenberg died peacefully at home in Munich on Sunday, February 1, 1976.

Over three decades and a new century later, the physics, the world, and the problems have undergone amazing advances and profound changes since Heisenberg's day. Even the terminology has taken on new meanings. Yet as much as change has occurred and even accelerated in recent years, many of the fundamentals, dilemmas and difficulties associated with physics and the social, political, and cultural sphere still bear the imprint of Werner Heisenberg and his generation.

In science, the quantum revolution unleashed by Einstein, Planck, Sommerfeld, and Bohr, and brought to fruition by Heisenberg, Pauli, Schrödinger, Dirac, and many others, is still the foundation of our understanding of events on the atomic and sub-atomic scales. The formulation of quantum mechanics during the 1920s, the development of relativistic quantum field theories, the discovery of nuclear forces, the study of these forces through high-energy collisions, and the search for a unification of the four forces of nature were all areas of research in which Heisenberg, together with his colleagues, were at the forefront of theoretical research. They made many of the initial breakthroughs and helped set much of the agenda for future work. The struggle continues today to understand the quantum mechanical behavior of events at the smallest levels, to unify these forces, and to study their properties through the collisions of high-speed elementary particles. The work has made enormous strides during recent decades, resulting in a unification of three of the four forces (except gravitation) in the so-called standard model. It is this progress that has recently inspired the highest energy accelerator ever built, the Large Hadron Collider, opened in 2008 at CERN near Geneva, Switzerland.

The impact of Heisenberg's uncertainty, or indeterminacy, principle has been equally profound, yet even more puzzling. As part of the Copenhagen Interpretation of quantum mechanics, it has been from the start an object of much philosophical study and experimental research into its limits and validity, and a cause for metaphysical speculation, both scientific and non-scientific. Recently, however, experiments have led to a confirmation of the strange, related phenomenon of quantum entanglement. This has led in turn to efforts to make practical use of entanglement in such areas as quantum encryption, telecommunications, and super-powerful quantum computers. If realized, they would add to the many applications of quantum mechanics already changing our lives, from lasers and medical imaging to the transistors of the current digital revolution.

During the slightly more than 74 years of his life from 1901 to 1976, Heisenberg lived in Germany through two lost world wars, three revolutions (1918, 1933, 1945), and four very different political regimes—the Wilhelmine monarchy, the Weimar democracy, the Third Reich, and the democratic Federal Republic. As in quantum mechanics, Heisenberg and others of his generation were on the early forefront of the encounter of contemporary scientists and citizens with the social, political, and moral dilemmas of an age marked by authoritarian rulers and, at times, dictatorial and genocidal regimes. As in physics, his life and thinking and behavior provide enduring lessons for us all.

Heisenberg, the exceptionally brilliant physicist, was in many ways, in social terms, a rather average representative of the educated academic upper social stratum in Germany. The upheavals of his early years—military defeat, economic collapse, a failed monarchy, and anxiety about the future—may have been more extreme than elsewhere, but these events are not unique to his time and place. We have seen them in our own times, in less intense measure, during the cold war, failed military ventures, sudden economic downturns, and the devastating attacks of September 11, 2001.

The response of Heisenberg and others to the onset of the Third Reich betrayed perhaps an understandable naïveté about the nature of the new regime, a susceptibility to political manipulation, and a propensity for self-serving rationality in accommodation to the regime. These may be excusable to some extent in that this was the first full-scale experience with the onset and growth of a regime that quickly became the behemoth of Nazi state terror and genocide. If Heisenberg and others had no historical precedents to warn them of where events were headed, we do not have that excuse, nor should we ever lose sight of the lessons of his failures and, at times, his successes as similar political tendencies inevitably reappear in our time and in the future.

Similarly, the Heisenberg story warns us of the potential dangers and dilemmas of scientific research in service to the state and to other benefactors of science, in times of peace and war. It is ironic, but perhaps no coincidence, that the scientific heads of the main German and Allied nuclear weapons programs during World War II were both theoretical physicists, ill-prepared administrators, and the products of the best education and culture their respective nations could offer. The forces and rationale and purposes that would drive people such as these to create nuclear weapons, or even to contemplate and work toward such a goal under a regime at war, may have seemed, or been, reasonable at the time. Although the horrors of gas warfare were known, the even greater horrors of nuclear weapons were made manifest only in 1945.

Now that we do know what concerted scientific research can produce and how eager modern states are for the acquisition of the power that flows from new weaponry, the moral, ethical, and professional dilemmas facing us all today are sometimes even more acute than those in Heisenberg's day. Yet the lessons of Heisenberg and others in their early confrontations with all of these issues in days past can help us to understand and to shape our own responses as we move beyond his era into the uncertainty and anxiety of today, and tomorrow.

NOTES

Abbreviations of frequently cited references and source locations are listed below. Archival references in the notes are followed by call numbers.

Many of the published papers of the scientists discussed here have been reprinted in their collected works. All of Heisenberg's published writings are now available in the ten volumes of Heisenberg, *Collected Works* (HCW). These volumes also include a number of previously unpublished materials, most notably all of his secret wartime nuclear research reports. In the notes, the references to Heisenberg's writings also include the HCW location. A complete bibliography of all of Heisenberg's published writings, including all translations and reprints, with cross references to HCW, is available in WH, *Biblio*.

All of Heisenberg's private correspondence with members of his family cited here is in his private papers (HP). Letters to his parents through March 1945 have been published in WH, *Briefe*. Unless otherwise noted, all of the letters to his parents cited in the notes may be found in either or both of these sources. Letters to his wife from 4 May 1945 to 20 January 1946 are currently available online at *http://werner-heisenberg.physics.unh.edu/e-Farm-Hall.htm*.

Unless otherwise indicated, published versions of all letters cited from and to Wolfgang Pauli may now be found in the appropriate volume of Pauli, *Wissenschaftlicher Briefwechsel* (PWB), the contents of which are arranged chronologically.

Most translations of original German and French quotations are by the author. Published English translations have been compared with the originals and edited as necessary.

ABBREVIATIONS USED IN THE NOTES

ADJ	Archiv der Deutschen Jugendbewegung, Burg Ludwigstein
AHQP x, y	Archive for History of Quantum Physics, AIP, CHP and elsewhere; x = microfilm number, y = section
AIP, CHP	American Institute of Physics, Center for History of Physics, College Park, MD
AP	*Annalen der Physik*
AS, *Atombau*	Arnold Sommerfeld, *Atombau und Spektrallinien*, multiple editions (Braunschweig: F. Vieweg und Sohn, 1919 ff.)
BA Koblenz	Bundesarchiv, Koblenz, Germany
BAW	Bayerische Akademie der Wissenschaften, archive, Munich
BCW x, y	Niels Bohr, *Collected Works*, 12 vols., Finn Aaserud, current ed. (Amsterdam: North-Holland, 1972-2007); x = volume, y = pages.
BDC	Berlin Document Center, U.S. Army, Berlin, now BA Koblenz
Beyerchen, *Scientists*	Alan D. Beyerchen, *Scientists under Hitler: Politics and the Physics Community in the Third Reich* (New Haven: Yale Univ. Press, 1977)
BGC	Bohr General Correspondence, in NBA
BHStA	Bayerisches Hauptstaatsarchiv, Allgemeines Archiv, Munich
BMS	Bohr Manuscripts, in NBA and on microfilm in AHQP
BP	Hans Bethe Papers, Library, Cornell University, Ithaca, NY
BPC	Bohr Private Correspondence, in NBA
BR, *Nuclear Forces*	Laurie M. Brown and Helmut Rechenberg, *The Origin of the Concept of Nuclear Forces* (New York: Taylor and Francis, 1996)
BSB	Bayerische Staatsbibliothek, Handschriftenabteilung, Munich

BSC x, y	Bohr Scientific Correspondence, in NBA and AHQP; x = microfilm, y = section
DC, *Diss.*	David C. Cassidy, "Werner Heisenberg and the Crisis in Quantum Theory, 1920-1925," (doctoral thesis, Purdue University, 1976)
DC, *JRO*	David C. Cassidy, *J. Robert Oppenheimer and the American Century* (New York, 2005; Baltimore: Johns Hopkins Univ. Press, 2009)
DJ–x, y	Third Reich Documents, Group 11: German Atomic Research, microfilms of materials compiled by David Irving (Wakefield, UK: Microform Academic Publishers), x = microfilm, y = frame numbers
EA	Einstein Archive, The Hebrew University, Jerusalem
EB, *Briefw.*	Albert Einstein, Hedwig Born, and Max Born, *Briefwechsel 1916–1955* (Munich: Nymphenburger Verlagshandlung, 1969)
EH, *Recoll.*	Elisabeth Heisenberg, *Inner Exile: Recollections of a Life with Werner Heisenberg*, S. Cappellari and C. Morris, trans. (Boston: Birkhäuser, 1984)
ES, *Briefw.*	Albert Einstein and Arnold Sommerfeld, *Briefwechsel*, A. Hermann, ed. (Basel: Schwabe, 1968).
ETH	Eidgenössische Technische Hochschule, Handschriftenabteilung, Zurich
Goudsmit, *Alsos*	Samuel A. Goudsmit, *Alsos* (New York: Henry Schuman, 1947; latest reprint Woodbury, NY: AIP Press, 1996)
GP	Samuel A. Goudsmit Papers, in AIP, CHP
HA	Heisenberg Archive, Max Planck Institute for Physics and Astrophysics, Munich
HCW x, y, z	WH, *Gesammelte Werke / Collected Works*, multiple vols., ed. W. Blum, H. Rechenberg et al. (Berlin: Springer-Verlag, and Munich: Piper-Verlag, 1985–1993); x = series, y = volume, z = pages
Hentschel, *Anthology*	Klaus Hentschel, ed., *Physics and National Socialism: An Anthology of Primary Sources,* A. M. Hentschel, trans. (Basel: Birkhäuser, 1996)
Hermann, *Jahrhundert*	Armin Hermann, *Die Jahrhundertwissenschaft: Werner Heisenberg und die Physik seiner Zeit* (Stuttgart: Deutsche Verlags-Anstalt, 1977)
HP	Heisenberg Private Papers, in care of Heisenberg family
HSPS	*Historical Studies in the Physical and Biological Sciences*
IfZ	Institut für Zeitgeschichte, Munich
Jammer, *Conceptual*	Max Jammer, *The Conceptual Development of Quantum Mechanics* (New York: McGraw-Hill, 1966)
Jammer, *Philosophy*	Max Jammer, *The Philosophy of Quantum Mechanics* (New York: John Wiley and Sons, 1974)
JM	Christa Jungnickel and Russell McCormmach, *Intellectual Mastery of Nature: Theoretical Physics from Ohm to Einstein,* 2 vols. (Chicago: Univ. of Chicago Press, 1986)
LC	Library of Congress, Manuscript Division, Washington, DC, microfilm collection of captured German war documents
LNN	*Leipziger Neueste Nachrichten*
MNN	*Münchner Neueste Nachrichten*
MPG	Archiv zur Geschichte der Max Planck-Gesellschaft, Berlin-Dahlem
MR x, y	Jagdish Mehra and Helmut Rechenberg, *The Historical Development of Quantum Theory,* multiple vols. (New York: Springer-Verlag, 1982-2001); x = volume, y = pages
NARA	National Archives and Records Administration, Washington, DC and vicinity
NBA	Niels Bohr Archive, Niels Bohr Institute, Copenhagen
Nwn	*Die Naturwissensachaften*
Pais, *Inward*	Abraham Pais, *Inward Bound: Of Matter and Forces in the Physical World* (Oxford: Clarendon Press, 1986)
Pais, *Subtle*	Abraham Pais, *"Subtle Is The Lord...": The Science and the Life of Albert Einstein* (Oxford: Clarendon Press, 1982)
PBl	*Physikalische Blätter*
PR	*Physical Review*
PRS	*Proceedings of the Royal Society of London*, series A

PWB	Wolfgang Pauli, *Wissenschaftlicher Briefwechsel mit Bohr, Einstein, Heisenberg u. a.*, 4 vols., Karl von Meyenn, current ed. (Berlin: Springer-Verlag, 1979–2005)
PZ	*Physikalische Zeitschrift*
RAC	Rockefeller Archive Center, North Tarrytown, New York
SAM	Stadtarchiv Munich
SAW, MPK	Sächsische Akademie der Wissenschaften, mathematisch-physikalische Klasse
SN	Sommerfeld Nachlass, Deutsches Museum, Munich
SPK	Staatsbibliothek Preussischer Kulturbesitz, Berlin
SStA	Sächsisches Staatsarchiv Dresden
StAM	Staatsarchiv Munich
UA	Universitätsarchiv, or University Archive
UB	Universitätsbibliothek, or University Library
Walker, *Nazi Science*	Mark Walker, *Nazi Science, Myth, Truth, and the German Atomic Bomb* (New York: Plenum Press, 1995).
Walker, *Nuclear Power*	Mark Walker, *German National Socialism and the Quest for Nuclear Power 1939–1949* (New York: Cambridge Univ. Press, 1989)
WH, *Biblio.*	*Werner Heisenberg: A Bibliography of His Writings*, second rev. ed., David C. Cassidy, comp. (Island Park, NY: Whittier, 2001), online at *http://www.aip.org /history/heisenberg/bibliography/contents.htm*
WH, *Briefe*	WH, *Liebe Eltern! Briefe aus kritischer Zeit 1918 bis 1945*, Anna Maria Hirsch-Heisenberg, ed. (Munich: Langen Müller, 2003)
WH, *PB*	WH, *Physics and Beyond: Encounters and Conversations* (New York: Harper and Row, 1971), translation of *Der Teil und das Ganze* (1969) by A. J. Pomerans
WH-Wirtz, *FIAT*	WH and Karl Wirtz, "Grossversuche zur Vorbereitung der Konstruktion eines Uranbrenners," in *FIAT Review of German Science 1939–1946, Nuclear Physics and Cosmic Rays*, part 2, Walther Bothe and Siegfried Flügge, eds. (Wiesbaden, 1948), 143–165, reprinted HCW B, 419–441
ZN	*Zeitschrift für Naturforschung*
ZP	*Zeitschrift für Physik*
ZStA Potsdam	Zentrales Staatsarchiv Potsdam

Chapter 1. The Early Years

1. August Heisenberg, autobiographical sketch, in *Geistiges und künstlerisches München in Selbstbiographien*, ed. W. Zils (Munich: Max Kellerers Verlag, 1913), 156–160. Heisenberg's ancestors are discussed by Jochen Heisenberg, "Die Vorfahren von Werner Heisenberg," in *Werner Heisenberg 1901–1976: Schritte in die neue Physik*, ed. H. Rechenberg and G. Wiemers (Beucha: Sax-Verlag, 2001), 11–15; and in *Werner Heisenberg 1901–1976*, ed. C. Kleint, H. Rechenberg and G. Wiemers (Leipzig: Sächsische Akademie, 2005), 23–29.

2. August Heisenberg to Erich Petzet, 8 Jan 1913 (BSB, E. Petzetiana IVb).

3. Hartmut Kaelble, *Vierteljahrschrift für Soziologie und Wirtschaftsgeschichte, 60* (1973), 41–71.

4. This has been argued by Fritz K. Ringer, *The Decline of the German Mandarins: The Academic Community 1890–1933* (Cambridge, MA: Harvard Univ. Press, 1969).

5. August Heisenberg, "Studien zur Textgeschichte des Georgios Akropolites" (Ph.D. thesis, Munich, 1893).

6. Surmised from August Heisenberg to Erich Petzet, 26 Oct 1895 (in note 3).

7. August Heisenberg to his father, 13 Apr 1897 (HP).

8. Report of Staatsministerium, 17 Apr 1893, in "Acta des Königlichen Staats-Ministerium des Inneren für Kirchen und Schul-Angelegenheiten. Dr. Heisenberg August" (BHStA, MK 17732).

9. Nickolaus Wecklein, "Zeising: Adolf Z.," *Allgemeine Deutsche Biographie*, vol. 55 (1910), 404–411; Otto Hagenmaier, *Der Goldene Schnitt. Ein Harmoniegesetz und seine Anwendung* (Munich: Heinz Moos, 1963).

10. An outline of Wecklein's life was provided by his school colleague Johannes Melber, in "Geheimer Hofrat Dr. Nik. Wecklein," *Bayerische Blätter für das Gymnasialschulwesen, 69* (1927), 88–102.

11. Ibid., as well as Wecklein to E. Halm, 27 Sep 1875 and 19 Jan 1877 (BSB, Halmiana IX).

12. Note 1.

13. "Frequenz der humanistischen Gymnasien . . . am Schlusse des Schuljahres 1901/02," *Blätter für das Gymnasial-Schulwesen, 38* (1902), 661–663.

14. Dieter Schäfer, *Der Weg der Industrie in Unterfranken* (Würzburg: Stürtz, 1970).

15. Franz Dölger, "Die Byzantinisten der Akademie. Jakob Philipp Fallmerayer, Karl Krumbacher, August Heisenberg," in *Geist und Gestalt. Biographische Beiträge zur Geschichte der Bayerischen Akademie der Wissenschaften vornehmlich im zweiten Jahrhundert ihres Bestehens* (Munich: Beck, 1959), 139–157, on 153; Ernst Reisinger, recollection of August Heisenberg as teacher, in *Meine Jugend in Alt-Schwabing* (Munich: Franzis-Verlag, 1952), pp. 68–71.

16. "Qualifikationsliste. K. Lateinschule Lindau," 3 May 1898 (in note 8).

17. For instance, Geneviève Bianquis, "La femme allemande à l'époque moderne (du XVIIe au XXe siècle)," in *Histoire mondiale de la femme*, ed. Pierre Grimal, vol. 4 (Paris: Nouvelle Librairie de France, no date), 253–290; and *Frauen in der Geschichte,* ed. Annette Kuhn und Gerhard Schneider, (Düsseldorf: Schwann, 1979).

18. *Jahres-Bericht über das Kgl. Alte Gymnasium zu Würzburg,* for 1901/02 to 1909/10.

19. Bibliography in Franz Dölger, "August Heisenberg. Geboren 13. November 1869, gestorben 22. November 1930," *Jahresbericht über die Fortschritte der klassischen Altertumswissenschaft, 241* (1933), 25–55, on 43–55.

20. August Heisenberg, *Grabeskirche und Apostelkirche: Zwei Basiliken Konstantins*, 2 vols. (Leipzig, 1908); and "Acta des K. Acad. Senats der Ludwig-Max.-Universität München. Betreffend Dr. August Heisenberg" (UA Munich).

21. Kaspar Hammer, "Qualifikationsliste. Altes Gymnasium, 1902" (in note 8).

22. Interview with Carl Friedrich von Weizsäcker, Starnberg, 30 April 1982.

23. George L. Mosse, *Nationalism and sexuality: Respectability and Abnormal Sexuality in Modern Europe* (New York: Howard Fertig, 1985).

24. Reports in August's personnel file (note 8).

25. J. L. Heilbron, *The Dilemmas of an Upright Man: Max Planck as Spokesman for German Science* (Berkeley: Univ. of California Press, 1986), 4.

26. See Carl Schorske, *Fin-de-Siècle Vienna: Politics and Culture* (New York: Vintage, 1980); Deborah Coen, *Vienna in the Age of Uncertainty: Science, Liberalism, and Private Life* (Chicago: Univ. of Chicago Press, 2007).

27. WH to his parents, 11 Jan 1928 (HP).

28. "Double Dialogue with Werner Heisenberg," 1974, HCW, C3, 464–486, on 475, interview by Vintila Horia.

29. WH, "Ordnung der Wirklichkeit," ca. 1942, HCW, C1, 218–306; and WH, "Naturwissenschaftliche und religiöse Wahrheit," 1973, HCW, C3, 422–439.

30. Interview with Elisabeth Heisenberg, Göttingen, Feb 1982.

31. Carl Friedrich von Weizsäcker, "Heisenbergs Entwicklung seit 1927," in Weizsäcker and B. L. van der Waerden, *Werner Heisenberg* (Munich: Hanser, 1977), pp. 25–40, on 40.

32. This accords with Mosse's assertions regarding the nuclear family, note 23.

33. Note 30.

34. Ibid.

35. Interview of WH by T. S. Kuhn, 30 Nov 1962 (AHQP).

36. "Zeugnisnoten-Protokoll des K. Maximilians-Gymnasiums in München, 1913/14, Klasse IIIA," quoted by A. Hermann, *Werner Heisenberg: In Selbstzeugnissen und Bilddokumenten* (Reinbek: Rowohlt, 1976), 8.

37. Interview with Fr. Gottfried Simmerding, former youth comrade, Munich, 1982.

38. EH, *Recoll.*, 13.

39. K. Maximilians-Gymnasium München, "Zensur, Werner Heisenberg," class 4A, 1914/15 and class 5A, 1915/16 (Archive, Maximilians-Gymnasium, Munich).

40. WH to Koji Kigoshi, 5 Mar 1971 (HA).

41. Note 30.

42. WH, "Theory, Criticism and a Philosophy," 1968, HCW, C2, 423–438, on 438.

43. Note 10.

44. August Heisenberg's active involvement in gymnasium political affairs is indicated by "Mitteilung über die Gymnasiallehrer-Vereinigung München," *Blätter für das Gymnasial-Schulwesen, 41* (1905), 554–556, and by his speeches to teachers' meetings in *Bericht über die XXIV. Generalversammlung des Bay. Gymnasiallehrervereins, München, 4 April bis 6 April 1907*, 18–20, on 18, supplement to *Blätter für*

das Gymnasial-Schulwesen, 43 (1907); and in *Bericht über die XXV. Generalversammlung . . . 15 April bis 17 April 1909*, supplement to *Blätter für das Gymnasial-Schulwesen, 45* (1909). The issues of the period are described by Johannes Guthmann, *Der Bayerische Lehrer- und Lehrerinnenverein: Seine Geschichte*, vol. 2 (Munich: Oldenbourg, 1959).

45. "Rede von Professor Werner Heisenberg," 1971, HCW, C5, 433–435.

46. Theodor Preger, "Karl Krumbacher. geb. 23. September 1856, gest. 12. Dezember 1909," *Blätter für das Gymnasial-Schulwesen, 46* (1910), 78–79.

47. Dr. Otto Crusius to Akademischer Senat, 21 Dec 1909 (in personnel file, note 20).

48. Ibid.

49. August Heisenberg to Akad. Senat, 9 Jan 1910 (ibid.).

Chapter 2. The World at War

1. This according to file card on August Heisenberg (SAM), and *Adressbuch für München und Umgebung* (Munich, 1911).

2. See Jeffrey Gaab, *Munich: Hofbrauhaus and History—Beer, Culture, and Politics* (New York: Peter Lang, 2006).

3. Oscar Brunn, *Plan von München*, 1911 (Graphiksammlung, Stadtmuseum, Munich).

4. Alois Wagner, *Zu meiner Zeit. Ein Bubenleben in Schwabing 1904 bis 1918* (Munich: Süddeutscher Verlag, 1980).

5. The Heisenberg family also acquired an automobile.

6. Heisenberg's primary school records, once in his private papers, could not be found. The school he attended is surmised from note 1 and "Bericht über die Städtischen Volks- und Mittelschulen Münchens für das Geschäftsjahr 1911. Erstattet von Stadtschulrat Dr. Sg. Kerschenstein" (SAM, Schulamt 2236).

7. Hans Scharold, *100 Jahre Maximilians-Gymnasium. Ein Beitrag zur Geschichte des Gymnasiums in Bayern* (Munich, 1949), 2.

8. "Erste Klasse A," K. Maximilians-Gymnasium in München, *Jahresbericht für das Schuljahr 1911/12* (Munich, 1912), 39–40.

9. Johannes Melber, ed., *Die Schulordnung an den höheren Lehranstalten Bayerns nach der königlichen Verordnung vom 30. Mai 1914* (Munich: Lindauersche Universitätsbuchhandlung, 1914).

10. A classical caricature of a gymnasium teacher is Heinrich Mann's *Professor Unrat* (Frankfurt, 1905).

11. Maximilians-Gymnasium, *Jahresbericht*, 1911/12, 47.

12. During the middle years, 1914–1917, Herr Wolff was at the front, then assigned to the lower grades.

13. Texts for each subject are listed in the *Jahresberichte* [annual report], and in Melber (note 9).

14. Maximilians-Gymnasium, "Zeugnisnoten-Protokoll," quoted by A. Hermann, *Werner Heisenberg: In Selbstzeugnissen und Bilddokumenten* (Reinbek: Rowohlt, 1976), 8.

15. Maximilians-Gymnasium, *Jahresbericht*, 1912/13, 46, which includes the poem.

16. "München am 1. Mobilmachungstag," *MNN, 67*, no. 393 (3 August 1914), Morgenblatt, 3, dated 2 Aug.

17. Note 4, 43.

18. WH, *Epoca, 2* (1964), 31, reprinted HCW, C4, 21.

19. For example, Fritz K. Ringer, *The Decline of the German Mandarins: The Academic Community 1890–1933* (Cambridge, MA. Harvard Univ. Press, 1969); Konrad H. Jarausch, *Students, Society, and Politics in Imperial Germany: The Rise of Academic Illiberalism* (Princeton: Princeton Univ. Press, 1983).

20. August Heisenberg, "Die jüngste Entwicklung der Sprachfrage in Griechenland," *Internationale Wochenschrift für Wissenschaft, Kunst und Technik, 5* (1911), 685–702; Karl Krumbacher, *Das Problem der neugriechischen Schriftsprache* (Munich, 1902).

21. Klaus Schwabe, *Wissenschaft und Kriegsmoral. Die deutschen Hochschullehrer und die politischen Grundfragen des Ersten Weltkrieges* (Göttingen: Musterschmidt, 1969).

22. See, e.g., "Aufrufe und Aeusserungen der 'Intellektuellen'," *Die Eiche, 3*, no. 2 (April 1915), 94–196; and Hermann Kellermann, *Der Krieg der Geister. Eine Auslese deutscher und ausländischer Stimmen zum Weltkriege 1914* (Weimar, 1915).

23. August Heisenberg, *Der Philhellenismus einst und jetzt* (Munich: Beck, 1913), founding address to Deutsch-Griechische Gesellschaft, 8 Dec 1912.

24. August Heisenberg, "Die Zukunft Griechenlands," *Süddeutsche Monatshefte, 12* (1915), 939–947; "Griechenland und die Mittelmeerfrage," *Panther, 5* (1917), 349–356; *Neugriechenland* (Leipzig: Teubner, 1919).

25. First work in note 24, 942.

26. "Kriegserlebnisse eines Münchner Universitätsprofessors [Heisenberg]," *MNN, 68,* no. 283 (6 June 1915), 4, and no. 291 (10 June 1915), 3.

27. August Heisenberg to Otto Crusius, 2 Feb 1915 (BSB, Crusiana).

28. Ibid., 24 Apr 1915.

29. WH, interview by T. S. Kuhn, 30 Nov 1962 (AHQP), p. 1. No record of wounds or of even a relapse of rheumatism has been found.

30. Based on Maximilians-Gymnasium, *Jahresberichte.*

31. For example, Karl Theodor Heigel (President, Bay. Akad. Wiss.), "An die akademische Jugend!" *Süddeutsche Monatshefte, 11* (1914), 776–779.

32. (Ernst Kemmer), "Krieg und Schule," Maximilians-Gymnasium, *Jahresbericht,* 1914/15, 36.

33. Ernst Kemmer, "Schule, militärische Jugenderziehung und vaterländischer Hilfsdienst," Maximilians-Gymnasium, *Jahresbericht,* 1916/17, 26–29, on 27.

34. "Kriegsgefahr und Lebensmittelmarkt," *MNN, 67,* no. 389 (1 August 1914), Vorabend-Ausgabe, 3.

35. Note 4, 48; and Johannes Melber, "Geheimer Hofrat Dr. Nik. Wecklein," *Bayerische Blätter für das Gymnasialschulwesen, 69* (1927), 88–102.

36. Note 33, 28.

37. "Jungmannen auf Land," *MNN, 71,* no. 192 (17 April 1918), Morgen-Ausgabe, 2; "Mittelschüler im landwirtschaftlichen Hilfsdienst," *MNN, 71,* no. 238 (12 May 1918), 3.

38. Maximilians-Gymnasium, "Jahreszeugnis Werner Heisenberg, 1917/18, 7. Klasse" (HP); according to Jugendwehr records, Heisenberg was appointed Gruppenführer in September 1918 (Communication from Dr. Heyl, Archivdirektor, BHStA Kriegsarchiv).

39. Note 29, 3.

40. WH to his father, 15 May 1918.

41. WH to his parents, May 1918.

42. WH, *PB,* 2.

43. Based on reports from the front in *MNN.*

Chapter 3. The Gymnasium Years

1. WH, *ZP, 8* (1922), 273–297, rec. 17 Dec 21, reprinted HCW, A1, 134–158.

2. Yearly grade reports for WH (Maximilians-Gymnasium, Archive, Munich), their emphasis. Further quotes below are from the same source.

3. "Reifezeugnis" for WH, 15 Jul 1920 (HP).

4. Johannes Melber, report on examination of WH (Maximilianeum-Stiftung, Archiv, Munich).

5. WH interview by T. S. Kuhn, 30 Nov 1962 (AHQP); and WH, centenary lecture at Maximilians-Gymnasium, 1949, reprinted HCW, C5, 395–408.

6. Quotes from ibid., both works.

7. Ibid.

8. Albert Einstein, *Über die spezielle und die allgemeine Relativitätstheorie. Gemeinverständlich* (Braunschweig: Vieweg, 1917). Recalled by WH, "Begegnungen und Gespräche mit Albert Einstein," 1974, publ. HCW, C4, 202–216.

9. Johannes Melber, ed., *Die Schulordnung an den höheren Lehranstalten Bayerns nach der Königlichen Verordnung vom 30. Mai 1914* (Munich: Lindauersche Universitätsbuchhandlung, 1914), 92.

10. WH, note 8, 202.

11. Interview, note 5, 2.

12. Interview, note 5, 2.

13. Note 4.

14. The professor made the most of the opportunity. August Heisenberg, "Dialekte und Umgangssprache im Neugriechischen," Bay. Akad. Wiss., *Jahrbuch 1918,* 1–26, Festrede on 29 May 1918; and "Phonographische Fixierung von Sprach- und Gesangsproben bei griechischen (kriegsgefangenen) Truppen in Görlitz ausgeführt von Prof. August Heisenberg 1917/1919" (BAW, VII 466).

15. Grade report for 1915–16, note 2.

16. Fermat claimed he could prove the theorem, but he died without revealing the secret.

17. L. Kronecker, "Über die Auflösung der Pellschen Gleichung mittels elliptischer Functionen," 1863, and "Zur Theorie der elliptischen Functionen," 1883–1889, reprinted in *Leopold Kronecker's Werke,* ed. K. Hensel, vol. 4 (Leipzig: Teubner, 1929), 219–225 and 345–495.

18. WH, in *Ensemble 2* (Munich: Oldenbourg, 1971), 228–243; reprinted HCW, C1, 369–384, on 369; and interview, note 5, 2.

19. WH, notebook (HA).

20. Paul Bachmann, *Zahlentheorie*, 5 vols. (Leipzig: Teubner, 1892ff).

21. Acquisition list, Library, Maximilians-Gymnasium, Munich.

22. Johann Kleiber and Adalbert Grüttner, *Kleiber-Nath: Physik für die Oberstufe*, 7th edition (1916); 8th ed. (1918).

23. WH, *PB*, 2.

24. Plato, *Timaeus*, para. 55–56.

25. WH, *PB*, 8.

26. Note 5, second work.

27. WH to Pauli, 24 Nov 1925 (PWB).

28. Interview with C. F. von Weizsäcker, April 1982.

29. Maximilians-Gymnasium, *Jahresbericht*, 1919/20.

30. Foundation documents, in *125 Jahre Maximilianeum 1852 bis 1977* (Munich: UNI-Druck, 1977).

31. Note 4.

32. "Die Mitglieder des Maximilianeums von 1852 bis 1977," in note 30, 55–111.

33. *125 Jahre Maximiliansgymnasium München. Rückblick-Ausblick. Eine Dokumentation* (Munich: Maximilians-Gymnasium, 1974), 113.

34. Recalled in Felix Andreas Wittmann to Dr. Riedl, 1972 (Maximilianeum-Stiftung, Archive, Munich).

Chapter 4. The Battle of Munich

1. August Heisenberg to Karl Krumbacher, 26 Jun 1892 (BSB Krumbacheriana I).

2. The following draws upon, among others, Allan Mitchell, *Revolution in Bavaria, 1918–1919: The Eisner Regime and the Soviet Republic* (Princeton: Princeton Univ. Press, 1965); Richard Grunberger, *Red Rising in Bavaria* (New York: St. Martin's Press, 1973); Michael Doeberl, *Entwicklungsgeschichte Bayerns*, vol. 3 (Munich, 1931); and Karl Schwend, *Bayern zwischen Monarchie und Diktatur* (Munich, 1954).

3. This is argued by Mitchell, note 2.

4. Mitchell, note 2, 218; and Wolfgang Treue, *Die deutschen Parteien: Vom 19. Jahrhundert bis zur Gegenwart* (Frankfurt am Main: Ullstein, 1975).

5. Dr. Gustav Wyneken, "An die Schüler der höheren Schulen!" *MNN, 72*, no. 100 (3 March 1919), 4.

6. [No first name] Dresler, "An die Schüler," *MNN, 72* , no. 168 (12/13 April 1919), 4–5, and in other newspapers.

7. Account of remarks by Heisenberg in *Die neue Seite, 3. MPZ, Stammesmitteilungen, 2* (1962), 2–3.

8. Interview with Fr. Gottfried Simmerding, Munich, March 1982, former youth comrade of Heisenberg's and a resident of the city at the time.

9. WH, *PB*, 7.

10. Robert G. L. Waite, *Vanguard of Nazism: The Free Corps Movement in Postwar Germany, 1918–1923* (Cambridge, MA: Harvard Univ. Press, 1952). For example, Rudolf Höss, *Kommandant in Auschwitz: Autobiographische Aufzeichnungen*, ed. M. Broszat (Munich: DTV, 1963).

11. The most detailed account of the invasion and its planning is provided in Kriegsgeschichtliche Forschungsanstalt des Heeres, ed., *Darstellungen aus den Nachkriegskämpfen deutscher Truppen und Freikorps*, vol. 4: *Die Niederwerfung der Räteherrschaft in Bayern 1919* (Berlin, 1939). Notice is taken, however, of the period when this was written.

12. Ibid. and "Die Beteiligung der II. Marine-Brigade," *MNN, 72*, no. 171 (5 May 1919), Abend-Ausgabe, 3.

13. Oven's orders are printed in note 11, 218–219.

14. Grunberger, note 2, 144.

15. "Verluste," in note 11, 209–212. Ernst Toller wrote an insightful autobiography, *Eine Jugend in Deutschland* (Amsterdam, 1933; reprinted Reinbek: Rowohlt, 1983).

16. Interview of WH by T. S. Kuhn, 30 Nov 1962 (AHQP); and WH, address for centenary of Maximilians-Gymnasium 1949, reprinted HCW, C5, 395–408.

17. *MNN* for the period.

18. "Allgemeine Studentenversammlung an der Universität," *MNN, 72*, no. 181 (10/11 May 1919); also "Aufruf der Universität zum Eintritt in die Freikorps," *MNN, 72*, no. 177 (8 May 1919), Abend-Ausgabe, 3.

19. See, among others, Lucy S. Dawidowicz, *The War against the Jews 1933–1945* (New York: Holt, Rinehart and Winston, 1975).

20. WH, address, note 16, 397; WH, interview, note 16; WH, *PB*, 7.

21. EH, *Recoll.*, chap. 1.

22. WH, "Wissenschaft als Mittel zur Verständigung unter den Völkern," 1946, reprinted HCW, C5, 384–394, on 385.

Chapter 5. Finding His Path

1. WH, *PB*, 1.

2. Walter Z. Laqueur, *Young Germany: A History of the German Youth Movement* (London: Routledge and Kegan Paul, 1962).

3. Willibald Karl, *Jugend, Gesellschaft und Politik im Zeitraum des Ersten Weltkriegs* (Miscellanea Bavarica Monacensia, vol. 48) (Munich: Stadtarchiv München, 1973).

4. Interview with Rev. Wolfgang Rüdel, March 1982; and Rüdel, "Erinnerungen 27.1.1962" (poem), *Die Neue Seite. 3. MPZ. Stammesmitteilungen*, 2 (1962), between 20 and 21.

5. Histories of the German Pathfinders, including the New Pathfinders, are provided by Günther Bandick, "Ursprung und geistige Entwicklung der deutschen Pfadfinderbewegung bis 1933" (doctoral diss., Universität Hamburg, 1955); and Karl Seidelmann, *Die Pfadfinder in der deutschen Jugend-geschichte*, Teil 1: *Darstellung* (Hannover: Hermann Schroedel, 1977).

6. Seidelmann, note 5.

7. Interview, note 4.

8. Schlenk had also served with Heisenberg in the agricultural service, as recalled in Carl Zenker to WH, 6 August 1943 (HA).

9. Interview with Fr. Gottfried Simmerding, March 1982.

10. WH, *PB*, 10.

11. Ibid.

12. Franz Ludwig Habbel and Ludwig Voggenreiter eds., *Schloss Prunn. Der deutsche Pfadfindertag von 1919* (*Der Weisse Ritter*, 2, 1. Beiheft) (Regensburg: Der Weisse Ritter-Verlag, 1919).

13. Such views are described by numerous authors, among them George L. Mosse, *The Crisis of German Ideology: Intellectual Origins of the Third Reich* (New York: Schocken Books, 1964/1981); Peter Gay, *Weimar Culture* (New York: Harper and Row, 1968); Paul Forman, "Weimar culture, causality and quantum theory, 1918–1927," *HSPS, 3* (1971), 1–115.

14. Fritz Ludwig Habbel, "Vorrede zu der Zeitschrift 'Der Aufbau,' 1918," reprinted in *Die Pfadfinder in der deutschen Jugendgeschichte*, Teil 2,1: *Quellen und Dokumente aus der Zeit bis 1945*, ed. Karl Seidelmann (Hannover: Schriedel, 1980), 51–52, his emphasis.

15. Ernst Kemmer, "Windsbach," *Der Weisse Ritter, 3* (1921), 189–191, on 189.

16. Fritz Ludwig Habbel, "Unser Pfadfindertum (1919)," in note 14, 52–55, on 54.

17. Karl Ettinger, statement, in note 12, 39–40.

18. [No first name] Steidle, statement, ibid., 14–18.

19. Fritz Ludwig Habbel, Karl Sonntag, and Ludwig Voggenreiter, "Ein Geleitwort," *Der Weisse Ritter, 2* (1919), 4–6, on 6. The journal started with volume 2. Its predecessor, *Der Aufbau*, was designated volume 1.

20. This is also asserted by George L. Mosse, *Nationalism and Sexuality: Respectability and Abnormal Sexuality in Modern Europe* (New York: Howard Fertig, 1985), 47.

21. See Mosse, note 13, 173–174.

22. Michael H. Kater, *Studentenschaft und Rechtsradikalismus in Deutschland 1918–1933* (Hamburg: Hoffmann und Campe, 1975).

23. WH to Kurt Pflügel, 21 October 1923 (Pflügel papers, private).

24. Note 9.

25. WH, *PB*, 54.

26. EH, *Recoll.*, 25–26.

27. Simmerding, diary of their travels (his private papers).

28. Note 9.

29. WH, *PB*, 11.

30. Note 12, 41.

31. Note 29.

32. WH to his father, 16 Aug 1919.

33. Based on materials in Nachlass Seidelmann (ADJ).

34. Wolfgang Rüdel to WH, 15 Feb 1923 (HP).

35. Mosse, note 13, 181, quotes one source reporting in 1926 that the New Pathfinders admitted "only those Jews . . . who are more Nordic than the Nordic Aryans. Secretly they are made fun of all the same."

36. Hans Blüher, *Secessio Judaica: Philosophische Grundlegung der historischen Situation des Judentums und der antisemitischen Bewegung* (Berlin: Der Weisse Ritter-Verlag, 1922).

37. Note 22.

38. Hurt to Völkel, 24 April 1921 (Nachlass Seidelmann, ADJ). However, Walter's brother Franz did not help matters.

39. Karl Sonntag, recollections, in *Die Neue Seite. 3. MPZ Stammesmitteilungen, 4* (1967), 77–86. Publication referenced below as *Die Neue Seite.*

40. Sonntag to Habbel, 19 Dec 1921; and Martin Völkel, "Rundbrief," 29 March 1922 (both Nachlass Seidelmann, ADJ). Homosexual activity was a crime under the Weimar constitution.

41. Völkel to Habbel, 26 May 1922; Sonntag to Voggenreiter, 9 April 1925; and Völkel-Sonntag correspondence (all Nachlass Seidelmann, ADJ).

42. Plans for camp activities in Karl Sonntag, "Gaubrief," April 1924; and "Sommerlager, Lagerordnung," 1925 (Nachlass Seidelmann, ADJ).

43. For example, [Martin Völkel], "Der grosse Häuptling spricht," *Die Spur in ein deutsches Jugendland, 2* (March 1924), 155.

44. Interview note 4.

45. Interview with Friedrich Hund, February 1982.

46. Annie Heisenberg to Kurt Pflügel, 4 August (1922), in *Die Neue Seite, 2* (1962), between 20 and 21.

47. Diary of trip (Simmerding papers, private); and WH, *PB*, 52.

48. Interview with Elisabeth Heisenberg, February 1982; the Rüdel brothers to WH, 22 August 1920 (HP).

49. WH to his father, 15 May 1918.

50. WH, interview session 1 (AHQP).

51. Note 9.

52. Rolf Wägele, "Werner und der jüngere Jungstamm," *Die Neue Seite, 2* (1962), 27–31; and Sonntag, remarks, ibid., 8.

53. WH, "Wissenschaft als Mittel zur Verständigung unter den Völkern," 1946, reprinted HCW, C5, 384–394, on 385.

54. WH, *Magnum, 35* (1961), 39, reprinted HCW, C4, 19.

55. WH, *PB*, 18–19.

56. For example, WH to Renée Weber, and to Fritjof Capra, 1970s (HA). WH, "Ordnung der Wirklichkeit," 1942, publ. HCW, C1, 218–306.

57. Ruth Nanda Anshen, "World Perspectives: What This Series Means," in WH, *PB*, 249–260, on 249 and 252.

58. Karl Seidelmann, sunrise lecture to the Jungstamm on Easter Sunday. 1931 (Nachlass Seidelmann, ADJ).

59. Note 9.

60. WH, fragment of autobiographical sketch, ca. 1934 (Archiv für Geschichte der Naturforschung und Medizin, Deutsche Akademie der Naturforscher, Leopoldina, Halle).

Chapter 6. Sommerfeld's Institute

1. Born to Sommerfeld, 5 March 1920 (SN); and Neils Born and WH, "La mecanique des quanta," 1927 Solvay Congress report, reprinted HCW B, 58–96.

2. In November 1921 Wilhelm Brüchner, university student, founded the Storm Division, SA, of the Nazi Party and populated it with many of his classmates, among them Rudolf Hess. In Karl Schwend, *Bayern zwischen Monarchie und Diktatur* (Munich, 1954).

3. Arno Seifert, "In den Kriegen und Krisen des 20. Jahrhunderts," in *Ludwig-Maximilians-Universität. Ingolstadt-Landshut-München 1472–1972*, ed. Laetitia Böhm and Johannes Spörl (Berlin: Duncker und Humblot, 1972), 315–362.

4. "Studentenversammlung in der Universität," *MNN, 73*, no. 23 (17/18 Jan 1920), 2; "Münchner Studentenschaft und Arco-Prozess," *MNN, 73*, no. 24 (19 Jan 1920), 4; "Eine verhinderte Vorlesung," *MNN, 73*, no. 29 (22 Jan 1920, morgen), 3.

5. Based on *Chronik der Ludwig-Maximilians-Universität* for the period. The proportion of women students remained constant at about 10 percent.

6. Aloys Fischer, *Die Wirtschaftliche Lage der Studentenschaft Münchens und die Bedeutung für die Studentenfürsorge* (Munich: Verlag des Vereins Studentenhaus München, 1921).

7. Based on "Beamtenbesoldungsgesetz," *Gesetz- und Verordnungsblatt für die Freistaat Bayern, 1920,* 275–322; and "Gesetz zur Abänderung des Beamtenbesoldungsgesetzes," ibid., *1923,* 1–3.

8. Erich Simon, "Der Haushalt eines höheren Beamten," *Jahrbücher für Nationalökonomie und Statistik, 119* (1922), 425–432.

9. WH to his mother and to his father, both Jena, 23 Sep 1921.

10. WH to his mother, Berlin, 28 Sep 1921.

11. Richard Willstätter, *Aus meinem Leben. Von Arbeit, Musse und Freunden,* 2nd ed. (Weinheim: Verlag Chemie, 1949/19), 302.

12. Max Born, *Mein Leben. Die Erinnnerungen des Nobelpresiträgers* (Munich: Nymphenburg, 1975), 268.

13. WH, "Begegnungen und Gespräche mit Albert Einstein," 1974, HCW, C4, 202–216. Most of the letters are published in ES, *Briefw.*

14. Report in *PZ, 21* (1920), 649–668.

15. Summaries of the backgrounds and attitudes of Lenard and Johannes Stark are offered in Beyerchen, *Scientists,* chaps. 5 and 6; and Walker, *Nazi Science,* chaps. 1 and 2.

16. JM, vol. 2, 286–287.

17. Ibid., recounting an episode in 1912.

18. WH to his mother, Jena, 21 Sep 1921.

19. The article, published in *Weltbühne,* is reprinted in ES, *Briefw.,* 89–90; Einstein to Sommerfeld, 27 Sep and 9 Oct 1921, ibid.

20. For example, E. Amaldi, recollection, in Enrico Fermi, *Collected Papers (note e memorie),* vol. 1 (Chicago: University of Chicago Press, 1962), 811.

21. Wolfgang Pauli, "An Hermann Weyl zum 6. Nov. 1955," quoted in PWB, vol. 1, 33, n. 6.

22. WH, *PB,* 27.

23. One discussion of the mathematical-physical seminar is provided by JM, vol. 1, 78–107.

24. Based on *Personalstand der Ludwig-Maximilians-Universität München,* Winterhalbjahr 1920/21.

25. WH, *PB,* 16.

26. WH, interview by T. S. Kuhn, I (AHQP), 5.

27. Ulrich Benz, *Arnold Sommerfeld. Eine wissenschaftliche Biographie* (Stuttgart: Wissen. Verlagsgesellschaft, 1975).

28. Klein's strategy is described by Karl-Heinz Manegold, *Universität, Technische Hochschule und Industrie* (Berlin, 1970).

29. Kultusminister Matt to Univ. Senat, 3 Aug 1920 (Acta des K. Akad. Senats. Das physicalische Cabinet betr. UA München; Personalakt Wilhelm Wien Littera E, Abt. II, Fascikel 698, UA München); and Wilhelm Wien, "Ein Rückblick," in *Aus dem Leben und Wirken eines Physikers* (Leipzig: Barth, 1930), 1–50.

30. Arnold Sommerfeld, "Das Institut für theoretische Physik," in *Die wissenschaftlichen Anstalten der Ludwig-Maximilians-Universität zu München,* ed. K. A. von Müller (München: Oldenbourg und Wolf, 1926), 290–292; and "Academie der Wissenschaften. Mathematisch-physikalisches Cabinett—Sammlung, nun seit 1909 Institut für theoretische Physik," Band II: 1853–1927 (Acta des K. Staatsministeriums des Innern für Kirchen und Schul-Angelegenheiten. BHStA, Munich. MK 11317).

31. Wilhelm Röntgen, "Bericht der Kommission für die Wiederbesetzung der ordentlichen Professur für theoretische Physik," 20 Jul 1905 (Arnold Sommerfeld, Personalakt, E II-N, UA München). Röntgen and the foundation of the Munich Institute for Theoretical Physics are discussed in JM vol. 2, 274–287.

32. Einstein to Sommerfeld, 14 Jan 1922, ES, *Briefw.,* 98.

33. Quoted by JM, vol. 2, 284.

34. Ludwig-Maximilians-Universität, *Vorlesungsverzeichnis,* for entire period.

35. Born to Sommerfeld, 13 May 1922 (SN); Arnold Sommerfeld, "Vorwort," in AS, *Atombau,* first ed.

36. Note 26.

37. Peter Paul Ewald, "Sommerfeld als Mensch, Lehrer und Freund," in *Physics of the One- and Two-Electron Atoms,* ed. Fritz Bopp and H. Kleinpoppen (Amsterdam: North-Holland, 1969), 8–16.

38. WH to his mother, 23 Sep 21.

39. Wolfgang Pauli, "Relativitätstheorie," *Encyklopädie der mathematischen Wissenschaften*, vol. 5, part 2 (Leipzig: Teubner, 1921), 539–775.

40. Arnold Sommerfeld, "Zwanzig Jahre spectroscopischer Theorie in München," *Scientia*, Nov–Dec 1942, 123–130.

41. Matriculation "Verzeichnis" for WH, WS 1920/21 and SS 1921 (UA München). The rest were lost in World War II.

42. Matriculation "Verzeichnis" for Pauli, WS 1918/19 (ibid.).

43. Note 26.

44. Detlev Richardt, report of WH's remarks at Lindau meeting of Nobel Prize winners, 1974, in *Phys. Blätter, 30* (1974), 30–37.

45. August Heisenberg to WH, 2 Nov 1922.

46. Born to Sommerfeld, 5 January 1923 (SN).

47. Note 26.

48. WH, *PB*, 25–26.

49. Note 46.

Chapter 7. Confronting the Quantum

1. For example, Jammer, *Conceptual*; Paul Forman, "The Environment and Practice of Atomic Physics in Weimar Germany: A Study in the History of Science," (Ph.D. diss., University of California, Berkeley, 1967); DC, *Diss.;* Daniel Serwer, *HSPS, 8* (1977), 189–256; the encyclopedic MR; and a host of specialized studies. Some further details may be found in DC, *Uncertainty*: The Life and Science of Werner Heisenberg (New York: W. H. Freeman, 1992).

2. Niels Bohr, address 1913, in BCW 2, 283–301; J. L. Heilbron and T. S. Kuhn, "The Genesis of Bohr's Atom," *HSPS, 1* (1969), 211–290.

3. Arnold Sommerfeld, *AP, 51* (1916), 1–94, 125–167.

4. Arnold Sommerfeld, *PZ, 17* (1916), 491–507; Peter Debye, *ibid.*, 507–512; Sommerfeld to Bohr, 4 Sep 13 (BSC 7, 3).

5. AS, *Atombau*, first ed., vii; and *Nwn 8* (1920), 61–64.

6. Ibid., 64.

7. Univ. Munich, *Jahresberichte*, [annual reports] for the years in question.

8. *Inventory of Sources for History of Twentieth Century Physics*, Office for History of Science and Technology, University of California, Berkeley.

9. T. S. Kuhn et al., *Sources for History of Quantum Physics: An Inventory and Report* (Philadelphia: American Philosophical Society, 1967).

10. DC, *Diss.*, tables on 8 and 9.

11. This is based on Forman, J. L. Heilbron, S. Weart, "Physics *circa* 1900: Personnel, Funding, and Productivity of the Academic Establishments," *HSPS, 5* (1975), 1–185, tables on 12 and 31.

12. Arnold Sommerfeld, *AP, 63* (1920), 221–263. Paschen to Sommerfeld (AHQP 33,1) and Back to Sommerfeld (AHQP 29,5) indicate that they regularly supplied data, but none of the available letters contains the data in question. They were probably on enclosures that have since been lost.

13. Notebook (HA). The following is based on DC, *HSPS, 10* (1979), 187–224; and Paul Forman, *HSPS, 2* (1970), 153–261.

14. WH, interview session I (AHQP).

15. WH, *PB*, 35.

16. Arnold Sommerfeld, *ZP, 8* (1922), 257–272, rec. 12 Dec 1921.

17. Indicated in Heisenberg's letters to Alfred Landé in those months (AHQP 6, 2).

18. WH to his mother, Jena, 21 Sep 1921, and to his father, Jena, 23 Sep 1921.

19. WH to Pauli, 19 Nov 1921.

20. WH, *ZP, 8* (1922), 273–297, in *HCW*, A1, 134–158; Sommerfeld, note 16.

21. Sommerfeld to Einstein, 11 Jan 1922, in ES, *Briefw.*

22. Bohr to Landé, 15 May 1922 (BSC 4,3).

23. Niels Bohr, "On the Application of Quantum Theory to Atomic Structure: Part I—The Fundamental Postulates," 1922, BCW 3, 455–500.

24. See, among others, Forman, note 1; and Paul Forman, *Isis, 64* (1973), 151–180.

25. For example, Sommerfeld to Carlsberg Foundation, 26 Oct 1919, quoted in Peter Robertson, *The Early Years: The Niels Bohr Institute 1921–1930* (Copenhagen: Akademisk Forlag, 1979), 34–35.

26. Bohr to Sommerfeld, 30 Apr 1922, in BCW 3, 691–692.

27. Neils Bohr, "Sieben Vorträge über die Theorie des Atombaus" (Bohr MSS 10), publ. in English translation in BCW 4, 341–419. The lectures were delivered 12–22 June 1922.

28. WH to Karl and Helen Heisenberg, 15 Jun 1922 (HP).

29. Bohr, lecture 3, 14 June 1922, in BCW 4, 372–387, reference to Kramers at end of lecture, 370–371. Kramers's paper is *ZP, 3* (1920), 199–223.

30. Interview with Friedrich Hund, who was in attendance, Göttingen, April 1982. Kramers's work and Heisenberg's objection are discussed in MR 2 and in Max Dresden, *H. A. Kramers: Between Tradition and Revolution* (New York: Springer-Verlag, 1987), 124–132.

31. WH, *PB*, 37–42; WH, "Quantum Theory and Its Interpretation," 1967, in HCW C2, 345–346.

32. Note 28.

33. WH to his parents, 15 Jun 1922.

34. Neils Bohr, lecture 5, 20 June 1922, in *BCW* 4, 391.

35. Note 33.

Chapter 8. Modeling Atoms

1. "Akten des Rektorats der Universität München. Dr. Arnold Sommerfeld" (UA Munich, Personalakt E II-N).

2. Arnold Sommerfeld and WH, *ZP, 10* (1922), 393–398, rec. 3 Aug 22, in HCW, A1, 159–164; and *ZP, 11* (1922), 131–154, rec. 26 Aug 1922, in HCW, A1, 165–188.

3. WH, "Nichtlaminare Lösungen der Differentialgleichungen für reibende Flüssigkeiten," 1922, in HCW, B, 23–26.

4. Franz Ludwig Habbel, "Die Aussenpolitik der deutschen Pfadfinderbewegung," *Der Weisse Ritter*, 7 (1927), 60–80, on 66. For the physicists, see Paul Forman, *Isis, 64* (1973), 150–180.

5. Habbel, note 4, 61.

6. WH to his father, 11 Nov 1922.

7. Martin Völkel, draft of circular letter marked "streng vertraulich" (top secret), 28 Feb 1922 (Nachlass Seidelmann, ADJ).

8. Karl Sonntag, circular letter, Bavaria District, 11 Mar 1922 (Nachlass Seidelmann, ADJ).

9. Habbel, note 4; Kurt Pflügel, "Finnlandfahrt," *Die neue Seite, 3. MPZ Stammesmitteilungen*, 2 (1962), 13–18.

10. [WH], "Der Kampf um die Überfahrt," *Die Spur in ein deutsches Jugendland*, 3 (1924), 37–39.

11. *Die neue Seite, 3. MPZ Stammesmitteilungen*, 2 (1962), 18–22, reprint of note 10.

12. Kurt Pflügel, "Wie wir 3 Enten schossen und nur eine erbeuteten," *Die Spur in ein deutsches Jugendland, 4* (1925/26), 134–135.

13. Elis J. Huetin to WH, Helsingfors, 1 Oct 1923 (HA); WH to his mother, 13 Nov 1923.

14. WH to Bohr, Zell am Ziller, 31 Aug 1925 (BSC 11,2).

15. WH to his parents, Leipzig, 17 Sep 1922.

16. Philipp Lenard, *Ueber Aether und Uräther. Mit einem Mahnwort an deutsche Naturforscher* (Leipzig: Hirzel, 1922). This episode is further discussed in Beyerchen, *Scientists*, 91–93.

17. Note 15.

18. Handbill (SPK, Sammlung Darmstädter, F1e 1908 (7): Einstein).

19. WH, *PB*, 43–45.

20. WH to his parents, 19 Sep 1922; and WH to Pauli, 29 Sep 1922.

21. Heisenberg still referred to the speaker as "Einstein" in WH, *PB*, 44, written in 1969.

22. Max Born, *My Life: Recollections of a Nobel Laureate* (New York: Charles Scribner's Sons, 1975), 212.

23. [Felix Klein?], "Universität Göttingen—Philosophische Fakultät I. Mathematisch physikalisches Seminar" (UA Göttingen, "Zur Geschichte des Königlichen Math.-phys. Seminar"). Göttingen mathematics and physics are discussed by Lewis Pyenson, *The Young Einstein: The Advent of Relativity* (Bristol: Adam Hilger, 1985), 101–193.

24. Publ. in EB, *Briefw.*

25. Born, note 22, 199–200; and "Personalakt Max Born" (UA Göttingen).

26. "Chronik der Georg August-Universität für die Rechnungsjahre 1921/23," Universitätsbund Göttingen, *Mitteilungen, 6* (1923), Heft 1/2; Born, note 22, 210.

27. Born to Walther Gerlach, 11 May 1921 (AHQP 19,1).

28. 1,257 of 3,263 students studied one of the sciences. *Amtliches Namenverzeichnis, Georg August-Universität zu Göttingen, Winterhalbjahr 1922/23.*

29. Born to Sommerfeld, 5 Jan 1923 (SN).

30. WH to Helen and Karl Heisenberg, 12 Jun 1922 (HP).

31. *The University of Goettingen* (Göttingen: Verein für Fremdenverkehr, ca. 1926), 18–19.

32. Born to Einstein, 7 Apr 1923, in EB, *Briefw.*

33. WH to his brother, 6 Dec 1922 (HP).

34. Eberhard Rüdel to WH, 23 Nov 1922 (HP).

35. WH to his mother, 7 Dec 1922.

36. Götz von Selle, *Die Georg-August-Universität zu Göttingen 1737–1937* (Göttingen, 1937); August Tecklenburg, *Göttingen: Die Geschichte einer deutschen Stadt* (Göttingen, 1930); Albrecht Saathoff, *Geschichte der Stadt Göttingen seit der Gründung der Universität* (Göttingen, 1940).

37. WH to his father, 5 and 16 Nov 1922, and to his mother, 20 Nov, 1 and 7 Dec (1922).

38. WH to his mother, 1 Dec 1922.

39. Note 22, 211.

40. WH to his father, 16 Nov 1922; WH to Sommerfeld, 15 Jan (1923) (AHQP 83,H).

41. WH to his brother, 6 Dec 1922 (HP), and to his mother, 7 Dec (1922).

42. Born to Sommerfeld, 5 Jan 1923 (SN).

43. WH to his father, 5 Nov 1922.

44. Note 22, 211.

45. WH to his father, 16 Nov 1922. The following is based upon DC, *Diss.* For technical details, the reader is referred to this and to MR 2.

46. Max Born and Wolfgang Pauli, *ZP, 10* (1922), 137–158.

47. Neils Bohr, *ZP, 13* (1923), 117–165, rec. 15 Nov 22; H. H. Kramers, *ZP, 13* (1923), 312–341, on 339, rec. 31 Dec 1922.

48. WH to Sommerfeld, 28 Oct 1922 (AHQP 83,H); WH to Landé, 15 Sep 1922 (AHQP 6,2).

49. Ibid., first letter.

50. Arthur Sommerfeld, *Journal of Optical Society of America, 7* (1923), 509–515. Nevertheless, Sommerfeld had to withdraw the model a year later when it proved unstable.

51. Max Born, *Nwn, 11* (1922), 677–678, dated 27 Jun 1922.

52. WH to his father, 16 Nov 1922; note 22, 202.

53. Neils Born and WH, *ZP, 14* (1923), 44–55, rec. 16 Jan 23, in HCW, A1, 189–200. The hypothesis was replaced in 1925 by Pauli's "exclusion principle."

54. WH to Sommerfeld, 4 Jan 1923 (AHQP 83,H).

55. WH to Sommerfeld, 15 Jan 1923 (AHQP 83,H).

56. WH to Bohr, 2 Feb 1923 (BSC 9,2).

57. Their results appeared in Max Born and WH, *ZP, 16* (1923), 229–243, rec. 11 May 1923, in HCW, A1, 201–215.

58. Born to Bohr, 4 Mar 1923, in BCW 4, 669.

59. WH to Pauli, 19 Feb 1923 and 26 Mar 1923; also note 58.

60. Pauli to Sommerfeld, 6 Jun 1923; Bohr to Landé, 3 Mar 1923 (AHQP 4,1).

61. Neils Born, *Nwn, 11* (1923), 537–542, on 542.

Chapter 9. Channeling Rivers, Challenging Causality

1. Eberhard Rüdel to WH, 12 Jun 1923 (HP).

2. Reynolds's work is outlined in MR 2, 53.

3. WH, "Über Stabilität und Turbulenz von Flüssigkeitsströmen" (doctoral thesis, Munich, 1923, UB Munich); revised version, *AP, 74* (1924), 577–627, in HCW, A1, 31–81. Technical reviews of Heisenberg's work on the problem and its background are offered by Olivier Darrigol, "Turbulence in 19th-century Hydrodynamics," *HSPS, 32* (2002), 207–262; and MR 2, 49–63.

4. WH, *PZ, 23* (1922), 363–366, in HCW, A1, 27–30. Prandtl, however, expressed objections in an appended note, 30.

5. *Promotions-Ordnung der Philosophischen Fakultät (II. Sektion) der Ludwig-Maximilians-Universität München* (Munich, 1922).

6. "Protokoll, Promotion des Herrn Werner Heisenberg" (UA Munich, Call no. OCI 49p).

7. Ibid. It is possible that the exams were held separately.

8. "Protokoll, Promotion des Herrn Wolfgang Pauli" (UA Munich, OCI 49p).

9. WH, interview session 1 (AHQP).

10. Ibid. and note 6.

11. Max Born, *My Life: Recollections of a Nobel Laureate* (New York: Charles Scribner's Sons, 1975), 213.

12. *Jahrbuch der Ludwig-Maximilians-Universität München, 1919–1925* (Munich, 1925).

13. WH to his parents, 29 Nov 1923.

14. Note 9.

15. WH to Kurt Pflügel, 24 Nov 1923 (Pflügel Papers).

16. "Bekanntmachung über das Diensteinkommen der Hochschulprofessoren," *Gesetz- und Verordnungsblatt für den Freistaat Bayern, 1923*, 313; "Gesetz über die Festsetzung der Teuerungszuschläge der Staatsbeamten," ibid., 7.

17. "Teuerungszahlen der Gemeinden vom Juli 1923 bis Januar 1925," *Statistisches Jahrbuch für das deutsche Reich, 44* (1924/25), 261.

18. Hans Guradze and Karl Freudenberg, in *Jahrbücher für Nationalökonomie und Statistik, 121* (1923), 354–355.

19. Note 17; and "Lebensmittelpreise im Kleinhandel in einigen deutschen Städten im Durchschnitt 1913/14 und vom Juli 1923 bis Januar 1925," *Statistisches Jahrbuch für das deutsche Reich, 44* (1924/25), 262.

20. Wilhelm Wien, "Ein Rückblick," 1927, in *Aus dem Leben und Wirken eines Physikers* (Leipzig: Barth, 1930), 1–50, on 47.

21. Born to Einstein, 7 Apr 1923 (EB, *Briefw.*).

22. Ibid.; and Personalakt Max Born (UA Göttingen).

23. Brigitte Schröder-Gudehus, *Minerva, 10* (1972), 537–570; Paul Forman, *Minerva, 12* (1974), 39–66; and, for example, Friedrich Schmitt-Ott, "Die Kulturaufgaben und das Reich," *Internationale Monatsschrift für Wissenschaft, Kunst und Technik, 11* (1919), 450–459 ; Georg Schreiber, *Die Not der deutschen Wissenschaft und der geistigen Arbeiter* (Leipzig: Quelle und Meyer, 1923).

24. Forman, note 23.

25. Notgemeinschaft, *Bericht I* (1922), 38.

26. See Forman, note 23.

27. Records of Elektrophysikausschuss der Notgemeinschaft (BA Koblenz, R73), and Steffen Richter, *Forschungsförderung in Deutschland 1920–1936.* (Düsseldorf: VDI-Verlag, 1972).

28. WH to his parents, 20 Nov 1923.

29. Ibid.; Richter, note 27; and Born to Universitätskurator, 18 Jul 25, in Personalakt Dr. Heisenberg (UA Göttingen, 4/Vc 317).

30. Ibid., third source, and Courant to Universitätskurator, 30 Oct 1925 (ibid.). On September 1, 1924, the Rentenmark was replaced by the Reichmark, of equal value and same abbreviation.

31. Notgemeinschaft, *Bericht V*, 85. The statement was probably written by Planck.

32. WH to Kurt Pflügel, 21 Oct 1923 (Pflügel Papers).

33. Ibid.

34. Kurt Pflügel to WH, 24 Oct 1923 (HP).

35. WH to Kurt Pflügel, 31 Oct 1923 (Pflügel Papers).

36. The dramatic effect was lost on the farmers near Göttingen, who astutely observed when it was all over: "Hei lebet noch!" (He's still living!). WH to Kurt Pflügel, 24 Nov 1923 (Pflügel Papers).

37. Interview with Fr. Gottfried Simmerding, Munich, 17 Mar 82. The story is unconfirmed.

38. Kurt Pflügel to WH, 12 Nov 1923 (HP); and WH to his mother, 13 Nov 1923.

39. WH to Kurt Pflügel, 24 Nov 1923 (Pflügel Papers).

40. Wolfgang Rüdel to WH, 22 Jan 1924 (HP).

41. WH to Kurt Pflügel, 24 Nov 1923.

42. Alfred Landé, *ZP, 15* (1923), 189–205; Ernst Back, *ZP, 15* (1923), 206–243.

43. WH, *ZP, 26* (1924), 291–307, rec. 13 Jun 1924, HCW, A1, 289–305. For technical details see Daniel Serwer, *HSPS, 8* (1977), 189–256; MR 2, 106–124; John Hendry, *Centaurus, 25* (1981), 189–221; and DC, *Diss.*

44. WH to Pauli, 9 Oct 1923.

45. WH to his mother, 7 Nov 1923.

46. WH to his father, 29 Nov 1923.

47. Bohr to WH, 31 Jan 1924 (BSC 11,2).

48. Pauli to Landé, 14 Dec 1923) to Bohr, 21 Feb 1924, and to Kramers, 19 Dec 1923.

49. Pauli to Bohr, 11 Feb 1924.

50. WH to his parents, 15 Mar 1924.

51. Ibid., and Bohr to Rose, 16 Apr 1924 (BGC).

52. WH to his parents, 20, 23, and 27 Mar 1924.

53. Ibid.

54. WH to his parents, 27 Mar 1924.

55. Postdoctoral physicist Slater was on a Sheldon Traveling Fellowship. He stayed in Copenhagen until April 1924, according to data compiled by Peter Robertson, *The Early Years: The Niels Bohr Institute 1921–1930* (Copenhagen: Akademisk Forlag, 1979), 156–159.

56. Niels Bohr, H. A. Kramers, and John C. Slater, *Phil. Mag., 47* (1924), 785–802, dated January 1924, reprinted BCW 5, 101–118; Kramers, *Nature, 133* (1924), 673–676, dated 25 March 1924.

57. WH, *Nwn, 17* (1929), 490–496, in HCW, B, 109–115.

58. Ladenburg, *ZP, 4* (1921), 451–468.

59. Slater to Kramers, 8 Dec 23 (AHQP 8, 10); John C. Slater, *Nature, 113* (1924), 307–308.

60. First work, note 56, 796. Slater did not concur with the full theory: John C. Slater, *Nature, 116* (1925), 278; and *PR, 25* (1925), 395–428.

61. WH to his mother, 31 May 1924.

62. Rose to Bohr, 2 Jun and 10 Jun 1924 (RAC, IEB series 1, Box 50, W. Heisenberg file); Bohr to Rose, 5 Jul 24 (RAC, IEB Denmark, Box 2146, Folder 403).

63. WH to his mother, 5 Jun 1924.

64. WH to his parents, 8 Jun 1924.

65. WH to Pauli, 8 Jun 1924.

66. Einstein to Born, 29 Apr 1924 (EB, *Briefw.*).

67. Max Born, "Über Quantenmechanik," *ZP, 26* (1924), 379–395, translated in *Sources of Quantum Mechanics*, B. L. van der Waerden, ed. and trans. (New York: Dover, 1967), 181–198, on 189.

68. WH to Landé, 6 Jul 1924 (AHQP 6, 2).

69. Bohr to WH, 5 Jul 1924 (BSC 11, 2).

70. Personalakt Dr. Heisenberg (UA Göttingen, 4/Vc 317). Heisenberg's certificate (venia legendi) for the field of theoretical physics was officially conferred on 10 Oct 1924 (Certificate in HA).

Chapter 10. Entering the Matrix

1. WH, *ZP, 33* (1925), 879–893; reprinted HCW, A1, 382–396; English translation in *Sources of Quantum Mechanics*, B. L. van der Waerden, ed. and trans. (New York: Dover, 1967), 261–276. All quoted translations subject to slight editing.

2. Overall histories of this development include the classic work by Jammer, *Conceptual;* as well as portions of A. Pais, *Inward*; Olivier Darrigol, *From C-numbers to Q-numbers: The Classical Analogy in the History of Quantum Theory* (Berkeley, 1993); Daniel Serwer, *HSPS, 8* (1977), 189–256; Max Dresden, *H. A. Kramers: Between Tradition and Revolution* (New York: Springer-Verlag, 1987); MR, vols. 2 and 3; and DC, *Diss.*

3. WH, *Nwn, 17* (1929), 490–496, in HCW, B, 109–115 on 111.

4. The founding of the institute is described in Peter Robertson, *The Early Years: The Niels Bohr Institute 1921–1930* (Copenhagen: Akademisk Forlag, 1979).

5. WH to his parents, 27 Mar 1924.

6. Note 4, 94 and 106–109.

7. WH was a "Privatdozent," one who lectured "privately" (that is, without state appointment) while awaiting a vacant teaching chair.

8. WH to his parents, 19 Sep 1924, and WH to his mother, 25 Sep 1924; WH to Sommerfeld, 18 Nov 1924 (AHQP 83, H). The level of Heisenberg's active Danish is uncertain. Although his Copenhagen lecture notes are in Danish, Heisenberg always responded in German to Bohr's letters in Danish.

9. Dresden, note 2, 252–276, is an excellent summary of the constellation of relationships among these *primae donnae* in Copenhagen.

10. WH to Bohr, 5 Jul 1924; also 25 Aug 1924, 4 Sep 1924 (BSC 11, 2).

11. WH to his parents, 19 Sep 1924, and to Wallace Lund, 27 Jan 1925 (HA).

12. H. A. Kramers, *Nature*, *114* (1924), 310–311.

13. H. A. Kramers and WH, *ZP*, *31* (1925), 681–708, rec. 5 Jan 25; reprinted HCW, A1, 354–381; English in van der Waerden, note 1, 223–251. This theory, its background, and implications are discussed more extensively in Dresden, note 2; John Hendry, *Centaurus, 25* (1981), 189–221; MR 2; and DC, *Diss.*, chap. 6.

14. WH, "Quantum theory and its interpretation," 1967, in HCW, C2, 345–361, on 351.

15. Kramers and WH, note 13, and van der Waerden, note 1, 234

16. WH to Landé, 15 Jun 1924 (AHQP 6,2).

17. Pauli and Heisenberg met in Munich on 8 Jan 1925. Pauli's criticisms are indicated in WH to Bohr, 8 Jan 1925 (BSC 11,2) and WH to Pauli, 28 Feb 25. Pauli's position in the period 1924–25 is well summarized by Daniel Serwer, *HSPS, 8* (1977), 189–256, and by John Hendry, *The Creation of Quantum Mechanics and the Bohr-Pauli Dialogue* (Hingham, MA: Reidel, 1984).

18. Pauli to Sommerfeld, 6 Dec 1924.

19. WH to his parents, 15 Mar, 20 Mar, and 3 Apr 1925.

20. WH, *ZP*, *32* (1925), 841–860, rec. 10 Apr 1925; reprinted HCW, A1, 306–325.

21. Bohr to WH, 18 Apr 1925 (BSC 11,2).

22. Recalled by WH in "Erinnerungen an die Zeit der Entwicklung der Quantenmechanik," 1960, HCW, C2, 263–270. He still believed hydrogen to be the least problematic for quantum theory. WH to Landé, 18 Feb 1925 (AHQP 6,2).

23. Indicated in WH to Bohr, 16 May 1925 (BSC 11,2), and WH to Kronig, 8 and 20 May 1925 (AHQP 16,6).

24. WH to Bohr, 16 May 1925 (BSC 11,2).

25. Pauli to Kronig, 21 May 25 (AHQP 16,10).

26. WH to Pauli, 24 Jun 1925.

27. WH to Pauli, 9 July 1925.

28. Note 26.

29. Max Born and Pascual Jordan, *ZP*, *33* (1925), 479–505, rec. 11 June 1925, 493, n. 1.

30. WH to his parents, 15 May 1925.

31. WH to Kronig, 5 Jun 1925 (AHQP 16,6).

32. Ibid.

33. Note 31; WH to Bohr, 8 Jun 1925 (BSC 11,2).

34. WH to B. L. van der Waerden, private communication, recounted in van der Waerden, note 1, 25.

35. Indicated by WH to Pauli, 21 and 24 Jun 1925, 4 Jul 1925.

36. WH to Pauli, 24 Jun 1925.

37. Ibid.

38. WH to his father, 30 June 1925.

39. This sequence according to Born, *My Life: Recollections of a Nobel Laureate* (New York: Charles Scribner's Sons, 1978), 216–217; Born's comment in EB, *Briefw.*, 125; and an unpublished chronology prepared by B. L. van der Waerden (in Born folder, AHQP).

Chapter 11. Awash in Matrices, Rescued by Waves

1. Pauli to Kramers, 27 July 1925; Pauli to Kronig, 9 Oct 1925.

2. Born to Schrödinger, 16 May 1927 (AHQP 41,7).

3. Max Born and Pascual Jordan, *ZP, 34* (1925), 858–888, on 858.

4. The development of matrix mechanics is summarized, along with a more extensive discussion of technical details, in Jammer, *Conceptual*; Olivier Darrigol, *From C-numbers to Q-numbers: The Classical Analogy in the History of Quantum Theory* (Berkeley, 1993); Helge Kragh, *Quantum Generations: A History of Physics in the Twentieth Century* (Princeton, 1999); and MR 3, 4.

5. Note 3, 858.

6. Dirac interview, session 1, 1 Apr 1962 (AHQP). Dirac's background and the technical details of his work in this period are described in the works cited in note 4.

7. Max Born, WH, Pascual Jordan, *ZP, 35* (1926), 557–615, rec. 16 Nov 1925; reprinted HCW, A1, 397–455. Again, for technical details, see note 4.

8. Born to Bohr, 10 Oct 1925, in BCW 5, 311–312.

9. Franck substituted for Born as director of the theoretical physics institute, while Hund and Heisenberg split Born's teaching duties. The lecture required only minimal preparation, since Born left his extensive lecture notes for Heisenberg. Ibid.; Dekan Courant to University Kurator, 30 Oct 1925 (Personalakt Dr.

Heisenberg, UA Göttingen, 4/Vc 317); and *Verzeichnis der Vorlesungen auf der Georg August-Universität zu Göttingen während des Wintersemesters 1925/26.*

10. Wolfgang Pauli, *ZP, 36* (1926), 336–363; Paul Dirac, *PRSL, A110* (1926), 561–569.

11. D. C. Cassidy, *HSPS, 37* (2007), 247–269.

12. Pauli to Bohr, 12 Mar 1926. See Ralph Kronig, "The Turning Point," in *Theoretical Physics in the Twentieth Century: A Memorial Volume to Wolfgang Pauli*, eds. M. Fierz and V. F. Weisskopf (New York: Interscience, 1960), 5–39. The invention of spin is discussed by Daniel Serwer, *HSPS, 8* (1977), 189–256; E. Rüdinger and K. Stolzenburg, "Introduction," BCW 5, 219–240; and MR 3.

13. WH and Pascual Jordan, *ZP, 37* (1926), 263–277, rec. 16 Mar 26; reprinted HCW, A1, 516–530.

14. Note 12.

15. Wolfgang Pauli, *ZP, 31* (1925), 373–385, and 765–783, J. L. Heilbron, "The Origins of the Exclusion Principle," *HSPS, 13* (1983), 261–310.

16. Schrödinger's wave mechanics papers appeared in *AP,* 1926, and are reprinted in Erwin Schrödinger, *Gesammelte Abhandlungen*, vol. 3 (Wien, 1984), English trans.: Schrödinger, *Collected Papers on Wave Mechanics*, 2nd ed. (New York: Chelsea, 1978). Histories of his work (and life) include: W. Moore, *Schrödinger: Life and Thought* (New York: Cambridge Univ. Press, 1989); Linda Wessels, "Schrödinger's route to wave mechanics," *Studies in History and Philosophy of Science, 10* (1979), 311–340, Helge Kragh, "Erwin Schrödinger and the Wave Equation: The Crucial Phase," *Centaurus, 26* (1982), 154–197; as well as in MR 5.

17. The hypothesis was later experimentally confirmed by electron diffraction experiments, interpreted as involving the diffraction of electron waves.

18. Sommerfeld to Pauli, 3 Feb 1926 (SN).

19. Pauli to Jordan, 12 Apr 1926.

20. Max Born, *ZP, 37* (1926), 863–867, rec. 25 Jun 1926, on 864. Max Born and Norbet Wiener, *ZP, 36* (1926), 177–187. His enthusiasm is recalled in Born to Schrödinger, 16 May 1927 (AHQP 41,7).

21. WH to Jordan, 28 Jul 1926 (AHQP 18,2).

22. WH to Jordan, 8 Apr 1926, ibid.; Born to Schrödinger, 16 May 1927 (AHQP 18,2).

23. WH to Pauli, 18 Sep 1925.

24. WH to Pauli, 16 Nov 1925.

25. Schrödinger, note 16 (Engl. trans.), 45, his emphasis.

26. Ibid., 59, slight changes; Schrödinger to Wien, 18 Jun 1926, excerpted in Wilhelm Wien, *Aus dem Leben und Wirken eines Physikers* (Leipzig, 1930), 72–73.

27. Schrödinger, *AP, 79* (1926), 734–756, on 734; Schrödinger, note 29 (Engl. trans.), 59ff.

28. WH to Pauli, 8 Jun 1926.

29. WH, *Nwn, 14* (1926), 989–994; reprinted HCW, B, 52–57.

30. Mara Beller, "The Genesis of Interpretations of Quantum Physics, 1925–1927," (PhD diss., University of Maryland, 1983); Beller, *Isis, 74* (1983), 469–491; and Beller, *Quantum Dialogue: the Making of a Revolution* (Chicago: Univ. of Chicago Press, 1999). See also John Hendry, *The Creation of Quantum Mechanics and the Bohr-Pauli Dialogue* (Hingham, MA: Reidel, 1984); A. I. Miller, "Redefining Anschaulichkeit," in *Physics as Natural Philosophy: Essays in Honor of Laszlo Tisza on His 75th Birthday*, eds. A. Shimony and H. Feshbach (Cambridge, MA, 1982), 376–411; E. MacKinnon, "The Rise and Fall of Schrödinger's Interpretation," in *Studies in the Foundations of Quantum Physics*, ed., P. Suppes (1980); and, for technical details, MR 5, 2.

31. Bohr to WH, 18 Nov 1925; Bohr to Born, 25 Nov 1925 (BSC 9,2)

32. Pauli to WH, 19 Oct 1926.

33. Trowbridge to Bohr, April 1926 (BGC).

34. WH to Bohr, 5 Apr 1926; Bohr to WH, telegram April 1926 (BSC 11,2).

35. Courant to Bohr, 24 Apr 1926, quoted in MR 4, 280; discussed in WH to his parents, 29 Apr 1926.

36. WH to "Herr Professor" [Max von Laue], 19 Apr 1926 (SPK, Darmstädter-Sammlung, F1a(5)1926).

37. WH to his parents, 29 Apr 1926. Only Planck was missing on vacation. Courant began immediate efforts to create a chair in Göttingen for Heisenberg. Courant, Dean of science faculty, to University Kurator, 8 Jun 1926 (Personalakt Dr. Heisenberg, UA Göttingen, 4/Vc 317).

38. WH to his parents, 14 May 1926, and to his mother, 20 Nov 1926.

39. August Heisenberg to Bohr, 17 July 1926 (BSC 11,2); Bohr to August Heisenberg, 4 Aug 1926 (BSC 11,2).

40. WH to Born, 26 May 1926 (AHQP 18,2); WH to his parents, 14 May 1926.

41. Ibid.

42. WH to his parents, 11 May 1926; lecture notes in HA and on AHQP 45,5.

43. Pauli to Schrödinger, 24 May 1926.

44. Pauli to Schrödinger, 22 Nov 1926.

45. WH to Pauli, 28 Jul 1926.

46. Register of lectures, "Münchener physikalisches Mittwochscolloquium" (AHQP 20). Schrödinger lectured on 23 and 24 July 1926, which despite the name of the colloquium were Friday and Saturday.

47. Recalled in WH, *PB*, 73.

48. Wien to Schrödinger, 20 Aug 1926; Schrödinger to Wien, 25 Aug 1926, excerpted in Wien, note 26.

49. WH to Jordan, 28 Jul 1926 (AHQP 18,2).

50. Note 29, 993–994.

51. WH to Jordan et al., 29 Oct and 24 Nov 1926 (AHQP 18,2).

52. WH, *ZP, 40* (1926), 501–506, rec 6 Nov 1926; reprinted HCW, A1, 472–477.

53. WH to Pauli, 4 Nov 1926.

54. Schrödinger delivered a lecture to the Fysisk-Forening in Copenhagen on 4 Oct 1926, titled "Grundlagen der undulatorischen Mechanik [Foundations of Wave Mechanics]," minute book of Fysisk-Forening, 1908–1946 (AHQP 35,4).

55. WH, "Quantum Theory and its Interpretation," 1967, in HCW, C2, 345–361, on 356.

56. Especially, Schrödinger to Joos, 17 Nov 1926 (AHQP 41,8); Schrödinger to Kramers, 19 Nov 1926 (AHQP 41,8); Schrödinger to Wien, 21 Oct 1926; Schrödinger to Bohr, 23 Oct 1926 (in BCW 6, 459–461); Bohr to Schrödinger, 2 Dec 1926 (BCW 6, 462–463); and others.

57. Born, *ZP, 38* (1926), 803–827, rec. 21 Jul 1926. Born's interpretation is examined in greater detail in Beller, diss., note 30; Jammer, *Conceptual*; and MR 5, 2.

58. Schrödinger, note 16 (Engl. trans.), 60.

59. Quoted in Jammer, *Conceptual*, 288.

60. Apparently attributed to Bohr, in Schrödinger to Bohr, note 56.

Chapter 12. Determining Uncertainty

1. WH, *ZP, 43* (1927), 172–198, rec. 23 Mar 1927; reprinted in HCW, A1, 478–504; English translation in *Quantum Ttheory and Measurement*, eds. John A. Wheeler and Wojciech Zurek (Princeton: Princeton Univ. Press, 1983), 62–84.

2. WH, *Forschungen und Fortschritte, 3* (1927), 83; reprinted HCW, C1, 21.

3. Writings on the uncertainty principle and the Copenhagen interpretation range from the extremely erudite to the popular to New Age fantasy. Some general historical works include Jammer, *Conceptual* and *Philosophy*; and Mara Beller, "The Genesis of Interpretations of Quantum Physics, 1925–1927," (PhD thesis, University of Maryland, 1983). Don Howard, "Who Invented the 'Copenhagen Interpretation'? A Study in Mythology," at *http://www.nd.edu/~dhoward1/Copenhagen%20Myth%20A.pdf*, argues that the full interpretation arose only in the 1950s. See also James T. Cushing, *Quantum Mechanics: Historical Contingency and the Copenhagen Hegemony* (Chicago: Univ. of Chicago Press, 1994); Beller, *Quantum Dialogue: The Making of a Revolution* (Chicago: Univ. Chicago Press, 1999); Ana Rioja, "Los origenes del principio de indeterminación," *Theoria, 22* (1995), 117–143; and the more popular-oriented Andrew Whitaker, *Einstein, Bohr, and the Quantum Dilemma: From Quantum Theory to Quantum Information* (New York: Cambridge Univ. Press, 2006).

4. Herman Weyl, *Gruppentheorie und Quantenmechanik* (Leipzig, 1928), 67. I thank the Quantum History Project reading group at the Max Planck Institute for History of Science, Berlin, for bringing this to my attention.

5. WH Note 2. Heisenberg did not use the German word for "uncertainty" (*Unsicherheit*); he used the word for "imprecision" (*Ungenauigkeit*). Nor did he yet call any of this a "principle." See D. C. Cassidy, "Answer to Question #62: When Did the Indeterminacy Principle Become the Uncertainty Principle?" *American Journal of Physics, 66* (1998), 278–279. The "uncertainty relations," as subsequently defined and expressed in mathematical symbols, may be given as follows: the uncertainty (as a standard deviation) in the measurement of the position (q), Δq, and the uncertainty in the measured momentum (p), Δp, at the same instant are related by the expression $\Delta p\, \Delta q \geq h/4\pi$, where h is Planck's constant. A similar "uncertainty relation" holds for the uncertainty in the measured energy, ΔE, and time, Δt: $\Delta E\, \Delta t \geq h/4\pi$. Heisenberg originally gave only approximate expressions for these relations.

6. Note 2.

7. Acausality and indeterminism have a long history. See, among others, Ernst Cassirer, *Determinismus und Indeterminismus in der modernen Physik* (Göteborg: Elanders Boktryckeri, 1937); Stephen G. Brush, "Irreversibility and Indeterminism: Fourier to Heisenberg," *Journal of the History of Ideas, 37* (1976), 603–630; Paul A. Hanle, "Indeterminacy before Heisenberg: The Case of Franz Exner and Erwin Schrödinger," *HSPS, 10* (1979), 225–269.

8. WH, note 1, 197.

9. WH, *ZP, 40* (1927), 501–506, rec. 6 Nov 1926; in HCW, A1, 472–477; WH, *ZP, 41* (1927), 239–267, rec. 22 Dec 1926; in HCW, A1, 551–579; and WH to Pauli, 5 Feb 1927.

10. Results of a literature survey reported in A. B. Kozhevnikov and O. I. Novik, *Analysis of Informational Ties Dynamics in Early Quantum Mechanics (1925–1927)* (Moscow: Academy of Sciences, 1987).

11. WH to Pauli, 16 May 1927.

12. Wiener to Sommerfeld, 28 Nov 1926 (SN).

13. Sommerfeld to Wiener, 3 Dec 1926 (SN).

14. Philosophische Fakultät, Protokolle, book VII, 337 (UA Leipzig).

15. Max Born, *ZP, 40* (1927), 167–192, rec 16 Oct 1926.

16. Paul Dirac, *PRS, A113* (1927), 621–641, rec 2 Dec 1926; Jordan, *ZP, 40* (1927), 809–838, rec 18 Dec 1926.

17. Pauli to Schrödinger, 12 Dec 1926; Jordan to Schrödinger May 1927, incorrectly dated 1926 (AHQP 41,8).

18. Pauli to WH, 19 Oct 1926.

19. WH to Pauli, 28 Oct 1926, his emphasis.

20. WH to Pauli, 23 Feb 1927; and WH, note 1.

21. WH to Pauli, 23 Nov 1926.

22. WH, note 1, 197; repeated in WH, note 2. Because this paragraph is not integral to his paper, it may represent Heisenberg's "capitulation" to the acausal demands of the cultural milieu prevailing in his homeland, in accord with Paul Forman, "Weimar Culture, Causality and Quantum Physics," *HSPS, 3* (1971), 1–115.

23. Pascual Jordan, *Nwn, 15* 4 Feb 1927, 105–110.

24. This was argued by Mara Beller, *Archive for History of Exact Sciences, 33* (1985), 337–349. Heisenberg acknowledged some of his debts in WH, note 1, 173–174. WH, note 1, 176, cites Jordan's paper.

25. WH, *PB,* 63.

26. Only Heisenberg's side of the correspondence has been found: WH to Einstein, 16 and 30 Nov 1925, 18 Feb 1926 (EA). Elisabeth Heisenberg said that Allied Forces removed a packet of letters, including Einstein letters, from their Urfeld home in 1945. Interview with the author, 1982.

27. During the early 1980s Einstein's Princeton house still contained the writings of the mentioned authors, some in editions extending back to this period (author's observation).

28. Einstein to Schrödinger, 26 Apr 1926, in K. Przibram, *Briefe zur Wellenmechanik* (Vienna: Springer Verlag, 1963), 26. There is no indication that Einstein already knew of Schrödinger's equivalence proof, which arrived at the publisher several weeks later.

29. WH, *PB,* 63, and WH, interview session 1 (AHQP).

30. WH, note 1, 179–184.

31. WH to his parents, 17 Feb 1927. Bohr's departure was imminent.

32. Bohr to Pauli, 25 Mar 1927.

33. WH to Pauli, 4 Apr 1927; WH, *Die physikalischen Prinzipien der Quantentheorie,* German of his Chicago lectures, summer 1929 (Leipzig, 1930), 15–16; Engl. reprinted HCW, B, 117–166; German, HCW, B, 167–170. A more complete derivation than that given by Heisenberg may be found in Jammer, *Philosophy,* 64–65. The microscope argument and Bohr's response have been analyzed by Scott Tanona, "Uncertainty in Bohr's Response to the Heisenberg Microscope," *Studies in History and Philosophy of Science, Part B, 35* (2004), 483–507.

34. WH to his parents, 16 and 30 May 1927; WH to Pauli, 16 May 1927.

35. WH, *PB,* 79.

36. WH, note 1, "Nachtrag zur Korrektur," 197–198.

37. WH to Pauli, 16 May 1927; Niels Bohr, "The Quantum Postulate and the Recent Development of Atomic Theory," in *Atti del Congresso Internazionale dei Fisici 11–20 Settembre 1927, Como-Pavia-Roma,* vol. 2 (Bologna, 1928), 565–588, lecture delivered 16 Sep 1927; reprinted in BCW 6, 113–136. Bohr's derivation of the microscope experiment is on 573.

38. Ibid. Bohr's ideas are summarized here more directly than in the original paper.

39. "Discussione sulla comunicazione Bohr," in *Atti del Congresso Internazionale dei Fisici 11–20 Settembre 1927, Como-Pavia-Roma* (Bologna, 1928), 589–598; reprinted BCW 6, 137–146; WH's remark on 593–594.

40. WH to his parents, Copenhagen, 22 June 1927. Debye had practically decided on Leipzig by the spring of 1927, "Prof. Debye Berufg. nach Leipzig" (Archive, Schweizerischer Schulrat, in ETH).

41. WH to Bohr, 18 Jun 1927 (BPC).

42. Bayerisches Staatsministerium to University Senat, Munich, 20 Jun 1927 (A. Sommerfeld, Personalakten E II-N, UA Munich); and Sommerfeld to Philosophische Fakultät II, 3 Dec 1925 (Univ. München, Lehraufträge, vol. 2, BHSA, MK 11303).

43. Sommerfeld to WH, 17 Jun 1927 (SN).

44. WH to Sommerfeld, 21 Jun 1927 (SN); WH to his parents, 22 Jun 1927.

45. WH to Bohr, 21 Aug 1927 (BPC).

46. Communication of 20 Mar 1979 from Dr. Schindler, concerning records of Schweizerischer Schulrat, ETH. American interest in Heisenberg is indicated by Trowbridge to Bohr, 5 Oct 1927 (BSC 11,2).

47. WH to Bohr, 5 Dec 1927 (BSC 11,2).

48. Note 14, 352. The lecture was probably "Erkenntnistheoretische Probleme der modernen Physik" (HA), published HCW, C1, 22–28.

Chapter 13. Reaching the Top

1. The zeal is described by J. L. Heilbron, "The Earliest Missionaries of the Copenhagen Spirit," *Revue d'histoire des sciences*, 38 (1985), 195–230.

2. WH to Bohr, 21 Aug 1927 (BPC).

3. Pauli to Bohr, 17 Oct 1927.

4. Max Born, comment during the "Discussione sulla comunicazione Bohr," following Bohr's Como lecture, in *Atti del Congresso Internazionale dei Fisici 11–20 Settembre 1927, Como-Pavia-Roma*, vol. 2 (Bologna, 1928), 589–598, on 589; reprinted BCW, 6, 137–146.

5. Max Born and WH, "La mécanique des quanta," in *Electrons et photons. Rapports et discussions du 5. Conseil de Physique*, ed. Institut International de Physique Solvay (Paris, 1928), 143–181, on 143, reprinted HCW, B, 58–96.

6. Ibid., 178.

7. This has been argued by Heilbron, note 1; James T. Cushing, *Quantum Mechanics: Historical Contingency and the Copenhagen Hegemony* (Chicago: Univ. of Chicago Press, 1994); Mara Beller, *Quantum Dialogue: The Making of a Revolution* (Chicago: Univ. of Chicago Press, 1999);

8. Pauli to Kramers, 27 Jul 1925.

9. See Jagdish Mehra, *The Solvay Conferences on Physics: Aspects of the Development of Physics since 1911* (Dordrecht: D. Reidel, 1975).

10. Among the many works on the debate are Jammer, *Philosophy*, 108–158; Pais, *Subtle*, 440–449; Arthur Fine, *The Shaky Game: Einstein, Realism, and the Quantum Theory* (Chicago: Univ. of Chicago, 1986); Don Howard, "Complementarity and Ontology: Niels Bohr and the Problem of Scientific Realism," (Ph.D. diss., Boston University, 1979); Edward MacKinnon, *Scientific Explanation and Atomic Physics* (Chicago: Univ. of Chicago, 1982); Bohr, "Discussion with Einstein on Epistemological Problems in Atomic Physics," in *Albert Einstein: Philosopher-Scientist*, ed. P .A. Schilpp, vol. 1 (Evansville, IL: Library of Living Philosophers, 1949), 199–241.

11. Einstein to Born, 4 Dec 1926 (EB, *Briefw.*), his emphasis.

12. Einstein to Sommerfeld, 9 Nov 1927 (ES, *Briefw.)*; Pauli to Weyl, 11 Jul 1929.

13. WH to his parents, Brussels, 29 Oct 1927.

14. This has been argued by Paul Forman, "*Kausalität, Anschaulichkeit*, and *Individualität*, or How Cultural Values Prescribed the Character and the Lessons Ascribed to Quantum Mechanics," in *Society and Knowledge: Contemporary Perspectives in the Sociology of Knowledge*, eds. Nico Stehr and Volker Meja (New Brunswick, NJ: Transaction Books, 1984), 333–347.

15. See WH, *Biblio.*

16. These are collected in HCW, C1.

17. Forman, note 14, attributes this to the German cultural milieu.

18. Born, *ZP, 37* (1926), 863–867.

19. Note 4.

20. Forman, note 14, and "Weimar Culture, Causality and Quantum Physics," *HSPS, 3* (1971), 1–115.

21. WH, in *Forschungen und Fortschritte, 3* (1927), 83; reprinted HCW, C1, 21.

22. WH, "Erkenntnistheoretische Probleme der modernen Physik," MS (1928?), publ. in HCW, C1, 22–28, on 28. This paper and Heisenberg's overall philosophical position have been analyzed by Patrick A. Heelan, *Quantum Mechanics and Objectivity: A Study of the Physical Philosophy of Werner Heisenberg* (The Hague: Martinus Nijhoff, 1965).

23. Respectively: HCW, C1, 45, 29–32, 39, 49.

24. Heisenberg-Schlick correspondence, 1930–1932 (Schlick Papers, Amsterdam). I am grateful to Anne Kox for a copy of this correspondence.

25. HCW, C1, 49.

26. Bohr's papers on complementarity are published in BCW, 6.

27. Heisenberg recalled these conversations in WH, *PB*, 117–124; author's interview with Carl Friedrich von Weizsäcker, 30 Apr 1982.

28. Carl Friedrich von Weizsäcker, *ZP, 70* (1931), 114–130.

29. Quoted by Fine, note 10, 34.

30. Albert Einstein, Boris Podolsky, and Nathan Rosen, *PR, 47* (1935), 777–780; reprinted in *Quantum Theory and Measurement*, eds. John A. Wheeler and Wojciech Zurek, (Princeton: Princeton Univ. Press, 1983), 138–141. Fine, note 10, argues for Podolosky's authorship.

31. Pauli to WH, 15 Jun 1935.

32. Further historical and philosophical discussions include Jammer, *Philosophy*, 159–251; MacKinnon, note 10, 338–348; and Fine, note 10.

33. Condon, quoted in Jammer, *Philosophy*, 189–190; Pauli to WH, 15 Jun 1935.

34. WH, "Ist eine deterministische Ergänzung der Quantenmechanik möglich?" (AHQP 45, 11), publ. PWB, 2, 409–418, enclosed in WH to Pauli, 2 Jul 1935; Bohr to WH, 2 Jul 1935 (BSC 20,2), enclosing manuscript of Bohr, *PR, 48* (1935), 696–702; reprinted Wheeler and Zurek, note 30, 145–151.

35. WH to Bohr, 28 Aug 1935; Bohr to WH, 10 and 15 Sep 1935; and WH to Bohr, 29 Sep 1935 (BSC 20,2).

36. WH, note 34, 417. Heisenberg referred to G. Hermann, *Die naturphilosophische Grundlagen der Quantenmechanik* (Berlin, 1935). Hermann's views, expounded before the EPR argument, and Heisenberg's later use of them are discussed by Jammer, *Philosophy*, 207–211.

37. WH, address delivered to University of Vienna, 27 Nov 1935, reprinted HCW, C1, 108–119.

38. WH, note 34, 410.

39. Einstein, "Remarks Concerning the Essays Brought Together in This Co-operative Volume," in Schilpp, ed., note 10, vol. 2, 665–688, on 666–667; his parenthetical remarks.

40. WH, *Physics and Philosophy: The Revolution in Modern Science* (New York: Harper and Row, 1958), 144 and 129; reprinted HCW, C2, 3–201.

41. Physicist J. H. Van Vleck, quoted by K. R. Sopka, *Quantum Physics in America 1920–1935* (New York: Arno Press, 1980).

42. This was argued by J. L. Heilbron, "La fisica negli Stati Uniti subito prima della meccanica quantistica," in *Fisici e società negli anni '20* (Milan, 1980), 135–158. The rise of American theoretical physics is explored by S. S. Schweber, "The Empiricist Temper Regnant: Theoretical Physics in the United States 1920–1950," *HSPS, 17* (1986), 55–98.

43. Schweber, ibid.

44. K. T. Compton, *Nature, 139* (1937), 238–239.

45. List of past and future invited physicists, locations, salaries, and lengths of stay in H. M. Randall to W. F. G. Swann, 19 Feb 1929 (Swann Papers, Am. Philos. Society, Philadelphia).

46. The salaries ranged from $1,500 for Hund to $6,000 for Heisenberg. By comparison, the annual salary offered Heisenberg a year earlier to teach full time at Columbia University was $10,000. These were all large sums at the time.

47. Arnold Sommerfeld, *Atombau und Spektrallinien, Wellenmechanischer Ergänzungsband* (Braunschweig: Springer-Verlag, 1929).

48. WH to Sommerfeld, 6 Feb 1929 (AHQP 83,H).

49. WH to Bohr, 1 Mar 1929 (BSC, 11,2).

50. WH to Bohr, Chicago 16 Jun 1929 (BSC, 11,2).

51. WH to Pauli, 20 Jul 1929. Heisenberg's and Dirac's travels are described by L. Brown and H. Rechenberg, "Paul Dirac and Werner Heisenberg—A Partnership in Science," in *Reminiscences about a Great Physicist: Paul A. M. Dirac*, eds. B. N. Kursunoglu and E. Wigner (Cambridge: Cambridge Univ. Press, 1987), 117–162.

52. David C. Cassidy, *American Journal of Physics, 66* (1998), 278–279.

53. WH, *The Physical Principles of the Quantum Theory*, trans. C. Eckart and F. C. Hoyt (Chicago: Univ. of Chicago Press, 1930), x; reprinted HCW, B, 117–166.

54. Max Born and Pascual Jordan, *Elementare Quantenmechanik* (Berlin: Springer, 1930); Sommerfeld, note 47. Wolfgang Pauli, in *Handbuch der Physik*, 2nd ed., vol. 24, part 1 (Berlin, 1933), 83–272. Section A1, 83–90, is titled "Unbestimmtheitsprinzip und Komplementärität." [Indeterminacy principle and complementarity.] A partial list of subsequent textbooks is provided by Jammer, *Philosophy*, 59.

55. For instance, E. C. Kemble, *The Fundamental Principles of Quantum Mechanics* (New York: McGraw-Hill, 1937), 5, written after EPR.

56. I am grateful to Silvan S. Schweber for informing me of this recollection.

Chapter 14. New Frontiers

1. WH, *PB*, 93.

2. L. Weickmann, "Nachruf auf Otto Wiener," SAW, MPK, *Berichte, 79* (1927), 107–123.

3. Wentzel to Sommerfeld, 26 May 1927 (SN).

4. Note from Ulrich, dated 31 May 1932, on letter of Debye to Saxon Culture Ministry, 26 May 1932 (Physikalisches Institut der Universität Leipzig, vol. 2, 1929–1939, SStA, Ministerium für Volksbildung, 10230/27).

5. Floor plans in Otto Wiener, "Das Physikalische und das Theoretisch-Physikalische Institut," *Festschrift zur Feier des 500 jährigen Bestehens der Universität Leipzig*, vol. 4, part 2 (Leipzig: Hirzel, 1909), 24–69.

6. Debye to Ministerialrat von Seydewitz, 19 Sep 27, and Debye to Notgemeinschaft, 20 Jul 1928 (Debye Papers, MPG).

7. WH to his parents, 5 and 9 Nov 1927; WH to his parents, 13 Dec 1927; Akten des Rentamtes (UA Leipzig).

8. WH to his parents, 9 Nov 1927; Sommerfeld to WH, 15 Nov 1927 (SN).

9. WH to parents, 7 May 1928, and Wentzel to Sommerfeld, 12 May 1928 (SN). The three students were Rudolf Peierls, Carl Eckart, and W. V. Houston, an American visitor.

10. WH to his parents, 27 Jun 1928.

11. Data derived from *Vierteljahrshefte zur Statistik des Deutschen Reiches, 37* (1928), *38* (1929), *39* (1930); Debye to Staatsministerium, 27 Feb 1931 (in note 4).

12. List compiled by Dr. Michael Eckert, to whom I am grateful for a copy.

13. The establishment and maintenance of this network is indicated by the travels of the assistants and students and by the correspondence of their teachers, especially WH to Pauli, 1 Aug 1929.

14. Interview with Friedrich Hund, Göttingen, April 1981; and Hund's "Wissenschaftliches Tagebuch" (Deutsches Museum, Handschrifenabteilung),

15. C. I. Zahn to S. A. Goudsmit, 24 Feb 1933, quoted in PWB 2, 148.

16. WH to Mansel Davies, 16 Jan 1970, publ. in Davies, "Peter Joseph Wilhelm Debye," Royal Society of London, *Biographical Memoirs, 16* (1970), 175–232, on 221.

17. Interview with Hund, 1981.

18. WH to Bohr, 23 Jul 1928 (BSC, 11,2).

19. WH to his mother, 25 Nov 1928 and 6 Dec 1928; WH to his parents, 27 Jun 1928 and 16 Oct 1928.

20. Felix Bloch, "Reminiscences of Heisenberg and the Early Days of Quantum Mechanics," *Physics Today, 29* (issue 12) (December 1976), 23–27, on 27.

21. Quoted by Bernhard Schweitzer, *Die Universität Leipzig 1409–1959* (Tübingen: J. C. B. Mohr, 1960), 13–14.

22. Interview with Carl Friedrich von Weizsäcker, 30 April 1982.

23. WH to his parents, 25 Feb 1930.

24. Interview with Hund, 1981.

25. Hans Driesch, *Lebenserinnerungen* (Basel: Ernst Reinhardt Verlag, 1951), 200.

26. Gerhard Ritter, *Carl Goerdeler und die deutsche Widerstandsbewegung* (Stuttgart: Deutsche Verlags-Anstalt, 1954).

27. Draft of nomination, 23 May 1930 (Debye Papers, MPG), and SAW, MPK, *Berichte, 82* (1930).

28. WH, SAW, MPK, *Phys. Berichte, 83* (1931), 3–9; del. 9 Jan 1931, reprinted HCW, A2, 116–131.

29. Notes 24 and 22.

30. Histories of quantum field theory include: Pais, *Inward*; S. S. Schweber, *QED and the Men Who Made it* (Princeton: Princeton Univ. Press, 1994); MR 4; Helge Kragh, *Dirac: A Scientific Biography* (New York: Cambridge Univ. Press, 2005); and Olivier Darrigol, "Les Débuts de la Théorie quantique des Champs (1925–1948)" (doctoral thesis, Université de Paris I [Panthéon-Sorbonne], 1982).

31. Pascual Jordan and Wolfgang Pauli, *ZP, 47* (1928), 151–173.

32. Paul Dirac, *PRS, A117* (1928), 610–624. For details, see the works in note 30.

33. The problem of infinities in field theory and efforts to resolve them are explored by Alexander Rueger, *HSPS, 22* (1992), 309–337.

34. Paul Dirac, *PZ, 29* (1928), 561–563, delivered (probably in English), 22 June 1928; conference program in Debye to Fermi, 9 Jun 1928 (Debye Papers, MPG); Pauli to Bohr, 16 Jun 1928.

35. WH to Pauli, 3 May 1928.

36. Arnold Sommerfeld, *Nwn, 15* (1927), 825–832. A fuller account of the electron theory of metals is provided by L. Hoddeson, G. Baym, and M. Eckert, *Reviews of Modern Physics, 59* (1987), 287–327.

37. Bloch, "Über die Quantenmechanik der Eletronen in Kritallgittern," doctoral diss., completed 2 Jul 1928, indicated in "Promotionen," (UB Leipzig), published *ZP, 52* (1928), 555–600.

38. Peierls recalls his work in Leipzig in Rudolf Peierls, *Bird of Passage* (Princeton: Princeton Univ. Press, 1985), 32–40. See Cathryn Carson, "The Peculiar Notion of Exchange Forces—I: Origins in Quantum Mechanics, 1926–1928," *Studies in History and Philosophy of Modern Physics, 27* (1996), 23–45.

39. WH, *ZP, 49* (1928), 619–636, rec. 20 May 1928, reprinted HCW, A1, 580–597.

40. WH, *Metallwirtschaft, 9* (1930), 843–844, reprinted HCW, B, 167–168.

41. Pauli to Bohr, 16 Jan 1929; Pauli to O. Klein, 18 Feb 1929; WH to Jordan, 22 Jan 1929 (Nachlass Jordan, SPK).

42. WH and Wolfgang Pauli, *ZP, 56* (1929), 1–61, rec. 19 Mar 1929, reprinted HCW, A2, 8–68; and *ZP, 59* (1930), 168–190, rec. 7 Sep 1929, reprinted HCW, A2, 69–91.

43. J. Robert Oppenheimer, *ZP, 55* (1929), 725–737, rec. 6 May 1929; and *PR, 35* (1930), 461–477, rec. 12 Nov 1929.

44. Neils Bohr, *Journal of Chemical Society of London, 1932*, 349–384, reprinted BCW, 6, 371–408.

45. Bohr to Dirac, 29 Aug 30 (BSC 18,4).

46. WH to Pauli, 20 Jul 1929.

47. WH to Dohr, 26 Feb, 10 Mar, 23 Mar 1930; Bohr to WH, 18 Mar 1930 (BSC 20,2). See B. Garazzo and Helge Kragh, *American Journal of Physics, 63* (1995), 595–605; Helge Kragh, *Reviews of the History of Science, 48* (1995), 401–434.

48. WH, *ZP, 65* (1930), 4–13, reprinted HCW, A2, 106–115, on 106.

49. WH to his parents, 30 May and 12 Jul 1930.

50. WH to his father, 4 Nov 1930.

51. WH to his parents, 7 Nov 1930; WH to his mother, 15 Dec 1930.

52. WH to Bohr, 18 Mar 1937 (BSC, 20,2).

53. WH, *ZP, 69* (1931), 287–297, reprinted HCW, A1, 598–608; *PZ, 32* (1931), 737–740, reprinted HCW, A1, 627–630; *AP, 13* (5th series) (1932), 430–452, reprinted HCW, A2, 250–272.

54. WH to Bohr, 27 Jul 1931 (BSC, 20,2), containing a discussion of various applications.

55. WH, *Berliner Tageblatt*, 25 Dec 1931, reprinted in HCW, C1, 48–49.

56. Heisenberg, Bohr, and the route to the neutron-proton model of the nucleus has been explored by, among others, Joan Bromberg, *HSPS, 3* (1972), 307–341; and BR, *Nuclear Forces*.

57. Bohr, in *Convegno di fisica nucleare, Ottobre 1931-IX* (Rome: Reale Accademia d'Italia, 1932), 119–130.

58. WH to Bohr, 3 and 23 Feb (32); Bohr to WH, 24 Feb 1932 (BSC, 20,2).

59. Chadwick to Bohr, 24 Feb 1932 (BSC). Heisenberg also learned of the discovery about the same time, that is, after meeting at the ski hut.

60. WH, *ZP, 77* (1932), 1–11, rec. 7 Jun 1932, reprinted HCW, A2, 197–207; *ZP, 78* (1932), 156–164, HCW, A2, 208–216; *ZP, 80* (1933), 587–596, HCW, A2, 217–226.

61. WH to Bohr, 20 Jun 1932 (BSC, 20,2).

62. WH, report to Solvay Congress, Brussels, October 1933, reprinted HCW, B, 179–225; WH, ZP, 96 (1935), 473–484, reprinted HCW, A2, 227–238; and report in *Peter Zeeman, 1865–1935* (The Hague, 1935), 108–116, reprinted HCW, B, 238–246.

63. Interview with Carl Friedrich von Weizsäcker, 30 April 1982.

Chapter 15. Into the Abyss

1. "Ein Ermächtigungsgesetz," *LNN*, no. 32 (1 February 1933), 1, editorial.

2. For example, Franz L. Neumann, *Behemoth: The Structure and Practice of National Socialism, 1933–1944* (New York: Harper and Row, 1966); Karl Dietrich Bracher, *The German Dictatorship*, trans. from the German by Jean Sternberg (New York: Holt, Rinehart and Winston, 1970); Hannah Arendt, *The Origins of Totalitarianism*, part 3: *Totalitarianism* (San Diego: Harcourt, Brace, Jovanovich, 1951); Fritz K. Ringer, *The Decline of the German Mandarins: The German Academic Community, 1890–1933* (Cambridge, MA: Harvard Univ. Press, 1969); Hans Mommsen, *Beamtentum im Dritten Reich (Schriftenreihe der Vierteljahrshefte für Zeitgeschichte, 12)* (Stuttgart: DVA, 1966); Kurt Sontheimer, *Antidemokratisches Denken in der Weimarer Republik: Die politischen Ideen des deutschen Nationalismus zwischen 1918 und 1933* (Munich: Nymphenburger Verlag, 1962/1968); Ian Kershaw, *The Nazi Dictatorship: Problems and Perspectives of Interpretation* (London: Edward Arnold, 1985).

3. "Deutsche Hochschullehrer für Adolf Hitler," *Der Führer*, 6 November 1932, signed by 55 academics; "Die deutsche Geisteswelt für die Liste I. Erklärung von 300 deutschen Universitätslehrern," *Völkischer Beobachter* (3 March 1933), 1; "Appell der deutschen Wissenschaft an die Welt," *Frankfurter Zeitung, 78* (13 November 1933), 4, and *Braunschweiger Landeszeitung* (13 November 1933); "Die deutsche Wissenschaft ruft auf: Für Adolf Hitler—Ja!" *Frankfurter Zeitung, 79* (19 August 1934), 1, with 68 signatures; Friedrich Schmidt-Ott (former head of Notgemeinschaft), "Für die Erneuerung unseres gesamten Kulturlebens," *LNN*, no. 225 (13 August 1934), 2.

4. "Ein Hilferuf für die deutsche Wissenschaft," *LNN*, no. 315 (10 November 1932), 3.

5. Planck's position and compromises are discussed by J. L. Heilbron, *The Dilemmas of an Upright Man: Max Planck as Spokesman for German Science* (Berkeley: Univ. of California Press, 1986).

6. This is based on election results as reported in *LNN*. Leipzig votes by precinct (*Gemeinde*) are given in Karl Dietrich Bracher, *Stufen der Machtergreifung* (part 1 of *Die nationalsozialistische Machtergreifung*) (Frankfurt am Main.: Ullstein, 1983), 180.

7. "43 Personen ins Konzentrationslager Dachau gebracht," *LNN*, no. 226 (14 August 1933), 3. These arrests and the early camps are discussed by Martin Broszat, "Nationalsozialistische Konzentrationslager 1933–1945," in *Anatomie des SS-Staates*, ed. Hans Buchheim et al., vol. 2 (Munich, 1967), 11–133, es15–22.

8. "Gesetz zur Wiederherstellung des Berufsbeamtentums. Vom 7. April 1933," *Reichsgesetzblatt*, no. 34 (1933), 175–177; "Gesetz gegen die Überfüllung deutscher Schulen und Hochschulen. Vom 25. April 1933," ibid., no. 43, (1993) 225.

9. Sarah Gordon, *Hitler, Germans and the "Jewish Question"* (Princeton: Princeton Univ. Press, 1984). Detlev J. K. Peukert, *Inside Nazi Germany: Conformity, Opposition and Racism in Everyday Life*, Richard Deveson, trans. (New Haven: Yale Univ. Press, 1982), 58, finds inconsistency in popular attitudes toward the persecution of Jews.

10. Gerhard Ritter, *Carl Goerdeler und die deutsche Widerstandsbewegung* (Stuttgart: Deutsche Verlags-Anstalt, 1954), 64.

11. Statement in *New York World-Telegram*, 11 Mar 1933, reprinted in *Einstein on Peace*, ed. Otto Nathan and Heinz Norden (New York: Schocken Books, 1960), 211.

12. *Einstein on Peace*, note 11; and Einstein to George Ellery Hale, 26 Apr 1933 (EA, 19–386).

13. "Boykott bei strenger Disziplin," *LNN*, no. 92 (2 April 1933), 1.

14. Von Laue to Einstein, 26 Jun 1933 (EA, 16-095); Planck to Einstein, 19 Mar 33 (EA, 19-388). Einstein's expulsion from the Berlin Academy and Planck's reactions to the situation are discussed by Heilbron, note 5.

15. Einstein to von Laue, 26 May 1933 (EA, 16-089).

16. E. Heymann, Academy secretary, press statement of 1 Apr 1933, in *Albert Einstein in Berlin, 1913–1933*, eds. Christa Kirsten and Hans-Jürgen Treder, vol. 1 (Berlin: Akademie-Verlag, 1979), 248. See Walker, *Nazi Science*, chap. 4.

17. This was Arnold Berliner's assessment in Berliner to Einstein, 1 Jan 1934 (EA, 7-038).

18. "Gesetz zur Wiederherstellung" (note 8), para. 3 and 4. Further discussed by Beyerchen, *Scientists*, and elsewhere.

19. Kopfermann to Bohr, 23 May 1933 (BSC, 22,2). Bohr was at that time in the States seeking support for refugee organizations. The changing definition of non-Aryans is discussed by Raul Hilberg, *The Destruction of the European Jews* (Chicago: Univ. of Chicago Press, 1967), esp. chap. 4.

20. WH to his mother, 6 Oct 1933.

21. Born to Einstein, 2 Jun 1933 (EB, *Briefw.*); Born to Heisenberg, 11 Jun 1933 (HA); Born to Mrs. Schrödinger, 15 May 1933, quoted in PWB 2, 207; and Born, *My Life: Recollections of a Nobel Laureate* (New York: Charles Scribner's Sons, 1975).

22. The decimation of Göttingen physics is described by Beyerchen, *Scientists*; Peter Kröner, *Vor fünfzig Jahren: Die Emigration deutschsprachiger Wissenschaftler 1933–1939* (Münster: Gesellschaft für Wissenschaftsgeschichte, 1983), 53–55, lists a total of 50 academics who had emigrated from Göttingen by 1939; of these 32 were in the natural sciences.

23. WH, reviews of Planck's *Postivismus und reale Aussenwelt*, 1931, reprinted HCW, C4, 238; *Wege zur physikalischen Erkenntnis*, 1933, reprinted HCW, C4, 239; and *Die Physik im Kampf um die Weltanschauung*, 1935, reprinted HCW, C4, 240.

24. As indicated in Planck to WII, 26 Apr 1933 (HA), Heisenberg had requested a meeting on 20 April.

25. "Neue Unruhen an der Universität Leipzig," *LNN*, no. 335 (30 November 1932), 3; and recollections of W. Goetz, *Historiker in meiner Zeit* (Köln, 1958).

26. For example, the authorities used paragraph 4 of the civil service law to force a music professor into retirement for signing a pacifist petition and supporting dismissed Jewish colleagues. Hans Driesch, *Lebenserinnerungen* (Leipzig, 1951), 270–272.

27. Heilbron, note 5, 150. Also recalled by WH, *PB*, 150–151.

28. Warren Weaver (an official of the Rockefeller Foundation), diary entry for 6 Dec 34, after a visit with Leipzig pro-rector and geophsicist Ludwig Weickmann (RAC Record Group 1.1, Series 717 Germany, Box 14, folders 127–137). Other professionals faced a similar dilemma. The case of newspaper publishers is discussed by Oron J. Hale, *The Captive Press in the Third Reich* (Princeton: Princeton Univ. Press, 1964), 78.

29. Notation by Staatssekretär Lammers on thank-you note from Planck to Hitler of 9 May 1933 (BA Koblenz, Reichskanzlei, Bd. R43 II/1227a). I thank Dr. Werner of the archive for this information.

30. Planck, "Mein Besuch bei Adolf Hitler," *Physikalische Blätter, 3* (1947), 143, a one-paragraph statement; Einstein to Ludwik Silberstein, 20 Sep 1934 (carbon copy, EA, 19-404).

31. Heisenberg's 1969 recollection of a meeting with Planck, during which he was informed of Planck's meeting with Hitler (WH, *PB*, 149–154), places the Heisenberg/Planck meeting in 1933, but the recollected events in Leipzig surrounding his visit actually occurred in 1935. It is possible, though unlikely, that Planck visited Hitler again in 1935. The account here relies upon documentation from that period placing the visit in 1933.

32. WH to Born, 2 Jun 1933, typed excerpt in a letter from Born to Ehrenfest, 11 Jun 1933 (Ehrenfest Papers, AHQP, EHR 18,5), excerpted in PWB, 2, 168.

33. WH to Bohr, 30 Jun 1933 (BSC, 20,2).

34. Kopfermann to Bohr, 23 May 1933 (BSC, 22,2).

35. Planck to WH, 18 Jun 1933 (HA). The petition and Courant's reaction are discussed, along with the differing reactions of Franck and Born, by Beyerchen, *Scientists* and (save for the petition) by Constance Reid, *Courant in Göttingen and New York: The Story of an Improbable Mathematician* (New York: Springer-Verlag, 1976).

36. Note 32.

37. Born to Ehrenfest, 11 Jun 1933 (Ehrenfest Papers, AHQP, EHR 18,5).

38. Born to WH, 11 Jun 1933 (HA).

39. Planck to WH, 18 Jun 1933 (HA).

40. Ibid. and Arnold Eucken to WH, 5 Jul 1933 (HA).

41. WH, affidavit for Prof. Achelis, 26 Jun 1946 (HA), and WH to his mother, 22 Jun and 31 Jul 1933.

42. Robin E. Rider, *HSPS, 15* (1984), 107–176; Donald Fleming and Bernard Bailyn, eds., *The Intellectual Migration: Europe and America, 1930–1960* (Cambridge, MA: Harvard, 1969); Norman Bentwich, *The Rescue and Achievement of Refugee Scholars: The Story of Displaced Scholars and Scientists, 1933–1952* (The Hague: Martinus Nijhoff, 1953); Leonard Dinnerstein, *America and the Survivors of the Holocaust* (New York: Columbia, 1982).

43. Born to WH, 4 Jul 1933 (HA).

44. Born to Sommerfeld, 1 Sep 1933 (SN).

45. Bohr to WH, 17 Aug 1933 (BSC, 20,2); Bohr to Pauli, 25 Aug 1933.

46. WH to his mother, 8 Sep and 6 Oct 1933.

47. Von Laue's activities are indicated by Laue to WH, 10 May 1933 (HA); Sommerfeld to Laue, 19 May 1933 (SN); and Ladenburg to Bohr in Pasadena, 24 May 1933 (BSC 17,3). Bohr's work on behalf of refugees in 1933 and 1934 is discussed by Finn Aaserud, *Redirecting Science: Niels Bohr, Philanthropy, and the Rise of Nuclear Physics* (Cambridge: Cambridge Univ. Press, 1990), chap. 5.

48. Heisenberg's role as conduit is indicated by WH to Bohr, 14 Oct 1933 (BSC 20,2), quoted in part by Hermann, *Jahrhundert*, 116.

49. Beyerchen, *Scientists*; Hermann, *Jahrhundert*; Einstein to von Laue, 23 Mar 1934 (EA, 16-102); Planck to von Laue, 22 Mar 1934 (Max von Laue papers, 1964-6/136a,b, Sondersammlungen, Deutsches Museum, Munich).

50. Planck to Ministerialdirektor Gerulis, 30 Aug 1933, forwarded to Ministerialrat Achelis (Archiv der Akademie der Wissenschaften, Berlin); and Planck to von Laue, 11 Sep 1933 (1964-6/138a,b, Sondersammlungen, Deutsches Museum, Munich). The situation surrounding the funeral service is described by Beyerchen, *Scientists*, and Heilbron, note 5.

51. Franz Senger, ed., *Reichs-Habilitations-Ordnung: Amtliche Bestimmungen über den Erwerb des Dr. habil. und der Lehrbefugnis an den deutschen wissenschaftlichen Hochschulen* (Berlin: Wiedemannsche Verlagsbuchhandlung, 1939), first promulgated 13 Dec 1934, revised 1 Oct 1938. Meitner's career is described by Ruth Sime, *Lise Meitner: A Life in Physics* (Berkeley: Univ. of California Press, 1997).

52. WH to his mother, 17 Sep 1933. See also Planck to von Laue, 11 Sep 1933, note 50. Weyl's wife was Jewish. Schrödinger's resignation is discussed by Walter Moore, *Schrödinger: Life and Thought* (Cambridge: Cambridge Univ. Press, 1989).

53. Planck to von Laue, 11 Sep 1933 (Laue papers, note 49).

54. WH to his mother, 6 Oct 1933.

55. Planck to WH, 8 and 15 Nov 1933 (HA). The Schrödinger-Heisenberg correspondence on this subject has not been found.

56. Casimir to WH, 23 Oct 1933 (HA); and Kramers to WH, 15 Dec 1933 (HA).

57. WH to his mother, 6 Oct 1933.

58. Courant, writing from Göttingen, notified Heisenberg that he was following Born into temporary exile. Courant to WH, 26 Dec 1933 (HA).

59. WH to Franck, 10 Jan 1934 (Franck Papers, Library, University of Chicago).

60. Planck reiterated his ethics to von Laue, 22 Mar 1934 (1964-6/136a,b, Sondersammlungen, Deutsches Museum, Munich).

61. Brandi to WH, 12 Dec 1933 and 1 Feb 1934 (HP).

62. Ibid., and M. Reich, Dean of the math-naturwiss. Fakultät, to WH, 9 Feb 1934 (HA).

63. Debye to Clemens Schäfer, 14 Feb 1934 (carbon copy, Debye Papers, MPG).

64. Sauter to WH, 13 and 18 Apr 1934 (HA).

65. Ibid., 5 May 1934 (HA).

66. Pohl to WH, 13 Jun 1934 (HA).

67. On Sauter in Göttingen, see David C. Cassidy, in *Wissenschaft und Gesellschaft*, vol. 1, R. Rürup, ed. (Berlin: Springer-Verlag, 1979), 373–387.

68. Planck's admiration in letters to Heisenberg, 8 and 15 Nov 1933 (HA).

Chapter 16. Social Atoms

1. WH to his mother, 6 Oct 1933.

2. Ibid., 6 Feb 1938.

3. Ibid., 25 Jun 1933, his emphasis.

4. WH to his mother, 12 Jul 1934; WH to Margrethe Bohr, 31 Jan 1935 (BSC, 20,2), also indicated in WH to Bohr, 17 Jun 1934 (BSC, 20,2); WH to his mother, 17 Mar 1935.

5. This is emphasized, for example, by Hans Buchheim, "Die SS—das Herrschaftsinstrument," in *Anatomie des SS–Staates*, ed. Buchheim et al., vol. 1 (Munich: Deutscher Taschenbuch Verlag, 1967), 15-212, esp. 15-29.

6. Certification by Leipzig Universitätsrat Sperling to Kaiser Wilhelm-Institut für Physik in Berlin, 13 Apr 1943 (HA).

7. The law and the wording of each oath are given in *Reichsgesetzblatt*, no. 98 (22 August 1934), 783.

8. EH, *Recoll.*, 38.

9. Karl Seidelmann, lecture delivered East Sunday, 1931 (Nachlass Seidelmann, ADJ).

10. Oberstudiendirektor Wimmer (1. Vorsitz der Neupfadfinder) to Registergericht München, 15 July 1934, in Registerakten des Amtsgerichtes München in Sachen: Bayerischer Pfadfinderbund e.V. Landesverband zur Förderung der Pfadfinderbewegung (StAM AG 33157). The last year of the New Pathfinders is recounted by Franz Ludwig Habbel, "The Story of the German Youth Movement 1896–1933," MS 1947 (IfZ, MS 73). The last official meeting of Heisenberg's group is recalled by Rolf Wägele, manuscript of 15 May 1962 (HA). Some of the "boys" of course met privately after 1934.

11. Note 8.

12. Records and correspondence in BDC and HA.

13. WH, "Die aktive und die passive Opposition im Dritten Reich," 12 Nov 1947 (HA). This differs from the affidavit that Heisenberg submitted for Ernst von Weizsäcker: WH, affidavit, 3 Apr 1948, Weizsäcker defense exhibit no. 303 (NARA, microfilms M897, roll 119).

14. Heisenberg to S. A. Goudsmit, 5 Jan 1948 (GP; quoted in Hermann, *Jahrhundert*, 119).

15. Interview with Friedrich Hund, Göttingen, 28 November 1979.

16. Lion Feuchtwanger, *Exil: Roman* (Frankfurt am Main: Fischer Taschenbuch Verlag, 1979), 363. The political dilemmas of the apolitical German professors are explored by Reinhard Siegmund-Schultze, "The Problem of Anti-Fascist Resistance of 'Apolitical' German Scholars," in *Science, Technology and National Socialism*, eds. M. Renneberg and M. Walker, (New York: Cambridge Univ. Press, 1994), 312–323.

17. *LNN*, no. 80 (21 March 1933), 3; and "Hochschule im neuen Staat. Die Bewegung der Besucherzahl an der Universität Leipzig," *LNN* 11 (July 1934) (newspaper clipping collection, BA Koblenz, Zsg. 129/643). See Hans Peter Bleuel and E. Klinnert, *Deutsche Studenten auf dem Weg ins Dritte Reich: Ideologien, Programme, Aktionen, 1918–1935* (1967).

18. The victim was Leon Lichtenstein, professor of mathematics. Herbert Helbig, *Universität Leipzig* (Frankfurt am Main: Weidlich, 1961), 105; and "Aufruf der Deutschen Studentenschaft gegen den undeutschen Geist," *LNN*, no. 96 (12 April 1933), 3.

19. Ettore Majorana to his mother, Leipzig, no date (early 1933), quoted in English trans. in Leonardo Sciascia, *The Moro Affair and the Mystery of Majorana*, trans. by Sacha Rabinovitch (New York: Carcanet Press, 1987), 151–152.

20. L. W. Jones and Warren Weaver, diary entry of 29 May 1933, Leipzig (RAC, Record Group 1.1, Series 717 Germany, Box 14, folders 127–137).

21. Staatsminister des Inneren Fritsche to SS-Sturmführer Wolf Friedrich, 31 Jul 1933, and Friedrich to Ministerpräsident von Killinger, 12 Aug 1933 (IfZ, records of the NSDStB [National Socialist German Students' League], microfilm MA-228, frames 5024396 and 5024403).

22. Helbig, note 18. By the end of May 1933, 8 professors and 11 assistants had been dismissed. Report of Dean Weickmann to Jones and Weaver, note 20. Peter Kröner, *Vor fünfzig Jahren: Die Emigration deutschsprachiger Wissenschaftler 1933–1939* (Münster: Gesellschaft für Wissenschaftsgeschichte, 1983), 66–67, lists a total loss of 35 academics (without indication of rank).

23. For Dornfeld: Heisenberg to Universitäts-Rentamt, 14 Oct 1933, Akten des Universitäts-Rentamtes zu Leipzig betreffend: das theoretisch-physikalische Institut (UA Leipzig, RA 1407).

24. Philosophische Fakultät, Protokolle, book 8, 1928–1947 (UA Leipzig); and "Akten der philosophischen Fakultät zu Leipzig betr. Berufungsverfahren" (UA Leipzig, A2/21).

25. Hahn to von Seydewitz, 30 May 1933, in "Physikalisches Institut der Universität Leipzig, Band 2: 1929–1939" (SStA Ministerium für Volksbildung, 10230/27).

26. Debye to Rector Achelis, 26 Jun 1933 (ibid.); von Seydewitz, press release, 4 Jul 1933 (ibid.); Debye to Ministerium für Volksbildung, asking that Bewilogua be named as Sack's replacement, 30 Jun 1933 (ibid.); Swiss Consul to Sächsisches Aussenministerium, Dresden, 28 Jul 33 (ibid.).

27. Bloch to Bohr, 6 Apr 1933 (BSC, 17, 3); and Bloch interview with Charles Weiner, 15 Aug 1968 (AIP, CHP).

28. WH to Bohr, 30 Jun 1933, and Bohr to WH, Pasadena, 19 May 1933 (BSC, 20, 2); Bloch to Bohr, Rome, 10 Feb 1934 (BSC, 17,3).

29. Polanyi to WH, 19 Jun 1934 (HA), and WH to Polanyi, 9 Jul 1934 (Archive, Univ. of Chicago). I am grateful to William Lanouette for bringing the latter to my attention.

30. Kopfermann to Bohr, 23 May 1933 (BSC 22,2).

31. Sponer to WH, 25 Oct 1933 (HA).

32. WH to Ehrenfest, 2 Feb 1932 (Ehrenfest Papers, AHQP, EHR 21,4); Reinhold Fürth to WH, 5 Oct 1933 (HA); Kramers to WH, 15 Dec 1933 (HA). Beck's odyssey is recalled by Rudolf Peierls, *Bird of Passage: Recollections of a Physicist* (Princeton: Princeton Univ. Press, 1985), 39, and in the obituary by H. M. Nussenzveig, "Guido Beck," *Physics Today, 43*, no. 12 (1990), 89–90.

33. Note 24.

34. Kirchner to Hans Bethe, 20 May 1935 (Hans Bethe Papers, Library, Cornell Univ., Ithaca, NY, Box 3, 5-57-H-5). Robert Döpel replaced Kirchner in 1938.

35. Debye's assistant, Möbius, was listed as a representative of the Dozentenschaft at meetings of the Philosophical Faculty starting in 1937 (note 24, *first work*). Heisenberg's efforts in the situation are indicated in Heisenberg to Debye, 5 Nov 1934, and Debye to WH, carbon copy, 10 Nov 1934 (Debye Papers, MPG).

36. Gerhard Geissler, "Die Universität. Ein neues Semester beginnt mit neuen Aufgaben," *Neue Leipziger Zeitung*, no. 302 (29 October 1933) (newspaper clipping collection, BA Koblenz, Zsg. 129/643).

37. "Jahresbericht des Rektors Dr. Arthur Golf," in *Akademische Reden: Gehalten am 31. Oktober 1934 in der Aula der Universität Leipzig aus Anlass des 525. Jahrestages ihrer Gründung* (Leipzig: A. Edelmann, 1934), 3–17; and "Dienstplan der Studentenschaft," 18 October 1934 (records of Reichsstudentenführung, IfZ, microfilm MA-528, frames 5021537-8). See also, "Feierliche Gründung der Leipziger Dozentenschaft: Erziehung statt Bildung," *LNN*, no. 198 (17 July 1934), 2; and "Entwurf der Dozentenschaft der Universität Leipzig," Dec 1933 (records of NSDStB, IfZ, microfilm MA-228, frames 5024324-25).

38. Heinz Woltereck, "Lutherfeier an der Universität," *LNN*, no. 315 (11 Nov 1933), 3. According to Rector Golf (note 37, 14), the university received from the Saxon Education Ministry a new constitution on 1 Jan 1934, introducing the "Führerprinzip."

39. WH, *PB*, 141–149.

40. WH to Bohr, 30 Jun 1933 (BSC, 20, 2).

41. Official communication from H. Pleijel (Stockholm Academy) to WH, 9 Nov 1933, facsimile in A. Hermann, *Werner Heisenberg: In Selbstzeugnissen und Bilddokumenten* (Reinbek: Rowohlt, 1976), 53.

42. The rally is reported in "Appell der deutschen Wissenschaft an die Welt," *Frankfurter Zeitung, 78* (13 November 1933), 4; "Die deutsche Wissenschaft spricht ihr 'Ja'! Der N.S.-Lehrerbund und der 12. November," *Völkischer Beobachter*, Berlin, 46 (1 Nov 1933), Beiblatt: Aus der Bewegung, 1; and *Bekenntnis der Professoren an den deutschen Universitäten und Hochschulen zu Adolf Hitler und dem nationalsozialistischen Staat, überreicht vom NS Lehrerbund Deutschlands/Sachsen* (Dresden, 1934). The echo could even be heard abroad: "Leipzig: Professors Plead for Better Understanding of Germany," *New York Times*, 12 Nov 1933, section 3, 2.

43. Hedwig Goerlich to WH, 9 Nov 1946 (HA).

44. EH, *Recoll.*, 36–39.

45. WH to his mother, 11 Nov 1933. This was after the annual Luther celebration, described by Woltereck, note 38.

46. Kreisführer of Kreis IV, NSDStB, to Saxon Minister of Education, 19 Nov 1933 (IfZ, records of Deutsche Studentenschaft, microfilm MA-228, frame 5024355).

47. Born to Schrödinger, 6 Nov 1960, quoted in PWB 2, 228.

48. For instance, Arnold Kramish, *The Griffin* (Boston: Houghton Mifflin, 1986).

49. WH to Bohr, 27 Nov 1933 (BSC 20,2); WH to Born, Zurich, 25 Nov 1933, quoted in English trans. in Born, *My Life: Recollections of a Nobel Laureate* (New York: Charles Scribner's Sons, 1975).

50. Martin Wein, *Die Weizsäckers: Geschichte einer deutschen Familie* (Stuttgart: Deutsche Verlags-Anstalt, 1988).

51. WH to his mother, 8 Oct 1934.

52. Minutes of the faculty meeting on 8 May 1935, note 24.

53. Minutes of faculty meeting on 8 May 1935, note 24.

54. Van der Waerden to Armin Hermann, undated (1970s), quoted in Hermann, *Jahrhundert*, 117. Van der Waerden, a Dutch citizen, was told to stay out of German affairs (interview with van der Waerden, Zurich, 25 April 1985). There is no indication of a reprimand in Heisenberg's personnel file (UA Leipzig, PA 560).

55. WH to his mother, 9 Apr 1935. Heisenberg was listed as a member of the university senate from 1935 until he left Leipzig in 1942, in *Rektorenwechsel an der Universität Leipzig* (Leipzig: A. Edelmann, annual).

56. Berve to WH, 13 Mar 1946 (HA). Nevertheless, the Education Ministry named Berve pro-rector of the university in 1937 and rector in 1940. *Rektorenwechsel*, note 55.

57. Note 54.

58. WH, *PB*, 149.

59. Ibid., 149–154.

60. Since Heisenberg wrote this recollection, ibid., long after the end of the war, it is uncertain what sort of future world Planck and Heisenberg had in mind; or perhaps Heisenberg read this into his memory of their meeting.

61. WH to Born, 2 Jun 1933, typed excerpt in letter from Born to Ehrenfest, 11 Jun 1933 (Ehrenfest Papers, AHQP, EHR 18, 5).

62. WH to his mother, 5 Oct 1935.

Chapter 17. Of Particles and Politics

1. Patrick Blackett and Giuseppe Occhialini, *PRS, 139* (1933), 699–727.

2. Carl D. Anderson, *Science, 76* (1932), 238–239.

3. Paul Dirac, *PRS, 117* (1928), 610–624, and *118* (1928), 351–361.

4. Paul Dirac, *PRS, 126* (1930), 360–365; and *PRS, 133* (1931), 60–72. Historical accounts include Joan Bromberg, *HSPS, 7* (1976), 161–191; Pais, *Inward,* 346–352; and Helge Kragh, *Dirac: A Scientific Biography* (New York: Cambridge Univ. Press, 2005).

5. Note 1, 713ff.

6. According to the "Copenhagen registers and minute books" (AHQP 35,2), Heisenberg visited Bohr's institute from 10 Sep to 3 Oct 1933. The conference was held 14–20 Sep. Bohr to WH, 17 Aug 1933 (BSC, 20,2).

7. Wendell Furry and J. Robert Oppenheimer, *PR, 45* (1934), 245–262; Hans Bethe and Walter Heitler, *PRS, 146* (1934), 83–112.

8. WH to Sommerfeld, 9 Oct 1933 (AHQP 83,H).

9. WH to Paul Goerns, 18 Apr 1934 (HA); WH, lecture in *Stahl und Eisen, 54* (1934), 749–752; reprinted in HCW, C1, 92–95, delivered 2 Jun 1934.

10. Pauli to WH, 14 Jun 1934.

11. WH to Pauli, 16 Jun 1934.

12. Wolfgang Pauli and Viktor Weisskopf, *Helvetia Physica Acta, 7* (1934), 709–731, rec. 27 Dec 1934.

13. WH to Pauli, 11 Jul 1934.

14. Stark thus refused to attend Heisenberg's lecture to the mining engineers. Laue to Dr. W. F. Berg, 21 Aug 1934, transcript by Berg forwarded to Einstein (EA, 16-107); WH to Bohr, 17 Jun 1934 (BSC, 20,2).

15. Based on Laue to Einstein, 22 Aug 1934 (EA 16-105); Stark to Laue, 21 Aug 1934, transcripts (Nachlass Laue, MPG, Berlin, and Deutsches Museum, Munich); and Beyerchen, *Scientists*, 118 and 243.

16. Stark to Laue, note 15.

17. WH, *Nwn, 22* (1934), 669–675; reprinted HCW, C1, 96–101.

18. WH to his mother, 21 Sep 1934.

19. WH, "Grundfragen der modernen Physik," lectures delivered in winter semester 1934–35 (HA). The main section headings are: "Ausführliche historische Einleitung," "Die Relativitätstheorie" (both special and general theories), and "Die Quantentheorie."

20. Dr. Dalfarts (?) to WH, 4 Dec 1934 (HA).

21. Dr. Rosskothen to Rosenberg, and Rosenberg to Rosskothen, both in 1934, quoted (without exact dates or source) by Werner Haberditzl, *NTM* [Naturn. Technik, Med.], Beiheft (1963), 320–326; on 323.

22. Hans Euler and Bernhard Kockel, *Nwn, 23* (1935), 246–247.

23. WH to Pauli, 25 Apr 1935.

24. Hans Euler, "Über die Streuung von Licht an Licht nach der Diracschen Theorie," (diss., Leipzig, in UA Leipzig, Anmeldungsschein, Promotionen MI and MII); WH's evaluation, 1 Nov 1935 (UA Leipzig, Philosophische Fakultät, Promotionsakten); WH and Euler, *ZP, 98* (1936), 714–732, rec. 22 Dec 1935, reprinted HCW, A2, 162–180. The physics of Heisenberg's papers on positrons in this period is summarized by Pais, intro. HCW, A2, 95–105, and in Pais, *Inward,* 374–388.

25. Pauli to Kronig, 20 Nov 1935. The comparison is with Pauli to Kronig, 21 May 1925.

26. WH and Euler, note 24, 732.

27. WH, affidavit on Kockel's political past, 7 Jan 1947 (HA).

28. WH to Debye, 14 Aug 1936 (Debye Papers, MPG).

29. WH, *PB*, 176.

30. WH, affidavit on Berve's political past, 19 Nov 1946 (HA).

31. Hans Euler, "Erklärungen" in his doctorate file (UA Leipzig, Promotionsakten, Philosophische Fakultät).

32. Beyerchen, *Scientists*, 51–57.

33. Bernhard Rust, lecture at University of Berlin, 6 May 1933, in *Die nationalsozialistische Revolution 1933*, ed. Friedrichs (Berlin: Junker, 1937), 278–285.

34. H. Mehrtens and S. Richter, eds., *Naturwissenschaft, Technik und NS-Ideologie* (Frankfurt am Main: Suhrkamp, 1980); Karl-Heinz Ludwig, *Technik und Ingenieure im Dritten Reich* (Königstein/Ts.: Athenäeum-Verlag, 1979); and M. Renneberg and M. Walker, "Scientists, engineers and National Socialism," in *Science, Technology and National Socialism*, eds. Renneberg and Walker (New York: Cambridge Univ. Press, 1994), 1–29.

35. See Beyerchen, *Scientists*, 79–122; Walker, *Nazi Science*, chaps. 2, 3.

36. Statements for Stark's denazification trial: Einstein to Senatsvorsitzender Schliefer, 14 May 1949 (EA 22–366); and WH to Schliefer, 24 May 1949 (HA). Stark's trial has been discussed by Andreas Kleinert, *Sudhoffs Archiv, 67* (1983), 13–24. Lenard was not brought to trial because of his age.

37. Sommerfeld to Rector, University of Munich, to be forwarded to Bavarian Culture Ministry, 26 Jul 1937 (UA Munich, Sommerfeld, Personalakten, E II-N).

38. "Deutsche Physik" is described by Beyerchen, *Scientists*, 123–140; Richter, "Die Deutsche Physik,"in Mehrtens and Richter, note 34, 116–141; Walker, note 35, chap. 2.

39. Philipp Lenard, *Deutsche Physik*, 4 vols., vol. 1: *Einleitung und Mechanik* (Munich: J.F. Lehmanns Verlag, 1936), ix.

40. Stark to Lenard, 3 Feb 1933, and Lenard to Hitler, 21 Mar 1933, both published by Andreas Kleinert, *PBl, 36* (1980), 35–43; on 35.

41. Johannes Stark, *Nationalsozialismus und Wissenschaft* (Munich: Zentralverlag der NSDAP [Nazi Party], 1934), quotes on 13–14. Beyerchen, *Scientists*, 115–122, provides a further account of Stark's activities in this period.

42. Published as Johannes Stark, "Organisation der physikalischen Forschung," *Zs. für technische Physik, 14* (1933), 433–435.

43. See Dieter Hofffman, "Between Autonomy and Accomodation: The German Physical Society during the Third Reich," *Physics in Perspective, 7* (2005), 293–329; and the further studies in Dieter Hoffmann and Mark Walker, eds., *Physiker zwischen Autonomie und Anpassung: Die Deutsche Physikalische Gesellschaft im Dritten Reich* (Weinheim: Wiley-VCH, 2007).

44. For example, Beyerchen, *Scientists*, 115–122; Ludwig, note 34, 210–216; and the discussion of the "Fall Wildhagen," the attack on Stark's manager at the Deutsche Forschungsgemeinschaft [German Research Association, in Helmut Heiber, *Walter Frank und sein Reichsinstitut für Geschichte des neuen Deutschlands* (Stuttgart: Deutsche Verlags-Anstalt, 1966), 821–847.

45. Heiber, note 44, 796.

Chapter 18. Heir Apparent

1. WH to Sommerfeld, 18 Jan 1935 (SN); and WH to his mother, 9 Apr 1935.

2. Correspondence January to June 1935 (UA München, Akten des Rektorats, Personalakte EII-N); minutes of faculty meeting, 15 Feb 1935 (UA München, Sitzungsprotokolle, Phil. Fak. II. Sektion, OC-N1d).

3. Dean to Rector, 24 Mar 1935 (note 2, first source).

4. Documents in UA München, Nachfolgeakt Sommerfeld, OC-N 10a; and Sommerfeld to Debye, 7 Jul 1935 (Debye Papers, MPG).

5. "Politische Erziehung der Leipziger Studentenschaft," *LNN*, no. 302 (29 October 1933), 2.

6. Studentkowski, Akten-Notiz (SSA, Akten des Ministeriums für Volksbildung, 10230); and Debye to Sommerfeld, Berlin, 20 Sep 1935 (ETH, Hs 627:13).

7. Submission of 4 Nov 1935, transcription (SN).

8. Martin Kersten, "Richard Becker, 1887–1955," *PBl, 34* (1978), 379–382.

9. Most notably, Beyerchen, *Scientists*, chap. 8, and Wolfgang Schlicker, *Jahrbuch für Geschichte, 27* (1983), 109–142. See also, Hermann, *Jahrhundert*, 126–147.

10. Quoted by Hans Buchheim, "Die SS—das Herrschaftsinstrument," in *Anatomie des SS-Staates*, ed. Buchheim et al., vol. 1 (Munich: Deutscher Taschenbuch Verlag, 1967), 98–99.

11. "Philipp Lenard als deutscher Naturforscher, Rede zur Einweihung des Philipp-Lenards-Instituts in

Heidelberg am 13. Dezember 1938," *Nationasozialistische Monatshefte, 7* (February 1936), 106–112. The speech is signed by Lenard and Stark.

12. Oron J. Hale, *The Captive Press in the Third Reich* (Princeton: Princeton Univ. Press, 1964), 31–32.

13. Lenard and Stark, note 11, and Willi Menzel, "Deutsche Physik und jüdische Physik," *VB, 49*, no. 29 (29 January 1936), 7.

14. WH to his mother, 15 Feb 1936, reporting the meeting with Kölbl.

15. WH to Sommerfeld, 14 Feb 1936 (SN).

16. "German Science Goose-steps," editorial, *New York Times*, 12 March 1936, 20; WH, "Zum Artikel: Deutsche und jüdische Physik. Entgegnung von Prof. Dr. Heisenberg," *VB, 49*, no. 59 (28 Feb 1936), 6, reprinted HCW, C5, 10–11; English trans. in Hentschel, *Anthology,* 121–123.

17. WH to his mother, 28 Feb 1936.

18. *New York Times*, note 16.

19. Alfred Rosenberg, editorial comment preceding the Heisenberg-Stark exchange, note 16 and "Stellungnahme von Prof. Dr. J. Stark," *VB, 49*, no. 59 (28 Feb 1936), 6. The comment is also published in HCW, C5, editorial introduction, 4."

20. Studentkowski, "Aktenvermerk," in "Besetzung des Ord. Lehrstuhls für Experimentalphysik, Nachf. Debye," Band 2, Heft 2 1935–1937 (SStA, Akten des Ministeriums für Volksbildung, 10230), referring to a meeting with Mentzel on 2 March 1936.

21. Ibid.

22. Beyerchen, *Scientists,* 154.

23. Rudorf to WH, 30 Mar 1936, and to Prorector Golf, same date (UA Leipzig, Heisenberg, Personalakte, Call no. PA 560).

24. Undated form letter with Sommerfeld's name in the blank for the addressee, with typed signatures of M. Wien, H. Geiger, W. Heisenberg (SN); another copy (Friedrich Hund Papers, Göttingen); reprinted, without names of the signatories, HCW, C5, 12–13, trans. in Hentschel, *Anthology,* 137–139.

25. For example, Lenard to Dr. Wacker, 10 Jun 1936, transcription (Nachlass Lenard, private hands).

26. Geiger replaced Hertz at the Technical College Berlin, discussed further by D. C. Cassidy, in *Wissenschaft und Gesellschaft*, ed. R. Rürup, vol. 1 (Berlin: Springer-Verlag, 1979), 373–387.

27. Private communication from Dr. Helmut Fischer.

28. The memo is probably "Die Physik an den deutschen Hochschulen," no date or author (SN).

29. Zschintzsch, file copy of letter of 2 Oct 1936, "Dem Herrn Reichsminister weitergeleitet," in "Korrespondenz des Staatssekretärs Zschintzsch," vol. 3 (BA Koblenz, R21/203). Original in "Research: Wi Heisenberg" (BDC).

30. Sommerfeld to WH, 9 Nov 1936 (HA); WH to Pauli, 26 Nov 1936.

31. The expectation is discussed by D. C. Cassidy, *HSPS, 12* (1981), 1–39, and Peter Galison, *Centaurus, 26* (1982), 262–316.

32. Carl Friedrich von Weizsäcker, *ZP, 88* (1934), 612–625; E. J. Williams, *PR, 45* (1934), 729–730; L. D. Landau, *Physikalische Zeitschrift der Sowjetunion, 5* (1934), 761–764.

33. J. R. Oppenheimer, *PR, 47* (1935), 44–52; L.W. Nordheim, *PR, 49* (1936), 189–191. The work of the Oppenheimer group is explored in DC, *JRO*, chap. 10.

34. WH to Pauli, 26 May 1936.

35. Enrico Fermi, *ZP, 88* (1934), 161–177.

36. Heisenberg's work on Fermi-force nuclear physics is discussed by BR, *Nuclear Forces.*

37. Heisenberg, *ZP, 101* (1936), 533–540, received 8 June 1936; reprinted HCW, A2, 275–282;

38. Heitler, *Angewandte Chemie, 49* (1936), 690.

39. Anderson to Heitler, 21 May 1936 (Hans Bethe Papers, Library, Cornell Univ., Ithaca, NY, Box 3).

40. Carl D. Anderson and S. Neddermeyer, *PR, 50* (1936), 263–271, rec. 7 Jun 1936.

41. J. Robert Oppenheimer, *PR, 50* (1936), 389, abstract of paper delivered to American Physical Society, June 1936; Homi J. Bhabha and Walter Heitler, letter to editor, *Nature, 138* (July 1936), 401. They presented their complete theories in Carlson and Oppenheimer, *PR, 51* (1937), 220–231, and Bhabha and Heitler, *PRS, 159* (1937), 432–458.

42. WH, *Forschungen und Fortschritte, 12* (1936), 341–342, publ. 20 Sep 1936; reprinted HCW, C1, 122–123.

43. G. Herzog and W. Scherrer, *Naturwiss., 24* (1936), 718–720; H. Geiger, *Preussische Akademie der Wissenschaften, Vorträge und Schriften, 3* (1940), 1–33, esp. 16.

44. Pauli to WH, 9 Jun 1936.

45. "Harvard Visit Off for Reich Physicist. Dr. Heisenberg, Scheduled to Give a Paper at Exercises, is Serving 8 weeks in Army," *New York Times*, Sunday, 30 August 1936, Section 2, 8; WH to Goudsmit, 9 Apr 1936 (HA).

46. WH to his mother, 28 Aug and 4 Sep 1936. See also WH to Born, 3 Nov 1936 (BN).

47. Paul to WH, 26 Oct 1936.

48. WH to Born, 3 Nov 1936 (Nackass Born, SPK)

Chapter 19. The Lonely Years

1. WH, *PB*, 165.

2. WH, *Briefe*.

3. WH to his mother, 19 Mar 1936 and 30 Mar 1936.

4. *Die Weizsäcker-Papiere 1933–1950*, Leonidas E. Hill, ed. (Frankfurt am Main: Verlag Ullstein, 1974), 131, 631 (n. 13), 503 (n. 79); Adelheid to WH, 27 Aug 1937 (HP).

5. WH to his mother, 9 Apr 1936.

6. Ibid., 12 Nov 1936 and 12 Feb 1937.

7. WH, *PB*, 166.

8. EH, *Recoll.*, 47.

9. Dates based on WH to his mother, 6 and 12 Feb 1937.

10. Klaus Schwabe, *Vierteljahrshefte für Zeitgeschichte, 14* (1966), 105–188.

11. Cf. Erich Kuby, *Mein Krieg: Aufzeichnungen aus 2129 Tagen* (Munich: Nymphenburger Verlagsbuchhandlung, 1975).

12. Barbara Wood, *E. F. Schumacher: His Life and Thought* (New York: Harper and Row, 1984), 2, 6.

13. Interview with EH, Göttingen, 8–9 February 1982.

14. Note 8; WH to his mother, 12 Feb 1937.

15. Ibid., 6 Dec 1935.

16. "Hilfsaktion des WHW zum 30. Januar," *LNN*, no. 23 (23 January 1937), 1.

17. WH, *PB*, 166; WH to Elisabeth Schumacher, 23 Mar 1937 (HP).

18. Note 13.

19. Euler to WH, 13 Mar 1937 (AHQP 45,1), and Pauli to WH, 10 Mar 1937.

20. WH to his mother, 23 and 27 Mar 1937.

21. WH to his mother, 27 Mar 1937; Wilmanns (Dean of math.-naturwiss. Abteilung der Phil. Fak., Leipzig), notes of conversation with Hund, 8 Apr 1937 (UB Leipzig, Personalakte Heisenberg, PA 560).

22. WH to Bohr, 18 Mar 1937 (BSC 20, 2).

23. Ibid., and WH to Pauli, 26 Apr 1937.

24. Invitation and menu (HP).

25. Pauli to WH, 22 Feb 38.

26. Note 13. The recollection is by Anna Maria Hirsch-Heisenberg, "Erinnerungen an meinen Vater," in WH, *Briefe*, 349-393, on 349.

27. Pauli to WH, 19 Jan 1937; WH to Pauli, 21 Jan 1937; C. D. Anderson and S. H. Neddermeyer, *PR, 50* (1936), 263–271; Homi J. Bhabha and Walter Heitler, *PRS, 159* (1937), 432–458, rec. 11 Dec. 1936; J. F. Carlson and J. Robert Oppenheimer, *PR, 51* (1937), 220–231, rec. 8 Dec. 1936. For further discussion, see BR, *Nuclear Forces*; David C. Cassidy, *HSPS, 12* (1981), 1–39; and Peter Galison, *Centaurus, 26* (1982), 262–316. The cosmic-ray work of the Oppenheimer group is further discussed in DC, *JRO*, chap. 10.

28. Anderson and Neddermeyer, note 27, first work, 268.

29. Carlson and Oppenheimer, note 27, second work, 221.

30. Lothar W. Nordheim, Gertrude Nordheim, J. Robert Oppenheimer, Robert Serber, *PR, 51* (1937), 1037–1045, on 1038.

31. WH to Pauli, 16 Jan 1937.

32. WH to Pauli, 12 Jun 1937.

33. WH to Bhabha, 21 Jan 1937.

34. WH to Pauli, 26 Apr 1937.

35. Pauli to WH, 2 May 1937.

36. WH to Pauli, 12 Jun 1937.

37. WH to Bohr, 5 Jul 1937 (BSC, 20,2). S. Neddermeyer and C. D. Anderson, *PR, 51* (1937), 884–886.

38. WH to Kramers, 22 May 1937 (AHQP 10,3).

39. WH to his mother, 1 Apr, 17 Jun, and 10 Jul 1937; and EH, *Recoll.*, 47.

40. To his mother, 9 Jun 1937.

41. Quoted by EH, *Recoll.*, 47.

Chapter 20. A Faustian Bargain

1. "'Weisse Juden' in der Wissenschaft," *Das Schwarze Korps* (15 July 1937), 6, trans. in Hentschel, *Anthology*, 152–156. Discussions of this episode from varying perspectives may be found in Beyerchen, *Scientists*, 156–167; Hermann, *Jahrhundert*, 143–147; Walker, *Nuclear Power*, 61–66; EH, *Recoll.*, 47–70; and Wolfgang Schlicker, *Jahrbuch für Geschichte*, 27 (1983), 109–142. Heisenberg himself chose not to discuss it.

2. Ludwig Wesch file, in "Hauptamt Wissenschaft," (IFZ, microfilm MA-116/117).

3. Hans Buchheim, "Die SS—das Herrschaftsinstrument," in *Anatomie des SS-Staates*, ed. Hans Buchheim et al., vol. 1 (Munich: Deutscher Taschenbuch Verlag, 1967), 15–212.

4. Allied interrogation of d'Alquen, 16 Feb 1948 (NARA, microfilm M-1019, roll 2).

5. Beuthe to Wesch, 1 Nov 1937 (Ludwig Wesch papers, UA Heidelberg).

6. Kurt R. Grossmann, *Ossietzky: Ein deutscher Patriot* (Munich: Kindler-Verlag, 1963).

7. Sommerfeld to Kölbl, 26 Jul 1937 (UA Munich, PA Sommerfeld, E II-N); Kölbl to Bavarian State Minister for Instruction and Culture, 22 Oct 1937 (ibid.).

8. Hund to Reichsminister Rust, 20 Jul 1937, via Koebe; and Hund to Koebe, 16 Jul 1937 (both in UA Leipzig, PA Heisenberg, PA 560,); Debye to Hund, 22 Jul 1937; Hund to Debye, 21 Jul 1937 (both in Debye Papers, MPG).

9. WH to Sommerfeld, 14 Apr 1938 (AHQP 31,5, original in SN).

10. WH to Berve, 17 Jul 1937; Berve to Koebe, 19 Jul 1937 (UA Leipzig, Heisenberg file, PA 560).

11. Studentkowski, memo "Wegberufung von Prof. Heisenberg," 3 June 1937 (ibid.).

12. WH to Wacker, 28 Jul 1937 (HA); Wacker to Bavarian Ministry for Instruction and Culture; and Wacker to Sommerfeld, both 16 Nov 1937 (both in PA Sommerfeld, E II-N, UA Munich).

13. WH to Himmler, 21 Jul 1937, excerpted without location in Hermann, *Jahrhundert*, 144; cited by Beyerchen, *Scientists*, 254, n. 64, as being in HA; it has not been found.

14. Quoted by Beyerchen, *Scientists*, 160, from his interview with WH, 13 July 1971. See also EH, *Recoll.*, 54.

15. Heinrich Himmler, Terminbücher 1938–1939 (NARA, Microfilm 37A; original in BA Koblenz, NS19/1437); and Himmler's private papers (microfilm MA-320, IFZ Munich; originals in BA Koblenz).

16. WH to Himmler, 7 Nov 1937, cited by Beyerchen, *Scientists*, 254, n. 66, has not been found in HA. The paraphrase below follows Beyerchen, *Scientists*, 161.

17. WH to his mother, 27 Sep 1938.

18. WH to Sommerfeld, 31 Aug 1938 (HA); George B. Pegram to WH, 3 Jan and 31 Mar 1938 (HA).

19. Elisabeth Heisenberg to Annie Heisenberg, 1 Nov 1937 (HP).

20. For example, Hugo Dingler, *Zs. f. ges. Naturwissenschaft*, 3 (1937), 321–335; Dingler, "Pascual Jordan, 'Die Physik des 20. Jahrhunderts' [2nd. ed.]," ibid., 4 (1938/39), 389–393; Ludwig Bieberbach, *Deutsche Mathematik*, 1 (1936), 109.

21. Beyerchen, *Scientists*, 164.

22. Indicated by Friedrich von Faber, Dean of Phil. Fak. II, to Rekorat, Universität München, 29 Dec 38 (Lehrstuhl für theoretische Physik, Nachfolge Sommerfeld, OC-N 10a, UA Munich).

23. Note 5.

24. WH to his mother, 14 and 21 Nov 1937.

25. Interview with E. Heisenberg, Göttingen, April 1983.

26. *Catalogus Professorum: Der Lehrkörper der TH Hannover* (Hannover, 1956); and Juilfs to the author, 6 Aug 1983.

27. Juilfs to the author, 6 Aug 1983.

28. Ibid.

29. Report on Heisenberg, enclosed in letter from office of Reichsführer-SS to REM, 26 May 1939 (ZStA Potsdam, REM 2943, 370–1, transcription by Mark Walker). I am grateful to Mark Walker for providing this transcription.

30. WH, quotation from letter to Heisenberg from an unnamed "lower level SS-Führer," in WH to Sommerfeld, 14 Apr 1938 (SN and AHQP 31,5).

31. WH to his mother, 12 Jun 1938.

32. C. Wieselsberger, "Ludwig Prandtl," *Zs. f. technische. Physik,* 17 (1935), 25–27.

33. From the opening paragraph of Prandtl to Himmler, 12 Jul 1938 (initialed transcript copy, Nachlass Prandtl, UA Göttingen).

34. Himmler to WH, 21 Jul 1938 (facsimile in Goudsmit, *Alsos,* 119; transcript in SN). Heisenberg did teach his course "Spezielle Relativitätstheorie" once again in the Spring of 1940, utilizing his lecture notes from 1931–32 (HA), but he changed its name to the title of Einstein's original paper: "The Electrodynamics of Moving Bodies."

35. Himmler to Heydrich, 21 Jul 1938 (facsimile of certified copy in Goudsmit, *Alsos,* 116; transcript in "Research: Ahnenerbe, Heisenberg, Werner," BDC).

36. WH to Himmler, Fischen (Allgäu), 23 Jul 1938 (GP).

37. Heisenberg notified Sommerfeld in WH to Sommerfeld, 23 Jul 1938 (SN); Sommerfeld urged the REM to act, Sommerfeld to REM, 17 Oct 1938 (AHQP 33, 6).

38. Heisenberg served from 2 August to 15 October 1938, according to his "militaire Fragebogen," dated 9 Feb 40 (UA Leipzig, PA Heisenberg, PA 560).

39. WH to his mother, barracks near Sonthofen, 18 Sep 1938.

Chapter 21. One Who Could Not Leave

1. Sommerfeld to WH, 28 Feb 1939, carbon copy (SN).

2. Dean von Faber to rectorate, 29 Dec 1938 (UA Munich, Lehrstuhl für theoretische Physik, Nachfolge Sommerfeld, OC-N 10a).

3. WH to Sommerfeld, 5 Nov 1938 (SN); notice in *PZ, 39* (1938), 136; WH to his mother, 4 Nov 1938.

4. Hess to REM, 12 Jan 1939, referring to an earlier complaint (ZStA Potsdam, REM 2943).

5. Himmler to REM, 26 May 1939 (ibid.); WH to Sommerfeld, 15 Feb and 13 May 1939 (SN); Sitzungsprotokolle, Philosophische Fakultät, II. Sektion (UA Munich, OC-N 1d).

6. Recommendation of Müller by Dean von Faber, note 2.

7. WH to Sommerfeld, 13 May 1939 (SN).

8. WH to Sommerfeld, 27 Oct 40 (SN), and complaint by Müller to Dean, 11 Sep 40 (SN).

9. Note 7.

10. Dames to Geiger, Pohl and Gerlach, 1 Feb 1938; and Führer to series of physicists, 29 Jul 41 (both BA Koblenz, R21/500).

11. WH, *Gutachten- und Prüfungsprotokolle für Promotionen und Habilitationen (1929–1942),* 2nd edition, H. Rechenberg and G. Wiemers, eds. (Berlin: ERS-Verlag, 2002).

12. Lucy S. Dawidowicz, *The War against the Jews 1933–1945* (Toronto: Bantam Books, 1975/1986), 99.

13. David H. Buffum, U.S. Counsel in Leipzig, diplomatic report of 21 Nov 1938, published in *Nazi Conspiracy and Aggression,* vol. 7, ed. Office of U.S. Chief of Counsel for Prosecution of Axis Criminality (Washington: Government Printing Office, 1946), trial Doc. L-202, 1037–1041.

14. Ibid.

15. WH to his mother, 12 Nov 1938.

16. Elisabeth Heisenberg, interview with the author, Göttingen, 1983.

17. WH to his mother, 6 Apr 1939.

18. WH to his mother, 23 Jun 1939.

19. WH to his mother, 21 Nov 1937.

20. WH, "Die gegenwärtigen Aufgaben der theoretischen Physik," published in excerpt in *Frankfurter Zeitung,* (26 August 1937), and in *Scientia, 32,* 1938, reprinted HCW, C1, 133–141; WH, paper presented to Bologna conference, 18–21 Oct 1937, reprinted HCW, B, 256–259; and several others, listed in WH, *Biblio.*

21. WH, *AP, 32* (1938), 20–33, rec. 13 Jan 1938, reprinted HCW, A2, 301–314.

22. WH, paper presented at University of Geneva, 11–16 Oct 1937, reprinted HCW, B, 249–255.

23. Abraham Esau to REM, 13 Jul 1939, indicating earlier observations; and Hess to REM, 22 Jun 1939 (ZStA Potsdam, REM 2943).

24. WH to Bohr, 4 May and 14 Jun 1938; Bohr to WH, 13 Jun 1938 (BSC 20,2).

25. Yukawa, Physico-Mathematical Society of Japan, *Proceedings, 17* (1935), 48–57. For historical accounts of this development see V. Mukherji, *Archive for History of Exact Sciences, 13* (1974), 27–102; Peter Galison, *Centaurus, 26* (1983), 262–316; BR, *Nuclear Forces;* and D. C. Cassidy, *HSPS, 12* (1981), 1–39.

26. Yukawa, note 25, 57.

27. J. Robert Oppenheimer and R. Serber, *PR, 51* (1937), 1113; and Serber, *PR, 53* (1938), 211.

28. Kemmer, *PRS, 166* (1938), 127–153.

29. Herbert Fröhlich, Walter Heitler, and Nicholas Kemmer, *PRS, 166* (1938), 154–177, on 155; Kemmer, Cambridge Philosophical Society, *Proceedings, 34* (1938), 354–364; Homi J. Bhabha, *PRS, 166* (1938), 501–528.

30. Kemmer, note 28, 148, rec. 9 Feb 1938.

31. WH to his wife, Manchester [Mar 1938].

32. WH, *ZP, 110* (1938), 251–266, rec. 24 Jun 1938, reprinted HCW, A2, 315–330.

33. WH to Bohr, 4 May 1938 (BSC 20,2); Hans Euler, *ZP, 110* (1938), 692–716.

34. Hans Euler and WH, *Ergebnisse der exakten Naturwiss.*, 17 (1938), 1–69; reprinted HCW B, 262–330; WH, *AP, 33* (1938), 594–599, dated 6 Oct 1938, reprinted HCW, A2, 331–336.

35. WH to Pauli, 15 Jul 1938; Pauli to Peierls, 18 July, to Rubinowicz, 22 July, to WH, 15 Aug 1938.

36. Lectures delivered in Hamburg, 1 Dec 1938, reprinted HCW, B, 331–332; in Munich, 4 Dec 1938, abstract reprinted HCW, B, 333; and in Leipzig, 7–8 Jan 1938, abstract reprinted HCW, B, 334.

37. Hans Bethe and Lothar Nordheim, *PR, 57* (1940), 998–1006; J. G. Wilson, *PRS, 172* (1939), 517–529, and *PRS, 1974* (1940), 73–85; Blackett to WH, 10 Sep 1938 (AHQP 45,1).

38. C. G. and D. D. Montgomery, *Reviews of Modern Physics, 11* (1939), 255–265, on 257.

39. These views and the "failed revolution" against QED were discussed by Galison, note 25.

40. Walker, *Nazi Science*, chap. 6.

41. WH, interview session 10, 28 Feb 1963 (AHQP); J. Robert Oppenheimer, discussion remark, *Reviews of Modern Physics, 11* (1939), 264–266. The controversy is further discussed in DC, *JRO*, chap. 10.

42. WH to Sommerfeld, Fischen, 31 Aug 1938 (SN).

43. A. H. Compton to his son Arthur, 16 Jul 1937, and Compton to WH, 16 Jul 1937 (UA Chicago, Arthur H. Compton Papers, series 2). The earlier offer is indicated in K. T. Compton to WH, 31 Jul 29, carbon copy (UA Princeton, K. T. Compton papers).

44. WH, *PB*, 170–171.

45. Nevill Mott and Rudolf Peierls, "Werner Heisenberg, 5 December 1901–1 February 1976," *Biographical Memoirs of Fellows of the Royal Society, 23* (1977), 213–251, on 232.

46. WH to his mother, 23 Jan 1939.

47. WH, *PB*, 169–172.

48. Goudsmit, "Werner Heisenberg (1901–1976)," *Yearbook of the American Philosophical Society, 1976*, 74–80, on 76.

49. Max Dresden, discussion with the author, Stony Brook, NY, 1989.

50. WH, *PB*, 172; and WH to his wife, 28 Jul 1939 (HP).

Chapter 22. Warfare and Its Uses

1. Ernst von Weizsäcker, *Memoirs*, trans. John Williams (London: Victor Gollancz, 1951), 212.

2. WH to his mother, 4 Dec 1939.

3. Erich Bagge, Kurt Diebner, and Kenneth Jay, *Von der Uranspaltung bis Calder Hall* (Hamburg: Rowohlt, 1957), 22.

4. WH to Sommerfeld, 4 Sep 1939 (AHQP 31,5).

5. Bagge's diary, entry for 25 Sep 1939 (DJ-29, 106–143), and WH, "Militaire Fragebogen," dated 9 Feb 40 (UA Leipzig, Personalakten PA 560).

6. Quotes in *Einstein on Peace*, eds. Otto Nathan and Heinz Norden (New York: Schocken Books, 1960), 289. Histories of the Allied fission project include Richard Rhodes, *The Making of the Atomic Bomb* (New York: Simon and Schuster, 1986); Spencer Weart, *Scientists in Power* (Cambridge, MA: Harvard Univ. Press, 1979); Margaret Gowing, *Britain and Atomic Energy 1939–1945* (New York: St. Martin's Press, 1964); and Henry DeWolf Smyth, *Atomic Energy for Military Purposes* (Princeton: Princeton Univ. Press, 1946).

7. The best book-length overview of the German fission project is Walker, *Nuclear Power*. See also parts of Rhodes, note 6; Jeremy Bernstein, "Prologue," in *Hitler's Uranium Club: The Secret Recordings at Farm Hall* (Woodbury, NY: AIP Press, 1996), 1–53; and Mark Walker, "German Work on Nuclear Weapons," *Historia Scientiarum, 14* (2005), 164–181.

8. Siegfried Flügge, *Nwn, 27* (1939), 402–410.

9. WH, *PB*, 170.

10. WH, research report of 6 Dec 1939 to Army Ordnance, publ. HCW, A2, 378–396.

11. Neils Bohr and John Wheeler, *PR, 56* (1939), 426–450.

12. WH, note 10, 394.

13. Ibid., 396.

14. WH, second report to Army Ordnance, dated 29 Feb 1940, publ. HCW, A2, 397–418.

15. See D. C. Cassidy, in *Wissenschaft und Gesellschaft*, ed. R. Rürup, vol. 1 (Berlin: Springer-Verlag, 1979), 373–387.

16. WH, note 14, 397; K. H. Höcker, report of 20 Apr 1940 (DJ-29, 427–433); P. O. Müller, report of 29 Apr 1940 (DJ-29, 434–436); H. Bothe, report of 8 March 1941 (DJ-31, 117–127); and Bothe and Jensen, report of 1941, *ZP, 122* (1944), 749–497.

17. Carl Friedrich von Weizsäcker, report of 17 July 1940 (DJ-29, 451–455).

18. Edwin McMillan and Philip Abelson, *PR, 57* (1940), 1185–1186.

19. Nevill Mott and Rudolf Peierls, "Werner Heisenberg 5 December 1901–1 February 1976," *Biographical Memoirs of Fellows of the Royal Society, 23* (1977), 213–251, on 232. See also Paul Lawrence Rose, *Heisenberg and the Nazi Atomic Bomb Project* (Berkeley: Univ. of California Press, 1998).

20. Martin Heisenberg to the author, 1 Jul 1990; Thomas Powers, *Heisenberg's War: The Secret History of the German Atomic Bomb* (New York: Da Capo Press, 2000).

21. Rainer Karlsch, *Hitlers Bombe* (Munich: Deutsche Verlags-Anstalt, 2005). See also Karlsch and Mark Walker, "New Light on Hitler's Bomb," *Physics World*, June 2005, 15–18, on 18; and Walker, "Eine Waffenschmiede? Kernwaffen- und Reaktorforschung am Kaiser-Wilhelm-Institut für Physik," in *Gemeinschaftsforschung, Bevollmächtigte und der Wissenstransfer*, ed. H. Maier (Berlin: Wallstein, 2007), 352–394.

22. Debye to W. E. Tisdale, Rockefeller Foundation, 7 Oct 1939 (Debye Papers, MPG).

23. Warren Weaver, diary, entry for 6 Feb 1940, Berlin (RAC).

24. WH, *PB*, 172; and interview of WH by J. J. Ermenc, Urfeld, 29 Aug 1967, 75-page transcript (Papers of Gen. Leslie Groves, NARA Gift Collection).

25. German work in each of these areas is discussed by Walker, *Nuclear Power*.

26. WH to Sommerfeld, 29 Oct 1940 (SN); WH to R. Ortvay, 8 Dec 1941 (HA). Walker, *Nuclear Power*, 52–53, lists nine main groups by 1942. "Energiegewinnung aus Uran," a report to the HWA, dated February 1942 (Bagge Papers), 135, lists a total of 22 participating institutes at 12 different locations.

27. WH to Harteck, 29 Apr 1940 (DJ-29). Auer later asked Degussa to turn the oxide into metal powder and plates.

28. Harteck to WH, 15 Jan 1940 (DJ-29); WH to Harteck, 18 Jan 1940 (DJ-29). Robert and Klara Döpel and WH, report of 7 Aug 1940, HCW, A2, 419–426.

29. See Per F. Dahl, *Heavy Water and the Wartime Race for Nuclear Energy* (Bristol: Institute of Physics Publishing, 1999).

30. WH to Bagge in Paris, 6 June 1941 (HA); Bagge's diary, note 6, entries for 1941. For Joliot and French fission research during the war see Spencer Weart, *Scientists in Power* (Cambridge, MA: Harvard Univ. Press, 1979).

31. WH to his mother, letters in early 1940.

32. WH to his mother, 11 Nov 1940.

33. A list of courses offered by the Leipzig Physics Institute appeared annually in the *PZ*; Heisenberg's doctoral candidates are given in WH, *Gutachten- und Prüfungsprotokolle für Promotionen und Habilitationen (1929–1942)*, 2nd edition, eds. H. Rechenberg and G. Wiemers (Berlin: ERS-Verlag, 2002).

34. "Aktennotiz: Betrifft Prof. Dr. Hans Volkelt, Universität Leipzig, 15.6.42," in Hauptamt Wissenschaft, Amt Wissenschaftsbeobachtung u. -wertung (Erxleben) (IfZ, MA 129/9). Also recalled by former Leipzig professor Hans-Georg Gadamer, *Philosophical Apprenticeships*, trans. Robert R. Sullivan (Cambridge, MA: MIT Press, 1985), 93–102.

35. "Goerdeler und die Deportation der Leipziger Juden," *Vierteljahrshefte für Zeitgeschichte, 13* (1965), 338–339.

36. Heisenberg only refereed Gora's dissertation; his examiner was Hund (UA Leipzig, Promotionen, MII). Gora's story is based on WH to Edwin Gora, 3 Jan 1941 (Gora Papers, Providence, RI); Gora to Christian Kleint, 20 Feb 1985 (ibid.); interview with Prof. Gora, Boston, March 1985; and Edwin K. Gora, "One Heisenberg Did Save," *Science News, 109* (1976), 179.

37. Heisenberg recalled Euler and Grönblom in WH, *PB*, 176–179; and 38. This and other letters from Euler to WH are in HA.

38. E. Zeigner, friend of the Euler family, to WH, 1 Oct 1941 (HA); Mrs. Martha Euler to WH, 11 Apr 42; Oberleutnant Heppner to Martha Euler, 1 Dec 1941 (all HA); Marita Euler to WH, 9 Nov 46 (HP); WH to Marita Euler, 26 Nov 46 (HP).

39. WH, *B.O. Grönbloms wissenschaftliche Arbeiten*, 1943, reprinted HCW, C4, 45–61.

Chapter 23. A Copenhagen Visit

1. The technical features of the Leipzig and Berlin experiments are summarized by WH-Wirtz, *FIAT*. For Leipzig research, see also Christian Kleint, *Kernenergie, 29* (1986), 245–251; as well as Walker, *Nuclear Power*.

2 A detailed description of the building is given by WH, report dated March 1941, publ. HCW, A2, 432–462, on 435.

3. Reports on experiments B-1 to B-4 through Jan 1942, co-authored by WH, are publ. in HCW, A2.

4. R. and K. Döpel and WH, report of July 1942, HCW, A2, 536–543.

5. R. and K. Döpel and WH, report of 28 Oct 1941, HCW, A2, 481–498.

6. R. and K. Döpel and WH, report of 26–28 Feb 1942, HCW, A2, 526–528; also, W. Bothe, report of 7 Jan 1942 (DJ-30, 023–024).

7. Robert Döpel, report of 9 Jul 1942 (DJ-29, 539–548; and DJ-30, 298–303).

8. Robert and Klara Döpel and WH, note 4, 543; and WH-Wirtz, *FIAT*, 149.

9. D. Irwing interview with WH, 23 Oct 1965 (DJ-31, 526–567).

10. Fritz G. Houtermans report dated on cover and in Nachwort, October 1944, but includes report dated Aug 1941 on 35, quote 33 (DJ-30, 704–719).

11. Quoted by H. Rechenberg, "Einleitung," in WH, *Ordnung der Wirklichkeit*, ed. H. Rechenberg (Munich: Piper, 1989), 17; original in HA.

12. Surmised from WH, report to REM on his visit to Copenhagen, 23 Sep 1941 (ZStA Potsdam, REM 2942, Bl. 547).

13. Michael Frayn, *Copenhagen* (New York: Anchor Books, 1998).

14. WH, affidavit on 1941 visit to Copenhagen, manuscript and typescript, n.d. [1948], 2 pp. (HA). The official trial affidavit is Weizsäcker defense document no. 239 (NARA, microfilms M897, roll 119). He may have submitted it in connection with Carl Friedrich's interrogation (N/ RA, microfilms M1019, roll 78) and later testimony (NARA, microfilms M897, roll 10, 24 Jun 1948, vol. 25, 10007–10040). Use of the word *bomb* is anachronistic. It was not yet determined that the explosive could be delivered as a bomb. If it were too unwieldy for flight, it would have to be delivered by ship.

15. Previous accounts of the meeting include Walker, *Nuclear Power*, 223–228, as well as Walker *Nazi Science*, 144–151, and Paul Lawrence Rose, *Heisenberg and the Nazi Atomic Bomb Project* (Berkeley: Univ. of California Press, 1998), 154ff.

16. Rose, ibid., 155. On the new documents, see Rainer Karlsch and Mark Walker, "New Light on Hitler's Bomb," *Physics World*, June 2005, 15–18.

17. WH to B. L. van der Waerden, typed transcription by van der Waerden, 28 Apr 1948 (DJ-29, 1190–1191).

18. WH, *PB*, 170; and interview, note 9.

19. WH, affidavit, note 14; EH, *Recoll.*, 79.

20. "Energiegewinnung aus Uran," a report to the HWA, dated February 1942 (Bagge Papers). I thank Mark Walker for a copy of this report.

21. WH, affidavit, note 14.

22. On the Danish occupation and Bohr's institute see: William Dan Andersen, "The German Armed Forces in Denmark 1940–1943: A Study in Occupation Policy," (Ph.D. thesis, University of Kansas, 1972); Erich Thomsen, *Deutsche Besatzungspolitik in Dänemark 1940–1945* (Düsseldorf: Bertelsmann Verlag, 1971); Niels Bohr, affidavit, 20 Dec 1947, Weizsäcker defense exhibit no. 301 (NARA, microfilms M897, roll 119); Werner Best (occupation head after 6 Nov 1942), affidavit, 27 Aug 1947, Weizsäcker defense exhibit no. 302 (ibid.); WH, affidavit, 3 April 1948, Weizsäcker defense exhibit no. 303 (ibid.).

23. The US State Department allowed the foundation to continue its funding, despite the Nazi occupation.

24. Dr. Brauwiler, report in May 1940 to Reichsministerium für Volksaufklärung und Propaganda (BA Koblenz, R58/1091, Bl. 32–40).

25. Andersen, note 22, 194.

26. Plenipotentiary of German Reich in Denmark to Berlin Foreign Office, 27 Mar 1941 (Z Potsdam, REM 2943); C. F. von Weizsäcker to WH, 26 Mar 1941 (HA).

27. C.F. von Weizsäcker to Lamberts (DAAD), 22 Jul 1941; Foreign Office to REM, 2 Aug 1941 (ZStA Potsdam, REM 2943); C.F. von Weizsäcker to Niels Bohr, 15 Aug 1941 (BSC 26,2).

28. REM documents in August and September 1941 (ZStA Potsdam, REM 2943).

29. I thank Erik Rüdinger for Mrs. Bohr's remark.

30. Weizsäcker, note 26. The Copenhagen view of that visit is repeated in Arnold Kramish, *The Griffin* (Boston: Houghton Mifflin, 1986), 120.

31. Quoted by Margaret Gowing, *Britain and Atomic Energy 1939–1945* (New York: St. Martin's Press, 1964), 246.

32. "Notes on Meeting of Sub-Committee September 10, 1943 (R.C.T.)" (NARA, RG 77, file 334, British Interchange Sub-Committee).

33. See Matthias Dörries, ed., *Michael Frayn's "Copenhagen" in Debate* (Univ. of C. Berkeley: Office for History of Science and Technology, 2005), also containing Bohr's draft letters in facsimile, Danish transcription, and English translation, 101–179, currently available also at *http://www.nba.nbi.dk/release.html*.

34. Robert Jungk, *Heller als Tausend Sonnen* (Bern: Alfred Scherz Verlag, 1956); Danish: *Stærkere end tusind sole: atomforskernes skæbne* (Copenhagen, 1957); English: *Brighter than a Thousand Suns: A Personal History of the Atomic Scientists*, James Cleugh, transl. (New York: Harcourt Brace Jovanovich, 1958), 88 and 89.

35. Ibid., English, 103–104; WH to Jungk, 18 Feb 1957 (HA).

36. Bohr to WH, doc. 1, in Dörries, note 33, 109 and 111; Gerald Holton, "What is *Copenhagen* Trying to Tell Us?" ibid., 49–58. See also D. C. Cassidy, "New Light on *Copenhagen* and the German Nuclear Project," *Physics in Perspective, 4* (2002), 447-455.

37. Bohr to WH, docs. 11a and 7, in Dörries, note 33, 163, 137, 139.

38. Schumann to directors and researchers, 5 Dec 1941, quoted by Bagge et al., *Von der Uranspaltung bis Calder Hall* (Hamburg: Rowohlt, 1957), 28.

39. Report, note 20, 133.

40. List of lectures, 26–28 Feb 1942 (DJ-29, 998–1005).

41. List of lectures, 26 Feb 1942 (DJ-29, 705).

42. WH, "Die theoretischen Grundlagen für die Energiegewinnung aus der Uranspaltung," 26 Feb 1942, publ. HCW, A2, 517–521, quote on 518–519; translation: "A Lecture on Bomb Physics," William Sweet, *Physics Today* (August 1995), 27–30.

43. Walker, *Nuclear Power,* 58.

44. This campaign is described by Beyerchen, *Scientists,* 190–191; Karl-Heinz Ludwig, *Technik und Ingenieure im Dritten Reich* (Königstein/Ts.: Athenäeum-Verlag, 1979), 241–242; and Walker, note 41, 119–122. See also D. Hoffmann and M. Walker, eds., *Physiker zwischen Autonomie und anpassung: Die Deutsche Physikalische Gesellschaft im Dritten Reich* (Weinheim: Wiley-VCH, 2007).

45. Finkelnburg to WH, 6 May 1942 (HA); WH to Finkelnburg, 22 May 1942 (HA).

46. "Niederschrift über die Sitzung des Senats der KWG," 24 Apr 1942 (MPG).

Chapter 24. Ordering Reality

1. WH, untitled typescript (HA); published as "Ordnung der Wirklichkeit," in HCW, C1, 218–306; and as *Ordnung der Wirklichkeit* (Munich: Piper, 1989).

2. WH, "Die Goethe'sche und die Newton'sche Farbenlehre im Lichte der modernen Physik," Budapest, 28 Apr 1941, reprinted HCW, C1, 146–160; "Die Einheit des naturwissenschaftlichen Weltbildes," Zurich, 27 Nov 1942, published HCW, C1, 201–215; "Die Einheit des naturwissenschaftlichen Weltbildes," Leipzig, 26 Nov 1941, reprinted HCW, C1, 161–192; "100 Jahre Energiegesetz," radio lecture, Aug 1942, reprinted HCW, C1, 202–206. Heisenberg's world view has been discussed most recently by Gregor Schiemann, *Werner Heisenberg* (Munich: C. H. Beck, 2008), chap. 4.

3. WH, note 1, 218. All references are to the HCW edition.

4. Ibid., 226.

5. Ibid., 232. The "layers" are listed vertically on the page, but in reverse order.

6. WH, note 2, first work, 160.

7. WH, note 1, 304–306.

8. Ibid., 298 and 305.

9. Ibid., 304.

. The details of these changes are discussed by Walker, *Nuclear Power;* recollections by Speer, *Inside the Third Reich: Memoirs*, trans. R. and C. Winston (New York: Macmillan, 1970), 276; Reich Research Council documents (LC, Microfilm 107, File 12847).

11. See Kristie Macrakis, *Surviving the Swastika: Scientific Research in Nazi Germany* (New York: Oxford Univ Press, 1993), chaps. 7 and 8.

12. Göring, order of 8 Dec 1942 (DJ-29, 1031).

13. WH to Telschow, 11 Jun 1942 (HA); WH to KWG, 2 Jul 1942 (HA), on return of the institute to the KWG.

14. WH to dean of Philosophical Faculty, Leipzig, 9 Sep 1942 (UA Leipzig, Personalakten Heisenberg, PA 560); Head of Saxon Ministry for Education to Rector, University of Leipzig, 22 Sep 1942 (HA), approval; WH to H. Falkenhagen, 19 Jun 1942 (HA).

15. Opposition to Heisenberg's call is indicated by REM correspondence with the Party Chancellery in that period (BDC, REM files 5512-5518).

16. Telschow, Aktennotiz, 8 Feb 1943 (HA and DJ-29, 1049–1050), on meeting between Vögler, Mentzel, and Telschow. Esau to Mentzel, 5 Apr 1943 (DJ-29, 1060–1061), on financing.

17. EH, *Recoll.*, 91.

18. Published in part as "Eingabe an Rust," *PB, 3* (1947), 43–47. See Dieter Hoffmann, "Die Ramsauer-Ära und die Selbstmobilisierung der Deutschen Physikalischen Gesellschaft," in *Physiker zwischen Autonomie und Anpassung*, eds. D. Hoffmann and M. Walker (Weinheim: Wiley-VCH, 2007), 173–215.

19. Erxleben to Bechtold, Partei-Kanzlei, 9 Sep 1942 (IfZ, Hauptamt Wissenschaft, Heisenberg file, MA 116/5).

20. WH to Pascual Jordan, 31 Jul 1942 (HP). The date of the meeting is given in Dr. Borger to Partei-Kanzlei, 9 Sep 1942 (IfZ, MA 116/5).

21. Dr. Borger to Partei-Kanzlei, 9 Sep 1942 (IfZ, MA 116/5); Erxleben to Bechtold, 9 Sep 1942 (IfZ, MA 116/5).

22. The Seefeld meeting and the compromises reached are discussed in Beyerchen, *Scientists*, chapt. 9. Heisenberg's attendance is indicated in WH to Hotel Tiroler Weinstube, Seefeld, 29 Oct 1942 (HA); Bruno Thüring to WH, 17 Nov 1942 (HA).

23. REM to WH, 20 Feb 1943 (HA); Telschow, Aktennotiz, 8 Feb 1943 (DJ-29, 1049–1050 and HA); Dean Schwender to Rector, University of Leipzig, 10 Mar 1943 (UA Leipzig, Personalakten Heisenberg, PA 560).

24. Hitler, decree of 9 Jun 1942, RFR documents, note 10.

25. Quoted in Speer, note 10, 677, n. 24.

26. Ibid., 301.

27. Ibid., 120. Heisenberg's "pineapple" remark was off by about a factor of *ten*, if he referred to U-235. But it was close to the correct size if he referred to plutonium surrounded by a uranium tamper (which seems unlikely). Mark Walker, *Historia Scientiarum, 14* (2005), 164–181, notes that the earlier Leeb report cited a critical mass of 10-100 kg of U-235. There is no indication who made this calculation.

28. Quoted in Speer, note 10, 677, n. 26.

29. Based on R. Döpel, accident report of 9 Jul 1942 (DJ-29, 539–547).

Chapter 25. Professor in Berlin

1. WH, research report of 31 Jul 1942, publ. HCW, A2, 545–552; Bothe to WH, 18 Aug 1942 (HA), encouraging the use of plates.

2. WH to Bothe, 23 Oct 1942 (HA); Esau to Mentzel, 5 Apr 1943 (DJ-29, 1060–1061).

3. Fritz Bopp, Fischer, WH, Carl Friedrich von Weizsäcker, and Karl Wirtz, report on experiments B–3 to B–5 of 30 Oct 1942, publ. HCW, A2, 553–561. The last plates for B-6 and B-7 may not have been delivered until as late as 15 Jun 1944.

4. WH, ed., *Kosmische Strahlung: Vorträge gehalten im Max Planck-Institut, Berlin-Dahlem* (Berlin: Springer, 1943), reprinted HCW, B, 363–406.

5. Klaus Scholder, *Die Mittwochs-Gesellschaft: Protokolle aus dem geistigen Deutschland 1932 bis 1944* (n.p., n.d.[1982]), 305. Heisenberg's brother had already left Berlin for a position with Agfa in Bittersfeld, near Stuttgart.

6. Goedeler's role is described by Gerhard Ritter, *Carl Goedeler und die deutsche Widerstandsbewegung* (Stuttgart: Deutsche Verlags-Anstalt, 1954). The conspiracy is treated by Peter Hoffmann, *The History of the German Resistance 1933–1945*, trans. Richard Barry (Cambridge, MA: MIT Press, 1977).

7. Quoted by Scholder, note 5, 326.

8. EH, *Recoll.*, 98–99.

9. Scholder, note 5, 351–353.

10. Heisenberg only briefly recalled these painful events in WH, *PB*, 189–190.

11. WH, affidavit for Dr. Julius Hiby, 23 Nov 1947 (HA); Dr. Graun, Dozentenführer der Freien Forschungsinstitute, to WH, 20 Aug 1942 (HA); Helmut Joachim Fischer, "Feuerwehr für die Forschung," MS recollections, 1970 (IfZ). I thank Dr. Fischer for a copy.

12. WH to Sommerfeld, 8 Oct 1942 (SN); Sommerfeld to WH, 14 Oct 1942 (HA); Sommerfeld to W. Becker, Akademische Verlagsgesellschaft, 15 Oct 1942 (SN); Mentzel to von Laue, 22 May 1943 (HA).

13. Based on WH, *PB*, 188–189; WH to Karl and Helen Heisenberg, 1 Nov 1945 (HP); WH, affidavit for E. Sethe, 30 Jul 1946 (HA).

14. WH to NSDAP Ortsgruppe München, 11 Oct and 1 Nov 1943 (HA); WH to R. Döpel, 18 Dec 1943 (HA).

15. WH to Dr. Boseck, Sachbearbeiter in Turowsky's office, 26 Jun 1942 (HA); WH to Himmler, 4 Feb 1943 (HA).

16. Thüring to WH, 17 Nov 1942 (HA); also Thüring to WH, 20 Oct 1942 (HA), and WH to Thüring, 26 Oct 1942 (HA).

17. WH, *Zeitschrift für die gesamte Naturwissenschaft*, 9 (1943), 201–212, rec. 20 May 1943, reprinted HCW, C5, 14–25; R. Brandt, member of Himmler's personal staff, to WH, 15 Feb 1943 (HA).

18. WH to Vahlen, 10 Sep 1942 (HA).

19. WH to Dr. Gustav Borger, 11 Jun 1943 (HA). By then the Copernicus Prize was awarded by Hans Frank's Institut für deutsche Ostarbeit [Institute for German Eastern European Research].

20. Staff member of *Völkischer Beobachter* to WH, 27 Mar 1943 (HA); "Max Planck," *Das Reich, 1943*, 11 Apr 1943, 1; "Werner Heisenberg," *Das Reich, 1944*, 14 May 1944, 1.

21. Hans-Georg Gadamer, *Philosophical Apprenticeships*, trans. Robert R. Sullivan (Cambridge, MA: MIT Press, 1985), 99.

22. Foreign Office to REM, 27 Nov 1941 (ZStA Potsdam, REM 2943, Bl. 557).

23. Fritz von Twardowski, affidavit, 17 Apr 1948, Weizsäcker defense exhibit no. 208, Doc. 352 (NARS, Weizsäcker case, M897, Roll 119); also German Embassy, Prague, to Foreign Office, 6 Jan 1943 (ZStA Potsdam, REM 2943) on Budapest lectures by Heisenberg, Planck, and von Weizsäcker.

24. Hendrik Casimir, *Haphazard Reality: Half a Century of Science* (New York: Harper and Row, 1983), 209; WH, "Ordnung der Wirklichkeit," 1941–42, publ. HCW, C1, 218–306, on 305.

25. The propaganda value of lectures at these institutes is discussed by Walker, *Nazi Science*, chaps. 6 and 7.

26. Dr. Coblitz, director of Institut für deutsche Ostarbeit, Cracow, to WH, 25 May and 29 Sep 1943 (HA); WH to Coblitz, 1943 (HA); WH to Harteck, 8 Dec 1943 (HA).

27. Based on records in HA. These trips are also discussed by Walker, *Nuclear Power*, 105–118.

28. Telegrams from Best's office to Wehrmacht Headquarters, Sep to Oct 1943 (DJ-31, 1000–1002). See William Dan Andersen, "The German Armed Forces in Denmark 1940–1943: A Study in Occupation Policy," (Ph.D. thesis, University of Kansas, 1972); and Erich Thomsen, *Deutsche Besatzungspolitik in Dänemark 1940–1945* (Düsseldorf: Bertelsmann Verlag, 1971).

29. J. G. Crowther, *Science in Liberated Europe* (London: Pilot Press, 1949), 105–109.

30. Based on "Rapport over Begivenhederne under Besættelsen af Universitets Institut for teoretisk Fysik fra d. 6. December 1943 til d. 3. Februar 1944" (BGC). I thank Frederick Nebeker for translating this report. An historical account is offered by Stephan Schwarz, "On the Occupation and Release of Niels Bohr's Institute (6 Dec. 1943–3 Feb 1944)" MS, Copenhagen. I thank the author for a copy of his manuscript.

31. WH to his mother, 23 Jan 1944.

32. "Rapport," note 30; Hans von Euler to WH, 8 Jan 1944 (HA).

33. "Rapport," note 30. Bohr had burned any compromising correspondence before he left Copenhagen; communication from Erik Rüdinger.

34. This account is based on works in note 30.

35. REM to WH, via Biberbach, 1 Mar 1944 (HA).

36. Quoted by Crowther, note 29, 108.

37. Arthur Seyss-Inquart, "Report of the Situation and Developments in Occupied Territories of the Netherlands, 29 May–19 July 1940," translation of Doc. 997-PS, in *Nazi Conspiracy and Aggression*, vol. 3, ed. Office of U.S. Chief of Counsel for Prosecution of Axis Criminality (Washington, DC: GPO, 1946), 641–656, on 653.

38. Werner Warmbrunn, *The Dutch under German Occupation 1940–1945* (Stanford: Stanford Univ. Press, 1963), 146–147.

39. Casimir, note 24, 202.; Warmbrunn, note 38, 149.

Van Dellen, General Secretary, Dutch Education Ministry, to WH, 28 May 1943 (HA); REM to WH,

15 Jun 1943 (HA); WH to van Dellen, 21 Jun 1943 (HA). Kramers to WH, 29 Jul 1943, rec. 19 Aug 1943; WH to Kramers, 20 Aug 1943 (HA).

41. Scherrer to Kramers, 24 Jul 1943 (AHQP 13,3).

42. Dr. Plutzar, Hauptabteilung Wissenschaft, Volksbildung und Kulturpflege, to WH, 15 Sep 1943 (HA).

43. Based on WH, "Bericht über eine Reise nach Holland vom 18.–26.10.43," 10 Nov 1943 (HA); and testimony by Seyss-Inquart during his Nürnberg trial, 11 Jun 1946, in *Der Prozess gegen die Hauptkriegsverbrecher vor dem internationalen Militärgerichtshof Nürnberg 14. November 1945–1. Oktober 1946*, vol. 16 (Nürnberg, 1948), 14.

44. WH, "Bericht," note 43; WH to Kramers, 23 May 1944 (AHQP 12,5); Rosenfeld to WH, 10 Dec 1943 and 14 Apr 1944 (HA).

45. Rosenfeld, ibid.; Kramers to WH, 1 Dec 1943 (HA).

46. G. Kuiper to a Major Fischer, 30 Jun 1945 (University of Arizona Library, Kuiper Papers, Box 28). I thank Ronald Doel for informing me of this letter.

47. Walker, *Nuclear Power*, 113.

48. WH to Gregory Breit, 9 Jan 1951, carbon (HA).

Chapter 26. *Return to the Matrix*

1. WH to Geiger, 23 Oct 1942 (HA).

2. WH, *ZP, 120* (1943), 513–538, rec. 8 Sep 1942; and *ZP, 120* (1943), 673–702, rec. 30 Oct 1942, reprinted HCW, A2, 611–636, 637–666. WH, *ZP, 123* (1944), 93–112, rec. 12 May 1944, reprinted HCW, A2, 667–686. The fourth paper, unpublished then, is in HCW, A2, 687–698.

3. WH, *ZP, 110* (1938), 251–266, reprinted HCW, A2, 315–330; WH and Pauli, report for 8th Solvay Congress of 1939 (canceled), parts 2 and 3 by WH, published in HCW, B, 346–358.

4. WH to Sommerfeld, 19 Jun 1942 (SN); WH to Wick, 19 Jun 1942 (HA).

5. WH, note 2, first work. Heisenberg's S-matrix theory has been discussed by Pais, *Inward*; James T. Cushing, *Theory Construction and Selection in Modern Physics* (Cambridge: Cambridge Univ. Press, 1990); Inge Grythe, *Centaurus, 26* (1982/3), 198–203; Reinhard Oehme, introduction, HCW, A2, 605–610; and Helmut Rechenberg, "The Early S-matrix Theory and its Propagation (1942–1952)," in *Pions and Quarks: Particle Physics in the 1950s*, eds. L.M. Brown et al. (Cambridge: Cambridge Univ. Press, 1989), 551–578.

6. Pauli to Dirac, 21 Dec 1943.

7. For example, Max Dresden, *H. A. Kramers: Between Tradition and Revolution* (New York: Springer-Verlag, 1987), 453–458.

8. WH to Kramers, 31 Oct 1943 (AHQP 12,5).

9. Kramers to WH, 1 Dec 1943 (HA).

10. Kramers to WH, 12 Apr 1944 (AHQP 12,5 and HA).

11. WH-Møller correspondence (HA); Kramers, *Nederlands Tijdschrift voor Natuurkunde, 11* (1944), 134–140, dated 14 Apr 1944, reprinted Kramers, *Collected Scientific Papers* (Amsterdam: North-Holland, 1956), 838–844.

12. WH, *ZN, 1* (1946), 608–622, reprinted HCW, A2, 699–713; and paper in Cambridge, Engl., Dec. 1947, reprinted HCW, B, 444–449.

13. Pais, *Inward*, 497–505; Grythe and Rechenberg, note 5.

14. Carl Ramsauer, "Programm der Deutschen Physikalischen Gesellschaft für den Ausbau der Physik in Grossdeutschland," DPG, *Verhandlungen, 25* (1944), 1–6. See Dieter Hoffmann, "Die Ramsauer-Ära und die Selbstmobilisierung der Deutschen Physikalischen Gesellschaft," in *Physiker zwischen Autonomie und Anpassung*, eds. D. Hoffmann and M. Walker, (Weinheim: Wiley-VCH, 2007), 173–215.

15. Carl Ramsauer, address to German Acadany for Aeronautical Research, 3 Apr 1943 (DJ-31, 157–170); Ramsauer to WH, 21 Jul 1943 (HA).

16. Ernst Telschow, Aktennotiz (memo, 8 Feb 1943 (DJ-29, 1049–1050); Esau to Mentzel, 5 Apr 1943 (DJ-29, 1060–1061).

17. Esau to Mentzel, 28 Oct 1943 (DJ-29, 1082); Mentzel to Görnnert, Göring's office, 6 Nov 1943 (DJ-29, 1084); Speer to Görnnert, 17 Nov 1943 (DJ-29, 1088); Göring to Gerlach, 2 Dec 1943 (DJ-29, 1091).

18. WH, address published in *Probleme der Kernphysik* (Deutsche Akademie der Luftfahrtforschung, Schriften 1943/44) (Berlin, 1943), 29–36; reprinted HCW, A2, 570–575.

19. Indicated by Bäumker, Sec'y of Deutsche Akademie der Luftfahrtforschung, to WH, 17 Feb 1943 (HA)

20. WH, note 18, 30.

21. Graue, Leiter des Geschäftsführenden Beirates des RFR, notes on a meeting with Mentzel and other policy officials on 7 Sep 1943, dated 13 Sep 1943 (LC, Film 107, file 12853); OKW, Wehrersatzamt, Rundschreiben, 18 Dec 1943 (BA Koblenz, R26 III/108).

22. Max von Laue, *PB, 3* (1947), 424–425; EH, *Recoll.*, 90.

23. Report of Goudsmit's remarks on Heisenberg by Dietrich E. Thomsen, *Science News, 109* (1976), 157; Dresden, note 7, 458.

24. WH to Himmler's personal staff, 9 Mar 1943 (HA).

25. Kramers to WH, 12 Apr 1944 (HA); WH to Kramers, 4 May 1944 (AHQP 12,5).

26. WH to Coster, 16 Feb 1943 (GP and HA). Coster's letter to WH could not be found.

27. Goudsmit, *Alsos*, 48.

28. Goudsmit, "Werner Heisenberg (1901–1976), *Yearbook of the American Philosophical Society, 1976*, 74–80, on 78.

29. Mentzel to Görnnert in Göring's office, 8 Jul 1943 (DJ-29, 1077).

Chapter 27. One Last Attempt

1. WH, research report of 31 Jul 1942, publ. HCW, A2, 545–552. Bothe to WH, 30 Jul 1943 (HA).

2. Abraham Esau, report as of 31 Mar 1944, dated 21 Jul 1944 (DJ-29, 1102–1108).

3. WH, address published in *Probleme der Kernphysik* (Berlin, 1943), reprinted HCW, A2, 570–575, on 570.

4. Albert Vögler to WH, 18 Jan 1944 (HA); Beuthe to Mentzel, 22 Dec 1943 (DJ-29, 1095–1096).

5. Described (after the war) by WH-Wirtz, *FIAT*, 157.

6. Ibid., 156; and WH et al., report on B-7 of 3 Jan 1945, publ. HCW, A2, 595–601.

7. Laue to WH, 20 Aug 1943 (HA).

8. WH to Adolf Hornung, institute technician, 20 Jul 1943 (HA); Erich Bagge, diary, entries for Aug to Sep 1943 (DJ-29, 106–133); and list of 55 members of the institute, with titles and locations, as of 24 Jan 1944 (HA).

9. WH to the Mayor of Kochel, 3 Apr 1944 (HA); WH to Kurt Staun, Leipzig official, 24 Apr 1944 (HA).

10. EH, *Recoll.*, 93–94; WH to his mother, 5 Sep 1943, and to Uncle Karl, 1 Nov 1945 (HA).

11. WH to his mother, 19 May 1944.

12. Ibid., 2 Dec 1944.

13. Paul Rosbaud's activities are described by Arnold Kramish, *The Griffin* (Boston: Houghton Mifflin, 1986); Victor Weisskopf to J. Robert Oppenheimer, 1942, published by S. S. Schweber, in *Les Houches*, eds. B. S. DeWitt and R. Stora (Amsterdam: North-Holland, 1984), 37–220, on 126–128.

14. This account of Berg's activities is based on unpublished War Department reports of 5 Dec 1945, 25 Sep 1946, 30 Sep 1946, and Louis Kaufmann et al., *Moe Berg: Athlete, Scholar, Spy* (Boston: Little, Brown, 1974).

15. See Alsos mission reports and correspondence (NARA, microfilms M1109, 5 rolls); NARA, RG 77 (Chief of Engineers), 371.2 (Goudsmit Mission); as well as Goudsmit, *Alsos*; Boris T. Pash, *The Alsos Mission* (New York: Award House, 1969); and Leo James Mahoney, "A History of the War Department Scientific Intelligence Mission (Alsos), 1943–1945," (Ph.D. thesis, Kent State University, 1981).

16. Jozef Garlinski, *The Swiss Corridor: Espionage Networks in Switzerland during World War II* (London: Dent, 1981), 17.

17. Kaufman et al., note 14, 195. However, this story is not in the War Dept. reports cited in note 14.

18. Goudsmit, *Alsos*, 114.

19. Recalled in Gerlach to WH, 16 Apr 1946 (HA).

20. Gerlach, report on research from 1 Feb to 31 May 1944, n. d. (DJ-29, 1118–1122).

21. An official report of the B-8 experiment could not be found. The following account is based on technical data in WH-Wirtz, *FIAT*, 158–165.

22. Rainer Karlsch, *Hitlers Bombe* (Munich: Deutsche Verlags-Anstalt, 2005).

23. "Sitzungsbericht vom 21. August 1944 des Reichsmin. f. Rüstung u. Kriegsproduktion" (BA Koblenz, R26 III/92); Bormann, Rundschreiben, 3 Sep 1944, and Osenberg, Rundschreiben, 7 Sep 1944 (both BA Koblenz, R26 III/108).

24. WH to Helmut Volz, 24 Apr 1944 (HA).

25. WH-Wirtz, *FIAT*, 164.

26. Col. Lansdale, report to Gen. Groves, 5 May 1945 (NARA, M1109, roll 2).

27. "Organizational disposition," 15 Apr 1945 (NARA, M1109, roll 4). Paris served as Alsos headquarters, Aachen as Advance Base, North.

28. Groves to Army Chief of Staff, 23 Apr 1945 (NARA, M1109, roll 2).

29. Note 26.

30. Heisenberg described the last day of the war in WH to Fritz Schumacher, 6 May 1945 (GP); WH to Wolfgang Schadewaldt, 19 Jan 1946 (HA), WH to Karl and Helen Heisenberg, from Farm Hall, 1 Nov 1945 (HA); and in his recently published "Diary for the period April 15th to May 15th 1945," currently at *http://werner-heisenberg.physics.unh.edu/diary.htm*.

31. Col. Boris T. Pash to Chief, Military Intelligence Service, War Dept., "Subject: Alpine Operation," 18 May 1945 (NARA, M1109, roll 4); and Pash, note 15, 219–241. The "Munich Operation" is described in a report by Major R. C. Ham to Colonel Boris T. Pash, 12 May 1945 (NARA, M1109, roll 4).

Chapter 28. *Explaining the Project: Farm Hall*

1. Goudsmit, *Alsos*, 113.

2. Ibid., 112.

3. Max von Laue to his son, Theodor, 29 May 1945 (Nachlass von Laue, Deutsches Museum, 1976-20).

4. Max von Laue to his son, 26 May 1945 (Nachlass von Laue).

5. R. V. Jones, *The Wizard War: British Scientific Intelligence 1939–1945* (New York: Coward, MCann, and Geoghegan 1978), 481.

6. Leslie R. Groves, *Now It Can Be Told: The Story of the Manhattan Project* (New York: Plenum, 1962/1983).

7. A publication of the unedited manuscript: *Operation Epsilon: The Farm Hall Transcripts*, introduction by Charles Frank (Bristol: Institute of Physics Publishing, 1993); edited version: *Hitler's Uranium Club: The Secret Recordings at Farm Hall* (Woodbury, NY: American Institute of Physics, 1996), Jeremy Bernstein ed. and commentary; German translation: *Operation Epsilon: Die Farm-Hall-Protokolle oder die Angst der Alliierten vor der deutschen Atombombe*, ed. Dieter Hoffmann (Berlin: Rowohlt, 1993).

8. *Operation Epsilon*, note 7, first work, 33. All quotations are taken from this edition.

9. Max von Laue to his son, 7 Aug 1945 (Nachlass von Laue).

10. Bernstein, note 7.

11. Note 9.

12. *Operation Epsilon*, note 7, 76–77. A study of the Farm Hall deliberations is offered by Walker, *Nazi Science*, chap. 9.

13. WH, notebook (HA); "Appendix 2. 8 August 1945," in *Operation Epsilon*, note 7, 102–103; "Memorandum vom 7. August 1945," publ. HCW, C5 26, 27, excerpted in English in Groves, note 6, 336–337, without the supplementary comments.

14. Erich Bagge, diary entry for 10 Aug 1945, in Erich Bagge, Kurt Diebner and Kenneth Jay, *Von der Uranspaltung bis Calder Hall* (Hamburg: Rowohlt, 1957), 58.

15. Rittner's report, *Operation Epsilon*, note 7, 93–94.

16. Max von Laue to his son, 7 Aug and 22 Sep 1945 (Nachlass von Laue).

17. Note 7, *Operation Epsilon*, Weizsäcker, 78 and 76–77; Rittner's paraphrase of Heisenberg, 83.

18. Note 9.

19. Groves, note 6, 334.

20. Henry DeWolf Smyth, *Atomic Energy for Military Purposes: The Official Report on the Development of the Atomic Bomb under the Auspices of the United States Government, 1940–1945* (Princeton: Princeton Univ. Press, 1946), 223.

21. Philip Morrison, "Alsos: The Story of German Scientists," *Bulletin of the Atomic Scientists, 3* (1947), 354, 365.

22. Note 7, *Operation Epsilon*, 92.

Chapter 29. *Explaining the Project: The World*

1. WH, *Nwn, 33* (15 Dec 1946), 325–329, reprinted HCW, C5, 28–32; *Schwäbische Donau-Zeitung*, 2 Aug 1948, 3, reprinted HCW, C5, 33–34; *New York Times*, 30 January 1949, section 4, 8, reprinted HCW, C5, 41–42; *Die Welt, 5*, no. 28 (2 Feb 1950), 2, reprinted HCW, C5, 43. News report: Waldemar Kaempffert, *New York Times*, 26 October 1947, E9; Kaempffert, interview with WH, *New York Times* December 1948, 10, reprinted HCW, C5, 37–40;

2. Kaempffert, note 1, second work, quoting WH.

3. WH, note 1, first work. The same formulation is in WH to his Schumacher in-laws, 11 Feb 19

4. WH, *PB*, 218.

5. Goudsmit, *Alsos*; also Goudsmit, *Bulletin of the Atomic Scientists, 3* (1947), 64 and 67; *New York Times*, 9 November 1947, E8, letter of 29 Oct 1947; *New York Times*, 9 January 1949, section 4, 8, letter of 4 Jan 1949. Heisenberg's responses are cited in note 1.

6. Goudsmit, note 5, last work.

7. Goudsmit, *Alsos*, 121.

8. *Times* exchanges, notes 1 and 5; WH-van der Waerden correspondence (HA); WH-Weizsäcker correspondence (HP).

9. Kaempffert, note 1, first work; Goudsmit, note 5, third work; Henry Schuman to Einstein, 11 Nov 1947 (EA 12–176). Einstein responded that they may lie under the pressure of circumstances, Einstein to Schuman, 17 Nov 1947 (EA 12–177).

10. Max von Laue, *PB, 3* (1947), 424–425, and repeated by Heisenberg.

11. Karl Heisenberg to Bohr, 8 Jun 1946 (BSC 28,4); Karl Heisenberg to WH, 16 May 1946 (HP); WH to Goudsmit, 23 Sep 1947 (GP and BSC 28,4).

12. WH to B. L. van der Waerden, 28 Apr 1948 (HA); B. L. van der Waerden, "Aide-Mémoire," 12 Dec 1948 (BSC, 33,1).

13. WH to Goudsmit, 23 Sep 1947 (GP and BSC 28,4).

14. WH, "Die aktive und die passive Opposition im Dritten Reich," 12 Nov 1947 (HA); WH, "Ordnung der Wirklichkeit," 1941/42, publ. HCW, C1, 218–306.

15. Goudsmit to WH, 1 Dec 1947 (GP and BSC 28,4).

16. WH to Wirtz, FIAT.

17. Goudsmit, note 5, last work; Kaempffert, note 1, second work; Goudsmit to WH, 20 Sep 1948 (GP).

18. Goudsmit, *Alsos*, 115.

19. Carl Friedrich von Weizsäcker to WH, Chicago, 14 Oct 1949 (HP).

20. Goudsmit, "Werner Heisenberg (1901–1976)," *Yearbook of the American Philosophical Society, 1976*, 74–80, on 79. Subsequent historical research has supported his view.

21. Robert Jungk, *Heller als Tausend Sonnen* (Bern: Alfred Scherz Verlag, 1956); Danish: *Stærkere end tusind sole: atomforskernes skæbne* (Copenhagen, 1957); English: *Brighter than a Thousand Suns: A Personal History of the Atomic Scientists*, James Cleugh, trans. (New York: Harcourt Brace Jovanovich, 1958). Bohr's draft letters are published in *Michael Frayn's "Copenhagen" in Debate,* ed. Matthias Dörries (U.C. Berkeley: Office for History of Science and Technology, 2005), 101–179, and currently available at *http://www.nba.nbi.dk/release.html.*

22. Jungk, note 21, English, 88 and 103.

23. Ibid., English, 105, German, 112.

24. Ibid., 102–104.

25. Quoted by Arnold Kramish, *The Griffin* (Boston: Houghton Mifflin, 1986), 247.

26. Laue to Rosbaud, 4 Apr 59 (Nachlass von Laue, Deutsches Museum, 1976-20).

27. Quoted in WH to Hans Bethe, 27 Apr 64 (HA).

28. D. Irving, *The German Atomic Bomb: The History of Nuclear Research in Nazi Germany* (New York: Da Capo Press, 1967).

29. WH, *Bulletin of the Atomic Scientists, 24* (1968), no. 6, 34–35, German in *Frankfurter Allgemeine Zeitung*; reprinted HCW, C5, 50–52, on 52 ; WH, interview, *Der Spiegel, 21*, no. 27, 3 Jul 1967, reprinted HCW, C5, 45–48.

30. Eugene Rabinowitz, *Bulletin of the Atomic Scientists, 24* (1968), no. 6, 32–35.

31. Goudsmit, note 20, 80.

Chapter 30. The Later Years

1. Max von Laue to Theodor von Laue, 19 May 1946 (Nachlass von Laue, Deutsches Museum, 1976-20). The postwar mental state of German physicists is well examined by Klaus Hentschel, *The Mental Aftermath: The Mentality of German Physicists 1945–1949* (Oxford: Oxford Univ. Press, 2007).

2. WH to the Schumachers, 11 Feb 1946 (HP).

3. WH to Karl Heisenberg, 8 May 1946 (HP).

4. Ibid. and Johann Dieckmann to WH, 10 Nov 1946 (HA); Hund to WH, 1 May 1946 (HA); and drich Hund, "Wissenschaftliches Tagebuch," entry for 6 Apr 1945 (Hund papers, Göttingen). Note 3, and WH to Annaliese Clar, 18 Apr 1947 (HA); WH to his wife, 25 Jan 1946 (HP); WH to humacher, 6 May 1945 (GP).

6. Johann Dieckmann to WH, Osnabrück, 10 Nov 1946 (HA); Firma Hartmann to WH, Osnabrück, 26 Jun 1947; WH to Hartmann, 2 Jul 1947 (HA).

7. Interview with Elisabeth Heisenberg, Göttingen, 1982; WH to the Schumachers, Alswede, 11 Feb 1946 (HP); WH to his wife, 3 Jan 1946 (HP), currently publ. at *http://werner-heisenberg.physics.unh.edu/e-Farm-Hall.htm*.

8. WH to his wife, [6] Aug 1945, ibid.; WH to Karl and Helen Heisenberg, 1 Nov 1945 (HP); WH to Sommerfeld, 5 Feb 1946 (SN).

9. WH to Blackett, 5 Oct 1945 (HA); WH to the Schumachers, 11 Feb 1946 (HP); WH to his wife, 20 Jan 1946 (HP), and *http://werner-heisenberg.physics.unh.edu/e-Farm-Hall.htm*.

10. Bagge, diary entry for 3 Feb 1946, in Erich Bagge, Kurt Diebner, and Kenneth Jay, *Von der Uranspaltung bis Calder Hall* (Hamburg, Rowohlt, 1957); WH to Sommerfeld, 5 Feb 1946 (SN).

11. WH to Sommerfeld, 5 Feb and 29 Jun 1946 (SN); WH to Gerlach, 16 Jul 1946 (HA).

12. Sommerfeld to WH, 17 Feb 1946 (HA).

13. WH to B. Schweitzer, 17 Apr 1947, and to E. Sethe, 30 Jul 1946 (HA); M. Paul to WH, Hechingen, 3 Mar 1946; WH to Otto Hahn, Urfeld, 16 Aug 1946 (HA).

14. Sommerfeld to WH, 24 Sep 1947 (HA).

15. Interview with Friedrich Hund, Göttingen, 1983.

16. Armin Hermann, "Deutsche Wissenschaftspolitik und die Gründung von CERN," in *Wissenschaftsgeschichte heute*, Christian Hünemörder, ed. (Stuttgart: Steiner Verlag, 1987), 29–45.

17. WH, writings for Humboldt Foundation, reprinted HCW, C5, group G; WH, widely published talk to Göttingen students, 1946, reprinted HCW, C5, 384–394.

18. Law 25, dated 29 Apr 1946, in Felix Brandl, ed., *Das Recht der Besatzungsmacht* (Munich: Oldenbourg, 1947), 674–687.

19. Laue, note 1. Denazification policy is discussed by Tom Brower, *The Pledge Betrayed. America, Britain and the Denazification of Post-War Germany* (New York: Doubleday, 1981); Lutz Niethammer, *Entnazifizierung in Bayern* (Frankfurt, 1972); and Jeffrey Gaab, *Justice Delayed: the Restoration of Justice in Bavaria under American Occupation, 1945–1949* (New York: Lang, 1999).

20. Documents in HA; also discussed by Walker, *Nuclear Power*, 195–201.

21. British Research Branch, certificate for WH, 27 Feb 1947 (HA).

22. D. Cassidy, "Controlling German Science, I," *HSPS*, 24 (1994), 197–235.

23. Heisenberg's many papers, addresses, and interviews relating to German atomic energy policy in those years are reprinted in HCW, C5. Heisenberg's postwar activities as cultural figure and policy advocate are explored by Cathryn Carson, "Particle Physics and Cultural Politics: Werner Heisenberg and the Shaping of a Role for the Physicist in Postwar West Germany," (Ph.D. thesis, Harvard Univ., 1995); and Carson, "New Models for Science in Politics: Heisenberg in West Germany," *HSPS, 30* (1999), 115–172,

24. Heisenberg, the history of postwar German nuclear technology, and the scientists' opposition to nuclear weapons have been described by numerous authors. They include: Carson, note 23, chap. 4; Mark Cioc, *Pax Atomica: The Nuclear Defense Debate in West Germany during the Adenauer Era* (New York: Columbia Univ. Press, 1988); Michael Eckert, *HSPS, 19* (1988), 81–113, and *HSPS, 21* (1990), 29–58.

25. Heisenberg and the debates over German science policy in this period are explored by Thomas Stamm, *Zwischen Staat und Selbstverwaltung: Die deutsche Forschung im Wiederaufbau 1945–1965* (Köln, 1981); Maria Osietzki, *Wissenschaftsorganisation und Restauration: Der Aufbau außeruniversitärer Forschungseinrichtungen und die Gründung des westdeutschen Staates 1945–1952* (Köln: Böhlau Verlag, 1984); Cathryn Carson, "New Models for Science in Politics: Heisenberg in West Germany," *HSPS, 30* (1999), 115–171; and D. C. Cassidy, "Controlling German Science, II," *HSPS, 26* (1996), 197–239.

26. "Military Governors' Aide-Mémoire for the Parliamentary Council," 22 November 1948, reprinted in John Ford Golay, *The Founding of the Federal Republic of Germany* (Chicago: Univ. of Chicago Press, 1958), 263–264, on 263.

27. Kurt Zierold recalled his activities in *Forschungsförderung in drei Epochen: Deutsche Forschungsgemeinschaft—Geschichte, Arbeitsweise, Kommentar* (Wiesbaden: Franz Steiner Verlag, 1968).

28. The founding and history of the DFR are discussed by Cathryn Carson and Michael Gubser, "Science Advising and Science Policy in Post-war West Germany: The Example of the Deutscher Forschungsrat," *Minerva, 40* (2002), 147–179; and the works in note 25.

29. WH, Regener, Rein, and Zenneck to Carlo Schmid, 15 Dec 1948, letter and stenographic record of debate, printed in *Parlamentarischer Rat: Verhandlungen des Hauptausschusses* (Bonn, 1949), 30th session on 6 Jan 1949, reprinted in part, HCW, C5, 71.

30. Lists of members and other information in *Abschlussbericht des Deutschen Forschungsrats (DFR) über seine Tätigkeit*, ed. H. Eickemeyer (Munich: R. Oldenbourg, 1953).

31. WH, manuscript and discussion of lecture, dated 9 Mar 1949 (HA), publ. HCW, C5, 72–86. Art. 2, sect. 4 of the by-laws of the DFR, reprinted in *Abschlussbericht*, note 30, 82 and 85.

32. WH to Dr. Gerhard Hess, president of the University of Heidelberg, 8 May 51 (HA); and Col. Bertie Blount, head of the Research Branch of the British Control Commission, to WH, 3 May 1949 (HA). WH to Karl Geiler, president of the Notgemeinschaft, draft of 11 Jan 1951 (HA).

33. "Staatsabkommen der Länder der Amerikanischen, des Britischen und des Französischen Besatzungs-gebietes über die Finanzierung wissenschaftlicher Forschungseinrichtungen," Königstein, 1 Apr 1949, reprinted in *50 Jahre Kaiser Wilhelm-Gesellschaft und Max Planck-Gesellschaft 1911–1961* (Bonn: Max Planck-Gesellschaft, 1961), 227–231.

34. WH, "Denkschrift des DFR . . . ," Göttingen, 1 Sep 1949, reprinted HCW, C5, 87–91.

35. WH to J. Zenneck, 17 Dec 1949 (HA).

36. WH to Oberregierungsrat Dr. Rust in Bundeskanzleramt, 8 Jun 1950 (HA). WH memos, including "Erforschung und wirtschaftliche Nutzbarmachung der Atomenergie im Frieden," attached to memo of 19 Jun 1951 (HA).

37. Dr. E. Lehnartz, Vorsitzender des Hauptausschusses der Notgemeinschaft, to WH, official and unof-ficial letters of 17 Oct 1949 (HA).

38. Adenauer to WH, 11 May 1951 (HA).

39. Adenauer to WH, 2 Jul and 17 Jul 1951 (HA).

40. For example, the wide-ranging "Program for New Technologies."

41. WH, "Vorwort," in *Abschlussbericht*, note 30, reprinted HCW, C5, 115–116.

42. Carson, note 25, examines Heisenberg's dual roles as physicist and policy advocate in this period.

43. Heisenberg's papers on superconductivity are reprinted in HCW, A3. They are discussed by a co-worker of that period, H. Koppe, "Über Heisenbergs Arbeiten zur Supraleitung," ibid.

44. Heisenberg's papers on turbulence are reprinted in HCW, A1, group 1.

45. The history of renormalization and postwar particle physics has been discussed by a number of authors, among them Pais, *Inward*; Silvan S. Schweber, *QED and the Men who Made it* (Princeton: Princeton Univ. Press, 1994); and Val Fitch and Jonathan Rosner, "Elementary Particle Physics in the Second Half of the Twentieth Century," in *Twentieth Century Physics*, vol. 2, eds. L. M. Brown et al. (New York: AIP Press, 1995), 635–794.

46. Heisenberg's postwar papers on particle physics are reprinted in HCW, A3. He summarized his work through 1957 in HCW, B, 552–561. His work is explored by Carson, note 23, chap. 5.

47. WH, *ZN, 6a* (1951), 281–284, reprinted HCW, A3, 166–169.

48. WH, *ZN, 1* (1946), 608–622, reprinted HCW, A2, 699–713.

49. Heisenberg-Pauli correspondence, PWB, vol. 4, part A; recalled in WH, *PB*, 223–226.

50. WH and Wolfgang Pauli, "On the Isospingroup in the Theory of the Elementary Particles," mimeo-graph typescript, 1958, first publ. HCW, A3, 337–351.

51. Quoted in Weisskopf to Pauli, 7 March 1958. This episode is recounted by Hermann, *Jahrhundert*, and recalled by WH, *PB*, chap. 19.

52. Pauli, statement, 8 Apr 1958, in PWB, vol. 4B, 1137.

53. Heisenberg's paper, "Remarks on the Non-linear Spinor Theory with Indefinite Metric in Hilbert Space," and the ensuing discussion are reprinted in HCW, B, 563–570.

54. WH, *PB*, 235.

55. Communication from Manfred Schröder, a former student.

56. WH, *Physics and Philosophy: The Revolution in Modern Science* (New York: Harper and Row, 1958), The Gifford Lectures; German version reprinted HCW, C2, 1–201. Heisenberg's philosophy and its relation to Greek sources have been explored by Patrick A. Heelan, *Quantum Mechanics and Objectivity: A Study of the Physical Philosophy of Werner Heisenberg* (The Hague: Martinus Nijhoff, 1965); Heelan, *Zeitschrift für allgemeine Wissenschaftstheorie*, 6 (1975), 113–138, with reply by Heisenberg; and Gregor Schiemann, *Werner Heisenberg* (Munich: C. H. Beck, 2008), chap. 3.

57. WH, *Nwn, 63* (1976), 1–7, reprinted HCW, C3, 507–513; Engl. transl. "The Nature of Elementary ...cles," *Physics Today*, 29, no. 3 (1976), 32–39, reprinted HCW, B, 917–927, on 924.

WH, *PB*, 244.

...id, 247.

ACKNOWLEDGMENTS

SINCE *BEYOND UNCERTAINTY* BEGAN WITH *UNCERTAINTY*, I WOULD LIKE TO REITERATE MY gratitude to all of those who helped make *Uncertainty* possible. I would like especially to express my gratitude to the late Elisabeth Heisenberg and to Dr. Helmut Rechenberg, former director of the Heisenberg Archive in the Werner Heisenberg Institute of the Max Planck Institute for Physics and Astrophysics, Munich. Over the years since *Uncertainty*, I have benefitted greatly from comments, conversations, and exchanges of papers and perspectives with many of my friends and colleagues. I would like to express my sincere thanks to, among many others, Cathryn Carson, Michael Eckert, Paul Forman, Michael Frayn, Elizabeth Garber, Dieter Hoffmann, Gerald Holton, Don Howard, Thomas Powers, Helmut Rechenberg, Silvan S. Schweber, Suman Seth, Karl von Meyenn, Mark Walker, and Gerald Wiemers. I am very grateful to Jürgen Renn and the members of the international project History and Foundations of Quantum Physics, coordinated by the Max Planck Institute for the History of Science, Berlin, for stimulating discussions, comments, and meetings. I am also grateful to students and other readers of *Uncertainty* for their insightful comments and questions over the years.

I thank Dr. Rechenberg and the Werner Heisenberg Archive for many of the photographs appearing in this volume and for permission to include them. I am also grateful to the Emilio Segrè Visual Archives of the American Institute of Physics for the photographs indicating "Courtesy ESVA, AIP Niels Bohr Library." The photographs without credits are courtesy of the Heisenberg Archive.

I would like particularly to express my appreciation to Erika Goldman of Bellevue Literary Press for her continued encouragement and support of this book. Last but not least, I am eternally grateful to my spouse, Janet, for her loving support and her many good suggestions.

INDEX